ENGINEERING MECHANICS OF DEFORMABLE SOLIDS

# Engineering Mechanics of Deformable Solids

## A Presentation with Exercises

SANJAY GOVINDJEE

*Department of Civil and Environmental Engineering*
*University of California, Berkeley*

OXFORD
UNIVERSITY PRESS

OXFORD
UNIVERSITY PRESS

Great Clarendon Street, Oxford, OX2 6DP,
United Kingdom

Oxford University Press is a department of the University of Oxford.
It furthers the University's objective of excellence in research, scholarship,
and education by publishing worldwide. Oxford is a registered trade mark of
Oxford University Press in the UK and in certain other countries



First Edition published in 2013
Impression: 1



British Library Cataloguing in Publication Data
Data available

Library of Congress Cataloging in Publication Data
Library of Congress Control Number: 2012945096

ISBN 978–0–19–965164–1

Printed and bound by
CPI Group (UK) Ltd, Croydon, CR0 4YY

To my teachers who showed me the beauty of learning,
To my parents who led me to academia,
To Arjun and Rajiv for loving all things technical,
To Marilyn for always being there for all of us

# Preface

This text was developed for a Strength of Materials course I have taught at the University of California, Berkeley for more than 15 years. The students in this course are typically second-semester Sophomores and first-semester Juniors. They have already studied one semester of mechanics in the Physics Department and had a separate two-unit engineering course in statics, and most have also completed or are concurrently completing a four-semester mathematics sequence in calculus, linear algebra, and ordinary and partial differential equations. Additionally they have already completed a laboratory course on materials. With regard to this background, the essential prerequisites for this text are the basic physics course in mechanics and the mathematics background (elementary one- and multi-dimensional integration, linear ordinary differential equations with constant coefficients, introduction to partial differentiation, and concepts of matrices and eigen-problems). The additional background is helpful but not required. While there is a wealth of texts appropriate for such a course, they uniformly leave much to be desired by focusing heavily on special techniques of analysis overlaid with a dizzying array of examples, as opposed to focusing on basic principles of mechanics. The outlook of such books is perfectly valid and serves a useful purpose, but does not place students in a good position for higher studies.

The goal of this text is to provide a self-contained, concise description of the main material of this type of course in a modern way. The emphasis is upon kinematic relations and assumptions, equilibrium relations, constitutive relations, and the construction of appropriate sets of equations in a manner in which the underlying assumptions are clearly exposed. The preparation given puts weight upon model development as opposed to solution technique. This is not to say that problem-solving is not a large part of the material presented, but it does mean that "solving a problem" involves two key items: the formulation of the governing equations of a model, and then their solution. A central motivation for placing emphasis upon the formulation of governing equations is that many problems, and especially many interesting problems, first require modeling before solution. Often such problems are not amenable to hand solution, and thus they are solved numerically. In well-posed numerical computations one needs a clear definition of a complete set of equations with boundary conditions. For effective further studies in mechanics this viewpoint is essential, and thus the presentation, in this regard, is strongly influenced by the need to adequately prepare students for further study in modern methods.

Sanjay Govindjee
Berkeley and Zürich

# Contents

*Mechanics is the paradise of mathematical science,*
*because here we come to the fruits of mathematics*

**Leonardo da Vinci**

*Theory is the captain, practice the soldier*

**Leonardo da Vinci**

*Mechanics is not a spectator sport*

**Sanjay Govindjee**

# Introduction

# 1



The reliable design of many engineering systems is connected with their ability to sustain the demands put upon them. These demands can come in many forms, such as the high temperatures seen inside gas turbines, excessive forces from earthquakes, or simply the daily demands of traffic loading on a bridge. In this book we will exam the behavior of mechanical systems subjected to various load systems. In particular we will be interested in constructing theories which describe the deformation of mechanical systems in states of static equilibrium. With these theories we will examine and try to understand how common load-carrying systems function and what are their load-carrying limits. To come to a fundamental understanding of all load-carrying systems is a very large endeavor. Thus, we will in this introductory presentation restrict ourselves to classes of problems that are both accessible with an elementary level of analysis and at the same time are useful to everyday engineering practice. More specifically, we will consider the behavior of slender structural systems under the action of axial forces, torsional loads, and bending loads. The presentation will mainly focus on elastic systems, but on occasion we will discuss the behavior of plastically deforming systems.

The subject matter of this book is not unlike many other subjects in engineering science. At a high level, the elements of the theories we will develop and how they are manipulated and used are common to many areas of engineering. We will, like in these other areas, need to work with concepts that describe the measurable state of a system independent of its material composition or condition of equilibrium. We will encounter principles of conservation and balance that apply to all systems independent of their composition or measurable state. And lastly, we will be forced to introduce relations that account directly for a system's material composition. For us, these three concepts will be the concepts of kinematics, the science of the description of motion; statics, the science of forces and their equilibrium; and constitutive relations, the connection between kinematical quantities and the forces that induce them. If as one reads this book one pays careful attention to when these concepts are being introduced and utilized, then one will gain considerable advantage when the time comes to apply them to the solution of individual problems. It is easy to confuse which ideas have been combined to create a derived result, but if one can commit these to memory one will be much better served.

For the remainder of this introductory chapter we will restate some basic concepts from elementary physics that we will require throughout the reminder of the book. In particular, we will review some concepts of force systems and equilibrium. The reader who is already comfortable with such notions, can skip directly to Chapter 2 without any loss.

## 1.1 Force systems

Forces are the agents that cause changes in a system. In this book we are concerned with the deformation state of solids and thus we will be concerned with traditional forces – those that cause motion of a system. In general, forces are abstract quantities that cannot be directly measured. Their existence is confirmed only via the changes they produce. Notwithstanding, all of us have an intuitive feeling for forces, and we will take advantage of this and not dwell any further upon their philosophical aspects.

In order to move or deform a body or system one needs to apply forces to it. In this regard there are two basic ways in which one can apply a force:

(1) on the surface of a body with a surface traction, a force per unit area, or

(2) throughout the volume of a body with a body force, a force per unit volume.

Common examples of surface tractions would be, for example, drag forces on a vehicle, the pressure between your feet and the ground when you walk, or the forces between flowing water and turbine blades in a generator in a hydroelectric dam. Examples of body forces include gravitational forces and magnetic forces.

A fundamental postulate of mechanics further states that for every force system acting on a body there is an equal and opposite force system acting on the body which causes the forces. In various forms, this reaction force principle is known as Newton's Third Law.

### 1.1.1 Units

The units of force in the SI system (Le Système International d'Unités) are Newtons (N). This is a derived unit from those of mass, length, and time. 1 Newton is equal to 1 kilogram times 1 meter per second squared: $1\ \text{N} = 1\ \text{kg}\ 1\ \text{m/s}^2$. In the USCS (United States Customary System) the units of force are pounds-force (or simply pounds) and are denoted by the abbreviations lb or lbf. In this system 1 pound force is exactly equal to 1 pound-mass times 32.1740 feet per second squared: $1\ \text{lbf} = 1\ \text{lbm}\ 32.1740\ \text{ft/s}^2$. In the USCS the pound-mass (lbm) is defined exactly in terms of the kilogram – $0.45359237\ \text{kg} = 1\ \text{lbm}$. When using the USCS one should exercise some caution, due to the occasionally used unit of slug for mass. To accelerate a mass of 1 slug

1 foot per second squared requires a force of 1 pound-force. Surface tractions have dimensions of force per unit area, and body forces have dimensions of force per unit volume. Thus, for example, in the SI system traction has units of Newtons per meter squared, which is also known as a Pascal (Pa).

## 1.2 Characterization of force systems

When dealing with many engineering problems it is often inconvenient to treat all the details of a force system. In particular, it can be cumbersome to always take into account the fact that forces are distributed over finite areas and volumes. It is more convenient to have a characterization of a force system in terms of some vector (or scalar) quantities acting at a single point. This type of characterization of a force system is known as a statically equivalent force system. It is an equivalent description of a force system within certain assumptions. Most importantly, efficient analysis can be carried out using statically equivalent force systems and in many cases without any measurable loss of accuracy.

### 1.2.1 Distributed forces

Distributed forces are described in general by vector-valued functions over surfaces and volumes. For example, Fig. 1.1(a) shows a homogeneous rigid body of density $\rho$ under the action of a gravitational field with acceleration $g$ in the minus $z$-direction. The force system acting on the body is a body force which is given by a constant vector-valued function, $\boldsymbol{b} = -\rho g \boldsymbol{e}_z$. At each point in the body, the magnitude of $\boldsymbol{b}$ indicates the local force per unit volume acting on the material at that point, and the direction of $\boldsymbol{b}$ indicates the direction in which this force acts. As a second example, consider the dam shown in Fig. 1.1(b). The water applies a force which is distributed on the face of the dam. The force can be shown to vary linearly from the top of the dam to the bottom, and further shown to act perpendicular to the face. It can be described by the vector-valued function $\boldsymbol{t}(y, z) = \rho g(h - z)\boldsymbol{e}_x$, where $\rho$ is the density of the water and $g$ is the gravitational constant. The magnitude of $\boldsymbol{t}$ at each point $(y, z)$ on the face of the dam gives the local force per unit area acting on the material at that point, and the direction of $\boldsymbol{t}$ gives the direction of the force. In these examples, the form of the distributed force function is rather simple and can easily be used in analysis. However, to answer certain questions about these systems, knowing the details of the force distribution is not necessary and can make the analysis somewhat cumbersome.

A simple characterization of a force system can be had by considering only the total force given by the system. For example, for the rigid body in Fig. 1.1(a) this would correspond to the total weight of the body – the volume of the body times its density and the gravitational constant, $W = V\rho g$. For the dam in Fig. 1.1(b) we need to account for the fact that

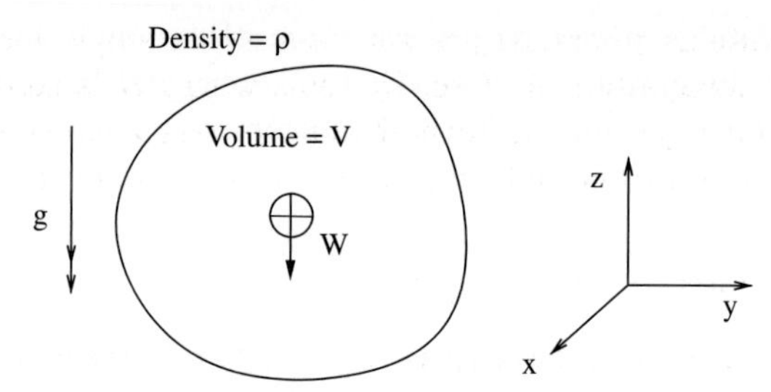


(a) Homogeneous body under the influence of gravity - a distributed body force system.

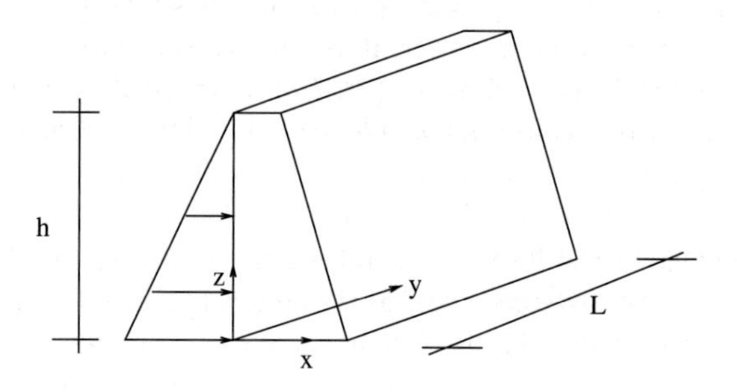


(b) Dam under the influence of a distributed surface force system.

**Fig. 1.1** Systems with distributed loads.

the traction is a function of position on the dam face. Local to each point is a different amount of force, and these amounts need to be summed. The appropriate mathematical device in this regard is integration, and the total resultant force will be $\boldsymbol{R} = \int_A \boldsymbol{t}(y,z)\,dydz = (1/2)h^2 L\rho g$. One way of thinking about the total resultant force is that it represents the (spatial) average of the force system times the area over which it acts. The characterization of a force system solely by its resultant force is quite useful, but in many situations too crude for effective analysis. One important refining concept is that of the first moment of a force distribution or simply moment. Moments of functions are defined relative to reference points which can be freely chosen. If we take as our reference point a point labeled $\boldsymbol{x}_o$, then an effective definition of the (first) moment of a surface force system about this point is given by $\boldsymbol{M}_o = \int_A (\boldsymbol{x} - \boldsymbol{x}_o) \times \boldsymbol{t}(\boldsymbol{x})\,dA$. This moment gives additional information about the spatial distribution of the force system.

**Remarks:**

(1) Taken together, $\{\boldsymbol{R}, \boldsymbol{M}_o\}$ give us two parameters to characterize a force system – at least approximately.
(2) If the force system involves body forces, then the appropriate definitions are $\boldsymbol{R} = \int_V \boldsymbol{b}(\boldsymbol{x})\,dV$ and $\boldsymbol{M}_o = \int_V (\boldsymbol{x} - \boldsymbol{x}_o) \times \boldsymbol{b}(\boldsymbol{x})\,dV$.
(3) Knowledge of $\{\boldsymbol{R}, \boldsymbol{M}_o\}$ is sufficient to fully characterize the effects of a force system acting on a rigid body.

(4) Force systems where $\boldsymbol{M}_o$ is zero are called single force systems or point force systems. Force systems where $\boldsymbol{R}$ is zero are called force couple systems. If both $\boldsymbol{R}$ and $\boldsymbol{M}_o$ are zero, then the force system is said to be self-equilibrated or in equilibrium.

### 1.2.2 Equivalent forces systems

The characterization of a force system in terms of a resultant force and a moment is dependent upon the choice of a reference point. Since the reference point is arbitrary, the representation is not unique. If we choose another reference point $\boldsymbol{x}_p = \boldsymbol{x}_o + \boldsymbol{a}$, where $\boldsymbol{a}$ is the vector from $\boldsymbol{x}_o$ to the new reference point $\boldsymbol{x}_p$, then it is easy to see that the new characterization of our force system is $\{\boldsymbol{R}, \boldsymbol{M}_p\}$, where $\boldsymbol{M}_p = \boldsymbol{M}_o - \boldsymbol{a} \times \boldsymbol{R}$. This new characterization is considered equivalent to the first.

**Remarks:**

(1) Note that if the force resultant, $\boldsymbol{R}$, is zero, then the first moment of the force system is independent of the reference point. In particular, this tells us that a force couple system will always be a force couple systems regardless of reference point.

(2) A force system which is in equilibrium will always be independent of the reference point.

(3) All the points along the line $\boldsymbol{x}(\mu) = \boldsymbol{x}_o + \mu \boldsymbol{n}_R$ have the same moment resultant characterization, where $\mu \in \mathbb{R}$ and $\boldsymbol{n}_R = \boldsymbol{R}/\|\boldsymbol{R}\|$; this is the locus of points through $\boldsymbol{x}_o$ in the direction of $\boldsymbol{R}$.

(4) The notion of equivalent must be carefully understood. The nomenclature stems from the study of rigid body mechanics. In the framework of rigid bodies, equivalent force systems have the exact same effect on a given rigid body. In a deformable body this is not true. What is true, however, is that if carefully chosen and interpreted, equivalent force systems will have almost the same effect on a given deformable body. We will see this more clearly later in the book.

---

**Example 1.1**

*Equivalent forces systems.* As an example, consider the bar shown in Fig. 1.2(a). The bar is loaded with a constant distributed load. In Fig. 1.2(b) one possible characterization of the force system is shown. It consists of a single force acting at a point a distance $b/2$ from the end of the bar. In Fig. 1.2(c) a second equivalent characterization of the force system is shown. All three forces systems are equivalent according to our definition of equivalence. If the bar is rigid, then all three force systems will have the exact same effect on the behavior of the bar. If, however, the bar is deformable then the three systems will all affect the bar in

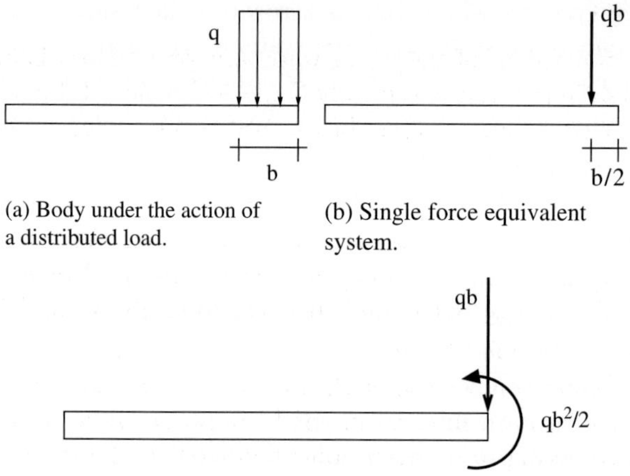

q

b

(a) Body under the action of a distributed load.

qb

b/2

(b) Single force equivalent system.

qb

$qb^2/2$

(c) Second characterization of the force system.

**Fig. 1.2** Three equivalent force systems.

different ways. This difference will manifest itself primarily in the region of length $b$ from the right-hand end of the bar. At distances greater than $b$ from the end of the bar, the effect of the three loading systems will be nearly identical.

**Remarks:**

(1) The load shown in Fig. 1.2(a) is an example of a parallel distributed loading system. That is, at every point where the load acts, the load points in the same (constant) direction in space. For such loading systems, one can always find a single force characterization of the loading (as shown in Fig. 1.2(b)). The point where this single force acts is the centroid of the loading. For example, if the distributed force is given as $\boldsymbol{t} = t\boldsymbol{e}$ over an area $A$ where $\boldsymbol{e}$ is a constant vector, then the single force equivalent load acts at the point $\boldsymbol{x}_R = \int_A t\boldsymbol{x}\, dA / \int_A t\, dA$. In the case where $t$ is also a constant, then $\boldsymbol{x}_R$ coincides with the geometric centroid of the area $A$, $\boldsymbol{x}_c = (1/A)\int_A \boldsymbol{x}\, dA$.

## 1.3 Work and power

The effect of forces is often characterized by the scalar concepts of work and power. The power of a force is defined as the scalar product of the force and the velocity of the material point where it acts: $\mathcal{P} = \boldsymbol{F} \cdot \boldsymbol{v}$. For distributed loads, the power is defined by $\mathcal{P} = \int_A \boldsymbol{t} \cdot \boldsymbol{v}\, dA$ and $\mathcal{P} = \int_V \boldsymbol{b} \cdot \boldsymbol{v}\, dV$. The work of a force system is defined as the time integral of the power over the time interval of application of the force system: $W = \int_I \mathcal{P}(t)\, dt$, where $I$ is some interval of time.

**Remarks:**

(1) The dimensions of energy, or work, are force times distance. In the SI system the unit is typically Joules (J), where 1 Joule is equal to 1 Newton times 1 meter. In the USCS the typical unit of energy is foot-pound (ft-lbf).

(2) The dimensions of power are energy per unit time or force times velocity. In the SI system the unit is typically Watts (W), where 1 Watt is equal to 1 Joule per second. In the USCS the typical unit of power is foot-pounds per second (ft-lbf/s), and 550 ft-lbf/s equals 1 horsepower (hp).

(3) The expression for the work done by a force can be rewritten in terms of a path integral in space. Suppose we have a body with a force acting on a given point and that at the start of the time interval of interest the point is located in space at $\boldsymbol{x}_P$. At the end of the time interval let us assume our point is located at $\boldsymbol{x}_Q$. In this case we can re-express the work of the force as:

$$W = \int_I \boldsymbol{F} \cdot \boldsymbol{v} dt = \int_I \boldsymbol{F} \cdot \frac{d\boldsymbol{x}}{dt} dt = \int_{\boldsymbol{x}_P}^{\boldsymbol{x}_Q} \boldsymbol{F} \cdot d\boldsymbol{x}. \tag{1.1}$$

In general, the value of this integral is dependent upon the path taken from $\boldsymbol{x}_P$ to $\boldsymbol{x}_Q$.

### 1.3.1 Conservative forces

A special class of important force systems are conservative forces. Conservative force systems are those where the path integral form of the work expression is path *independent.* In this case, it can be shown that the force $\boldsymbol{F}$ must be a gradient of a scalar function; i.e. we can always write (for conservative force systems) that

$$\boldsymbol{F} = -\frac{\partial V}{\partial \boldsymbol{x}}. \tag{1.2}$$

Further, we can also write that the work of the force is given by $W = -V(\boldsymbol{x}_Q) + V(\boldsymbol{x}_P)$.

**Remarks:**

(1) The function $V$ is called the potential of the loading system, or alternatively, the potential energy of the load.

(2) The introduction of the minus sign in eqn (1.2) is by convention.

(3) A common example would be the potential energy of a weight near the surface of the earth. The gravitational loading system is conservative and $V(\boldsymbol{x}) = Wz$, where $W$ is the weight of the object and $z$ is its elevation above the surface of the earth.

(4) The absolute value of the potential energy does not play any role in the way we use the concept of potential energy. It is easily observed that if we add an arbitrary constant to the value of the potential

energy, then it does not effect the force or the work expression – as the constant always drops out.

### 1.3.2 Conservative systems

In mechanics one also has the concept of conservative systems. Conservative systems are those that conserve their total energy; i.e. they do not dissipate energy. A common example of a conservative system is a gravitational pendulum. In such a system, the energy changes continuously from kinetic energy to potential energy of the mass of the pendulum, but at all times the sum of the two is constant. Later on, we will deal with deformable bodies under the action of various conservative loading systems. When we assume the bodies to be elastic (without dissipation) then we will be able to exploit the notion of energy conservation as long as we consider our system to be the deformable body plus the loading system.

## 1.4 Static equilibrium

Static equilibrium of a material system is a property of the material system and the force system acting upon it. A system is in a state of static equilibrium if all material points in the system have zero velocity and remain so as long as the force system acting on the material system does not change. Determining the conditions required for a static equilibrium is an important part of the analysis of many engineering systems, as many systems are designed to be in a state of equilibrium. To actually determine working relations, we will need to make a hypothesis about equilibrium.

### 1.4.1 Equilibrium of a body

The central axiom of mechanics states:

For a body to be in static equilibrium the resultant force and moment acting on every subset of the body must be equal to zero.

Thus, if we have a body $\Omega$, then for every subset $B \subset \Omega$ with a force system acting on $B$ which is characterized by $\{\boldsymbol{R}, \boldsymbol{M}_o\}_B$, we must have $\{\boldsymbol{R}, \boldsymbol{M}_o\}_B = \{\boldsymbol{0}, \boldsymbol{0}\}$. Another way of saying the same thing is that the resultant force system acting on any part of a body must be in equilibrium for the whole body to be in static equilibrium. This axiom takes as its origin Newton's Second Law.

### 1.4.2 Virtual work and virtual power

The concept of static equilibrium is very intuitive – the system in question is unchanging, i.e. static. The determination of static equilibrium given above is essentially vectorial in that for static equilibrium we must ensure that two vector valued quantities are zero (for every subset of a body). There is also an equivalent scalar test for equilibrium. It is known as the principle of virtual power or virtual work. The virtual power of a force system is defined as the power associated with the force system when the points of the body take on arbitrary velocities $\bar{\boldsymbol{v}}$. The virtual work of a force system is defined as the work associated with the force system when the points of the body take on arbitrary displacements $\bar{\boldsymbol{u}}$; i.e. $\bar{\mathcal{P}} = \boldsymbol{F} \cdot \bar{\boldsymbol{v}}$ or $\bar{\mathcal{W}} = \boldsymbol{F} \cdot \bar{\boldsymbol{u}}$. It can be shown that the virtual power (or work) of a force system is identically equal to zero for all virtual velocity fields (virtual displacement fields) if and only if the conditions for a static equilibrium are met. This statement is known as the *Principle of Virtual Power (Work)*. Because of the if-and-only-if condition, we can use the Principle of Virtual Power (Work) as an alternative means for studying the equilibrium of material systems. In the latter part of this book we will see how to exploit this principle and a closely related one to solve a number of different classes of problems.

## 1.5 Equilibrium of subsets: Free-body diagrams

The fact that all subsets of a system must be in equilibrium for a system to be in equilibrium leads to the concept of internal forces. Whenever we consider a subset of a body we note that on the boundary of this subset there must be surface forces due to its interaction with the rest of the system. These forces are called internal forces, and they can be characterized in terms of resultants which are simply called internal forces and moments. The boundaries of such a subset are often called section cuts, and the process of splitting a body along such an imaginary boundary is called making a section cut. Note that a section cut need not be straight. If we have a diagram of a part of a body that has been isolated from the remainder by section cuts and the diagram shows the forces and moments acting on the part, then we call the diagram a free-body diagram. As we shall see in this book, internal forces and moments are key to understanding how load-bearing bodies behave.

### 1.5.1 Internal force diagram

The internal force (moment) diagram for a body is a method of displaying information about a body that gives one a snapshot of how it carries its load. It is the graph of the internal force in the body parameterized by the location of a section cut. For our purposes we will adhere to the following sign convention: tension will correspond to positive internal forces, and compression will correspond to negative internal forces. This is predicated on the following convention: a force that acts in the direction of the outward normal to a section cut is considered positive; else it is negative.

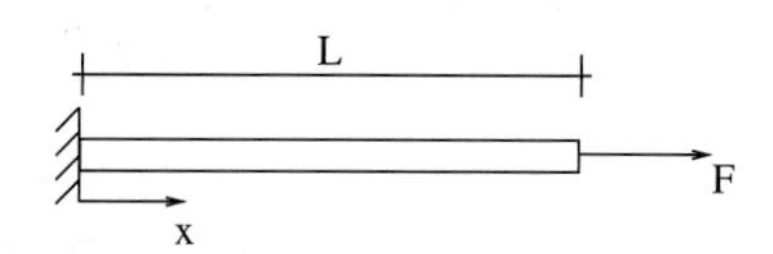


**Fig. 1.3** End-loaded bar.

## Example 1.2

*End-loaded bar.* Consider the bar shown in Fig. 1.3 that is loaded with a force $F$ in the $x$-direction and restrained from motion at the left end. Find the internal forces on a section cut located at $x = L/2$ if the entire system is to be in equilibrium.

*Solution*

Make a vertical section cut at $x = L/2$ and redraw the body in separated form. On the exposed (imaginary) surfaces place the unknown resultants (three in the two-dimensional case); see Fig. 1.4.

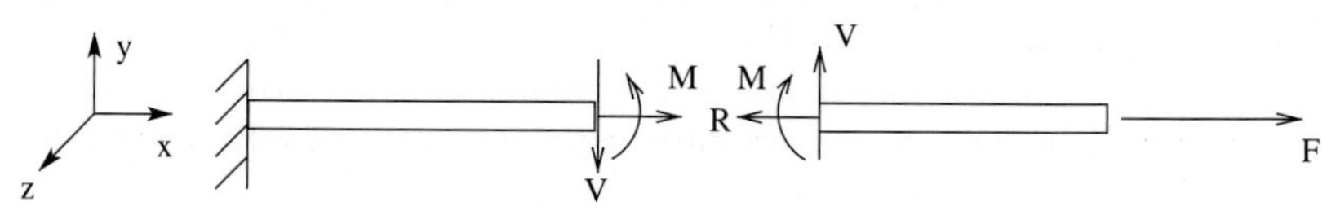


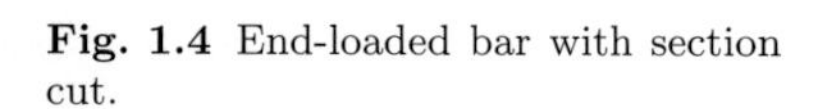

**Fig. 1.4** End-loaded bar with section cut.

Now apply our axiom for static equilibrium to the subset of the body shown on the right:

$$\begin{aligned} \sum F_x &= F - R = 0 \\ \sum F_y &= V \quad\;\;\; = 0 \\ \sum M_z &= -M \quad = 0 \end{aligned} \tag{1.3}$$

This implies

$$R(L/2) = F, \quad V(L/2) = 0, \quad \text{and} \quad M(L/2) = 0. \tag{1.4}$$

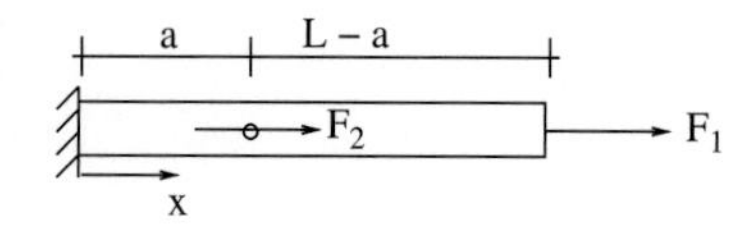


**Fig. 1.5** Bar with two point forces.

## Example 1.3

*Bar with two point forces.* Consider the bar shown in Fig. 1.5. Find $R(x)$ the axial internal force as a function of $x$ when the bar is in static equilibrium.

*Solution*

Make successive section cuts at various values of $x$, and sum the forces in the $x$-direction to obtain representative values of $R$. When one does this it is easy to see that all section cuts between $x = 0$ and $x = a$ give the same result, and all section cuts between $x = a$ and $x = L$ give the same result. The final solution is shown in Fig. 1.6.

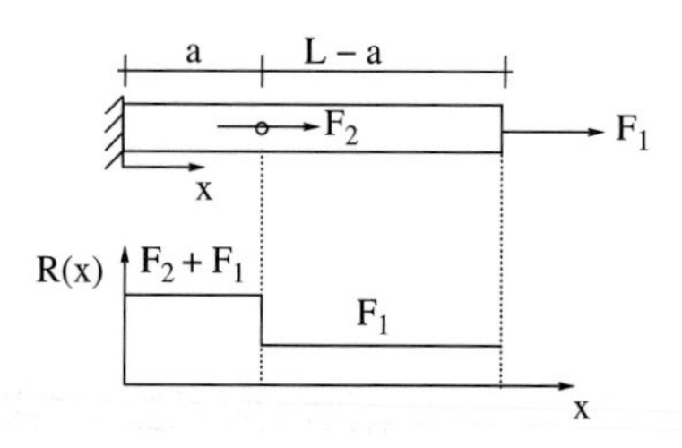


**Fig. 1.6** Internal-force diagram for bar with two point forces.

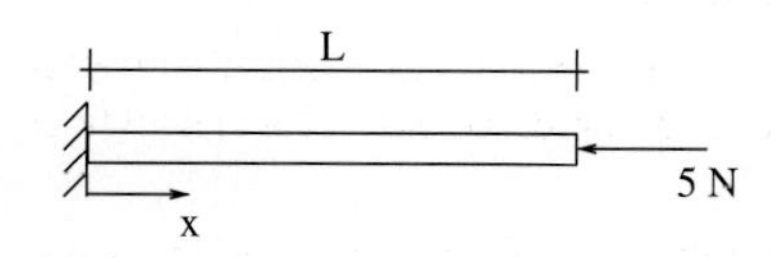


**Fig. 1.7** Compression of a bar.

## Example 1.4

*Compression of a bar.* Find the internal axial force $R(x)$ for the bar shown in Fig. 1.7, assuming static equilibrium.

*Solution*
One should first recognize that the internal forces will be a constant. This should be evident, because no matter where the section cut is made, the free-body diagram will look the same in terms of total loads. One such cut is made in Fig. 1.8.

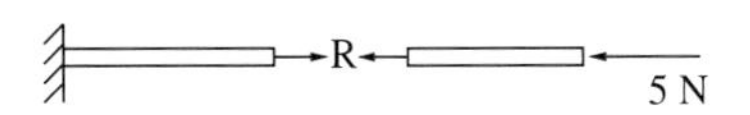


**Fig. 1.8** Compression of a bar with section cut.

Thus:

$$\sum F_x = 0 \quad \Rightarrow \quad R(x) = -5\,\mathrm{N} \tag{1.5}$$

**Remarks:**

(1) Note that one should in general draw the internal forces in the positive sense, and then let the signs take care of themselves. Here, the negative sign of $R(x)$ tells us that the bar is in compression at all points $x$.

## 1.6 Dimensional homogeneity

As we move forward and treat more and more complex problems, the concept of dimensional homogeneity becomes an important tool for double checking results. The concept of dimensional homogeneity simply states that all terms in an equation must have the same dimensions. Thus if one has an equation

$$a = b\,, \tag{1.6}$$

then for this equation to make sense the dimensions of $a$ and the dimensions of $b$ must be the same; for example, if $a$ is mass times length, then so must $b$. When numbers are used this also translates to the requirement that the units of $a$ must match the units of $b$. The requirement of dimensional homogeneity can be used at any stage in an analysis to check for errors, since all equations must be dimensionally homogeneous.

## Exercises

(1.1) Make a free-body diagram of your pencil, isolating it from the paper and your hand. Show appropriate forces and moments imparted by your hand and the paper. Omit ones that you think are zero.

(1.2) Make a free-body diagram of a bicycle, isolating it from the road and the rider. Omit support forces and moments that you think are zero.

(1.3) Make a free-body diagram of an airplane wing during flight. Isolate the wing from the fuselage and the aerodynamic loads. Omit forces and moments that you think are zero.

(1.4) What type of internal forces and moments do you think are present in the drive axle of a rear-wheel drive car?

(1.5) For the crane following assign appropriate variables to the dimensions and then provide expressions for the internal forces and moments at sections a–a and b–b.

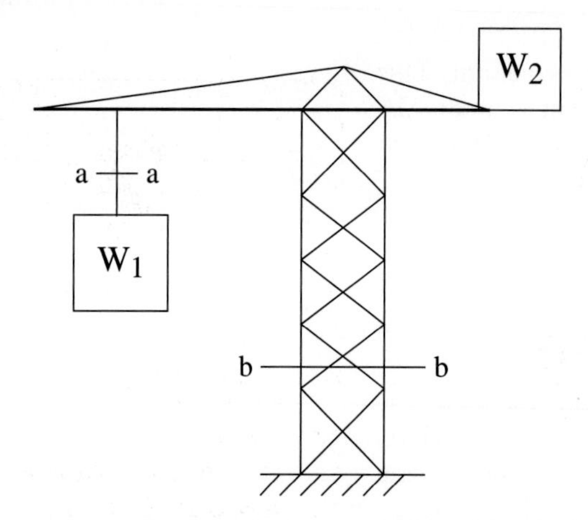


(1.6) For the following cantilever beam shown, determine the support reactions at the built-in end. Define any needed parameters for the analysis.

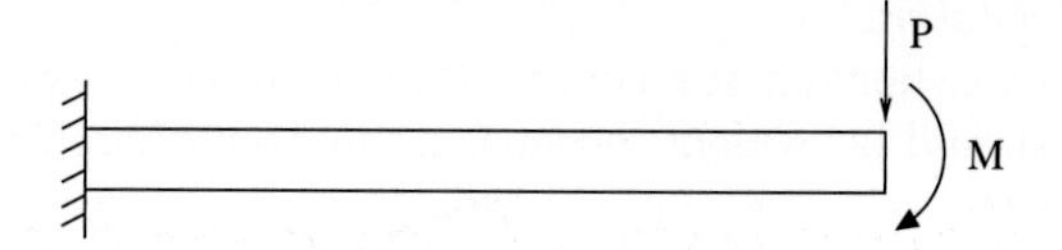


(1.7) Sketch the internal force/moment diagrams for Exercise 1.6.

(1.8) Sketch the internal force diagram for the rod shown below.

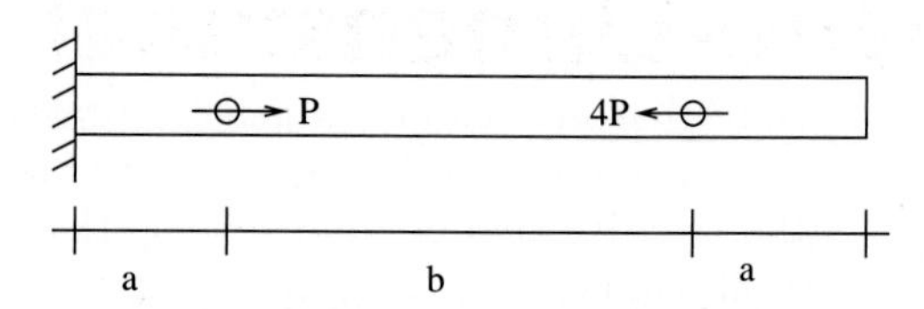

# Tension–Compression Bars: The One-Dimensional Case

# 2



One-dimensional tension–compression is a good place to begin to grasp the concepts of mechanics. It offers a relatively familiar and simple starting point for higher studies. In this chapter we will look at the essential features of mechanics problems. Our attention will be restricted to one-dimensional tension–compression problems so that we can concentrate on the fundamental concepts of mechanics. All of the remaining problem classes with which we will deal will mimic what we do here; the only real difference will be the geometric complexity. Our model system will be a bar with all loads applied in the $x$-direction and all motion occurring in the $x$ direction, where $x$ is the coordinate direction aligned with the long axis of the bar. Our goal will be a complete description of the bar's mechanical response to load. Before beginning, it is noted that even this very simple setting arises quite often in engineering and science. Figures 2.1 and 2.2 illustrate a few common examples.

## 2.1 Displacement field and strain

The first observation of deformable bodies is that when loads are applied to them they move; see Fig. 2.3. In one dimension the motion of a body is completely characterized by the $x$-displacement of the material particles. In general this displacement will be a function of the material particle which we will identify by its position. Thus the displacement field will by given by a function

$$u(x). \tag{2.1}$$

To each point $x$ in a one-dimensional body there is a corresponding displacement $u$. In Fig. 2.3, points at the wall on the left have zero displacement and those at the end where the load is applied move more.

The second observation is that loads are associated with differential motion. In Fig. 2.3, for example, the difference in motion at the two ends of the bar increases with increasing load. This differential motion $\Delta - u(L) - u(0)$ is closely related to the important concept of strain. In its simplest incarnation we say that the strain in a body is the change in

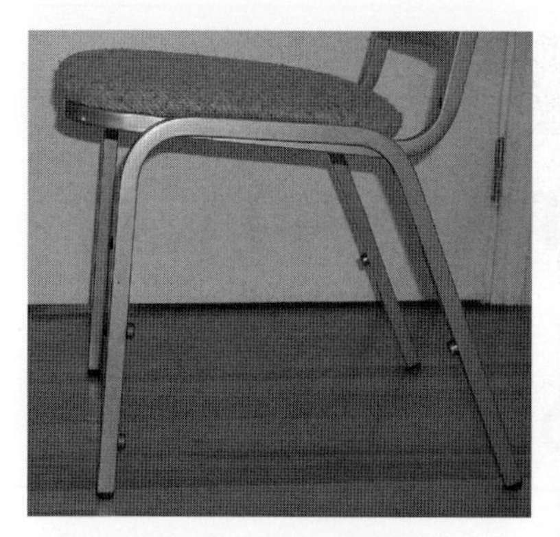

(a) Chair leg in axial compression.

(b) Deck support in compression with a point force applied in the middle.

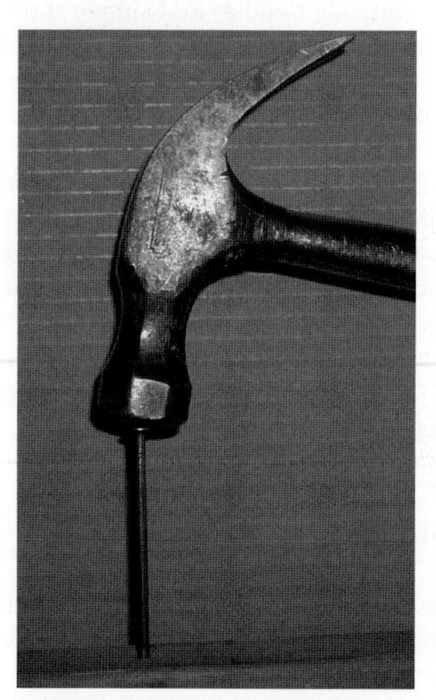

(c) A nail being struck by a hammer.

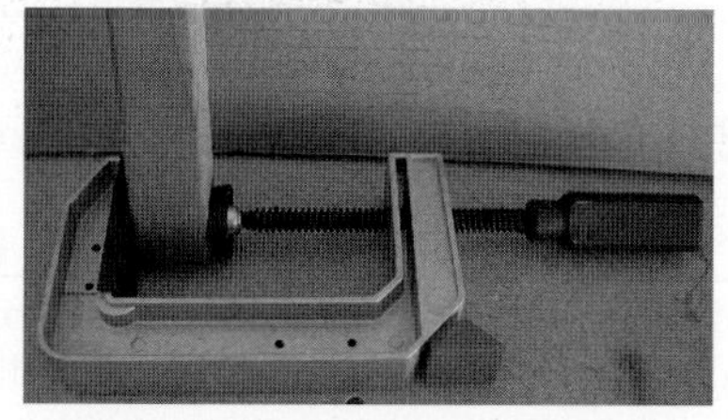

(d) A wood clamp. The tightening screw in the clamp is in compression.

**Fig. 2.1** Examples of physical systems where the mechanical loads are primarily in the axial direction.

length of a body divided by the length of the body. Mathematically this amounts to the engineer's mantra of *Delta ell over ell* as exemplified by Fig. 2.3. Physically, strain is a measure of relative deformation and is thus only non-zero when the motion is *not* rigid.

### 2.1.1 Units

The dimensions of strain are length per unit length. Thus strain is non-dimensional and does not require the specification of units. Since strains

(a) A roof strut that holds open a car hatch.

(b) Columns that support a freeway.

(c) Some of the structural elements of this front loader.

(d) A tree. The tree trunk is in compression from its own weight and the branches provide a distributed axial load.

**Fig. 2.2** Additional examples of physical systems where the mechanical loads are primarily in the axial direction.

are often of the order of $10^{-6}$ one often sees the notation $\mu$strain when discussing strain values; this is especially the case when dealing with metals in the elastic range. For other materials such as elastomers and biological materials, strains can easily be order 1.

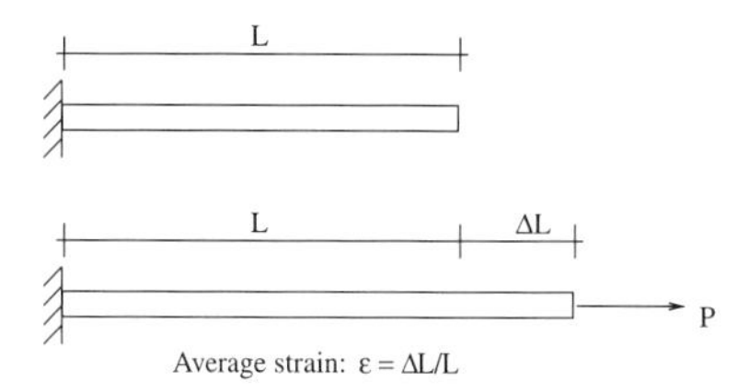


**Fig. 2.3** Definition of average strain.

## 2.1.2 Strain at a point

The definition of strain that we have given is quite common and useful in many contexts. However, it is somewhat limiting and we need to generalize it a bit even for one dimensional problems. It will be more useful to us to have a pointwise definition of strain – just as our definition of displacement is pointwise.

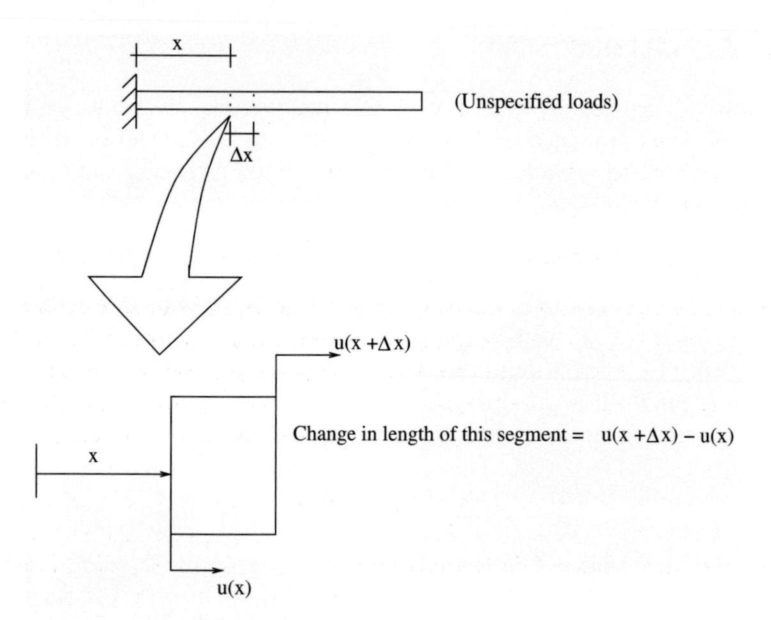


**Fig. 2.4** Construction for the derivation of pointwise strain in one dimension.

To obtain an expression for strain at a point, consider what happens to a segment of our body that starts at $x$ and ends at $x + \Delta x$. By looking at the change in length of this segment divided by its original length we will come to an expression for average strain in the $x$ direction for the segment itself; see Fig. 2.4. By taking the limit at $\Delta x \to 0$ we will then obtain an expression for strain at a point. The overall $x$-direction elongation of this particular segment is $u(x + \Delta x) - u(x)$. The average strain of the segment is then

$$\varepsilon = \frac{u(x + \Delta x) - u(x)}{\Delta x}. \tag{2.2}$$

If we take the limit as $\Delta x \to 0$ we will have a notion of strain at a point. The limit gives us

$$\varepsilon(x) = \frac{du}{dx}. \tag{2.3}$$

This is the definition of one-dimensional strain at a point.

**Remarks:**

(1) The sign convention for strain indicates that positive values of strain correspond to the elongation of material, and negative values correspond to the contraction of material.

(2) The strain-displacement relation $\varepsilon = du/dx$ will be our primary kinematic relation for one-dimensional problems.

## 2.2 Stress

Motion is caused by forces, and relative motion, strain, is properly related to stress. In one dimension we will restrict attention to forces that are acting parallel to the $x$-direction. Our working definition of stress will be:

The force per unit area acting on a section cut is the stress on the section cut.

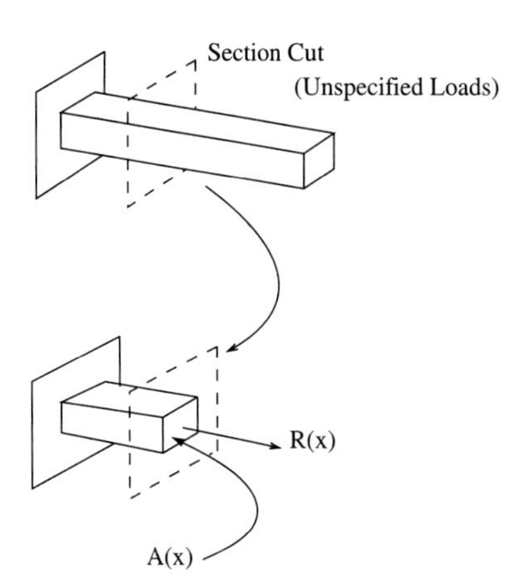


**Fig. 2.5** Construction for definition of stress.

In other words, the total force on a section divided by the area over which it acts is what we will define to be stress. This definition lends itself naturally to being a definition valid at a point along the bar, since the section cut itself is parameterized by its location along the bar (i.e. the $x$ coordinate), as is the internal force. Mathematically one has

$$\sigma(x) = \frac{R(x)}{A(x)}; \tag{2.4}$$

see Fig. 2.5, where $A$ is the cross-sectional area and $R$ is the internal force.

**Remarks:**

(1) By our sign convention for internal force we see that negative stresses correspond to compression and positive stresses to tension.

### 2.2.1 Units

The dimensions of stress are force per unit area. In the USCS the unit is most commonly pounds per square inch (psi) or thousand pounds per square inch (ksi). The international convention is commonly Newtons per square millimeter or million Newtons per square meter, also known as mega-Pascals (MPa). Note that $1\text{MPa} = 1\frac{\text{N}}{\text{mm}^2}$.

### 2.2.2 Pointwise equilibrium

Stresses are clearly related to forces; thus it should be possible to express equilibrium in terms of stresses. Doing so will bring us to an expression for equilibrium at a point in the bar as opposed to equilibrium of the overall bar (sometimes referred to as global equilibrium). Our method of arriving at such an expression will be similar to our construction for strains. Consider first a segment cut from our rod and apply the requirement of force equilibrium to it; see Fig. 2.6.

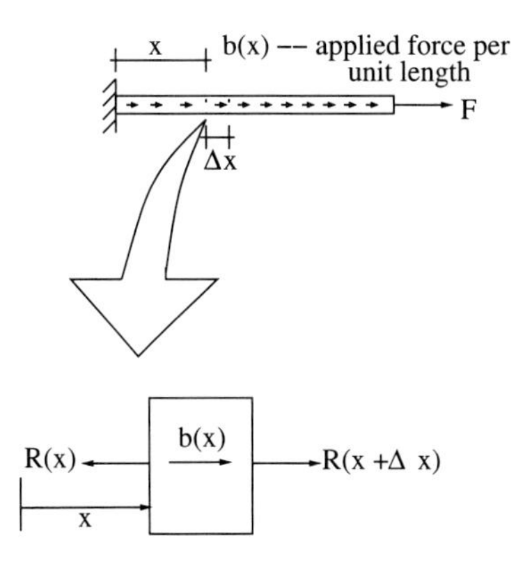


**Fig. 2.6** Construction for deriving the equilibrium equation in one dimension. Applied loads include distributed loads and point force.

$$\sum F_x = 0 = R(x + \Delta x) + b(x)\Delta x - R(x), \tag{2.5}$$

where $b(x)$ are applied loads per unit length along the length of the bar (units = force/length). If we divide through by $\Delta x$ and take the limit as $\Delta x \to 0$ we will arrive at a pointwise expression of equilibrium.

$$\frac{dR}{dx} + b = 0. \tag{2.6}$$

Utilizing our definition of stress we arrive at an expression of equilibrium at a point in terms of stresses:

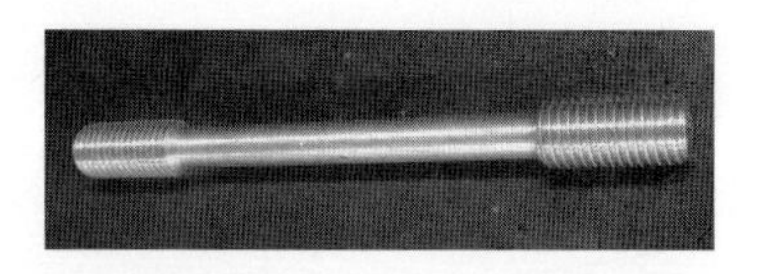

**Fig. 2.7** Test specimen for testing the tensile properties of Aluminum; the specimen is approximately 0.5 inches in diameter.

$$\frac{d}{dx}(\sigma A) + b = 0. \tag{2.7}$$

**Remarks:**

(1) The determination of the stress is made through the internal force field which can only be determined through the use of equilibrium.

## 2.3 Constitutive relations

**Fig. 2.8** Test specimen for testing the compressive properties of concrete; they are 6 by 12 inches.

Our equilibrium eqn (2.7) (equivalently eqn (2.6)) and our kinematic relation eqn (2.3) constitute two basic elements of any mechanical theory. To close the system of equations requires an expression connecting stresses to strains. The constitutive behavior of materials is the link between stress and strain. We will have occasion to use several different material models: linear elastic, non-linear elastic, elastic–plastic, and thermo-elastic. In general, constitutive properties (relations or equations) are determined from experimental investigation by taking test specimens and subjecting them to known stresses or strains and measuring the other quantity. Figures 2.7 and 2.8 show example test specimens that are used for such testing.

### 2.3.1 One-dimensional Hooke's Law

The simplest material law is Hooke's Law. It supposes a linear relation between stress and strain. In this setting, the slope of the stress–strain response curve for a material in a uniaxial test is called the Young's modulus; see Fig. 2.9. It is a material (constitutive) property; i.e. it is independent of equilibrium and kinematic considerations. Expressed in equation form, it says:

**Table 2.1** Young's moduli for some materials.

| Material | E (psi) | E (GPa) |
|---|---|---|
| Tungsten | $50 \times 10^6$ | 350 |
| Steel | $30 \times 10^6$ | 210 |
| Aluminum | $10 \times 10^6$ | 70 |
| Wood | $2 \times 10^6$ | 14 |

$$\sigma = E\varepsilon, \tag{2.8}$$

where $E$ is the symbol for Young's modulus. We will call this model linear elastic, since it represents elastic behavior and is linear in the variables that appear. Typical material properties are shown in Table 2.1.

In the presence of temperature changes the constitutive equation takes on an added term and can be expressed as

**Table 2.2** Coefficients of thermal expansion for some materials.

| Material | $\alpha$ ($\mu$strain / °C) |
|---|---|
| Iron | 12 |
| Diamond | 1.2 |
| Aluminum | 24 |
| Graphite | −0.6 |

$$\varepsilon = \frac{\sigma}{E} + \alpha\Delta T, \tag{2.9}$$

where $\alpha$ is a material property called the coefficient of thermal expansion, and $\Delta T$ is the change in temperature relative to some reference value; thus the last term ($\alpha\Delta T$) represents the thermal strain. We will call this model linear thermo-elastic. Typical values of $\alpha$ are given in Table 2.2.

### 2.3.2 Additional constitutive behaviors

The behavior described by Hooke's Law is known as linear elastic. The response of the stress to strain is linear and, further, upon reversal of the strain the unloading curve follows the loading curve; see Fig. 2.9. The linear thermo-elastic model has the same properties. Another common one-dimensional material response is elastic–plastic. Figure 2.10 shows the schematics of two such possible responses – i.e. one for a high-strength steel (HSS), and one for a low-carbon steel. Other examples of one-dimensional behavior are ceramics, which often are very brittle and possess a fracture stress and elastomers which can undergo very large strains without problem; see Fig. 2.11. The unloading curves for these last two materials follow the loading curves, and thus are termed elastic materials. The unloading curve for the elastic–plastic material is more complex. For our studies in this text we will work with an idealization of elastic–plastic behavior which is called elastic–perfectly plastic. It is shown in Fig. 2.12. Its characteristic features are that past the yield point the unloading curve does not follow the loading curve. Unloading from a state of yield gives rise to an elastic behavior but shifted with respect to the origin. Further compressive loading causes a reverse yield of the material.

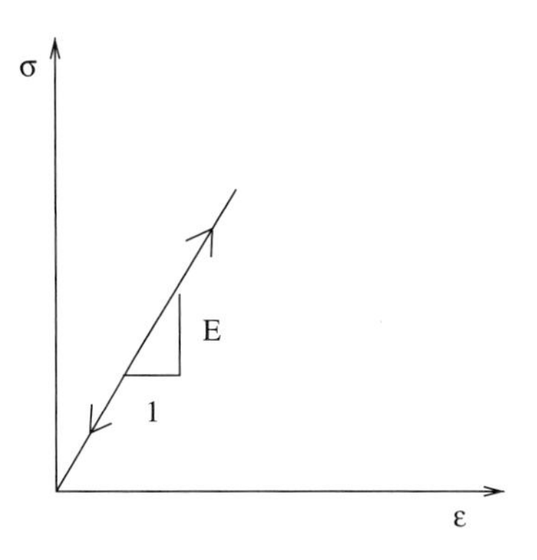

**Fig. 2.9** Linear elastic behavior.

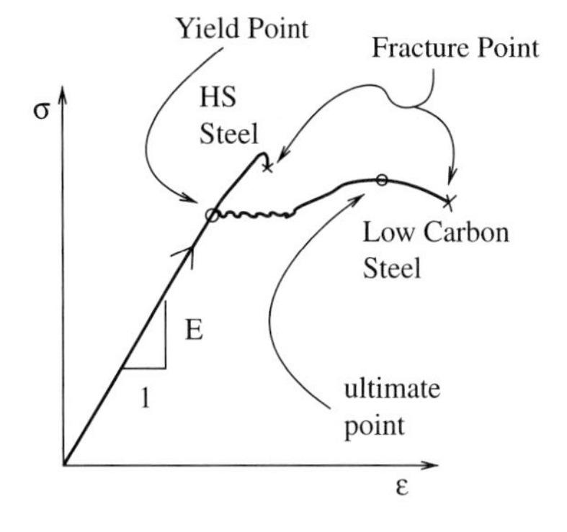

**Fig. 2.10** Schematic elastic–plastic response of a low-carbon steel and a high-strength steel.

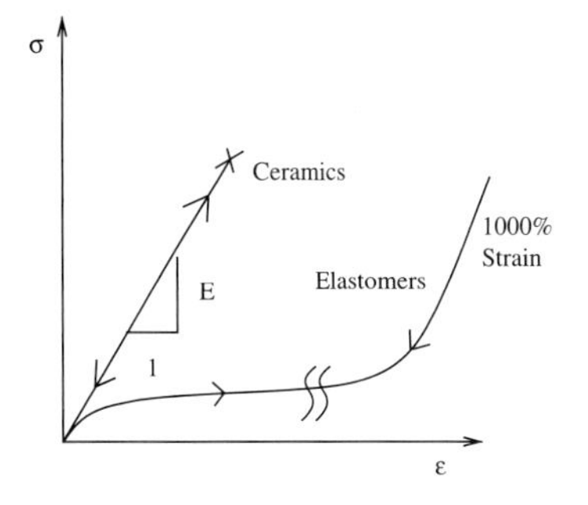

**Fig. 2.11** Schematic brittle–elastic behavior and elastomer behavior.

## 2.4 A one-dimensional theory of mechanical response

We have now defined the essential features of a one-dimensional mechanical system. The primary ingredients are a displacement field $u(x)$, a strain field $\varepsilon(x)$, a stress field $\sigma(x)$, a kinematic relation, an equilibrium expression, and a constitutive law. The basic one-dimensional problem we will deal with is this: Given the geometry of a one-dimensional body, the loads, and the boundary conditions, find the displacement, strain, and stress fields for a given constitutive specification. To solve such problems we will systematically apply the principles of kinematics, equilibrium, and constitutive relations.

### 2.4.1 Axial deformation of bars: Examples

As our first example application we will consider the bar shown in Fig. 2.13 that is loaded by a generic distributed load and some point forces. The question we will first consider is: What is the deflection at the end of the bar for a given set of loads? The central characteristic of our unknown is kinematic, and we are asked to relate it to the forces on the system. The only connection that we have between these two concepts is the constitutive relation. For this example we will assume a linear elastic material. Let us begin with kinematics. The strain in the bar is

$$\varepsilon = \frac{du}{dx}. \tag{2.10}$$

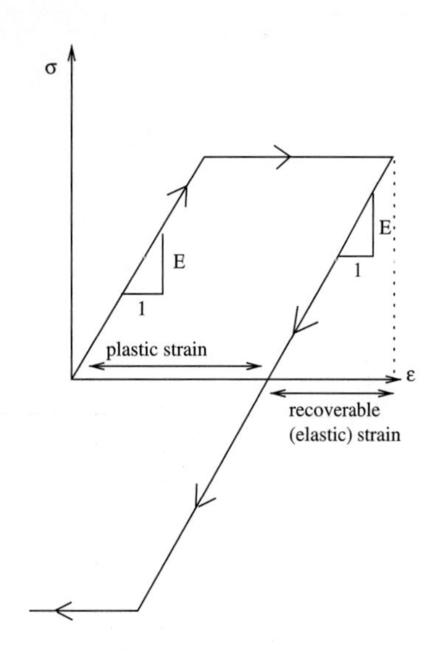


**Fig. 2.12** Idealized elastic–perfectly plastic behavior.

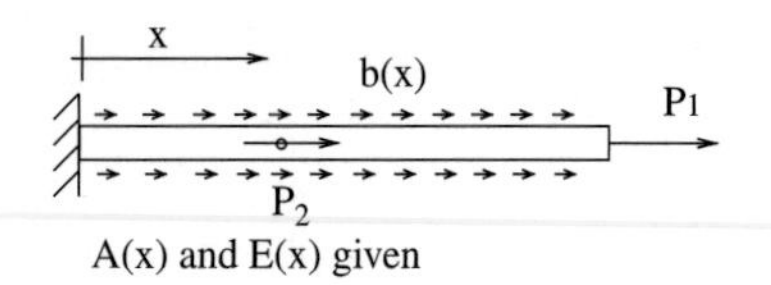


**Fig. 2.13** Generic one-dimensional bar loaded axially.

If one integrates this over the length of the bar, one will obtain the net change in length of the bar:

$$\Delta = u(L) - u(0) = \int_0^L \frac{du}{dx}\, dx = \int_0^L \varepsilon\, dx. \tag{2.11}$$

Since the bar is built-in at $x = 0$ we have the boundary condition $u(0) = 0$; note, the quantity of interest $\Delta = u(L)$. We can now apply our constitutive rule

$$\sigma(x) = E(x)\varepsilon(x) \tag{2.12}$$

to give

$$\Delta = \int_0^L \frac{\sigma}{E}\, dx. \tag{2.13}$$

Using the definition of stress

$$\sigma(x) = \frac{R(x)}{A(x)} \tag{2.14}$$

gives the "final result":

$$\Delta = \int_0^L \frac{R(x)}{A(x)E(x)}\, dx. \tag{2.15}$$

**Remarks:**

(1) So far we have used our kinematic relation, our constitutive relations, and our definition of stress. To actually use the final result we need to specify more precisely what the loads on the bar are so that we can apply equilibrium to determine the internal forces $R(x)$.

---

**Example 2.1**

*End-loaded bar.* For the end-loaded bar (see Fig. 2.3) find the deflection of the end of the bar.

*Solution*

An application of statics tells us that $R(x) = P$. If the bar is materially homogeneous (i.e. $E(x) = E$ a constant) and prismatic (i.e. $A(x) = A$ a constant) then we have that

$$\Delta = \frac{P}{AE}\int_0^L dx = \frac{PL}{AE}. \tag{2.16}$$

**Remarks:**

(1) The quantity $k = P/\Delta = AE/L$ is called the axial stiffness of the bar and represents the amount of force required to induce a unit displacement on the end of the bar.

(2) The quantity $f = \Delta/P = L/AE$ is called the flexibility of the bar and represents the amount of displacement that will result from the application of a unit force.

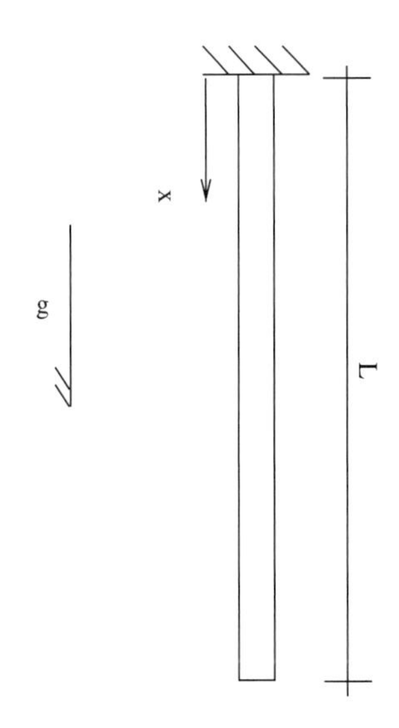


**Fig. 2.14** One-dimensional bar under the action of gravity.

### Example 2.2

*Gravity-loaded bar.* Assume our bar is acted upon by gravitational forces as shown in Fig. 2.14. Let the weight density of the bar be $\gamma$, which will have dimensions of force per unit volume. Assume $A$ and $E$ are constant. Find the deflection of the end of the bar.

*Solution*
The end deflection will be given by

$$\Delta = \int_0^L \frac{R(x)}{A(x)E(x)}\, dx = \frac{1}{AE}\int_0^L R(x)\, dx. \tag{2.17}$$

To determine the internal force we need to apply equilibrium. One way to find $R(x)$ is to solve the governing ordinary differential equation (ODE), eqn (2.6). At the moment let us not use this approach, but rather, take a direct approach of making section cuts and explicitly summing forces in the $x$-direction. Make a section cut at an arbitrary location $x$ and sum the forces on the lower section; see Fig. 2.15. This tells us that $R(x) = (L-x)A\gamma$. Inserting into the previous equation gives us

$$\Delta = \frac{1}{AE}\int_0^L R(x)\, dx = \frac{\gamma L^2}{2E}. \tag{2.18}$$

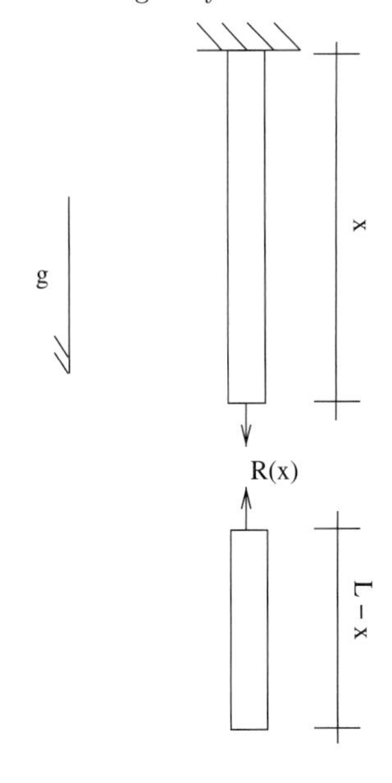


**Fig. 2.15** One-dimensional bar under the action of gravity with section cut.

**Remarks:**

(1) In coming to the last result we have used the boundary condition at $x = 0$, viz., $u(0) = 0$.

(2) To find the deflection at any other point in the bar we simply integrate from 0 to that point; i.e.

$$u(x) = u(x) - \underbrace{u(0)}_{0} = \frac{1}{AE}\int_0^x A\gamma(L-x)\, dx = \frac{\gamma}{E}(Lx - x^2/2). \tag{2.19}$$

(3) These past two examples rely on knowing what $R(x)$ is through the application of statics. This can only be achieved for statically determinate problems. When the problem is indeterminate the procedure we have followed still holds true, but there is an added complication to which one must attend.

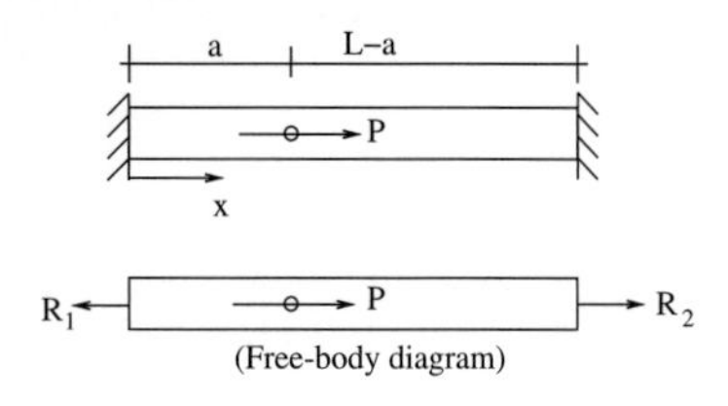


**Fig. 2.16** Statically indeterminate one-dimensional bar.

### Example 2.3

*Statically indeterminate bar with a point load.* Consider the bar as shown in Fig. 2.16. Assume $E$ and $A$ are constants and determine the internal force, stress, strain, and displacement fields.

*Solution*

The free-body diagram for the bar shows two unknown reactions, but there is only one meaningful equilibrium equation for us to use. Thus the problem is statically indeterminate (can not be determined solely from statics). A general procedure that will work for such problems (linear or non-linear) is:

(1) Assume that enough reactions are known so that the system is statically determinate.

(2) Solve the problem in terms of the assumed reactions.

(3) Eliminate the assumed reactions from the solution using the kinematic information associated with the location of the assumed reactions.

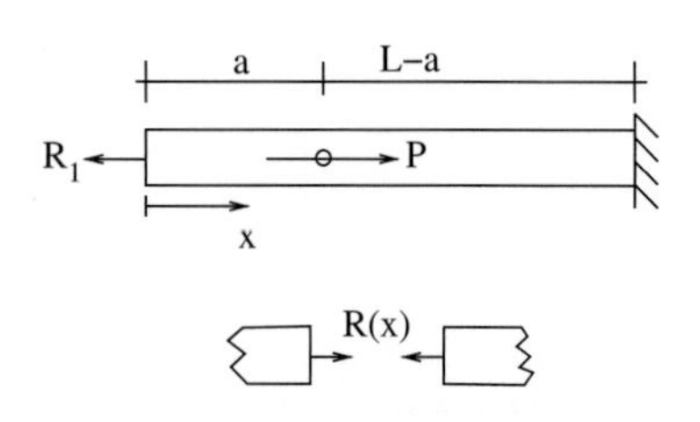


**Fig. 2.17** Statically indeterminate bar with assumed (known) reaction $R_1$.

For our bar let us apply this procedure by assuming we know the reaction at $x = 0$. Let us call this reaction $R_1$; see Fig. 2.17. We can now apply equilibrium in the horizontal direction to determine the internal forces in the bar, as shown in Fig. 2.18 (top). Dividing these by $A$ gives us the stresses (Fig. 2.18, middle) and further division by $E$ gives us the strains (Fig. 2.18, bottom). If we now integrate from 0 to $L$ we will find

$$u(L) - u(0) = \int_0^L \frac{R(x)}{AE}\, dx = \frac{R_1 a}{AE} + \frac{R_1 - P}{AE}(L - a). \tag{2.20}$$

We can now apply the known kinematic information about the system; viz., $u(0) = u(L) = 0$. This then gives an expression for $R_1$:

$$R_1 = P(1 - \frac{a}{L}). \tag{2.21}$$

This allows us to go back and correct our diagrams of the system to give an accurate picture of the state of the bar; see Fig. 2.19 where the displacement field is also plotted.

**Remarks:**

(1) For this problem we have taken a step-by-step approach, drawing the relevant functions that describe the state of the bar. We could have easily done this solely working with the appropriate equations. The process of going step-by-step with figures, however, has its merits in that it helps to emphasize the connections between the steps that are being executed in arriving at the final answer.

(2) A look at the diagrams also shows visually how the bar carries the load – tension to the left and compression to the right. Note that material motion is always to the right (positive $u$).

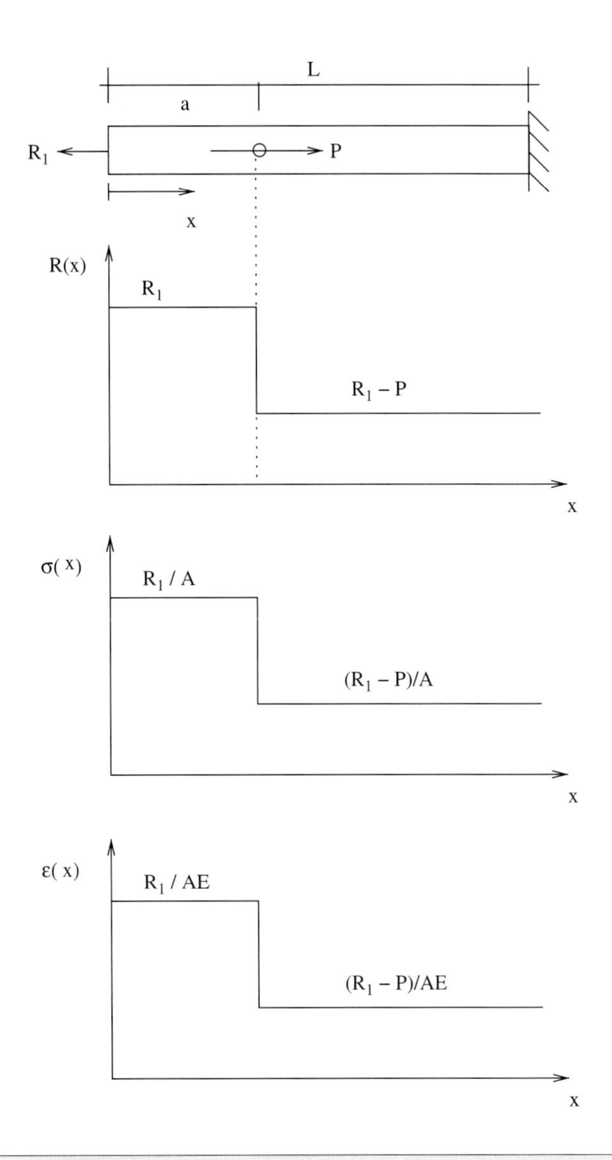


**Fig. 2.18** Internal force, stress, and strain diagrams.

### Example 2.4

*Indeterminate bar under a constant body force.* Consider the bar shown in Fig. 2.20. The bar shown is indeterminate and loaded by a distributed body force $b(x) = b_o$, a constant. Find the internal force, stress, strain, and displacement fields in the bar.

*Solution*
Since the bar is indeterminate with two unknown vertical reactions we will need to assume that one of them is known. Let us take the top reaction as known and call it $R_2$. Applying equilibrium to the bar with a section cut located at $x$ yields an expression for the internal force field as

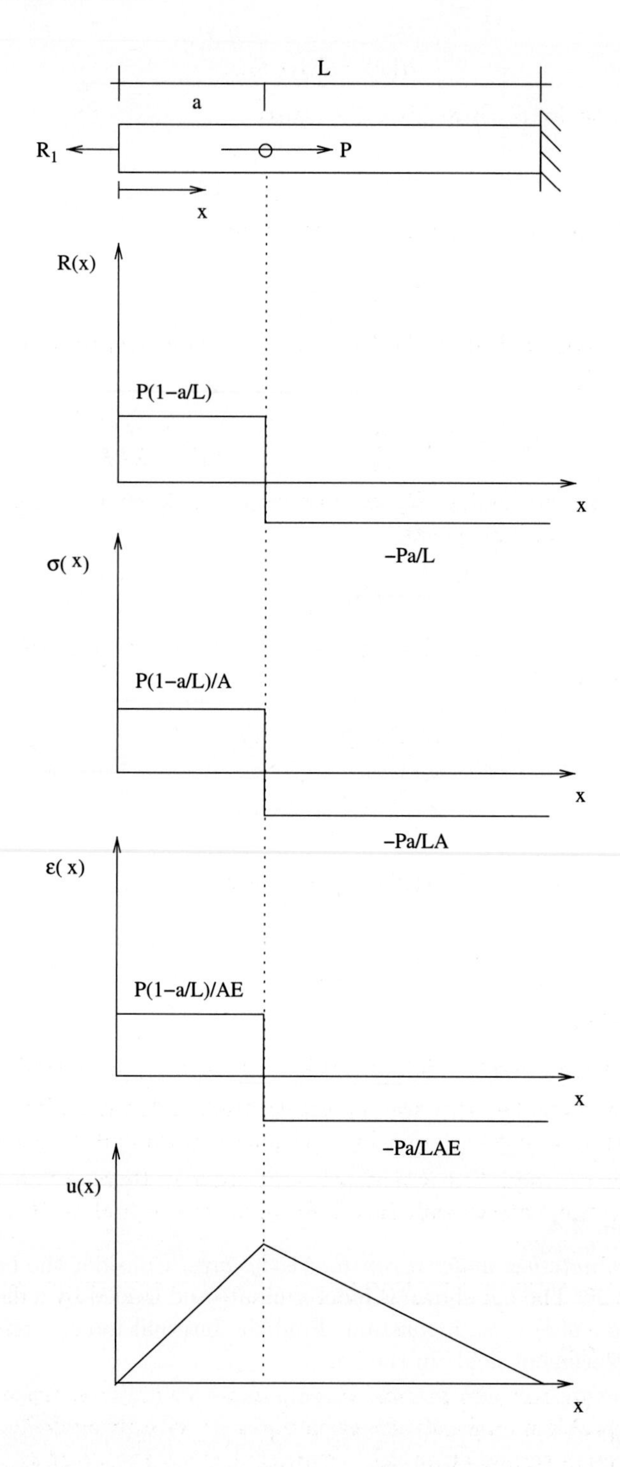


**Fig. 2.19** Corrected internal force, stress, strain, and displacement diagrams.

$$R(x) = R_2 - b_o x. \tag{2.22}$$

Dividing now by the area gives the stress

$$\sigma(x) = \frac{R_2 - b_o x}{A}. \tag{2.23}$$

Division by the modulus will give the strain

$$\varepsilon(x) = \frac{R_2 - b_o x}{AE}. \tag{2.24}$$

The displacement field in the bar is found by integrating the strain to give

$$u(x) - u(0) = \int_0^x \varepsilon(x)\, dx = \frac{R_2 x}{AE} - \frac{b_o x^2}{2AE}. \tag{2.25}$$

We can now determine $R_2$ by applying our kinematic information $u(L) = u(0) = 0$. This yields

$$R_2 = \frac{1}{2} b_o L. \tag{2.26}$$

We can now substitute this expression back into all our relations to obtain the final fields:

$$R(x) = b_o\left(\frac{L}{2} - x\right) \tag{2.27}$$

$$\sigma(x) = \frac{b_o}{A}\left(\frac{L}{2} - x\right) \tag{2.28}$$

$$\varepsilon(x) = \frac{b_o}{AE}\left(\frac{L}{2} - x\right) \tag{2.29}$$

$$u(x) = \frac{b_o}{2AE} x(L - x) \tag{2.30}$$

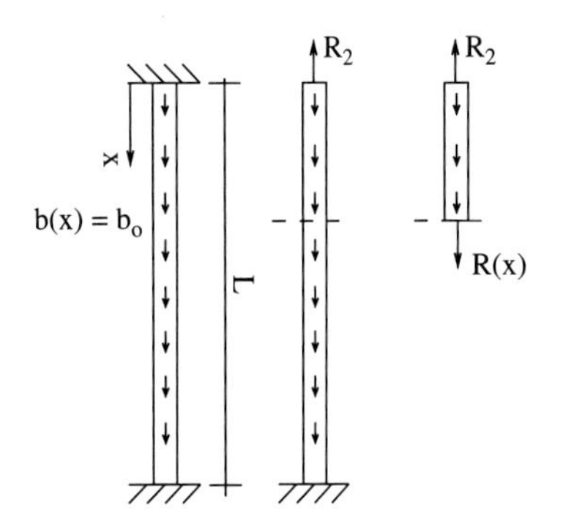


**Fig. 2.20** One-dimensional statically indeterminate bar under the action of a uniform body force.

**Remarks:**

(1) All the material points in the bar are moving downwards.

(2) The top half experiences a positive strain and is extending, and the bottom half is experiencing a negative strain and is contracting.

(3) The top half has a positive stress and is therefore in tension, and the bottom half has a negative stress and is therefore in compression.

**Example 2.5**

*Indeterminate bar with thermal loading and a point force.* Let us reconsider the problem originally shown in Fig. 2.16 with the added condition of a change in temperature $\Delta T$. Find $R(x)$, $\sigma(x)$, $\varepsilon(x)$, and $u(x)$.

*Solution*

Assume that the left-most reaction is known, and call it $R_1$. Equilibrium then tells us that

$$R(x) = \begin{cases} R_1 & x < a \\ R_1 - P & x > a\,. \end{cases} \tag{2.31}$$

Divide by the area to obtain the stress

$$\sigma(x) = \begin{cases} \frac{R_1}{A} & x < a \\ \\ \frac{R_1 - P}{A} & x > a\,. \end{cases} \tag{2.32}$$

Now apply the constitutive relation to obtain the strain. Note that we have to do more than simply divide the stress by the modulus. The thermal loading comes into play to give

$$\varepsilon(x) = \frac{\sigma(x)}{E} + \alpha\Delta T = \begin{cases} \frac{R_1}{AE} + \alpha\Delta T & x < a \\ \\ \frac{R_1 - P}{AE} + \alpha\Delta T & x > a\,. \end{cases} \tag{2.33}$$

The displacement field is then determined via integration to yield

$$\begin{aligned} u(x) - u(0) &= \int_0^x \varepsilon(x) \\ &= \begin{cases} \frac{R_1 x}{AE} + \alpha\Delta T x & x < a \\ \\ \frac{R_1 a}{AE} + \alpha\Delta T a + \frac{R_1 - P}{AE}(x-a) + \alpha\Delta T(x-a) & x > a. \end{cases} \end{aligned} \tag{2.34}$$

If we now apply our kinematic boundary conditions $u(0) = u(L) = 0$ we can determine $R_1$ as

$$R_1 = P(1 - a/L) - \alpha\Delta T A E. \tag{2.35}$$

This result can now be substituted back into the previous relations to give the final expressions for the unknown fields.

**Remarks:**

(1) Note the ease with which we were able to handle a problem with a change in temperature. The reason for this is that our solution procedure is designed to expose the interaction of the fundamental principles of equilibrium, kinematics, and constitutive relations. By keeping all the pieces separate in the process it was a simple matter to change the constitutive relation in the problem to the one appropriate for thermal loads.

### 2.4.2 Differential equation approach

The steps outlined in the examples help in seeing the interaction of the various pieces to problems in mechanics – viz., kinematics, equilibrium, and constitutive relations. From a mathematical viewpoint, however, there is no reason to introduce unknown reaction forces. In fact there is no reason to have separate solution procedures for statically determinate and indeterminate problems. To appreciate this point, let us concentrate on the linear elastic case.

There are three basic unknowns in our problems: stress, strain, and displacement. We have three relations governing them: equilibrium, strain-displacement, and the constitutive relation. If we wish we can combine these relations into a single one by first substituting the strain-displacement relation into the constitutive one to eliminate the strain; then we can substitute this new stress-displacement constitutive law into the equilibrium relation to yield:

$$\frac{d}{dx}\left(A(x)E(x)\frac{du}{dx}\right) + b(x) = 0. \tag{2.36}$$

This equation is a second-order ordinary differential equation for the displacement field. In the case where $A(x)E(x)$ is a constant it is a second-order ordinary differential equation with constant coefficients. To solve such an equation one needs two boundary conditions. We normally encounter two types of boundary conditions: displacement boundary conditions and force boundary conditions. Displacement boundary conditions will simply be the specification of the displacement at either end of the bar. Force boundary conditions will involve the specification of the rate of change of the displacement at either end of the bar. This stems from the fact that the force on a section is given by the strain times the modulus times the area. Recall that forces are related to the stresses by multiplication by the area, stresses are related to strains through the constitutive relation, and finally, strains are related to the displacements through the strain displacement expression.

As examples let us consider the problems of the previous section and first identify the distributed load function and the boundary conditions for each.

---

**Example 2.6**

*End-loaded bar revisited.* This problem was originally shown in Fig. 2.3. Here we have:

$$b(x) = 0 \tag{2.37}$$

$$u(0) = 0 \tag{2.38}$$

$$AE\frac{du}{dx}(L) = P, \tag{2.39}$$

i.e. no distributed load, fixed motion at $x = 0$, and an applied force $P$ at $x = L$.

---

### Example 2.7

*Gravity-loaded bar revisited.* This example was originally shown in Fig. 2.14; for this example we have

$$b(x) = \gamma A \tag{2.40}$$

$$u(0) = 0 \tag{2.41}$$

$$AE\frac{du}{dx}(L) = 0. \tag{2.42}$$

### Example 2.8

*Statically indeterminate bar with a point load revisited.* This example was originally shown in Fig. 2.16; for this example we have

$$b(x) = P\delta(x - a) \tag{2.43}$$

$$u(0) = 0 \tag{2.44}$$

$$u(L) = 0, \tag{2.45}$$

where $\delta(x - a)$ is a Dirac delta function located at $x = a$.

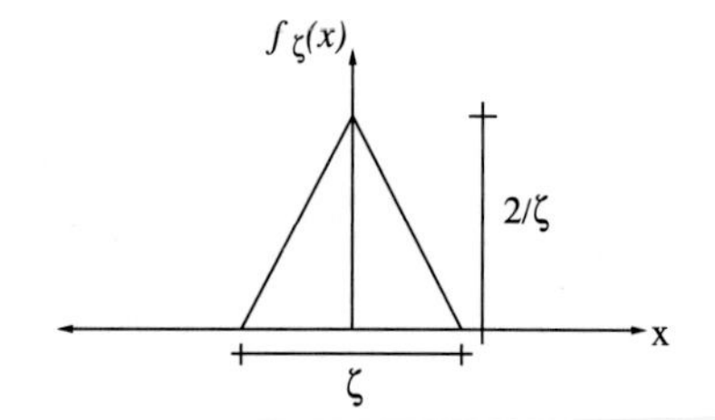


**Fig. 2.21** Distributed load representation of a localized force.

**Remarks:**

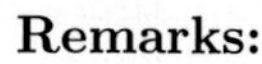

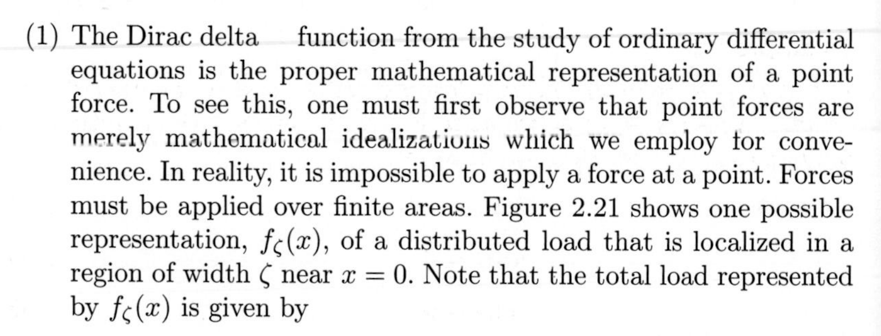

(1) The Dirac delta function from the study of ordinary differential equations is the proper mathematical representation of a point force. To see this, one must first observe that point forces are merely mathematical idealizations which we employ for convenience. In reality, it is impossible to apply a force at a point. Forces must be applied over finite areas. Figure 2.21 shows one possible representation, $f_\zeta(x)$, of a distributed load that is localized in a region of width $\zeta$ near $x = 0$. Note that the total load represented by $f_\zeta(x)$ is given by

$$\text{Total force} = \int_{-\zeta/2}^{\zeta/2} f_\zeta(x)\,dx = \frac{1}{2}\zeta\frac{2}{\zeta} = 1, \tag{2.46}$$

independent of $\zeta$. The idealization of a point force of magnitude 1 will then be given by $f_\zeta(x)$ in the limit as $\zeta$ goes to zero. We define this limit as $\delta(x)$; i.e.

$$\delta(x) = \lim_{\zeta \to 0} f_\zeta(x). \tag{2.47}$$

We call this function the Dirac delta function.

(2) As defined, the Dirac delta function has the following indefinite integration property

$$\int \delta(x)\,dx = H(x) + C, \tag{2.48}$$

where $H(x)$ is the Heaviside step function defined by:

$$H(x) = \begin{cases} 0 & x < 0 \\ 1 & x > 0 \,. \end{cases} \tag{2.49}$$

(3) It is also useful to introduce the the Macaulay bracket notation, where angle brackets have the following special meaning:

$$\langle x \rangle = \begin{cases} 0 & x < 0 \\ x & x \geq 0. \end{cases} \tag{2.50}$$

With these definitions one can deduce the following useful integration rules:

$$\int H(x)\, \mathrm{d}x = \langle x \rangle + C \tag{2.51}$$

$$\int \langle x \rangle^n \, \mathrm{d}x = \frac{1}{n+1} \langle x \rangle^{n+1} + C. \tag{2.52}$$

(4) Note that the definition we have introduced for the Dirac delta function also possesses the familiar property that for a continuous function $g(x)$,

$$\int_{0^-}^{0^+} g(x)\delta(x)\, \mathrm{d}x = g(0). \tag{2.53}$$

**Example 2.9**

*Indeterminate bar with thermal loading and a point load revisited.* This example uses the same setup as in Fig. 2.16 but with the addition of a temperature change. Here we have

$$b(x) = P\delta(x - a) \tag{2.54}$$

$$u(0) = 0 \tag{2.55}$$

$$u(L) = 0. \tag{2.56}$$

Note that our governing differential equation will need modification in this case, as eqn (2.36) was derived for a linear elastic material and not for a linear thermo-elastic material. The needed modification results in the following governing equation:

$$\frac{d}{dx}\left( A(x)E(x)\frac{du}{dx} - \alpha(x)A(x)E(x)\Delta T(x) \right) + b(x) = 0. \tag{2.57}$$

If there were force boundary conditions, they would also have to be suitably modified. For example, if we change the boundary condition at $x = L$ to an applied force of magnitude $P$, then we would write

$$AE\frac{du}{dx}(L) = P + AE\alpha\Delta T \tag{2.58}$$

for the boundary condition at $x = L$.

*Solution method*
A basic solution method for treating such equations, especially with non-constant coefficients, is to integrate both sides of the governing differential equation step by step using indefinite integration. The integration constants that will appear in this process can be eliminated by applying the boundary conditions. Note that the usual methods taught for ordinary differential equations can also be successfully applied here – viz., finding the homogeneous and particular solution and adding them together to give the general solution.

---

**Example 2.10**

*End-loaded bar.* Consider the end-loaded bar case given above, and find $u(x)$. Assume $A$ and $E$ are constants.

*Solution*
Begin with eqn (2.36) and integrate both sides twice, each time introducing a constant of integration.

$$\frac{d}{dx}\left(AE\frac{du}{dx}\right) = 0 \tag{2.59}$$

$$AE\frac{du}{dx} = C_1 \tag{2.60}$$

$$AEu = C_1 x + C_2. \tag{2.61}$$

To eliminate the constants of integration apply the boundary conditions.

$$u(0) = 0 \quad \Rightarrow \quad C_2 = 0 \tag{2.62}$$

and

$$AE\frac{du}{dx}(L) = P \quad \Rightarrow \quad C_1 = P. \tag{2.63}$$

Thus

$$u(x) = \frac{Px}{AE}. \tag{2.64}$$

Other quantities of interest such as strains, stresses, and reactions can be easily computed once $u(x)$ is known.

---

**Example 2.11**

*Statically indeterminate bar with a point load.* Consider the statically indeterminate bar with a point load given above and find $u(x)$. Assume $A$ and $E$ are constants.

*Solution*
Begin with eqn (2.36) and integrate both sides twice, each time introducing a constant of integration. Applying the rules for the Dirac delta function and its integrals, we find:

$$\frac{d}{dx}\left(AE\frac{du}{dx}\right) + b(x) = 0 \tag{2.65}$$

$$\frac{d}{dx}\left(AE\frac{du}{dx}\right) = -P\delta(x-a) \tag{2.66}$$

$$AE\frac{du}{dx} = -PH(x-a) + C_1 \tag{2.67}$$

$$AEu = -P\langle x-a\rangle + C_1x + C_2. \tag{2.68}$$

To eliminate the constants of integration, apply the boundary conditions.

$$u(0) = 0 \quad \Rightarrow \quad C_2 = 0 \tag{2.69}$$

and

$$u(L) = 0 \quad \Rightarrow \quad C_1 = P\frac{L-a}{L}. \tag{2.70}$$

Thus

$$u(x) = \frac{P}{AE}\left[-\langle x-a\rangle + \frac{L-a}{L}x\right]. \tag{2.71}$$

Other quantities of interest such as strains, stresses, and reactions can be easily computed once $u(x)$ is known.

## 2.5 Energy methods

Starting from our basic equations of kinematics, equilibrium, and material response is not always the most convenient route to answering a particular question. An alternative and very important set of methods which we will use throughout this book are energy methods. There are a variety of energy methods, but in this section we will concentrate on just one: conservation of energy. For our purposes we will restrict our attention to cases where the material of our (one-dimensional) systems is elastic. In this setting, any work we do on the material system is stored elastically:

$$W_{\text{in}} = W_{\text{stored}}. \tag{2.72}$$

In other words, we will assume that our material system does not dissipate energy. In this case $W_{\text{in}}$ is the energy lost from the loading system and by conservation it is stored in the material of our body as $W_{\text{stored}}$.

Consider now an elastic rod (not necessarily linear elastic) which we extend with an end-load. If we measure the displacement at the end of the rod, then we can make a plot of force versus displacement, as shown in Fig. 2.22. The work that we have done on the rod is the area under this curve:

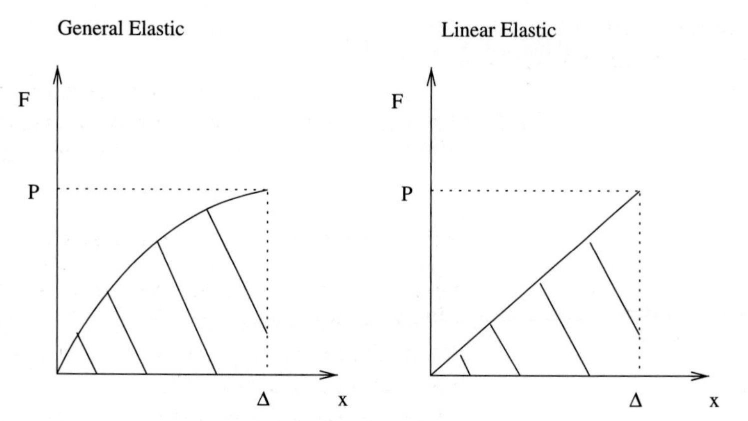


**Fig. 2.22** Force deflection curves for elastic rods.

$$W_{\text{in}} = \int_0^\Delta F(x)\,\mathrm{d}x. \tag{2.73}$$

If the material is linear elastic then the response will be linear and we find $W_{\text{in}} = \frac{1}{2}P\Delta$.

The energy stored in the material is exactly equal to this amount when the material is elastic. On a per unit volume basis, this stored energy is given by

$$\frac{W_{\text{stored}}}{AL} = \int_0^\Delta \frac{F(x)}{A}\frac{\mathrm{d}x}{L} = \int_0^\varepsilon \sigma(\varepsilon)\,\mathrm{d}\varepsilon. \tag{2.74}$$

Thus the energy stored (per unit volume) or strain energy density is given by

$$w = \int_0^\varepsilon \sigma(\varepsilon)\,\mathrm{d}\varepsilon. \tag{2.75}$$

In a linear elastic material this reduces to $w = \frac{1}{2}\sigma\varepsilon = \frac{1}{2}E\varepsilon^2 = \frac{1}{2}\sigma^2/E$. If one integrates this density over the volume of the material then we will come to an expression for the energy stored in the material:

$$W_{\text{stored}} = \int_V \frac{1}{2}\sigma\varepsilon\,\mathrm{d}V. \tag{2.76}$$

Using eqn (2.72) we find in the linear elastic case that

$$\frac{1}{2}P\Delta = \int_V \frac{1}{2}\sigma\varepsilon\,\mathrm{d}V. \tag{2.77}$$

What use is this relation? This is best seen by example.

---

**Example 2.12**

*Deflection of an end-loaded rod by conservation of energy.* Find the deflection of an end-loaded bar. Assume $A$ and $E$ to be constants.

*Solution*
Starting from eqn (2.77) we have

$$\frac{1}{2}P\Delta = \int_V \frac{1}{2}\sigma\varepsilon \, \mathrm{d}V \tag{2.78}$$

$$= \int_0^L \int_A \frac{1}{2}\sigma\varepsilon \, \mathrm{d}A \, \mathrm{d}x \tag{2.79}$$

$$= \int_0^L \frac{1}{2}\sigma\varepsilon A \, \mathrm{d}x \tag{2.80}$$

$$= \int_0^L \frac{1}{2}\left(\frac{P}{A}\right)\left(\frac{P}{AE}\right) A \, \mathrm{d}x \tag{2.81}$$

$$= \frac{1}{2}\frac{P^2 L}{AE} \tag{2.82}$$

If we now cancel $\frac{1}{2}P$ from both sides we find $\Delta = PL/AE$ – a result we had from before. Thus we see that by using conservation of energy it is possible to determine deflections in elastic systems. To see the real power of this method, consider the next example.

### Example 2.13

*Deflection of a two-bar truss.* Consider the two-bar truss shown in Fig. 2.23. Find the horizontal deflection of the truss at the point of application of the load.

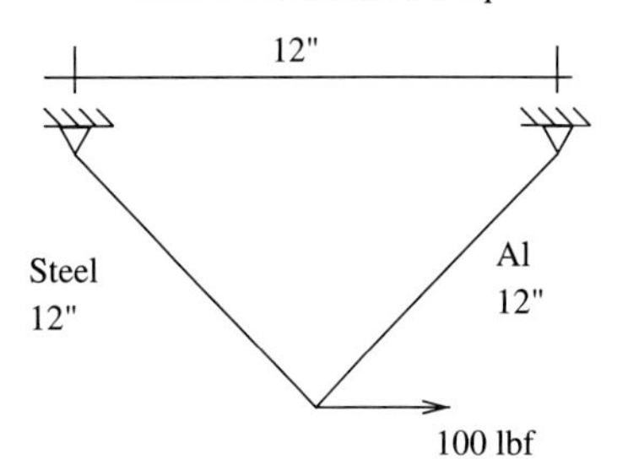


**Fig. 2.23** Two-bar truss.

*Solution*
Using energy conservation we have that

$$W_{\text{stored}} = W_{\text{Al}} + W_{\text{Steel}}$$
$$= \left(\frac{P^2 L}{2AE}\right)_{\text{Al}} + \left(\frac{P^2 L}{2AE}\right)_{\text{Steel}}. \tag{2.83}$$

From statics one has that $P_{\text{Steel}} = 100$ and $P_{\text{Al}} = -100$. Thus,

$$W_{\text{stored}} = 100^2 \left[\left(\frac{L}{2AE}\right)_{\text{Al}} + \left(\frac{L}{2AE}\right)_{\text{Steel}}\right]. \tag{2.84}$$

Setting this equal to the work done on the truss ($W_{\text{in}} = \frac{1}{2}100\Delta_H$) gives the final result:

$$\Delta_H = 100\left[\left(\frac{L}{AE}\right)_{\text{Al}} + \left(\frac{L}{AE}\right)_{\text{Steel}}\right] = 16 \times 10^{-5} \text{ inches.} \tag{2.85}$$

**Remarks:**

(1) Note that the method only gives the deflection in the direction of the applied load. No information is garnered about the vertical

deflection. Nonetheless, it is easy to see how this simple idea makes a somewhat complex problem rather tractable.

---

**Example 2.14**

*Elastic impact barrier.* Consider a rigid object of weight $W$ dropping from a height $h$ above an elastic bar of length $L$, area $A$, and modulus $E$ as shown in Fig. 2.24. Find the maximum force carried by the elastic bar before elastic release.

*Solution*

This system is conservative. The total initial energy of the system is $W(h+L)$. After the object drops it impacts the bar and deforms it. The maximum force will occur at maximum deformation, $\Delta$, which we take as positive in contraction for this problem. At the point of maximum deformation, the weight is momentarily at rest and has zero kinetic energy. At this state, the total energy of the system will be given by $W(L-\Delta)+\frac{1}{2}\frac{P^2L}{AE}$ – the potential energy of the weight plus the stored energy in the elastic bar. By conservation of energy we have that

$$W(h+L) = W(L-\Delta) + \frac{1}{2}\frac{P^2L}{AE}. \tag{2.86}$$

Noting that $\Delta = PL/AE$, we find through a little algebra that

$$P^2 - 2WP - 2WhAE/L = 0. \tag{2.87}$$

Solving shows

$$P = W[1 \pm \sqrt{1 + 2h/\Delta_s}], \tag{2.88}$$

where $\Delta_s = WL/AE$ is the static deflection of the bar.

There are two solutions to this problem, but one is physically meaningless – the one giving negative values of $P$. Note also that in this problem independent of $h$ and $\Delta_s$ the minimum force in the bar before elastic release is $2W$.

---

## 2.6 Stress-based design

The methods we have developed in this chapter can be easily incorporated into a stress-based design methodology for structural members that carry their loads axially (such as truss bars). Simply put, we can take our system of interest and apply our methodology to determine the stress field in the bar. Such information can then be used in simple stress-based design procedures when the allowable stresses for the materials of the body are known. The allowable stresses $\sigma_a$ are normally either the yield stress of the material $\sigma_Y$ or the ultimate stress of the material

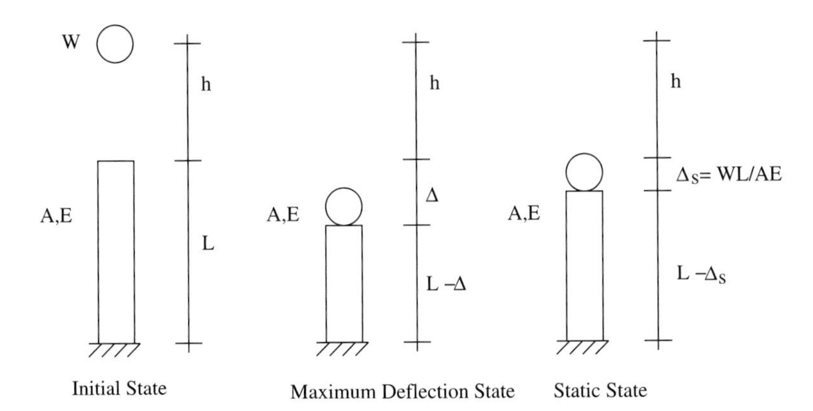


**Fig. 2.24** Elastic impact.

$\sigma_u$. To take into account uncertainties associated with the system under analysis (loading, geometry, material, and analysis model) stress-based design usually also introduces safety factors, $SF$. Simply put, the safety factor is the ratio of the allowable stress to the design stress $\sigma_d$ (the stress that is computed from the analysis):

$$SF = \frac{\sigma_a}{\sigma_d}. \tag{2.89}$$

Typical values for the safety factor vary from discipline to discipline and even within sub-disciplines. In mechanical engineering typical values are 3 on the yield stress and 4 on the ultimate stress. In civil engineering typical values are in the range 1.6–2.0 on the yield stress. In aerospace engineering the range is approximately 1.2–1.5 on the yield stress. Figure 2.25 shows an example of what happens if one exceeds the ultimate stress in a steel pipe. The pipe has split in two and displays evidence of extensive plastic yielding; this is inferred by noting the flaking-off of the whitewash which was painted on the pipe before the application of the load.

**Fig. 2.25** A steel pipe that has been pulled in tension beyond its ultimate stress point.

Design with safety factors is only one method of design. It is simple and rather common in certain industries, but there are also other design methodologies. In further design studies one will be exposed to these other methodologies.

## Chapter summary

- The essential elements that govern the analysis of a deformable mechanical system are its kinematics, its equilibrium, and its constitutive response. For an elastic tension–compression bar, the primary governing relations are:
- Strain-displacement relation

$$\varepsilon = \frac{du}{dx}$$

- Internal (resultant) force to stress relation
$$R = \sigma A$$
- Equilibrium relation
$$\frac{dR}{dx} + b = 0$$
- Hooke's Law
$$\sigma = E\varepsilon$$
- Differential equation for displacement
$$(AEu')' + b = 0$$
- Boundary conditions: fixed and forced
$$u = 0, \qquad AE\frac{du}{dx} = P$$
- For elastic material systems
$$W_{\text{in}} = W_{\text{stored}}$$
For a linear elastic system the strain energy density is $w = \frac{1}{2}\sigma\varepsilon$ and the work input is $\frac{1}{2}P\Delta$.

# Exercises

(2.1) A compression support is constructed by welding two round solid steel bars end-to-end as shown below. If the allowable stress in compression is 100 MPa, what is the allowable load $P_1$, if $P_2 = 150$ kN.

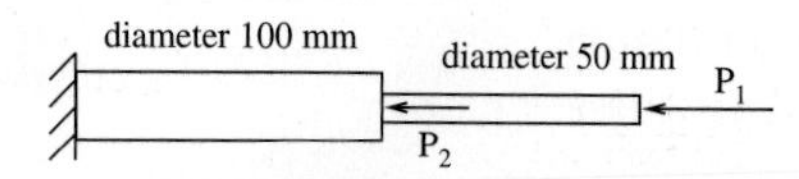


(2.2) Consider one of the concrete specimens shown in Fig. 2.8. In an experiment with applied end-forces a stress of 4.0 ksi was required to axially crush the cylinder. How much force did the testing machine exert on the specimen at crush?

(2.3) A 10-m long steel linkage is to be designed so that it can transmit 2 kN of force without stretching more than 5 mm nor having a stress state greater than 200 N/mm$^2$. If the linkage is to be constructed from solid round stock, what is the minimum required diameter?

(2.4) Estimate the amount of force it would take to (axially) crush *your* femur. Reasonable values to use for the Young's modulus are 17 GPa and for the compressive strength 190 MPa.

(2.5) Consider the Campanile (shown in the photograph). Introduce appropriate variables for the relevant dimensions, densities, etc., and derive an expression for the internal force field, $R(x)$. You may ignore the complexities associated with the openings for the bells, but make sure you account for the taper at the top.

(2.6) Consider a bar of length $L$ built-in at $x = L$ and subjected to a constant body force $b(x) = b_o$. Derive an expression for the cross-sectional area of the bar, $A(x)$, such that the stress in the bar is a constant (i.e. not a function of $x$). Assume $E$ to be a constant. You may assume that area at $x = L$ is given; i.e. assume $A(L)$ is given data.

(2.7) State which fundamental concept is represented by

$$\frac{dR}{dx} + b(x) = 0.$$

(2.8) Using the governing ordinary differential equation for the axial deformation of a bar, argue why the displacement field must be linear, independent of the boundary conditions for the bar, in the absence of any distributed body forces; i.e. for the case where $b(x) = 0$. Assume $AE$ is constant.

(2.9) Using the governing equation for the axial deformation of a bar, argue why the displacement field must be quadratic (independent of the boundary conditions for the bar) in the presence of a constant distributed body force; i.e. for the case where $b(x) = b_o$ a constant. Assume $AE$ is constant.

(2.10) Find $u(L)$ for the following bar. Assume $A$ and $E$ are constants.

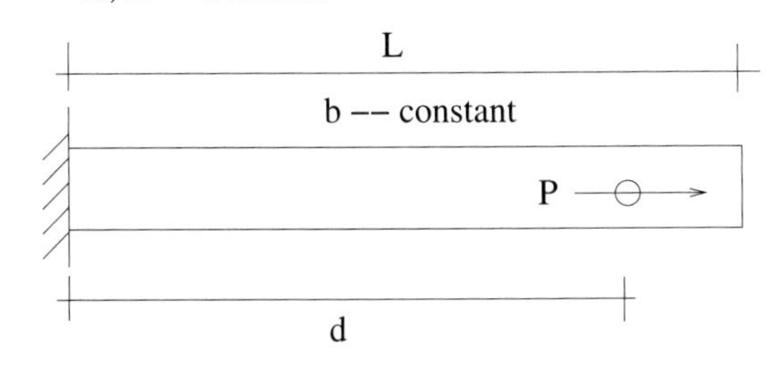


(2.11) Consider a bar which is built-in at $x = 0$, free at $x = L$, and loaded with a distributed load $b(x) = 10x$, where the constant 10 has units of N/mm$^2$. Find the displacement field $u(x)$ assuming that $E = 500$ N/mm$^2$, $A = 500$ mm$^2$, and $L = 500$ mm.

(2.12) For the bar shown determine the displacement field $u(x)$.

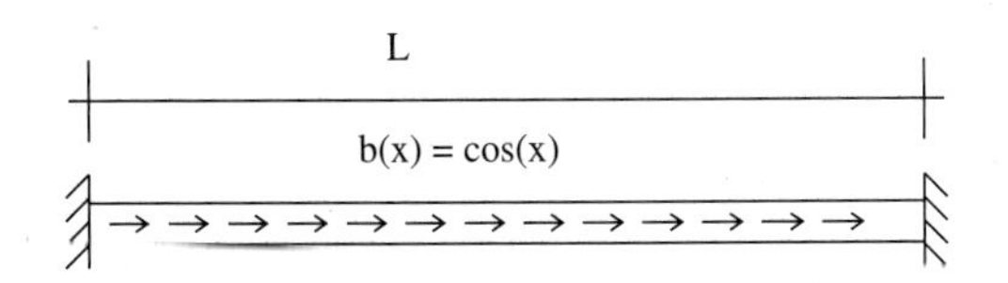


(2.13) The system shown has a linear spring support. State the relevant boundary conditions in terms of the kinematic variables and give an appropriate expression for the distributed load acting on the system.

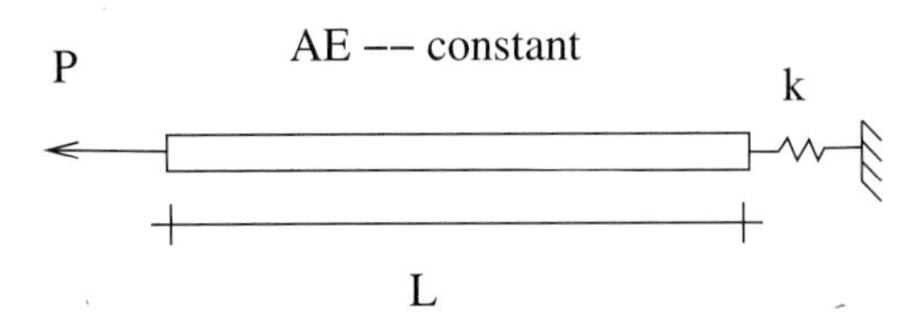


(2.14) A square rod with side length $a$ and span $L$ is to be used as an elastic spring. If the rod needs to be 10 mm long and the spring-constant needs to be 5 kN/mm, how big should $a$ be? Assume $E = 70$GPa.

(2.15) Consider a steel tape measure with cross-sectional area, $A = 0.0625$ inches squared, and length $L = 3,600$ inches at room temperature. How much error will occur if this tape measure is used on a hot day? Assume it is 130F and the coefficient of thermal expansion is $\alpha = 5 \times 10^{-6}$ 1/F. Does the error depend on the distance being measured?

(2.16) Consider the two loading cases shown on the left. By solving each case separately and adding the solutions together, show that "superposition" of the solutions is the solution of the case shown on the right. Show this for all the relevant field quantities ($u(x)$, $\varepsilon(x)$, $\sigma(x)$, $R(x)$).

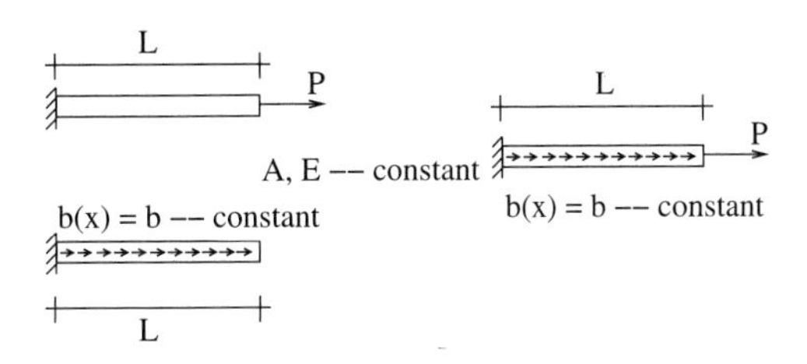


(2.17) Using a differential equation method, find the axial deflection in this spring-supported bar, $u(x)$.

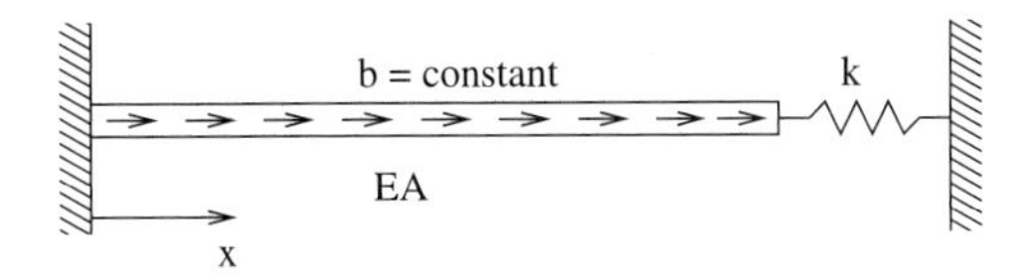


(2.18) For the linear elastic bar shown, determine the axial displacement as a function of $x$. Note that there is a distributed load and a point load.

AE -- constant

Distributed Load b(x) = Cx

x P

a L – a

(2.19) The bar shown below has a constant cross-sectional area, $A$, and is made of a non-linear elastic material whose constitutive relation is given by

$$\sigma = C\varepsilon^n,$$

where $C$ and $n$ are given material constants. Find the elongation of the bar in terms of the applied force, $P$, and the geometry.

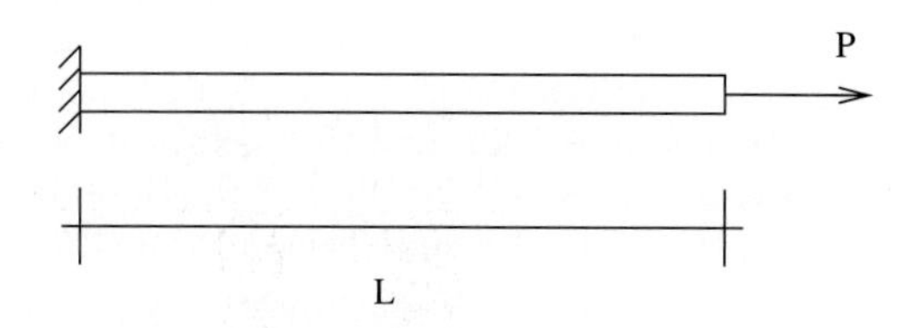


(2.20) Consider the bar in the previous exercise. If one doubles the load to $2P$, by what factor does the displacement on the end increase? For what value(s) of $n$ does the displacement double?

(2.21) You are given a prismatic bar with constant cross-sectional area $A$ and length $L$. The bar is functionally graded so that the Young's modulus is given as $E(x) = E_o + E_1 \frac{x}{L}$. Determine the reaction force at $x = L$ due to the load $P$. Express your answer in terms of $E_o = E(0)$, $E_{L/2} = E(L/2)$, $E_L = E(L)$, and the other given dimensions and load.

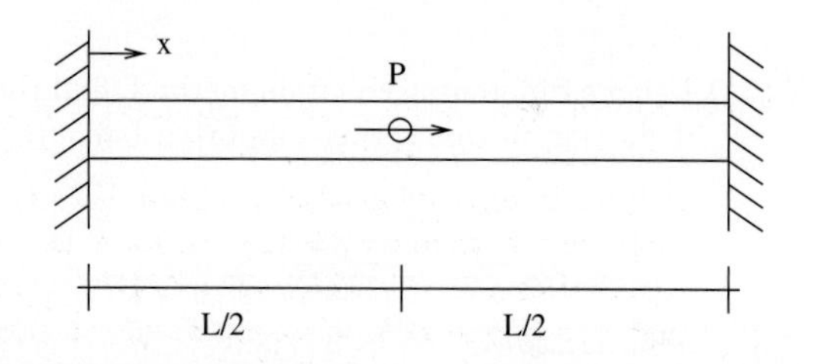


(2.22) The following bar has a linearly varying coefficient of thermal expansion. It is subjected to a gravitational acceleration and a temperature change. Find the support reaction at the top of the bar.

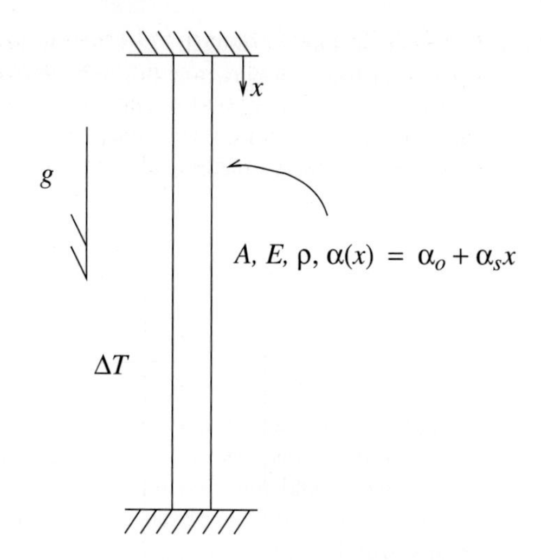


(2.23) Shown below is a slender bar with constant cross-sectional area and homogeneous linear elastic properties. Find an expression for $u(x)$, the axial displacement field in the bar. Accurately sketch $u(x)$, labeling the values of all critical points.

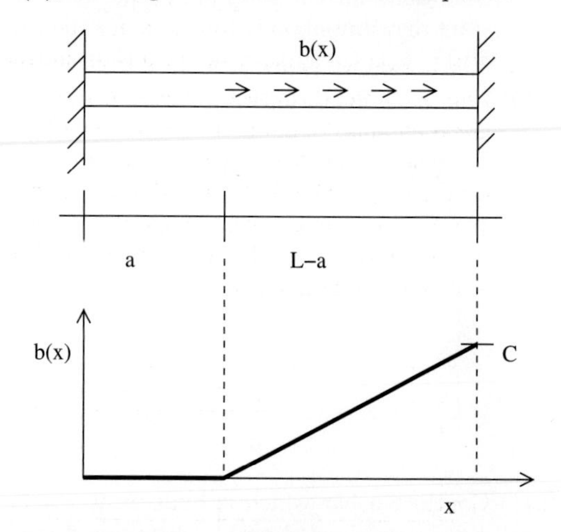


(2.24) Consider an elastic bar with constant Young's modulus, $E$, and cross-sectional area, $A$. The bar is built-in at both ends and subject to a spatially varying distributed axial load

$$b(x) = b_o \sin(\frac{2\pi}{L}x),$$

where $b_o$ is a constant with dimensions of force per unit length. Determine the largest **compressive** internal force in the bar.

(2.25) You are given a slender prismatic bar with constant cross-sectional area $A$, Young's modulus $E$, and length $L$. The bar is built in at both ends and acted upon only by gravity; assume the gravitational constant is $g$. If the mass density is $\rho$, determine $R(x)$, $\sigma(x)$, $\varepsilon(x)$, and $u(x)$. Make accurate plots of these functions, labeling all important points: maxima, minima, zero crossings, etc.

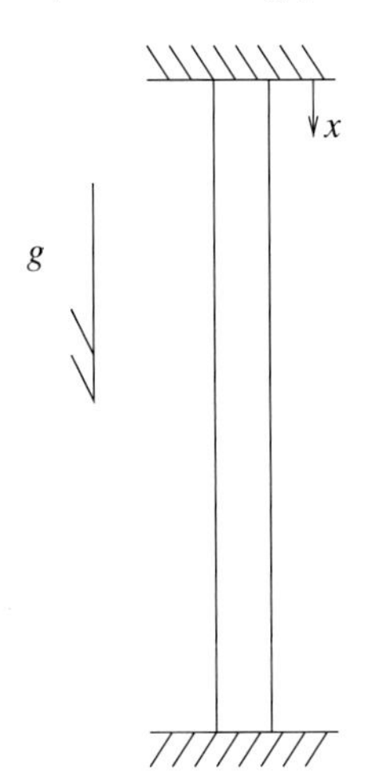


(2.26) The bar shown below is made of two pieces of metal that are welded together. The two pieces have the same cross-sectional area $A$ and Young's modulus $E$; however $\alpha_1 > \alpha_2$, where $\alpha_1$ and $\alpha_2$ are the coefficients of thermal expansion of the two pieces. For a given temperature change $\Delta T$, determine the internal force $R(x)$, the stress $\sigma(x)$, the strain $\varepsilon(x)$, and the displacement $u(x)$.

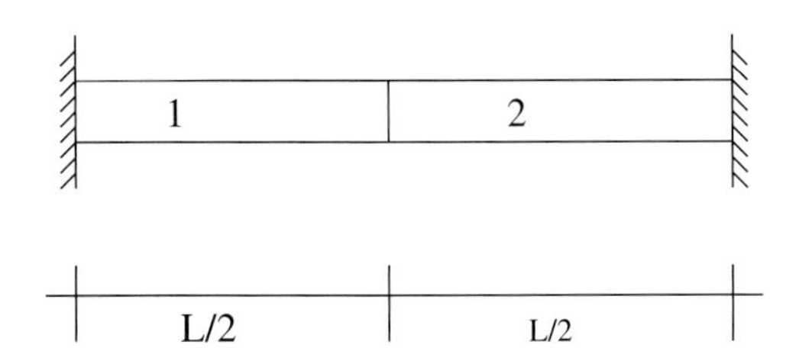


(2.27) The bar shown is built-in at the left and supported by a spring with spring constant $k$ at the right. List the boundary conditions at $x = 0$ and $x = L$. Find the expression for $u(x)$.

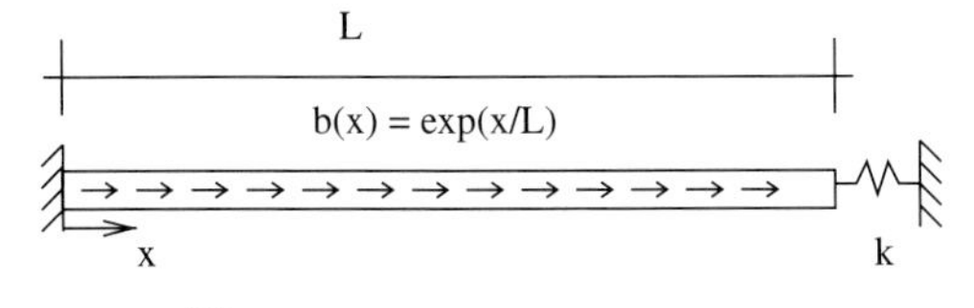


(2.28) Consider a spring-supported elastic bar with length $L$ and constant $AE$ that is subjected to a point force at $x = a$. Find $u(x)$.

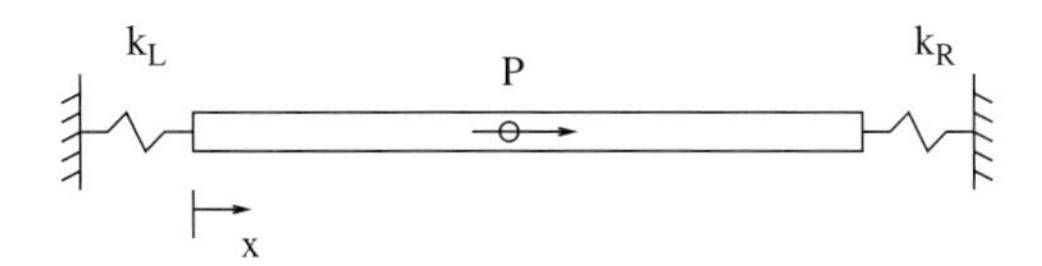


(2.29) A bolt with properties shown is inserted into a sleeve with the properties shown. A nut is threaded onto the other end and turned until it is just in contact with the sleeve. The bolt has $x$ threads per inch. How many turns of the bolt are required to initiate yield in the system if the bolt yields at $\sigma_Y^b$ and the sleeve at $\sigma_Y^s$.

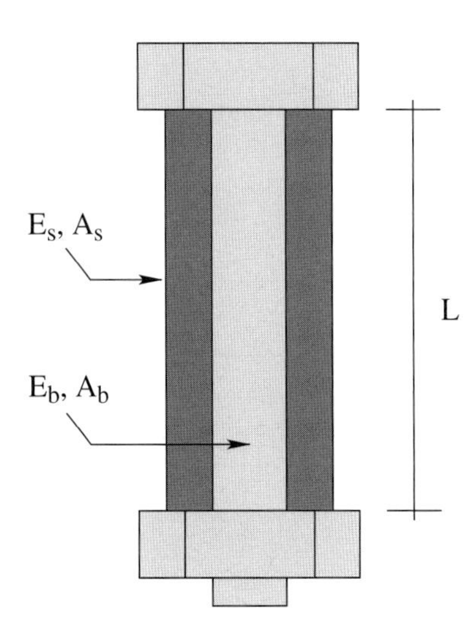


(2.30) One of the major new application areas for mechanics is in the design of devices that operate at the micron and nanometer scales. At these scales there are a number of interesting non-intuitive effects that arise. Consider a rod that spins about a post at a frequency $\omega$. If the material can support a stress of $\sigma_{\max}$ before failing, then for given dimensions and density there will be a corresponding maximum rate of spin ($\omega_c$). Let us see how this critical rate of spin depends on the scale of the device.

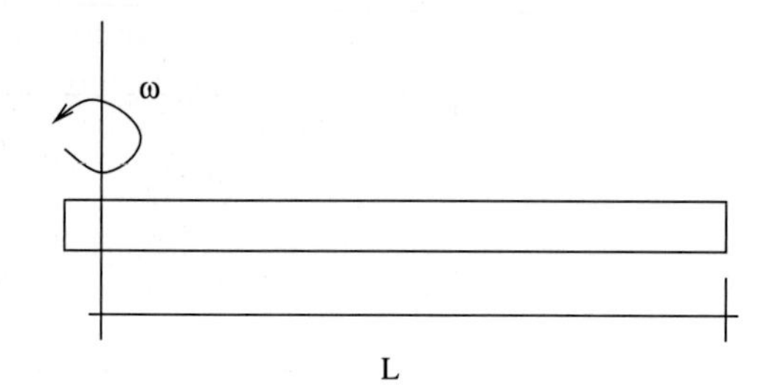


Let the length be given as $sL_o$, where $s$ is a scale factor and $L_o = 1$ m. Let the cross-sectional area be given as $s^2 A_o$ where $A_o = 25$ cm$^2$. Thus when $s = 1$, the scale of the rod is quite macroscopic (1 m long with a 20 to 1 aspect ratio). When $s = 10^{-6}$, we will be looking at a rod with nanoscale dimensions but the same aspect ratio. Plot $\omega_c$ versus $s$ for $s \in (10^{-6}, 1)$. Assume the material to be copper: $E = 128$ GPa, $\sigma_{\max} = 100$ MPa, $\rho = 8960$ kg/m$^3$. Your plot should have properly labeled axes etc.

(2.31) Consider a bar of length $L$ with constant $EA$ and constant density $\rho$. The bar is supported by a fixed pivot and spun about it at angular frequency $\omega$. Doing so produces a distributed body force $b(x) = A\rho\omega^2 x$, where $x$ is measured from the pivot. Find the maximum and minimum strains and their locations.

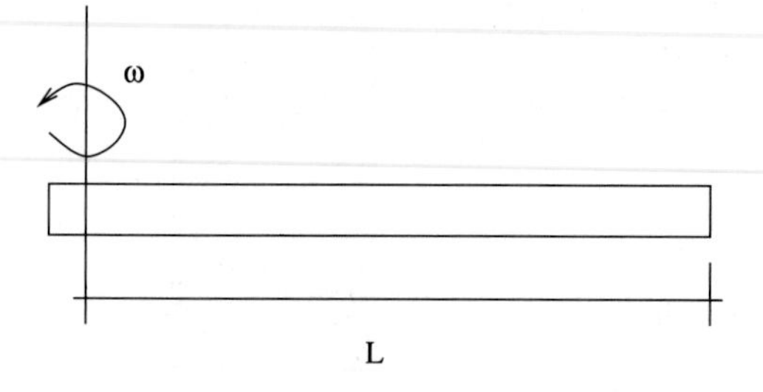


(2.32) For the system described in Exercise 2.31, find an expression for the maximum displacement and its location.

(2.33) For the two-bar truss shown, find an expression for the vertical deflection at the point of application of the load. Assume both truss bars are made from the same material with modulus $E$ and have constant cross-sectional areas $A$.

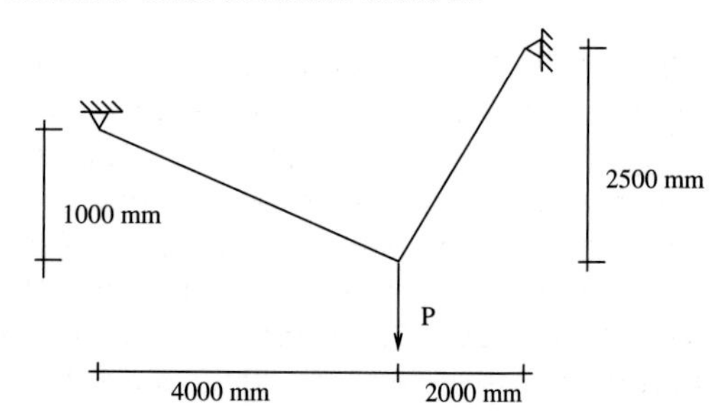


(2.34) How much energy does it take to compress a round bar of diameter 5 mm and length 300 mm to a length of 299.9 mm. Assume $E = 200$ GPa. Assume the load is applied at the ends of the bar with two opposing forces.

(2.35) Using an energy method verify the solution to Exercise 2.19. First, using your solution to Exercise 2.19, show that the work-in expression is given as $\frac{1}{n+1} P\Delta$, where $n$ is the material exponent, $P$ is the total applied load, and $\Delta$ is the total elongation. Second, equate the work-in to the work-stored. Now, using the notion of energy conservation, verify your elongation formula for Exercise 2.19.

# Stress

# 3



In the last chapter we presented an overview of mechanics through the one-dimensional problem of an axially loaded bar. The bar problem introduced us to the concepts of equilibrium, kinematics, and constitutive behavior in a deformable mechanical system; additionally we were exposed to the notions of displacement, strain, stress, and energy. In this chapter we will examine in detail the concept of stress in the general three-dimensional setting; this will be followed in the next two chapters by discussions of strain and constitutive relations in three dimensions.

We have already seen stress defined as force divided by area. This definition is quite basic and gives us a scalar that is more properly called average normal stress. In actuality, stress is a more complex concept than just force divided by area. Stress is really a *tensor* and not just a number. Without going into the details of defining a tensor, we can appreciate the distinction as something similar to the difference between the speed of an object (a scalar) and its velocity (a vector). If one knows the velocity of something $\boldsymbol{v}$, then its speed is $\|\boldsymbol{v}\|$. Velocity involves both speed (i.e. magnitude) and direction. A similar distinction holds for stress; it involves magnitudes and directions but in a more complex fashion than vectors.

## 3.1 Average normal and shear stress

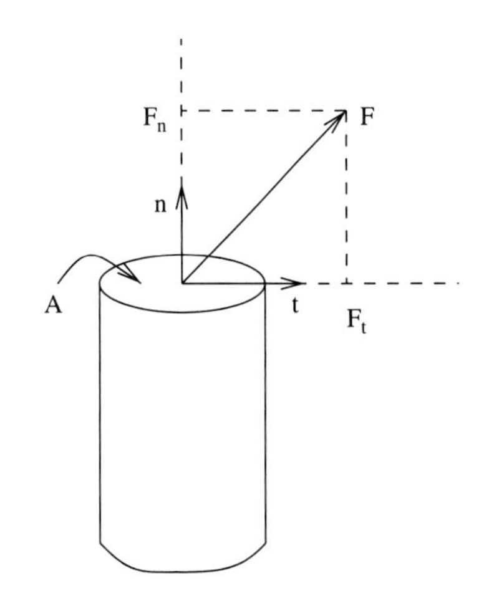


**Fig. 3.1** Section cut with generally oriented force.

As a first step to refining our understanding of stress, consider the body shown in Fig. 3.1. A force $\boldsymbol{F}$ acts on a section cut with area $A$. The force on the section cut is a vector, and it is not exactly clear how we should apply our previously developed definition of stress to compute the stress on the section cut. It proves convenient to first decompose $\boldsymbol{F}$ into its components normal and tangential to the section cut; i.e.

$$F_n = \boldsymbol{F} \cdot \boldsymbol{n} \tag{3.1}$$

$$F_t = \boldsymbol{F} \cdot \boldsymbol{t}, \tag{3.2}$$

where $\boldsymbol{n}$ and $\boldsymbol{t}$ are unit vectors normal and tangent to the section cut. We can now define *average normal stress* as

$$\sigma = \frac{F_n}{A} \tag{3.3}$$

and *average shear stress* as

$$\tau = \frac{F_t}{A}. \tag{3.4}$$

Already one can see that stress is a somewhat complex entity. It involves forces and more specifically their components. It also involves the area of the section cut (something that we will see can change).

### 3.1.1 Average stresses for a bar under axial load

To gain some appreciation for the last comment, consider the axially loaded bar shown Fig. 3.2 with cross-sectional area $A_o$. Under the given load ($\boldsymbol{F} = F\boldsymbol{e}_x$) the bar will be in a state of stress. To see what this state of stress is, make a section cut through the bar at some angle $\theta$. The normal to the section cut will be

$$\boldsymbol{n} = \sin(\theta)\boldsymbol{e}_x + \cos(\theta)\boldsymbol{e}_y \tag{3.5}$$

and a tangent to the section cut will be

$$\boldsymbol{t} = -\cos(\theta)\boldsymbol{e}_x + \sin(\theta)\boldsymbol{e}_y. \tag{3.6}$$

Note that $\boldsymbol{e}_x$ and $\boldsymbol{e}_y$ are the unit vectors in the $x$- and $y$-directions.[1] The components of the force normal and tangential to the section cut will be $F_n = F\sin(\theta)$ and $F_t = -F\cos(\theta)$. The area of the section cut will be $A_o/\sin(\theta)$. Applying our definitions from above we have for the average normal stress

[1] The unit vectors in the coordinate directions are also commonly denoted by $\hat{\imath}$ and $\hat{\jmath}$.

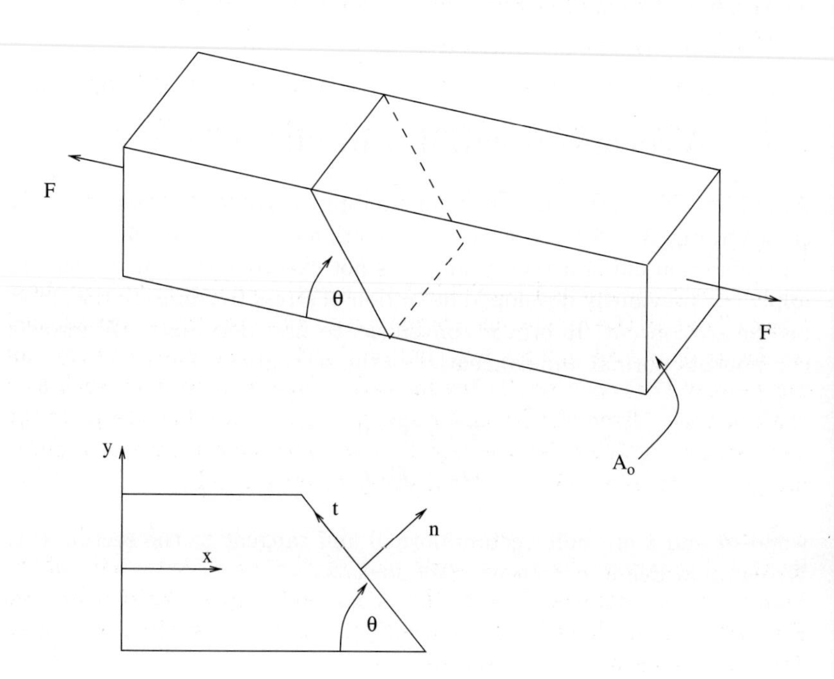


**Fig. 3.2** Bar with angled section cut.

$$\sigma = \frac{F}{A_o} \sin^2(\theta) \tag{3.7}$$

and for the average shear stress

$$\tau = -\frac{F}{A_o} \cos(\theta) \sin(\theta). \tag{3.8}$$

It should be clear that our values of average normal and shear stress strongly depend on the section cut we make, even though the state of load in the material is fixed.

This brings up the natural question: For which section cut angles are the stresses maximum? From a plot (see Exercise 3.2), one finds that the average shear stress will be extremal at an angle of $\theta = \pm\pi/4$ with a value of $\tau = F/2A_o$, and the average normal stress will be maximum at an angle of $\theta = \pi/2$ with a value of $\sigma = F/A_o$. The importance of these extremal values is that different materials are more sensitive to different kinds of stress. Loosely, one can can say that when designing with brittle materials such as cast iron and concrete one has to worry about tensile normal stresses, but when designing with ductile materials such as mild steel and aluminum one has to worry about shear stresses.

### 3.1.2 Design with average stresses

At this stage one can already perform some very basic engineering analysis of the stresses in a body. If the body is statically determinate we can define a simple procedure for stress analysis that is an extension of statics:

(1) Use equilibrium (statics) to find support reactions.

(2) Determine the internal forces[2] in the system.

(3) Resolve the internal forces into normal and tangential components.

(4) Compute the average normal and average shear stresses.

[2] Note that we have intentionally not mentioned internal moments here; to discuss the connection between internal moments and stress will require us to refine our definition of stress one more time.

**Example 3.1**

*Analysis of a shear key.* Figure 3.3 shows a gear on a drive shaft. In these types of system the keys which couple the gears to the shaft are designed as the weakest link. In this way, failure due to overloading destroys an inexpensive part such as the key instead of an expensive part such as a shaft or gear. Given the dimensions of the system and knowledge of the yield stress (in shear) for the key, find an expression for the maximum design torque when allowing for a safety factor on yield.

*Solution*

Start by isolating the shaft with half of the shear keys left intact. Summing the moments about the shaft center gives $2FR = M$; see Fig. 3.3 (bottom left). The force acts parallel to a surface with area $Lt$; thus the shear stress in the key is

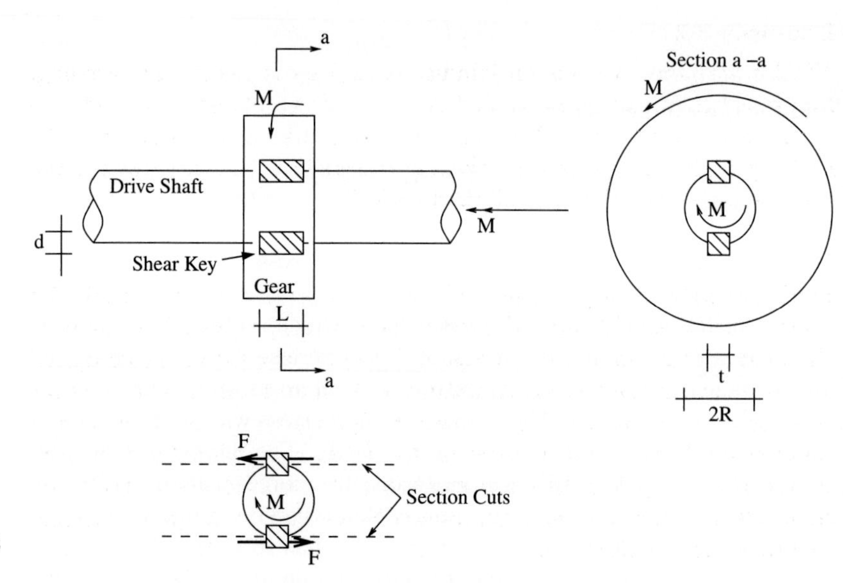


**Fig. 3.3** Gear affixed to a drive shaft by two shear keys.

$$\tau = F/Lt = M/2LRt. \tag{3.9}$$

Our safety factor and yield requirement imposes the restriction

$$\tau \cdot SF \leq \tau_Y. \tag{3.10}$$

This then implies

$$M \leq \frac{2LRt\tau_Y}{SF}. \tag{3.11}$$

To get a sense of order of magnitude of numbers, consider the following parameters (for a steel key):

$$\begin{aligned} \mathrm{SF} &= 3.0 \\ \tau_Y &= 20 \quad \text{ksi} \\ L &= 3/4 \quad \text{inches} \\ t &= 3/16 \quad \text{inches} \\ d &= 1/4 \quad \text{inches} \\ R &= 1/2 \quad \text{inches.} \end{aligned} \tag{3.12}$$

Inserting these values into our result gives $M \leq 938$ in-lbf. As a point of reference, a decent four-cylinder car has about a 100-hp engine with a peak torque output of 100 ft-lbf (equal to 1200 in-lbf).

**Example 3.2**

*Welded lap-joint.* A common joining technology is electric arc welding. Figure 3.4 shows a typical welded lap joint of two plates. The welds are designed to resist shear when the parts are pulled apart as shown. The limiting material property is given as the yield stress in shear for the weld material, $\tau_Y$. Find the allowed load $P_a$.

*Solution*

Find the loads on the weld by making section cuts as shown in Fig. 3.4 (bottom left). Equilibrium of forces tells us that the shear force on each the weld is $P/2$. To find the stress in the weld we need to divide by the area over which the force is transmitted. At first glance it appears that this area should be $Lw$. This, however, does not provide the maximal shear stress in the weld. One needs to consider additional section cuts to find the one giving the minimum area. This is as shown in the lower right of Fig. 3.4, which gives $Lt$, where $t = w/\sqrt{2}$ for a 45° weld. Putting this together one finds that:

$$\tau = \frac{P/2}{Lt}. \tag{3.13}$$

Thus,

$$P_a = 2\tau_Y Lt. \tag{3.14}$$

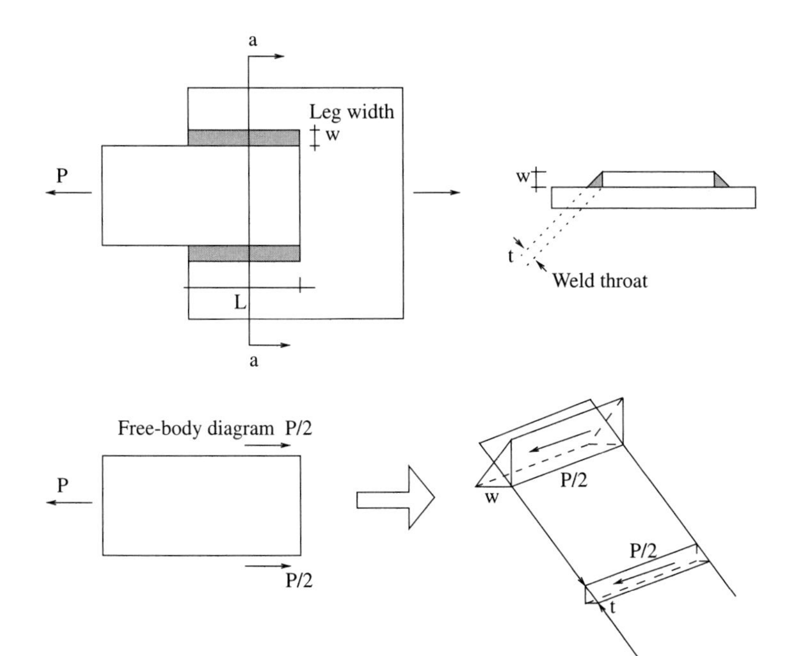


**Fig. 3.4** Welded lap joint with 45° fillet welds.

To get a sense of numbers, assume $L = 1$ inch, $w = 1/4$ inch, and $\tau_Y = 21$ ksi (typical for E70 electrodes a very common welding metal). Then, $P_a = 7.4$ kips. Note that 1 kip = 1,000 lbf.

## 3.2 Stress at a point

The stresses we have defined so far are average stresses on a section cut. This notion, however, is not sufficiently refined to allow us to have a complete picture of stress. For instance, our definition provides no connection yet between moments on a section cut and stresses. Consider the section cut shown in Fig. 3.6 with an internal moment acting on it. Let us now imagine that this moment is actually the result of a statically equivalent set of forces as shown on the right. There are an infinite number of statically equivalent force distributions for any given moment, but considering only one will suffice to motivate our main observation; viz., that the forces acting on a section cut in most cases will vary over the section cut. More generally, when one has an internal force and/or moment acting on a section cut, it is transmitted across the section cut by a distribution of forces that vary from point-to-point; see Fig. 3.5.

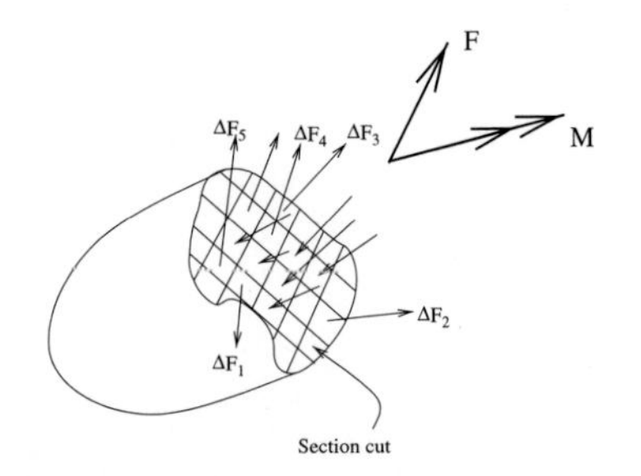


**Fig. 3.5** General distribution on a cross-section; total force is the sum of individual forces ($\Delta F_1$, $\Delta F_2$, etc.).

If we consider the section cut as shown in Fig. 3.5 and look at just one patch of material on that cut, we will find a force $\Delta \boldsymbol{F}$ acting on it. The sum over all the patches gives the total force on the cut:

$$\boldsymbol{F} = \sum_{\text{patches}} \Delta \boldsymbol{F}. \tag{3.15}$$

Similarly, the sum of the areas of the patches $\Delta A$ gives the area of the cut:

$$A = \sum_{\text{patches}} \Delta A. \tag{3.16}$$

Note that each patch can have a different area and force acting on it. For each patch define a normal stress and shear stress with our formulae from before:

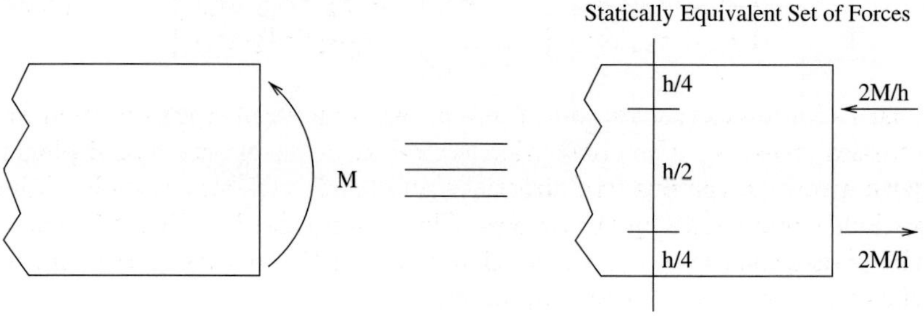

Statically Equivalent Set of Forces

M

h/4

h/2

h/4

2M/h

2M/h

**Fig. 3.6** Section cut with an internal moment.

$$\sigma = \frac{\Delta F_n}{\Delta A} \tag{3.17}$$

$$\tau = \frac{\Delta F_t}{\Delta A}. \tag{3.18}$$

In the limit as $\Delta A \to 0$ the patches will shrink to points, and we will recover a definition of stress at each point on the cross-section.

### 3.2.1 Nomenclature

At each point in a body, one can consider an infinite number of section cuts which pass through a particular point. This will lead to an infinite number of possible normal stresses and shear stresses at a point. Remarkably, knowing the normal stress and the shear stress on just three orthogonal planes passing through the point of interest permits one to know the stresses on any other section cut through the given point. In two-dimensional problems this is the case with only two orthogonal planes. In this regard, when stating the stress at a point there is a standard convention that is employed in choosing these orthogonal planes. We adorn our symbol $\sigma$ for stress with two subscripts:

$$\sigma_{\sqcup\sqcup}. \tag{3.19}$$

The first subscript is used to define the section cut orientation, and takes on the values $\{x, y, z\}$ or $\{1, 2, 3\}$. An '$x$' is used, for instance, if the section cut normal is in the $x$-direction. The second subscript is used to indicate the component of the force vector involved, and takes on values $\{x, y, z\}$ or $\{1, 2, 3\}$.

For example, the stress component $\sigma_{xx}$ corresponds to a normal stress on a section cut with normal in the $x$-direction. The symbol $\sigma_{12}$, for instance, corresponds to a shear stress in the 2-direction on a section cut with normal in the 1-direction. Quite often one will see the symbol $\tau$ used instead of $\sigma$ when the subscripts are not equal, and when the subscripts are equal one will often see the second subscript dropped. The convention for reporting the stress components at a point is to place them in a matrix using the following ordering:

$$\begin{bmatrix} \sigma_{xx} & \sigma_{xy} & \sigma_{xz} \\ \sigma_{yx} & \sigma_{yy} & \sigma_{yz} \\ \sigma_{zx} & \sigma_{zy} & \sigma_{zz} \end{bmatrix} \quad \text{or} \quad \begin{bmatrix} \sigma_{11} & \sigma_{12} & \sigma_{13} \\ \sigma_{21} & \sigma_{22} & \sigma_{23} \\ \sigma_{31} & \sigma_{32} & \sigma_{33} \end{bmatrix}. \tag{3.20}$$

Just as for forces (i.e. vectors) there is also a pictorial representation for stresses (tensors). The convention is to draw a set of orthogonal planes with arrows in the positive directions and to place the magnitude of the various components by the arrows. This is illustrated in Fig. 3.7, where the planes correspond to the section cuts and the arrows correspond to the directions of the force components.

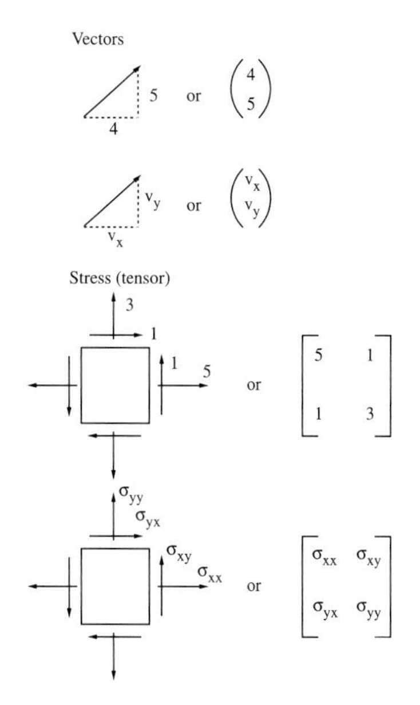


**Fig. 3.7** Vector and tensor drawing conventions.

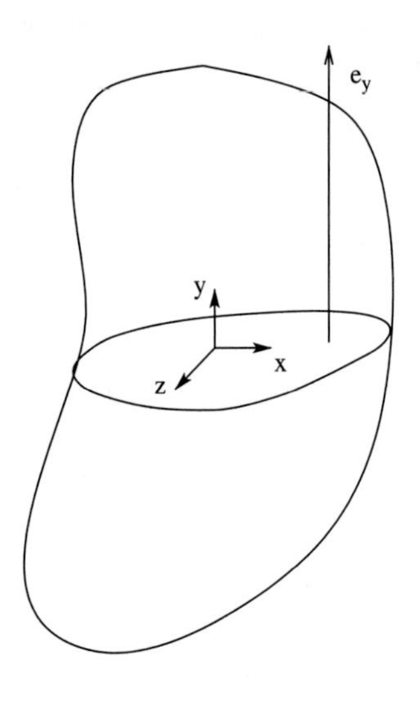


**Fig. 3.8** Body with section cut with normal $\boldsymbol{e}_y$.

### 3.2.2 Internal reactions in terms of stresses

With our established definitions we can now make precise the connection between internal forces on a section cut and the stresses on the section cut. Consider, for example, the section cut shown in Fig. 3.8 with normal $\boldsymbol{e}_y$. At each point on the cut we will find stresses $\sigma_{yy}$, $\sigma_{yx}$, and $\sigma_{yz}$; these stresses represent the force intensities (force per unit area) in the $y$, $x$, and $z$ directions, respectively. Thus to find the total force in the $y$ direction, one can simply integrate $\sigma_{yy}$ over the area of the cut:

$$F_y = \int_A \sigma_{yy}\, dA. \tag{3.21}$$

The total forces in the $x$ and $z$ directions on the section cut are given by

$$F_x = \int_A \sigma_{yx}\, dA \qquad \text{and} \qquad F_z = \int_A \sigma_{yz}\, dA. \tag{3.22}$$

To determine the total moments on the section requires us to define a point about which the moments will be computed. For convenience let us take this as the origin of our coordinate system which we will assume to be located on our cut. The stresses on our section cut give us three forces $\sigma_{yx} dA$, $\sigma_{yy} dA$, and $\sigma_{yz} dA$ on an elemental area $dA$ in the three coordinate directions $x$, $y$, and $z$. The force in the $y$-direction gives us moments about the $x$ and $z$ axes with lever arms $-z$ and $x$, respectively. The shear forces both contribute to the moment about the $y$-axis with lever arms of $z$ and $-x$. Adding the contributions from all the area elements on the section cut gives us

$$M_x = \int_A -z\sigma_{yy}\, dA \tag{3.23}$$

$$M_y = \int_A z\sigma_{yx} - x\sigma_{yz}\, dA \tag{3.24}$$

$$M_z = \int_A x\sigma_{yy}\, dA. \tag{3.25}$$

For section cuts with normals in the $x$ and $z$ directions similar arguments can be used to develop the appropriate expressions for the forces and moments.

*Forces and moments on arbitrary section cuts*

For section cuts with arbitrary normals the expressions are a little more complex to develop – though the principles are exactly the same. For the total force vector acting on a section cut with normal $\boldsymbol{n}$ and area $A$, one has:

$$\boldsymbol{F} = \int_A \boldsymbol{\sigma}^T \boldsymbol{n}\, dA, \tag{3.26}$$

where $\boldsymbol{\sigma}$ is the stress tensor and $\boldsymbol{\sigma}^T\boldsymbol{n}$ denotes the matrix-vector product of the transpose of the stress tensor with the normal vector. The moment acting on the section cut relative to a point $\boldsymbol{x}_o$ is given as:

$$\boldsymbol{M} = \int_A (\boldsymbol{x} - \boldsymbol{x}_o) \times \boldsymbol{\sigma}^T \boldsymbol{n}\, dA, \tag{3.27}$$

where $\boldsymbol{x}$ is the position vector and $\times$ is the cross product symbol.

These two relations imply that on a section cut with normal $\boldsymbol{n}$ the force per unit area, the traction $\boldsymbol{t}$, is given by $\boldsymbol{\sigma}^T \boldsymbol{n}$. This fact can be proved using an equilibrium argument. Consider the body shown in Fig. 3.9. From this body, local to a point, we cut out a small triangular wedge. Force balance on this wedge involves the balance between the surface forces (tractions) and the body forces. Let us assume without loss of generality that the stress field and the body force field are constant over the wedge. Further, let us assume that the body is two-dimensional with thickness $w$. Summing the forces on the wedge gives

$$\begin{aligned} 0 = \sum \boldsymbol{F} &= \Delta y w(-\sigma_{xx}\boldsymbol{e}_x - \sigma_{xy}\boldsymbol{e}_y) \\ &+ \Delta x w(-\sigma_{yy}\boldsymbol{e}_y - \sigma_{yx}\boldsymbol{e}_x) \\ &+ \sqrt{\Delta y^2 + \Delta x^2}\, w(t_x\boldsymbol{e}_x + t_y\boldsymbol{e}_y) \\ &+ \frac{1}{2}\Delta x \Delta y w(b_x\boldsymbol{e}_x + b_y\boldsymbol{e}_y). \end{aligned} \tag{3.28}$$

We can divide out the thickness from each term. Further, if we note that the $x$ component of the normal vector is $n_x = \Delta y/\sqrt{\Delta y^2 + \Delta x^2}$ and the $y$ component is $n_y = \Delta x/\sqrt{\Delta y^2 + \Delta x^2}$, then we can rewrite the force balance as

$$\begin{aligned} 0 &= n_x(-\sigma_{xx}\boldsymbol{e}_x - \sigma_{xy}\boldsymbol{e}_y) \\ &+ n_y(-\sigma_{yy}\boldsymbol{e}_y - \sigma_{yx}\boldsymbol{e}_x) \\ &+ (t_x\boldsymbol{e}_x + t_y\boldsymbol{e}_y) + \frac{1}{2}\frac{\Delta x \Delta y}{\sqrt{\Delta y^2 + \Delta x^2}}(b_x\boldsymbol{e}_x + b_y\boldsymbol{e}_y). \end{aligned} \tag{3.29}$$

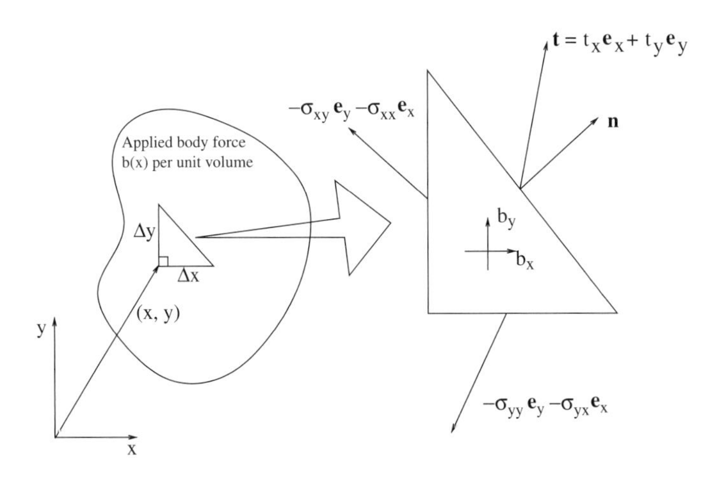


**Fig. 3.9** Wedge construction for Cauchy's Law.

If we now take the limit as $\Delta x, \Delta y \to 0$, then the body force term disappears and we are left with:

$$n_x(\sigma_{xx}\boldsymbol{e}_x + \sigma_{xy}\boldsymbol{e}_y) + n_y(\sigma_{yy}\boldsymbol{e}_y + \sigma_{yx}\boldsymbol{e}_x) = (t_x\boldsymbol{e}_x + t_y\boldsymbol{e}_y). \quad (3.30)$$

This can be expanded in matrix form to show that

$$\begin{bmatrix} \sigma_{xx} & \sigma_{xy} \\ \sigma_{yx} & \sigma_{yy} \end{bmatrix}^T \begin{pmatrix} n_x \\ n_y \end{pmatrix} = \begin{pmatrix} t_x \\ t_y \end{pmatrix}. \quad (3.31)$$

In compact form this reads $\boldsymbol{\sigma}^T\boldsymbol{n} = \boldsymbol{t}$; i.e. the transpose of the stress times the normal vector of the section cut is equal to the traction vector (force per unit area) on the cut. This result is known as *Cauchy's Law.* It also holds in three dimensions in the same form.

### 3.2.3 Equilibrium in terms of stresses

The discussion so far should give one the feeling that stresses are in some ways the representation of forces at a point. This notion leads to the natural question of whether or not we can expresses equilibrium in terms of stresses. There are two ways of approaching this question, and we will look at both. The first is the more common approach and provides some practice in utilizing a very basic method for developing many fundamental equations in engineering and science. The second approach is rather less common but mathematically more precise.

*Differential element approach*

To keep things simple we will carry out the derivations in two dimensions. Consider a body under load and look at a finite rectangular chunk of material that has a corner at the point $(x, y)$; see Fig. 3.10. The piece of material has dimensions $\Delta x$ and $\Delta y$. Let us start with force equilibrium

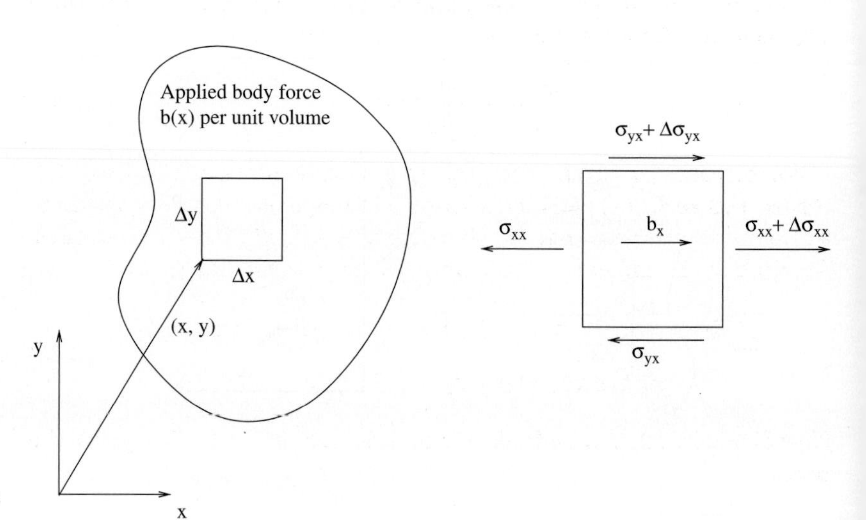


**Fig. 3.10** Two-dimensional body with applied body force $\boldsymbol{b}(x)$.

in the $x$-direction. The stresses that will contribute to forces in the $x$ directions will be $\sigma_{xx}$ and $\sigma_{yx}$. The $x$-component of the body force, $b_x$, will also contribute to the forces in the $x$-direction. If we assume that the stresses are relatively constant over the sides then we can add up the total force in the $x$ direction very easily. To do this, on the bottom and left sides we will assume the stresses to have the values of the stresses at the point $(x, y)$. On the right and top the stresses will have different values that we will define in terms of their change from those on the bottom and left. Multiplying the stresses by the areas over which they act gives the appropriate force values, and multiplying the body force by the volume over which it is distributed gives a force. Summing all the forces results in

$$\begin{aligned}\sum F_x &= (\sigma_{xx} + \Delta\sigma_{xx})t\Delta y - (\sigma_{xx})t\Delta y \\ &\quad + (\sigma_{yx} + \Delta\sigma_{yx})t\Delta x - (\sigma_{yx})t\Delta x \\ &\quad + b_x t\Delta x\Delta y = 0.\end{aligned} \tag{3.32}$$

In the last expression, $t$ represents the thickness of the two-dimensional body. It is only used to get the units correct; its exact value is not required, since it drops out of the equations. Now, divide through by $t\Delta x\Delta y$ to give

$$\frac{\Delta\sigma_{xx}}{\Delta x} + \frac{\Delta\sigma_{yx}}{\Delta y} + b_x = 0. \tag{3.33}$$

Taking the limit as $\Delta x, \Delta y \to 0$ gives the partial differential equation

$$\frac{\partial\sigma_{xx}}{\partial x} + \frac{\partial\sigma_{yx}}{\partial y} + b_x = 0. \tag{3.34}$$

Thus the statement "sum of the forces in the $x$-direction equals zero" is replaced by a partial differential equation. We can follow the same argument for the $y$ direction to obtain the relation:

$$\frac{\partial\sigma_{xy}}{\partial x} + \frac{\partial\sigma_{yy}}{\partial y} + b_y = 0. \tag{3.35}$$

For moment equilibrium we will take moments about the lower left corner. Figure 3.11 shows all the stresses components that will contribute to the moment. Note that it suffices to only include the shear stresses. Computing the moment about the lower left corner then gives:

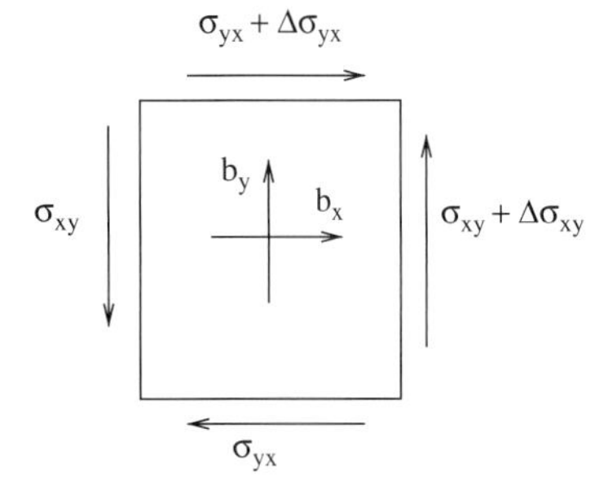


**Fig. 3.11** Small piece of material with stresses and body forces which contribute to moment about the z axis.

$$\begin{aligned}\sum M_z &= (\sigma_{xy} + \Delta\sigma_{xy})t\Delta y\Delta x - (\sigma_{yx} + \Delta\sigma_{yx})t\Delta x\Delta y \\ &\quad + b_x t\Delta x\Delta y(\Delta y/2) - b_y t\Delta x\Delta y(\Delta x/2) = 0.\end{aligned} \tag{3.36}$$

Now divide by $t\Delta x\Delta y$ and take the limit as $\Delta x, \Delta y \to 0$. This yields the result:

$$\sigma_{xy} = \sigma_{yx}. \tag{3.37}$$

So the standard force and moment equilibrium equations in two dimensions give rise to two partial differential equations and one algebraic equation when written in terms of stresses.

In three dimensions one has a very similar set of results; three partial differential equations and three algebraic relations:

$$\sum F_x = 0 \Rightarrow \frac{\partial \sigma_{xx}}{\partial x} + \frac{\partial \sigma_{yx}}{\partial y} + \frac{\partial \sigma_{zx}}{\partial z} + b_x = 0 \tag{3.38}$$

$$\sum F_y = 0 \Rightarrow \frac{\partial \sigma_{xy}}{\partial x} + \frac{\partial \sigma_{yy}}{\partial y} + \frac{\partial \sigma_{zy}}{\partial z} + b_y = 0 \tag{3.39}$$

$$\sum F_z = 0 \Rightarrow \frac{\partial \sigma_{xz}}{\partial x} + \frac{\partial \sigma_{yz}}{\partial y} + \frac{\partial \sigma_{zz}}{\partial z} + b_z = 0 \tag{3.40}$$

$$\sum M_x = 0 \Rightarrow \sigma_{yz} = \sigma_{zy} \tag{3.41}$$

$$\sum M_y = 0 \Rightarrow \sigma_{zx} = \sigma_{xz} \tag{3.42}$$

$$\sum M_z = 0 \Rightarrow \sigma_{xy} = \sigma_{yx}. \tag{3.43}$$

**Remarks:**

(1) The implication here is that the concept of equilibrium restricts how the stresses can vary from point to point in a body, and further, that the stress tensor is symmetric.

*Integral theorem approach*

Through the integral theorems of calculus one can also derive the above expressions. This method permits one to execute the argument in a mathematically cleaner manner. Note that there is nothing incorrect about what was done in the previous section. It is merely that we executed the argument without a detailed specification of a number of assumptions (which as it turns out are perfectly valid).

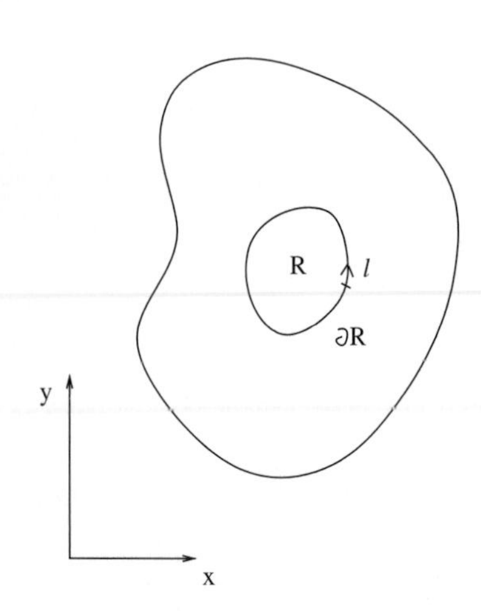


**Fig. 3.12** Two-dimensional body with part $R$ used for analysis.

Again we will do everything in two dimensions to reduce the amount of writing required. To simplify things we will ignore body forces. Consider first an arbitrary region, $R$, of material as shown in Fig. 3.12 with boundary curve $\partial R$. In equilibrium the total force on this region is zero; thus by eqn (3.26) we have that

$$\begin{pmatrix} 0 \\ 0 \end{pmatrix} = \begin{pmatrix} F_x \\ F_y \end{pmatrix} = \int_{\partial R} \begin{pmatrix} \sigma_{xx} n_x + \sigma_{yx} n_y \\ \sigma_{xy} n_x + \sigma_{yy} n_y \end{pmatrix} dl. \tag{3.44}$$

If we now apply the divergence theorem, we have that

$$\begin{pmatrix} 0 \\ 0 \end{pmatrix} = \int_R \begin{pmatrix} \frac{\partial \sigma_{xx}}{\partial x} + \frac{\partial \sigma_{yx}}{\partial y} \\ \frac{\partial \sigma_{xy}}{\partial x} + \frac{\partial \sigma_{yy}}{\partial y} \end{pmatrix} dxdy. \tag{3.45}$$

Since the region $R$ is arbitrary this implies by the localization theorem that the integrand must be equal to zero. Thus we recover our partial differential equations of equilibrium, eqns (3.34) and (3.35).

For moment equilibrium of region $R$ (about the origin) in two dimensions we have, using eqn (3.27):

$$0 = M_z = \int_{\partial R} x(\sigma_{xy}n_x + \sigma_{yy}n_y) - y(\sigma_{xx}n_x + \sigma_{yx}n_y)\,dl. \tag{3.46}$$

The application of the divergence theorem to this integral gives:

$$0 = \int_R \sigma_{xy} + x(\frac{\partial \sigma_{xy}}{\partial x} + \frac{\partial \sigma_{yy}}{\partial y}) - \sigma_{yx} - y(\frac{\partial \sigma_{xx}}{\partial x} + \frac{\partial \sigma_{yx}}{\partial y})\,dxdy. \tag{3.47}$$

By force equilibrium the second and fourth terms are zero, and we are left with the result that

$$0 = \int_R (\sigma_{xy} - \sigma_{yx})\,dxdy. \tag{3.48}$$

Noting that this has to hold for all regions $R$, an application of the localization theorem tells us that the integrand must be zero; i.e.

$$\sigma_{xy} = \sigma_{yx}. \tag{3.49}$$

## 3.3 Polar and spherical coordinates

So far we have described our mechanics in terms of Cartesian coordinates. While this is the easiest way of doing things it is not always the most convenient. In particular, one often finds it easier to use polar(cylindrical) or spherical coordinates. Figure 3.13 shows the definitions of these orthogonal coordinate systems. Note that while the definition of the cylindrical coordinate system is standard, the definition of the spherical coordinate system varies from book to book. For cylindrical coordinates one has

$$\begin{aligned} x_1 &= r\cos(\theta), & r &= \sqrt{x_1^2 + x_2^2}, \\ x_2 &= r\sin(\theta), & \theta &= \tan^{-1}(x_2/x_1), \\ x_3 &= z, & z &= x_3. \end{aligned} \tag{3.50}$$

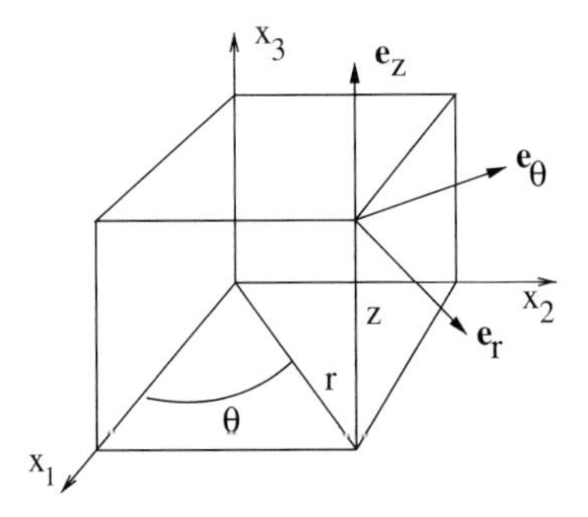


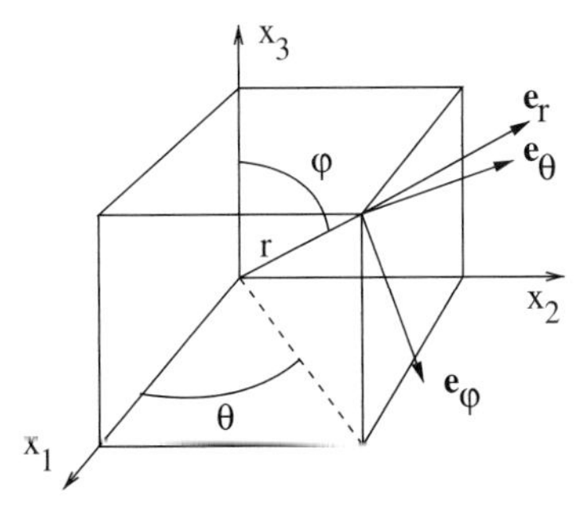


**Fig. 3.13** Definition of the cylindrical and spherical coordinate systems.

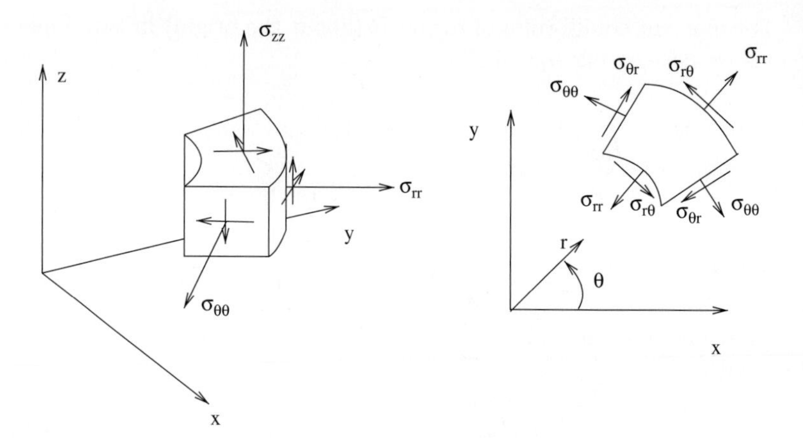


**Fig. 3.14** Cylindrical stresses. (Right) Three-dimensional representation with some stresses labeled. (Left) Two-dimensional case with all stresses labeled.

and for spherical coordinates one has

$$
\begin{aligned}
x_1 &= r\sin(\varphi)\cos(\theta), \qquad & r &= \sqrt{x_1^2+x_2^2+x_3^2}, \\
x_2 &= r\sin(\varphi)\sin(\theta), \qquad & \varphi &= \cos^{-1}\left(\frac{x_3}{\sqrt{x_1^2+x_2^2+x_3^2}}\right), \\
x_3 &= r\cos(\varphi), \qquad & \theta &= \tan^{-1}(x_2/x_1).
\end{aligned}
\tag{3.51}
$$

### 3.3.1 Cylindrical/polar stresses

Just as with Cartesian stresses we define polar stresses with reference to section cuts and force directions. For example, $\sigma_{r\theta}$ represents forces per unit area in the $\theta$-direction on a section cut with normal vector $\boldsymbol{e}_r$. Such a surface is a cylinder. Similar meaning can be ascribed to other stress components taking $\{r, z, \theta\}$ for subscripts. Figure 3.14 shows some of these stresses on the standard polar/cylindrical element cut from a solid.

The equilibrium equations in cylindrical coordinates can be derived using differential element arguments (where the shape of the differential element is the same as the integration volume in cylindrical coordinates). The result of such an exercise is that

$$
\begin{aligned}
\sum F_r &= \frac{\partial \sigma_{rr}}{\partial r} + \frac{1}{r}\frac{\partial \sigma_{r\theta}}{\partial \theta} + \frac{\partial \sigma_{rz}}{\partial z} + \frac{\sigma_{rr}-\sigma_{\theta\theta}}{r} + b_r = 0 \\
\sum F_\theta &= \frac{\partial \sigma_{\theta r}}{\partial r} + \frac{1}{r}\frac{\partial \sigma_{\theta\theta}}{\partial \theta} + \frac{\partial \sigma_{\theta z}}{\partial z} + \frac{2\sigma_{\theta r}}{r} + b_\theta = 0 \\
\sum F_z &= \frac{\partial \sigma_{zr}}{\partial r} + \frac{1}{r}\frac{\partial \sigma_{z\theta}}{\partial \theta} + \frac{\partial \sigma_{zz}}{\partial z} + \frac{\sigma_{zr}}{r} + b_z = 0
\end{aligned}
\tag{3.52}
$$

Note that moment equilibrium requires that $\sigma_{zr} = \sigma_{rz}$, $\sigma_{r\theta} = \sigma_{\theta r}$, and $\sigma_{z\theta} = \sigma_{\theta z}$. Thus just as with Cartesian stresses, the cylindrical/polar stresses are symmetric when placed in a matrix. In other words, the order of the subscripts does not matter.

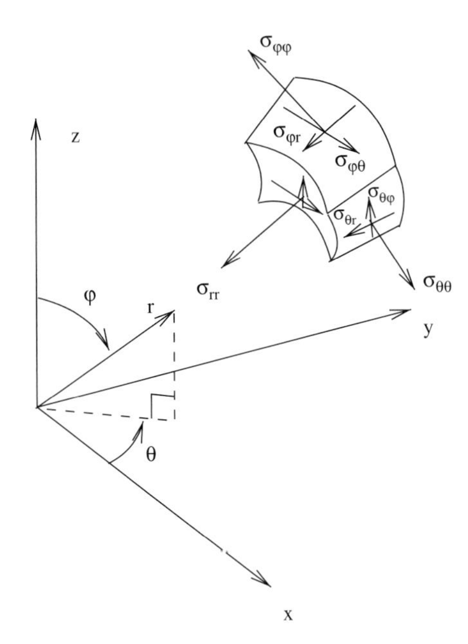


**Fig. 3.15** Spherical stresses. Three-dimensional representation with some stresses labeled on the standard spherical element.

### 3.3.2 Spherical stresses

A similar interpretation of stress in spherical coordinates also holds. The first subscript always refers to the normal vector of the section cut, and the second the direction of the force. Some of the stresses are labeled in Fig. 3.15. Applying equilibrium to a differential element yields the relations:

$$\begin{aligned}
\sum F_r &= \frac{\partial \sigma_{rr}}{\partial r} + \frac{1}{r}\frac{\partial \sigma_{r\varphi}}{\partial \varphi} + \frac{1}{r\sin(\varphi)}\frac{\partial \sigma_{r\theta}}{\partial \theta} \\
&\quad + \frac{2\sigma_{rr} - \sigma_{\varphi\varphi} - \sigma_{\theta\theta} + \sigma_{r\varphi}\cot(\varphi)}{r} + b_r = 0 \\
\sum F_\varphi &= \frac{\partial \sigma_{\varphi r}}{\partial r} + \frac{1}{r}\frac{\partial \sigma_{\varphi\varphi}}{\partial \varphi} + \frac{1}{r\sin(\varphi)}\frac{\partial \sigma_{\varphi\theta}}{\partial \theta} \\
&\quad + \frac{3\sigma_{\varphi r} + (\sigma_{\varphi\varphi} - \sigma_{\theta\theta})\cot(\varphi)}{r} + b_\varphi = 0 \\
\sum F_\theta &= \frac{\partial \sigma_{\theta r}}{\partial r} + \frac{1}{r}\frac{\partial \sigma_{\theta\varphi}}{\partial \varphi} + \frac{1}{r\sin(\varphi)}\frac{\partial \sigma_{\theta\theta}}{\partial \theta} \\
&\quad + \frac{3\sigma_{r\theta} + 2\sigma_{\varphi\theta}\cot(\varphi)}{r} + b_\theta = 0
\end{aligned} \tag{3.53}$$

Note that moment equilibrium requires $\sigma_{\varphi r} = \sigma_{r\varphi}$, $\sigma_{r\theta} = \sigma_{\theta r}$, and $\sigma_{\varphi\theta} = \sigma_{\theta\varphi}$. Thus just as with Cartesian and cylindrical/polar stresses, the spherical stresses are symmetric when placed in a matrix. In other words, the order of the subscripts does not matter here either.

## Chapter summary

- Average normal stresses on a section cut
$$\sigma = \frac{F_n}{A}$$
- Average shear stresses on a section cut
$$\tau = \frac{F_t}{A}$$
- Three-dimensional stress at a point is given by $\sigma_{\sqcup\sqcup}$, where the first index gives the direction of the section cut and the second the direction of the force. This holds for both Cartesian as well as non-Cartesian orthogonal coordinate systems.
- Cauchy's Law: the traction on a section cut in terms of the stresses and the normal vector to the section cut
$$\boldsymbol{t} = \boldsymbol{\sigma}^T \boldsymbol{n}$$
- Total force on a section cut
$$\boldsymbol{F} = \int_A \boldsymbol{\sigma}^T \boldsymbol{n}\, dA$$
- Total moment on a section cut
$$\boldsymbol{M} = \int_A (\boldsymbol{x} - \boldsymbol{x}_o) \times \boldsymbol{\sigma}^T \boldsymbol{n}\, dA$$
- Force equilibrium in terms of stresses is given by a set of partial differential equations.
- Moment equilibrium in terms of stresses is given by the requirement that the stress tensor be symmetric; i.e. shear stresses on orthogonal planes must be equal.

# Exercises

(3.1) Two pieces of wood are to be glued together using a lap joint. The shear strength of the glue is 10 $N/mm^2$. Find the maximum permissible load $F$.

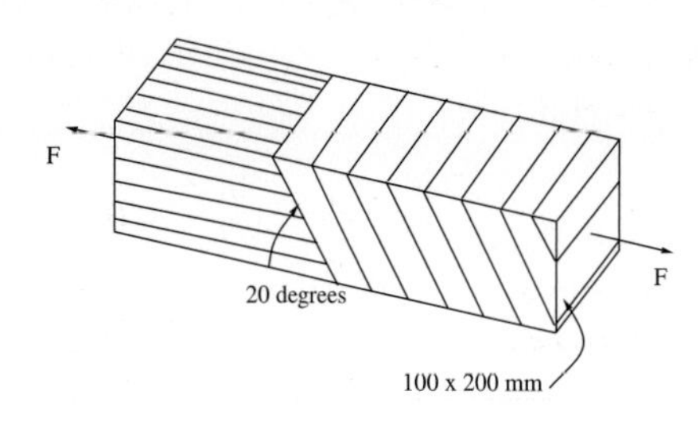


(3.2) Consider Fig. 3.2 and plot the average shear and normal stresses on the section as a function of angle $\theta \in [0, \pi]$. Normalize your stresses by $F/A_o$. Remember to properly label your graph.

(3.3) A common method of joining rods is to use a pin joint as shown. If the cross-sectional area of the pin is $A$ and the applied load is $F$, what is the maximum shear stress in the pin? (Hint: this configuration is known as double shear.)

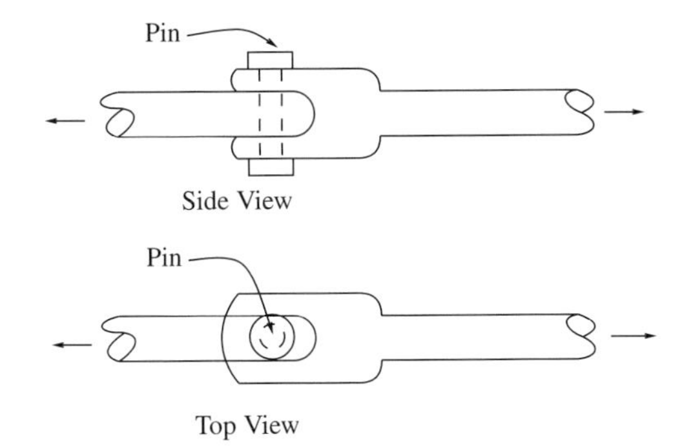


(3.4) The lap joint shown is made by gluing three pieces of metal together. The glue can support a shear stress of $\tau_{\max}$. Develop a formula for the maximum allowed load $P$. Assume a safety factor of 2.

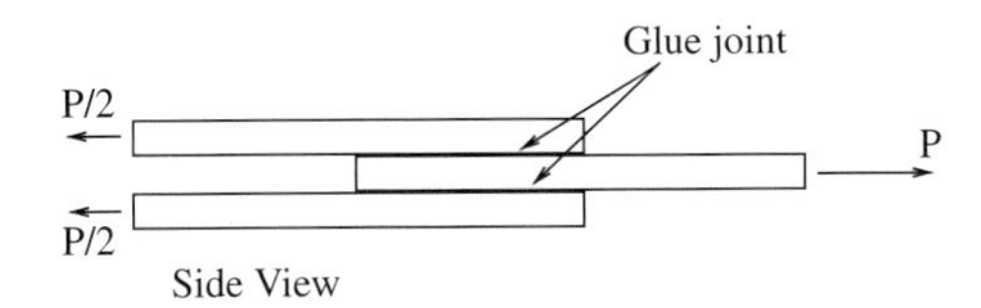


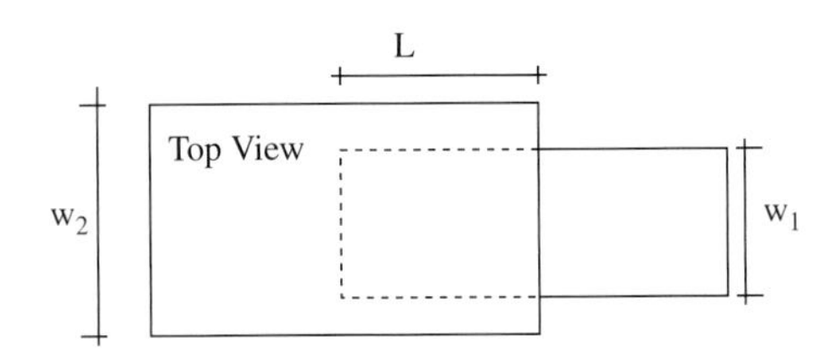


(3.5) Two truss bars are to be pinned to together to make a support frame. The load to be supported is a round barrel weighing 10,000 lbf. If the pin at $A$ is in double shear (see Exercise 3.3) with a 1 inch diameter, find the shear stress in the pin.

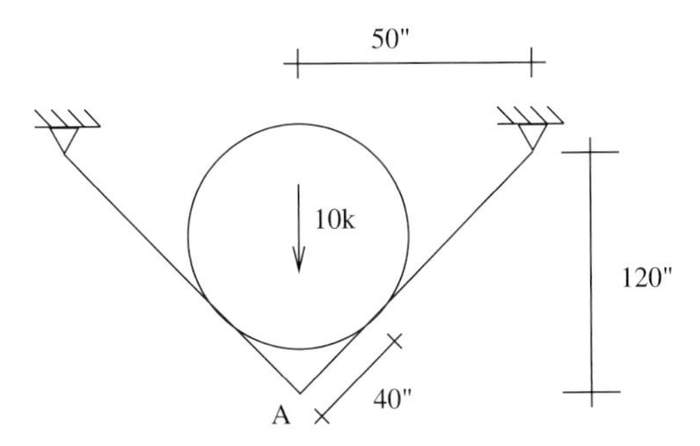


(3.6) Consider the frame that follows. If the pins at $A$, $B$, and $C$ are 3 cm in diameter and in double shear (see Exercise 3.3), find the shear stress in each pin?

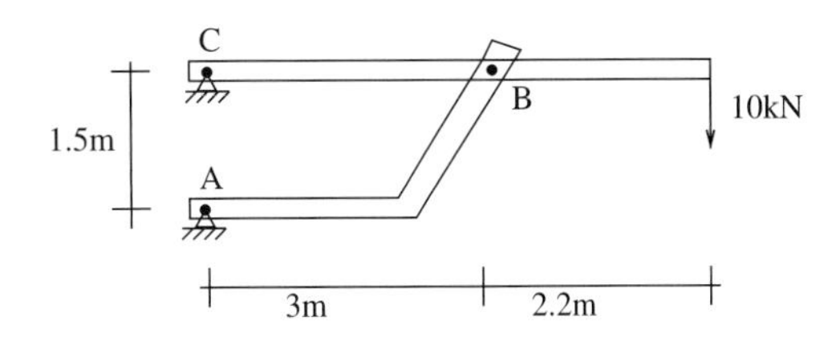


(3.7) Consider an arbitrary three-dimensional body with a section cut with normal $\boldsymbol{e}_z$. Give expressions for $F_x$, $F_y$, and $F_z$ on the cut in terms of the stress components.

(3.8) Given:

$$\boldsymbol{\sigma} = \begin{bmatrix} 10 & 12 & 13 \\ 12 & 11 & 15 \\ 13 & 15 & 20 \end{bmatrix} \text{ MPa}$$

at a point. What is the force per unit area at this point acting normal to the surface with unit normal vector $\boldsymbol{n} = (1/\sqrt{2})\boldsymbol{e}_x + (1/\sqrt{2})\boldsymbol{e}_z$? Are there any shear stresses acting on this surface?

(3.9) Consider a hydrostatic state of stress

$$\boldsymbol{\sigma} = \begin{bmatrix} 10 & 0 & 0 \\ 0 & 10 & 0 \\ 0 & 0 & 10 \end{bmatrix} \text{ MPa}$$

at a point. Consider an arbitrary section cut through this point. What is the force per unit area acting parallel to the section cut?

(3.10) You are given a body with no body forces and told that the stress state is given as:

$$\begin{bmatrix} 3\alpha x & 5\beta x^2 + \alpha y & \gamma z^3 \\ 5\beta x^2 + \alpha y & \beta x^2 & 0 \\ \gamma z^3 & 0 & 5 \end{bmatrix} \text{ psi},$$

where $(\alpha, \beta, \gamma)$ are constants with the following values: $\alpha = 1$ psi/in, $\beta = 1$ psi/in$^2$, and $\gamma = 1$ psi/in$^3$. Does this represent an equilibrium state of stress? Assume the body occupies the domain $\Omega = [0,1] \times [0,1] \times [0,1]$ (in inches).

(3.11) Show that force equilibrium in the $x$-direction implies

$$\frac{\partial \sigma_{xx}}{\partial x} + \frac{\partial \sigma_{yx}}{\partial y} = 0.$$

Assume plane stress and no distributed loads. Justify each step with a short *complete* sentence.

(3.12) Using a differential element argument derive eqn (3.35). Make sure you explain each step with a short *complete* sentence.

(3.13) Using a differential element argument derive eqn (3.40). Make sure you explain each step with a short *complete* sentence.

(3.14) Using a differential element argument derive eqn (3.42). Make sure you explain each step with a short *complete* sentence.

(3.15) Consider the section cut shown opposite. If the stress distribution on the cut is given by

$$\boldsymbol{\sigma} = \begin{bmatrix} -20y/h & 0 & 0 \\ 0 & 0 & 0 \\ 0 & 0 & 0 \end{bmatrix} \text{ ksi.}$$

What is the total force acting on the cut? What is the total moment acting on the cut (take moments about the origin)? Assume $h = 5$ inches and $b = 9$ inches.

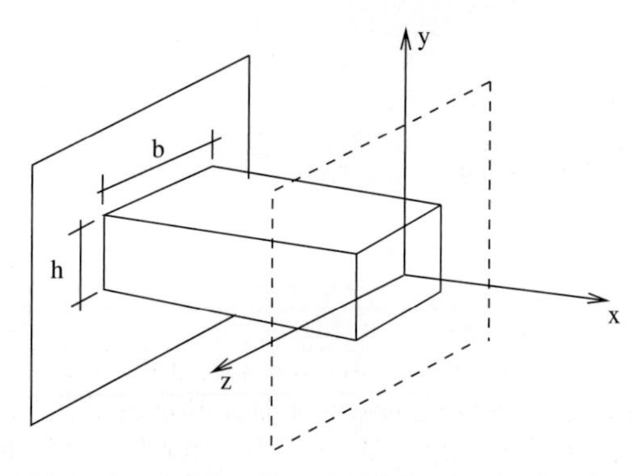


(3.16) Consider two-dimensional stress analysis where inertial forces cannot be ignored – i.e. the dynamic case. Using a differential element argument show that

$$\frac{\partial \sigma_{xx}}{\partial x} + \frac{\partial \sigma_{yx}}{\partial y} + b_x = \rho \frac{\partial^2 u_x}{\partial t^2},$$

where $\rho$ is the material density, $t$ is time, and $u_x$ is the material displacement in the $x$ direction.

# Strain

# 4



In this chapter we turn our attention to the concept of strain. We have already been exposed to this notion as the statement of change in length divided by length. We will try to refine this definition of strain to make it a pointwise definition in three dimensions in the same sense as we did with stress.

## 4.1 Shear strain

The notion of strain that we have seen already is more properly called normal strain. It is related to changes in length. There is one additional important strain concept associated with deforming bodies: changes in angle. Shear strain is the type of strain that is associated with changes in angle. The notion of shear strain takes into account "straining" motions that are not associated with length changes. In its simplest form it is defined by considering a motion of a rectangular body as shown in Fig. 4.1 and measuring the decrease in the right angle in the lower left corner. This definition will serve as our working definition of average shear strain, $\gamma$.

## 4.2 Pointwise strain

To refine our concept of strain, we will take our simple definitions of average normal and shear strain and make them pointwise definitions using differential constructions similar to the ones we have already seen. To keep things modestly simple we will consider only two-dimensional

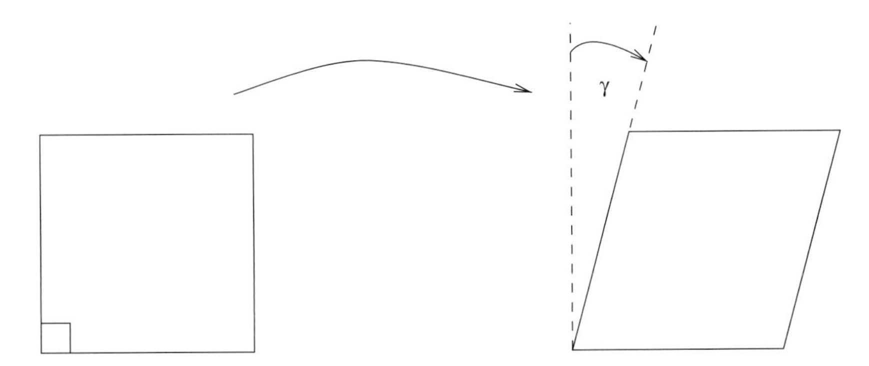


**Fig. 4.1** Shear strain, $\gamma$, definition.

bodies in the $x$-$y$ plane. As our body deforms every point in the body will be displaced. Let $u(x, y)$ be the displacement in the $x$-direction at a point $(x, y)$ and let $v(x, y)$ be the $y$-displacement at this same point.

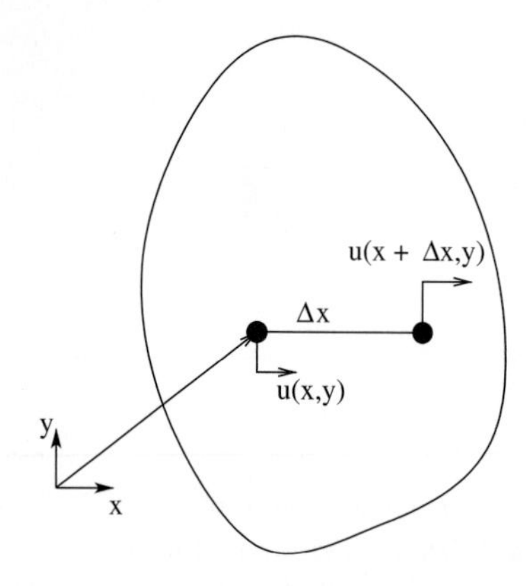


Fig. 4.2 Construction for $\varepsilon_{xx}$.

### 4.2.1 Normal strain at a point

To obtain an expression for normal strain at a point consider what happens to a line of material points that starts at $(x, y)$ and ends at $(x + \Delta x, y)$. By looking at the change in length of this line segment divided by its original length we will come to an expression for average normal strain in the $x$-direction. By taking the limit as $\Delta x \rightarrow 0$ we will then obtain an expression for normal strain in the $x$-direction at that point.

In Fig. 4.2 the overall $x$-direction elongation of this particular line segment is seen to be $u(x + \Delta x, y) - u(x, y)$. The average normal strain of the segment is then

$$\varepsilon = \frac{u(x + \Delta x, y) - u(x, y)}{\Delta x}. \tag{4.1}$$

If we take the limit as $\Delta x \rightarrow 0$ we will have a notion of normal strain at a point. The limit gives us

$$\varepsilon_{xx}(x, y) = \frac{\partial u}{\partial x}. \tag{4.2}$$

This is the definition of normal strain in the $x$ direction at a point.

By considering a small line of material points in the vertical direction of length $\Delta y$ we can develop an expression for pointwise normal strain in the $y$-direction. Applying our technique gives rise to

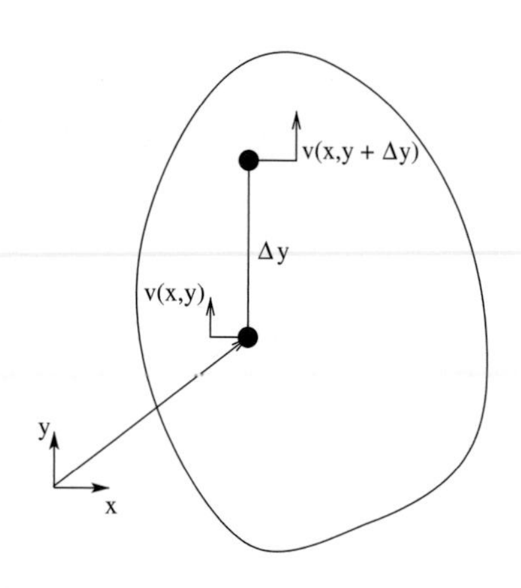


Fig. 4.3 Construction for $\varepsilon_{yy}$.

$$\varepsilon_{yy}(x, y) = \lim_{\Delta y \rightarrow 0} \frac{v(x, y + \Delta y) - v(x, y)}{\Delta y} = \frac{\partial v}{\partial y}; \tag{4.3}$$

see Fig. 4.3

**Remarks:**

(1) The sign convention for strain indicates that positive values of normal strain correspond to the elongation of material and negative values correspond to the contraction of material.

(2) The arguments above are somewhat imprecise. In each normal strain derivation we have accounted only for elongational motion of the end points of the line segments in the direction of interest. We have not taken into account what happens if the line segments also displace in the transverse direction. If we had, we would have found extra terms appearing in our expressions for the normal strains. These terms would have been non-linear in the derivatives of the displacements. In this book we will ignore such terms. The resulting theory that we will work with will then be called a small displacement or infinitesimal strain theory. The importance of these neglected terms occurs when the strains are large. To

make these phrases more precise, consider eqn (4.2); the hidden assumption we have made is that

$$\left| \frac{1}{2} \left[ \left( \frac{\partial u}{\partial x} \right)^2 + \left( \frac{\partial v}{\partial x} \right)^2 \right] \right| \ll \left| \frac{\partial u}{\partial x} \right|. \tag{4.4}$$

Such a situation will certainly occur if the first derivatives of the displacements are all small in comparison to unity. As it so happens this is true for a wide class of engineering systems – though certainly not all.

### 4.2.2 Shear strain at a point

To find an expression for shear strain at a point we can draw a right angle at the point of interest with leg lengths of $\Delta x$ and $\Delta y$; see Fig. 4.4. Upon deformation the point of interest as well as the two end-points will be displaced an amount $u$ and $v$-depending on the values of their coordinates. We now compute the change in angle between the three points; it will be composed of two contributions $\beta_1$ and $\beta_2$, as shown in Fig. 4.4 (bottom). As long as the deformation is not large we will have

$$\beta_1 = \frac{d}{c} = \frac{v(x + \Delta x, y) - v(x, y)}{\Delta x + u(x + \Delta x, y) - u(x, y)}. \tag{4.5}$$

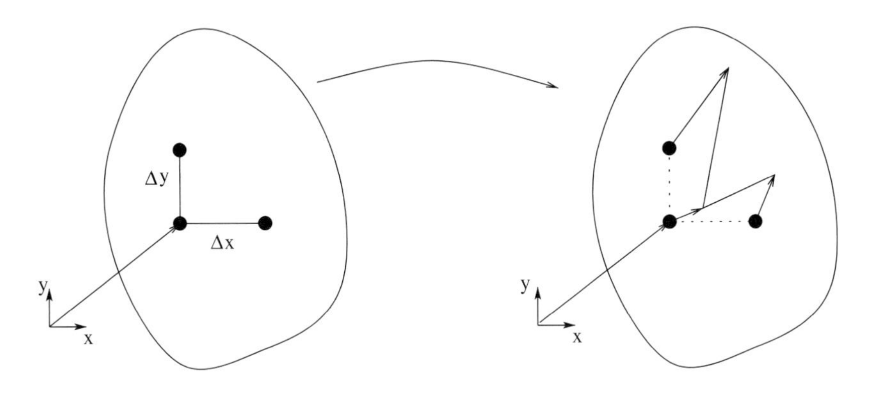


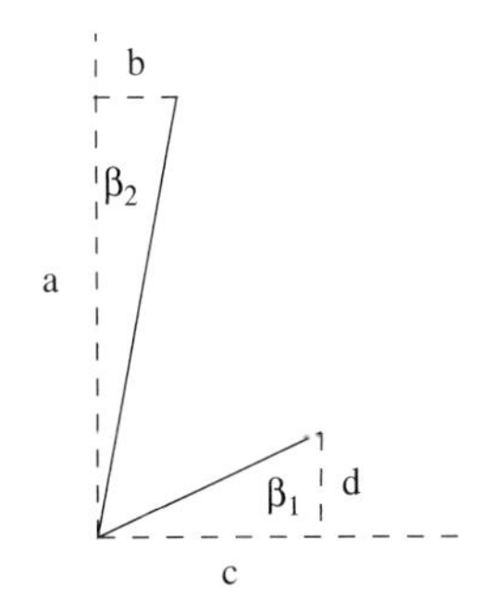


**Fig. 4.4** Construction for $\gamma_{xy}$.

Dividing the top and bottom by $\Delta x$ and taking the limit $\Delta x \to 0$ gives:

$$\beta_1 = \frac{\partial v/\partial x}{1 + \partial u/\partial x}. \tag{4.6}$$

In keeping with our previous assumptions, we will assume

$$\left|\frac{\partial u}{\partial x}\right| \ll 1; \tag{4.7}$$

thus,

$$\beta_1 \approx \frac{\partial v}{\partial x}. \tag{4.8}$$

A similar argument can be utilized to show that

$$\beta_2 = \frac{b}{a} \approx \frac{\partial u}{\partial y} \tag{4.9}$$

under the assumption that

$$\left|\frac{\partial v}{\partial y}\right| \ll 1. \tag{4.10}$$

Combining our results gives us our final small strain expression for the shear strain at a point in the $x$-$y$ plane:

$$\gamma_{xy} = \frac{\partial v}{\partial x} + \frac{\partial u}{\partial y}. \tag{4.11}$$

### 4.2.3 Two-dimensional strains

With the previous developments we now have expressions for the strains in two-dimensional problems; viz.,

$$\varepsilon_{xx} = \frac{\partial u}{\partial x} \tag{4.12}$$

$$\varepsilon_{yy} = \frac{\partial v}{\partial y} \tag{4.13}$$

$$\gamma_{xy} = \frac{\partial u}{\partial y} + \frac{\partial v}{\partial x}. \tag{4.14}$$

These relations define the components of the strain tensor. By convention, when reported, they are usually placed in a matrix in the following manner:

$$\boldsymbol{\varepsilon} = \begin{bmatrix} \varepsilon_{xx} & \frac{1}{2}\gamma_{xy} \\ \frac{1}{2}\gamma_{xy} & \varepsilon_{yy} \end{bmatrix}. \tag{4.15}$$

**Remarks:**

(1) The off-diagonal term $\frac{1}{2}\gamma_{xy}$ is often denoted as $\varepsilon_{xy}$. $\gamma_{xy}$ is known as the *engineering shear strain*, and $\varepsilon_{xy}$ is known as the *tensorial shear strain*. The distinction is very important, as they differ numerically by a factor of 2; i.e. $\varepsilon_{xy} = \frac{1}{2}\gamma_{xy}$.

**Example 4.1**

*Homogeneously strained plate.* Consider a thin (1-m x 1-m x 1-cm) steel plate which is loaded on one edge with a uniformly distributed 10-kN load and supported on the other end with a center pin and edge rollers, as shown in Fig. 4.5. Assume we make a measurement of the displacement field and find that $u(x, y) = (5 \times 10^{-6})x$ [m], and that $v(x, y) = (-0.15 \times 10^{-6})y$ [m]. Determine the strain field in the plate.

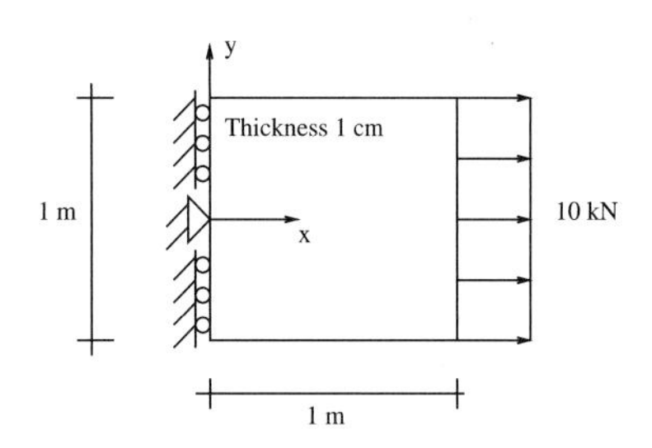


**Fig. 4.5** Homogeneously strained plate.

*Solution*
Knowing the displacement field the strain field can be determined by computing the appropriate derivatives of the displacement field.

$$\boldsymbol{\varepsilon}(x, y) = \begin{bmatrix} \varepsilon_{xx} & \frac{1}{2}\gamma_{xy} \\ \frac{1}{2}\gamma_{xy} & \varepsilon_{yy} \end{bmatrix} = \begin{bmatrix} 5 & 0 \\ 0 & -0.15 \end{bmatrix} \times 10^{-6}. \tag{4.16}$$

**Remarks:**

(1) The result tells us that the material elongates in the $x$-direction and contracts in the $y$-direction, and further, that these strains are homogeneous – are everywhere the same.

(2) The shear strain is everywhere zero, and this tells us that there are no angle changes between vertical and horizontal lines in the plate when the load is applied.

### 4.2.4 Three-dimensional strain

In three dimensions we can execute similar arguments to arrive at expressions for strain in three dimension. Here we will find three normal strains for the three coordinate directions and three shear strains – one for each coordinate plane. If we let $w(x, y, z)$ denote displacement in the $z$-direction we will have:

$$\varepsilon_{xx} = \frac{\partial u}{\partial x} \tag{4.17}$$

$$\varepsilon_{yy} = \frac{\partial v}{\partial y} \tag{4.18}$$

$$\varepsilon_{zz} = \frac{\partial w}{\partial z} \tag{4.19}$$

$$\gamma_{xy} = \frac{\partial u}{\partial y} + \frac{\partial v}{\partial x} \tag{4.20}$$

$$\gamma_{yz} = \frac{\partial v}{\partial z} + \frac{\partial w}{\partial y} \tag{4.21}$$

$$\gamma_{zx} = \frac{\partial w}{\partial x} + \frac{\partial u}{\partial z}. \tag{4.22}$$

The matrix convention for reporting the components of the strain tensor in three dimensions is:

$$\boldsymbol{\varepsilon} = \begin{bmatrix} \varepsilon_{xx} & \frac{1}{2}\gamma_{xy} & \frac{1}{2}\gamma_{zx} \\ \frac{1}{2}\gamma_{xy} & \varepsilon_{yy} & \frac{1}{2}\gamma_{yz} \\ \frac{1}{2}\gamma_{zx} & \frac{1}{2}\gamma_{yz} & \varepsilon_{zz} \end{bmatrix}. \tag{4.23}$$

## 4.3 Polar/cylindrical and spherical strain

Just as with stresses we can also define strains with respect to polar and spherical coordinate systems. Quantities such as $\varepsilon_{\theta\theta}$ represent normal strains in the $\theta$-direction at a point; a quantity such as $\gamma_{\theta z}$ would represent a change in angle in the $\theta$-$z$ coordinate plane passing through a point.

Using constructions like those in the Cartesian case one finds that

$$\begin{bmatrix} \varepsilon_{rr} & \frac{1}{2}\gamma_{r\theta} & \frac{1}{2}\gamma_{rz} \\ & \varepsilon_{\theta\theta} & \frac{1}{2}\gamma_{\theta z} \\ \text{sym.} & & \varepsilon_{zz} \end{bmatrix} = \begin{bmatrix} u_{r,r} & \frac{1}{2}\left(\frac{1}{r}\frac{\partial u_r}{\partial \theta} + r\frac{\partial(u_\theta/r)}{\partial r}\right) & \frac{1}{2}\left(\frac{\partial u_r}{\partial z} + \frac{\partial u_z}{\partial r}\right) \\ & \frac{1}{r}\left(\frac{\partial u_\theta}{\partial \theta} + u_r\right) & \frac{1}{2}\left(\frac{\partial u_\theta}{\partial z} + \frac{1}{r}\frac{\partial u_z}{\partial \theta}\right) \\ \text{sym.} & & \frac{\partial u_z}{\partial z} \end{bmatrix}. \tag{4.24}$$

$$\begin{bmatrix} \varepsilon_{rr} & \frac{1}{2}\gamma_{r\varphi} & \frac{1}{2}\gamma_{r\theta} \\ & \varepsilon_{\varphi\varphi} & \frac{1}{2}\gamma_{\varphi\theta} \\ \text{sym.} & & \varepsilon_{\theta\theta} \end{bmatrix} = \begin{bmatrix} u_{r,r} & \frac{1}{2}\left(\frac{1}{r}\frac{\partial u_r}{\partial \varphi} + r\frac{\partial(u_\varphi/r)}{\partial r}\right) & \frac{1}{2}\left(\frac{1}{r\sin(\varphi)}\frac{\partial u_r}{\partial \theta} + r\frac{\partial(u_\theta/r)}{\partial r}\right) \\ & \frac{1}{r}\left(\frac{\partial u_\varphi}{\partial \varphi} + u_r\right) & \frac{1}{2}\left(\frac{1}{r}\frac{\partial u_\theta}{\partial \varphi} + \frac{1}{r\sin(\varphi)}\frac{\partial u_\varphi}{\partial \theta}\right) \\ \text{sym.} & & \frac{1}{r\sin(\varphi)}\frac{\partial u_\theta}{\partial \theta} + u_r/r + u_\varphi\cot(\varphi)/r \end{bmatrix}. \tag{4.25}$$

## 4.4 Number of unknowns and equations

If we look over our theory as developed we find in three dimensions that there are six strains, three displacements and nine stresses – giving a total of 18 possible unknown quantities. In two dimensions we find that we have three strains, two displacements, and four stresses – giving nine possible unknown quantities. The total number of equations at hand in three dimensions is six equilibrium and six strain-displacement equations – for a total of twelve equations.[1] In two dimensions we will have three equilibrium and three strain-displacement equations – for a total of six equations. Since the theory is linear we are missing six equations for three-dimensional problems and three equations in two-dimensional problems. These missing equations are the constitutive equations. In the next chapter we will look at the generalization of Hooke's Law to two- and three-dimensional problems.

[1] When discussing the number of unknowns, many presentations explicitly assume that the stress tensor is symmetric. This reduces the number of unknown stresses in three dimensions to six and in two dimensions to three. Of course if one does this, then one should not count the moment equilibrium equations in the number of available equations.

## Chapter summary

- The general state of deformation of a body is given by the displacement field: $u(x, y, z)$, $v(x, y, z)$, $w(x, y, x)$.
- Normal strains at a point represent relative changes in length in particular directions; e.g. $\varepsilon_{yy}$ gives the relative change in length of a body in the $y$-direction.

$$\varepsilon_{xx} = \frac{\partial u}{\partial x}$$

$$\varepsilon_{yy} = \frac{\partial v}{\partial y}$$

$$\varepsilon_{zz} = \frac{\partial w}{\partial z}$$

- Shear strains at a point represent the decrease in angle between the coordinate lines at a given point; e.g. $\gamma_{zx}$ gives the decrease in angle between the $x$- and $z$-coordinate lines at a given point.

$$\gamma_{xy} = \frac{\partial u}{\partial y} + \frac{\partial v}{\partial x}$$

$$\gamma_{yz} = \frac{\partial v}{\partial z} + \frac{\partial w}{\partial y}$$

$$\gamma_{zx} = \frac{\partial w}{\partial x} + \frac{\partial u}{\partial z}$$

- Our interpretations of the strain components in Cartesian coordinates carry over to polar and spherical coordinates without change.

## Exercises

(4.1) Consider the two-dimensional body shown. The undeformed state of the body is shown on the left. After the application of load the body takes on the configuration shown on the right. What is the average shear strain, $\gamma_{xy}$, in the body?

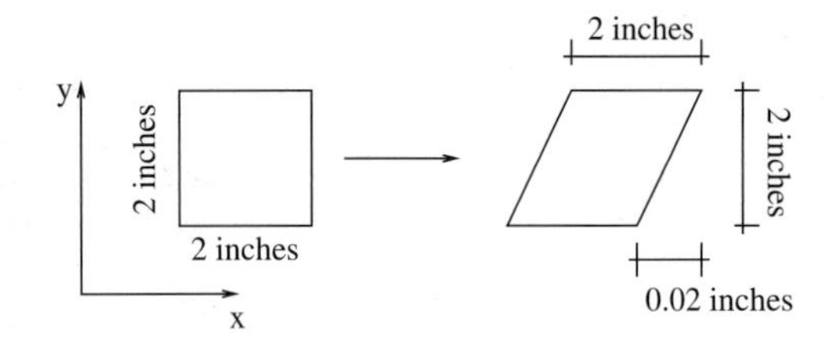


(4.2) Consider a two-dimensional body occupying the region $[0, 1] \times [0, 1]$ whose displacement field is given by $u = (4x^2 + 2) \times 10^{-4}$ and $v = (2x^4 + 3y^4) \times 10^{-4}$. What is the strain field for the body? Assume the numerical constants have consistent units.

(4.3) Consider a two-dimensional body occupying the region $[0, 1] \times [0, 1]$ whose displacement field is given by $u = (4x + 6y) \times 10^{-4}$ and $v = (3x + 5y) \times 10^{-4}$. What is the strain field for the body? Assume the numerical constants have consistent units.

(4.4) Consider a two-dimensional body occupying the region $[0, 1] \times [0, 1]$ whose displacement field is given by $u = (4y^2 + 6y) \times 10^{-4}$ and $v = (1x^2 + 2x) \times 10^{-4}$. What is change in angle between the $x$ and $y$ coordinate directions at the point $(0.5, 0.5)$. Assume the numerical constants have consistent units.

(4.5) Consider the strain field of Exercise 4.2. Sketch the deformed shape of a small square of material near the point $(0.2, 0.2)$.

(4.6) Consider the strain field of Exercise 4.3. Sketch the deformed shape of a small square of material near the point $(0.1, 0.2)$.

(4.7) Consider the strain field of Exercise 4.4. Sketch the deformed shape of a small square of material near the point $(0.2, 0.1)$.

(4.8) Shown below is a two-dimensional beam that is being bent by an applied moment, $M$. The motion of the material has been measured as $u_x = -\kappa yx$, $u_y = \kappa x^2/2$, where $\kappa$ is a given constant with dimensions of 1 over length. Find the strain field in the beam.

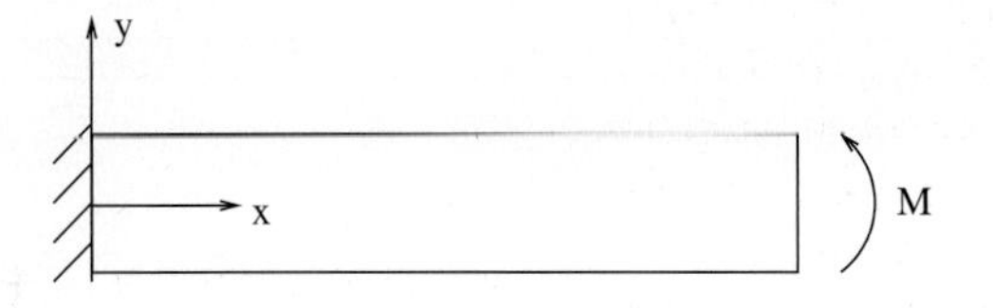


(4.9) Derive eqn (4.19).

(4.10) Derive eqn (4.21).

(4.11) Consider the ring of material shown below. Before deformation the ring has a radius of 50 mm. After deformation it has a radius of 51 mm. Assume that the ring is thin and that all the motion is in the radial direction. What is the hoop strain, $\varepsilon_{\theta\theta}$, in the ring?

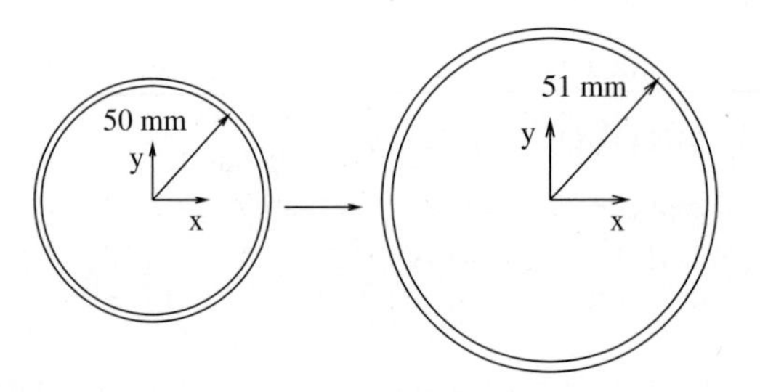


(4.12) Shown below is a solid circular rod of material. The bottom is clamped and a torque is applied at the top. The motion of the material has been measured in cylindrical coordinates as $u_r = 0$, $u_\theta = \alpha rz$, $u_z = 0$, where $\alpha$ is a given constant with appropriate dimensions. What is the strain field in the rod?

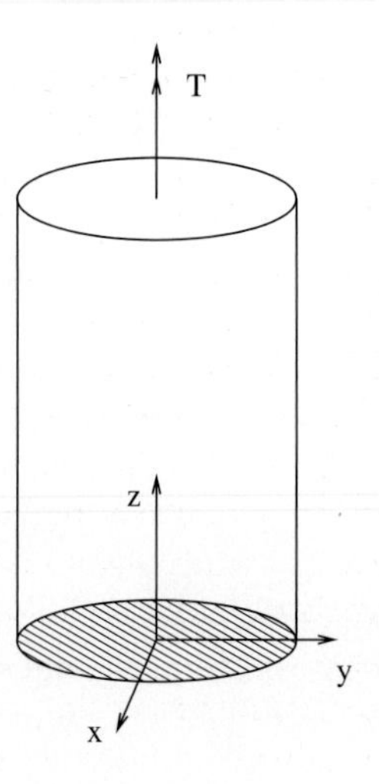


(4.13) Sketch a construction that allows one to derive the relation for $\varepsilon_{\theta\theta}$ in polar coordinates; i.e. draw a modified version of Fig. 4.2 from which one can show that $\varepsilon_{\theta\theta} = \frac{1}{r}[\partial u_\theta / \partial \theta + u_r]$.

(4.14) Use the construction of Exercise 4.13 to derive the relation for the hoop strain.

# Constitutive Response

# 5

In Chapter 2 we looked at some simple models of material behavior in one dimension. In this chapter we will extend these notions to two and three dimensions. Due to the complexity of constitutive relations in higher dimensions we will restrict our attention to isotropic linear elasticity and isotropic linear thermo-elasticity; these are the multi-dimensional counterparts to Hooke's Law and Hooke's Law with thermal effects.



## 5.1 Three-dimensional Hooke's Law

In three-dimensional isotropic[1] linear thermo-elasticity we have two basic material constants (Young's modulus, $E$, and Poisson's ratio, $\nu$) plus the coefficient of thermal expansion $\alpha$.

Young's modulus, we have seen, gives the relation of normal strain to normal stress (in a given direction). Poisson's ratio is an additional material constant that accounts for the fact that, generally, if one stresses a material in one direction there is a strain in the transverse direction. Consider, for example, Fig. 5.1 where a two-dimensional body has been stressed in the $x$-direction and a strain is generated in the $y$-direction. The stress $\sigma_{xx}$ induces a $\varepsilon_{yy}$ strain that is characterized by the relation $\varepsilon_{yy} = -\nu\sigma_{xx}/E$. A similar relation holds for induced strain in the $z$-direction due to normal stress in the $x$-direction. Likewise normal stresses in the $z$- and $y$-directions will induces Poisson effects in the $y$-, $x$- and $x$-, $z$-directions, respectively. Note that thermodynamic restrictions only allow $\nu \in [-1, \frac{1}{2})$. Common metals have $\nu = 0.3$; polymers tend to have $\nu \approx 0.45$ or slightly higher. Some structured foams have $\nu < 0$, which implies that they expand laterally when extended in one direction.

With Young's modulus and Poisson's ratio we can model the relation between normal strains and normal stress in an isotropic linear elastic solid. To fully characterize the response of such a solid we also need to specify the constitutive relations that govern the behavior of the material in shear. For the linear isotropic case at hand, this is given by relations of the form $\sigma_{xy} = G\gamma_{xy}$ for each shear stress–shear strain pair. Hence, shear strains are generated by the application of shear stresses. The shear modulus $G$ is a material parameter and is related to $E$ and $\nu$ as $G = E/(2(1+\nu))$; a proof of this will be given later in Chapter 9.[2]

Temperature changes in linear isotropic solids generally produce expansion and contraction that is uniform in space. Thus if one heats a solid and it expands, then the expansion is the same in all coordinate

[1] Isotropic means the mechanical properties are the same in all directions; this is the case for a wide variety of common engineering materials.

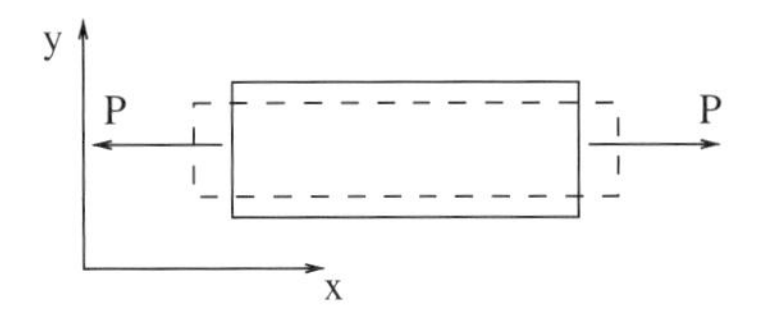


**Fig. 5.1** Geometric definition of Poisson's ratio.

[2] An alternative common symbol for $G$ is $\mu$.

directions. As such, for a temperature change $\Delta T$, the material strains as $\varepsilon_{xx} = \varepsilon_{yy} = \varepsilon_{zz} = \alpha \Delta T$, where $\alpha$ is the coefficient of thermal expansion.

If we now assume that each of the effects just described are independent of each other, then we can superpose them to find the overall constitutive relations for use when all are present. Doing so gives the strain tensor components as

$$\varepsilon_{xx} = \frac{\sigma_{xx}}{E} - \frac{\nu\sigma_{yy}}{E} - \frac{\nu\sigma_{zz}}{E} + \alpha\Delta T \tag{5.1}$$

$$\varepsilon_{yy} = \frac{\sigma_{yy}}{E} - \frac{\nu\sigma_{zz}}{E} - \frac{\nu\sigma_{xx}}{E} + \alpha\Delta T \tag{5.2}$$

$$\varepsilon_{zz} = \frac{\sigma_{zz}}{E} - \frac{\nu\sigma_{xx}}{E} - \frac{\nu\sigma_{yy}}{E} + \alpha\Delta T \tag{5.3}$$

$$\gamma_{xy} = \frac{\sigma_{xy}}{G} \tag{5.4}$$

$$\gamma_{yz} = \frac{\sigma_{yz}}{G} \tag{5.5}$$

$$\gamma_{zx} = \frac{\sigma_{zx}}{G}. \tag{5.6}$$

**Remarks:**

(1) The eqns (5.1)–(5.6) constitute the six equations which are needed for a complete description of the mechanical behavior of an isotropic linear elastic solid, as was discussed in Section 4.4.

---

**Example 5.1**

*Strain due to stress and temperature change.* Consider a linear thermoelastic body under the action of a temperature change of 1 F, where the stress state at a given point is known to be

$$\boldsymbol{\sigma} = \begin{bmatrix} 300 & 1200 & 0 \\ 1200 & -300 & 0 \\ 0 & 0 & 0 \end{bmatrix} \text{ psi.} \tag{5.7}$$

The material constants are $E = 30 \times 10^6$ psi, $G = 12 \times 10^6$ psi, $\nu = 0.25$, and $\alpha = 12 \times 10^{-6}$ 1/F. Determine the state of strain at this point and sketch the motion of the material in the vicinity of the point in the $x$-$y$ plane.

*Solution*

Using the thermoelastic Hooke's Law one finds

$$\varepsilon_{xx} = \frac{300}{E} - \frac{\nu}{E}(-300 + 0) + \alpha 1 \tag{5.8}$$

$$\varepsilon_{yy} = -\frac{300}{E} - \frac{\nu}{E}(300) + \alpha 1 \tag{5.9}$$

$$\varepsilon_{zz} = -\frac{\nu}{E}(300 - 300) + \alpha 1 \tag{5.10}$$

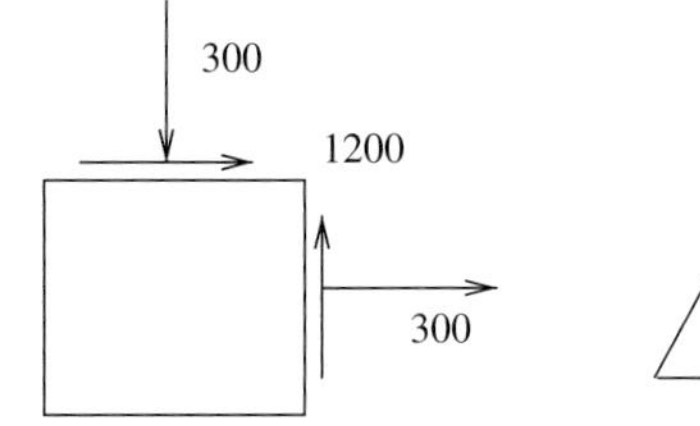


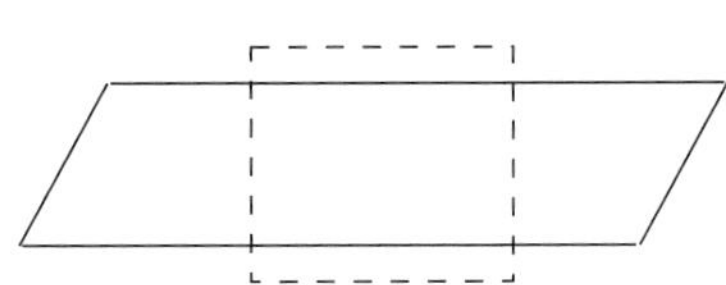


**Fig. 5.2** State of stress and resulting deformation.

$$\gamma_{xy} = \frac{1200}{G} \tag{5.11}$$

$$\gamma_{yz} = \frac{0}{G} \tag{5.12}$$

$$\gamma_{zx} = \frac{0}{G} \tag{5.13}$$

or in matrix form

$$\boldsymbol{\varepsilon} = \begin{bmatrix} 25 & 100 & 0 \\ 100 & -0.5 & 0 \\ 0 & 0 & 12 \end{bmatrix} \times 10^{-6}. \tag{5.14}$$

In the $x$-$y$ plane there is an extension in the $x$-direction and a relatively small contraction in the $y$-direction. In this plane the angle between the $x$- and $y$-coordinate axes is decreased by 200 $\mu$radians. This state of deformation is sketched in Fig. 5.2.

### 5.1.1 Pressure

It is often useful to talk about pressure in a solid. Pressure is defined to be the mean normal stress:

$$p = \frac{1}{3}(\sigma_{xx} + \sigma_{yy} + \sigma_{zz}). \tag{5.15}$$

In isotropic solids, pressures give rise to changes in volume, and this is commonly measured as volumetric strain (change in volume per unit volume) $\theta = \varepsilon_{xx} + \varepsilon_{yy} + \varepsilon_{zz}$. By adding eqns (5.1)–(5.3), one can show that

$$\theta = p/K, \tag{5.16}$$

where $K = E/(3(1-2\nu))$ is known as the bulk modulus, and thermal effects have been ignored.

**Remarks:**

(1) The mean normal stress, or pressure, is also commonly known as the "hydrostatic pressure".

(2) Beware that in some fields, such as fluid mechanics and geotechnical engineering, the pressure is defined as the negative of our definition, i.e. as $p = -\frac{1}{3}(\sigma_{xx} + \sigma_{yy} + \sigma_{zz})$. The convention which we adopt is common in solid mechanics.

### 5.1.2 Strain energy in three dimensions

The strain energy per unit volume in a linear elastic isotropic body (ignoring thermal effects) is given as

$$w = \frac{1}{2}(\sigma_{xx}\varepsilon_{xx} + \sigma_{yy}\varepsilon_{yy} + \sigma_{zz}\varepsilon_{zz} + \sigma_{xy}\gamma_{xy} + \sigma_{yz}\gamma_{yz} + \sigma_{zx}\gamma_{zx}). \quad (5.17)$$

This relation is a generalization of the relation derived in one dimension for the strain energy density in an axially loaded rod. The derivation follows directly along the same lines as the one in Section 2.5. Each term in eqn (5.17) can be observed to consist of matching pairs of like stresses and strains. Each pair contributes an additive contribution to the total strain energy density.

## 5.2 Two-dimensional Hooke's Law

When possible it is nice to reduce the dimensionality of a physical problem. Thus, if at all possible we try and take three-dimensional problems and turn them into two- and one-dimensional problems, since they are much easier to solve. This poses the question of how one takes a three-dimensional constitutive relation and turns it into one appropriate for one- and two-dimensional problems. There are many ways of doing this, but there are two very common sets of assumptions of which one should definitely be aware.

### 5.2.1 Two-dimensional plane stress

In plane-stress problems, we assume $\sigma_{zz} = \sigma_{yz} = \sigma_{zx} = 0$. This set of assumptions is useful near the free surface of a body where the normal to the surface is in the $z$-direction. It is also commonly used when the $z$-direction of the body is thin in comparison to the $x$- and $y$-directions. Note that this is done even if there are loads applied in the $z$-direction. While this may seem rather implausible, one can show from the three-dimensional theory of elasticity that it is a good assumption for a wide range of practical situations.

If we take these assumptions and re-evaluate the full three-dimensional theory we can shown that

$$\varepsilon_{xx} = \frac{\sigma_{xx}}{E} - \frac{\nu\sigma_{yy}}{E} + \alpha\Delta T \quad (5.18)$$

$$\varepsilon_{yy} = \frac{\sigma_{yy}}{E} - \frac{\nu\sigma_{xx}}{E} + \alpha\Delta T \quad (5.19)$$

$$\varepsilon_{zz} = -\frac{\nu\sigma_{xx}}{E} - \frac{\nu\sigma_{yy}}{E} + \alpha\Delta T \tag{5.20}$$

$$\gamma_{xy} = \frac{\tau_{xy}}{G} \tag{5.21}$$

$$\gamma_{yz} = 0 \tag{5.22}$$

$$\gamma_{zx} = 0. \tag{5.23}$$

**Remarks:**

(1) Equations (5.18), (5.19), and (5.21) constitute the three needed equations to complete the mechanical description for a two-dimensional problem as described in Section 4.4. Equation (5.20) is an extra equation that is not needed in a two-dimensional description of mechanical response. However, after a two-dimensional problem has been solved it can be used after the fact to compute the (Poisson) strain in the $z$-direction.

### 5.2.2 Two-dimensional plane strain

In plane strain we assume $\varepsilon_{zz} = \varepsilon_{yz} = \varepsilon_{zx} = 0$. This is basically the same as assuming that there is no displacement in the $z$-direction, $w = 0$, and that nothing varies in the $z$-direction, $\frac{\partial}{\partial z}(\cdot) = 0$. These assumptions are appropriate when the body in question is restrained from motion in the $z$-direction. This is something that happens often when the $z$-direction dimension of the body is quite large in comparison to the $x$- and $y$-dimensions. If we insert these assumptions into the three-dimensional theory we can show that

$$\varepsilon_{xx} = \frac{(1-\nu^2)\sigma_{xx}}{E} - \frac{\nu(1+\nu)\sigma_{yy}}{E} + (1+\nu)\alpha\Delta T \tag{5.24}$$

$$\varepsilon_{yy} = \frac{(1-\nu^2)\sigma_{yy}}{E} - \frac{\nu(1+\nu)\sigma_{xx}}{E} + (1+\nu)\alpha\Delta T \tag{5.25}$$

$$0 = \frac{\sigma_{zz}}{E} - \frac{\nu\sigma_{xx}}{E} - \frac{\nu\sigma_{yy}}{E} + \alpha\Delta T \tag{5.26}$$

$$\gamma_{xy} = \frac{\sigma_{xy}}{G} \tag{5.27}$$

$$0 = \sigma_{yz} \tag{5.28}$$

$$0 = \sigma_{zx}. \tag{5.29}$$

**Remarks:**

(1) Equations (5.24), (5.25), and (5.27) constitute the three needed equations to complete the mechanical description for a two-dimensional problem as described in Section 4.4. Equation (5.26) is an extra equation that is not need in a two-dimensional description of mechanical response. However, after a two-dimensional problem

has been solved it can be used after the fact to compute the (constraint) stress in the $z$-direction.

## 5.3 One-dimensional Hooke's Law: Uniaxial state of stress

The last common assumption that is often made is that of a uniaxial state of stress. The assumption here is that $\sigma_{xx}$ is the only possible non-zero stress. This assumption is useful for thin slender bodies under the action of axial or bending loads. Applying this assumption gives

$$\varepsilon_{xx} = \frac{\sigma_{xx}}{E} + \alpha\Delta T \tag{5.30}$$

$$\varepsilon_{yy} = -\frac{\nu\sigma_{xx}}{E} + \alpha\Delta T \tag{5.31}$$

$$\varepsilon_{zz} = -\frac{\nu\sigma_{xx}}{E} + \alpha\Delta T \tag{5.32}$$

$$\gamma_{xy} = 0 \tag{5.33}$$

$$\gamma_{yz} = 0 \tag{5.34}$$

$$\gamma_{zx} = 0. \tag{5.35}$$

**Remarks:**

(1) For the one-dimensional rod problems with axial loads we have used the result of this assumption: viz., eqn (5.30).

(2) Equations (5.31) and (5.32) can be used after the completion of a one dimensional analysis to compute the (Poisson) strains in the transverse directions.

## 5.4 Polar/cylindrical and spherical coordinates

The preceding relations also hold for other orthonormal coordinate systems such as polar/cylindrical and spherical coordinates. One simply needs to make the substitutions $(x, y, z) \rightarrow (r, \theta, z)$ or $(x, y, z) \rightarrow (r, \varphi, \theta)$.

## Chapter summary

- For linear elastic isotropic solids, the normal strain in a given coordinate direction is composed of a term directly related to the load in that direction plus terms associated with Poisson contraction due to loads in the other two coordinate directions and any thermal strains:

$$\varepsilon_{xx} = \frac{\sigma_{xx}}{E} - \frac{\nu\sigma_{yy}}{E} - \frac{\nu\sigma_{zz}}{E} + \alpha\Delta T$$
$$\varepsilon_{yy} = \frac{\sigma_{yy}}{E} - \frac{\nu\sigma_{zz}}{E} - \frac{\nu\sigma_{xx}}{E} + \alpha\Delta T$$
$$\varepsilon_{zz} = \frac{\sigma_{zz}}{E} - \frac{\nu\sigma_{xx}}{E} - \frac{\nu\sigma_{yy}}{E} + \alpha\Delta T$$

- For linear elastic isotropic solids, the shear strains are only related to the corresponding shear stresses:

$$\gamma_{xy} = \frac{\sigma_{xy}}{G}$$
$$\gamma_{yz} = \frac{\sigma_{yz}}{G}$$
$$\gamma_{zx} = \frac{\sigma_{zx}}{G}$$

- The two principal assumptions for two-dimensional problems are plane stress $\sigma_{zz} = \sigma_{zx} = \sigma_{zy} = 0$ and plane strain $\varepsilon_{zz} = \varepsilon_{zx} = \varepsilon_{zy} = 0$.
- The common elastic constants are Young's modulus $E$, Poisson's ratio $\nu$, bulk modulus $K$, shear modulus $G$, and the coefficient of thermal expansion $\alpha$.
- Pressure is defined as the mean normal stress:

$$p = \frac{1}{3}(\sigma_{xx} + \sigma_{yy} + \sigma_{zz})$$

- The volume strain is given as $\theta = \varepsilon_{xx} + \varepsilon_{yy} + \varepsilon_{zz}$.
- The strain energy density in a linear elastic body is composed of sums of corresponding stress–strain pairs:

$$w = \frac{1}{2}(\sigma_{xx}\varepsilon_{xx} + \sigma_{yy}\varepsilon_{yy} + \sigma_{zz}\varepsilon_{zz} + \sigma_{xy}\gamma_{xy} + \sigma_{yz}\gamma_{yz} + \sigma_{zx}\gamma_{zx})$$

# Exercises

(5.1) What happens in the transverse direction if one puts a rod in tension when the Poisson's ratio is negative?

(5.2) What set of strains will be generated by the following stresses:

$$\begin{bmatrix} 50 & -10 & 0 \\ -10 & 20 & 8 \\ 0 & 8 & 0 \end{bmatrix} \text{MPa.}$$

Assume $E = 100$ GPa, $\nu = 0.3$.

(5.3) What set of stresses is required to generate the following set of strains:

$$\begin{bmatrix} 5 & 1 & -4 \\ 1 & 2 & 7 \\ -4 & 7 & -10 \end{bmatrix} \mu\text{strain}.$$

Assume $E = 100$ GPa, $G = 45$ GPa.

(5.4) Consider a square steel plate under the following plane state of stress:

$$\begin{bmatrix} 300 & 1200 & 0 \\ 1200 & -300 & 0 \\ 0 & 0 & 0 \end{bmatrix} \text{psi}.$$

Further, the plate is subjected to a temperature rise of 10F. Assume $E = 30 \times 10^6$ psi, $G = 12 \times 10^6$ psi, $\alpha = 6.5 \times 10^{-6}$ 1/F, and find the strain state of the plate. Once you have determined the strain state, accurately sketch the deformed shape of the plate assuming that it occupies the region $[0, 10] \times [0, 10]$ (inches $\times$ inches).

(5.5) Consider the two-dimensional body, shown below, that is to be homogeneously stressed in the $x$-direction until its length (in the $x$-direction) has been increased by 1%. What value of $\sigma_{xx}$ is required if one assumes (a) two-dimensional plane-strain and (b) two-dimensional plane-stress? Assume the material is ABS (Acrylonitrile Butadiene Styrene, a common engineering thermoplastic) with $E = 3.2 \times 10^5$ psi, $\nu = 0.35$.

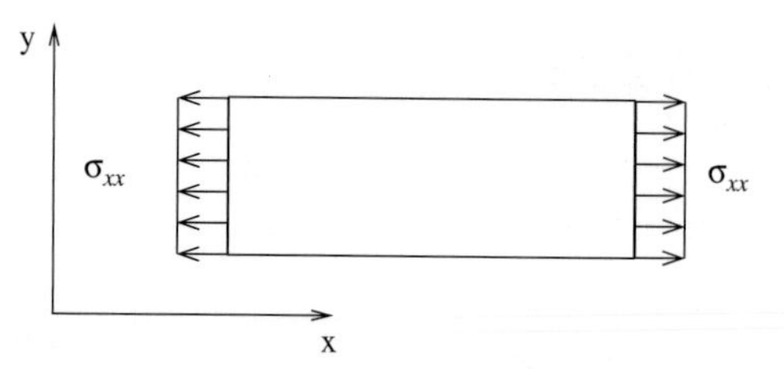


(5.6) Consider a three-dimensional thermo-elastic body that is restrained so that all strains are zero. What temperature change would be required to produce a compressive state of stress $\sigma_{xx} = \sigma_{yy} = \sigma_{zz} = -20$ ksi. Assume $\alpha = 12\mu$strain/F, $E = 30 \times 10^6$ psi, $\nu = 0.3$.

(5.7) Consider a rubber sphere of radius 5 cm. How much pressure would be required to squeeze the sphere down to a radius of 4.9 cm? Assume $E = 1$ MPa and $\nu = 0.49$.

(5.8) How much total strain energy is stored in the sphere in Exercise 5.7? Make a plot of strain energy in the sphere as a function of radius for radii from 5.1 cm to 4.9 cm.

(5.9) Compute the stress field associated with the strain field in Exercise 4.12.

(5.10) Explain why a material is said to be "incompressible" when $\nu$ approaches $\frac{1}{2}$.

(5.11) A foam rod of length $L = 50$ mm and diameter $d = 10$ mm has a Poisson's ratio $\nu = -0.7$ and a Young's modulus $E = 10$ MPa. The rod is to be inserted into a hole with diameter $\delta = 9$ mm. To do this the rod is first axially compressed and then slid into the hole after which the axial compression is released. How much axial compression stress is required for the insertion step? After the axial compression is released, what is the stress state in the rod? Assume the plate is rigid and that there is no friction between the rod and the plate.

(5.12) Write the stress components as a function of the strain components for the case of two-dimensional plane strain.

(5.13) Write the stress components as a function of the strain components for the case of two-dimensional plane stress.

# Basic Techniques of Strength of Materials

# 6



The complete set of equations that describe three-dimensional (elastic) mechanical systems are rather involved, and their solution requires modestly sophisticated mathematical and numerical methods. In engineering mechanics one employs carefully chosen assumptions to reduce the governing system of equations to a more practical and easier-to-solve set. The principal techniques of engineering mechanics are based upon kinematic assumptions and stress-distribution assumptions. These two sets of assumptions when combined with an appropriate constitutive model render the generally complex problem of determining the stress, strain, and displacement state of a body tractable. In other words, by a careful selection of assumptions we can turn a problem that involves the solution of coupled partial differential and algebraic equations into something quite achievable. In fact, we already performed such an exercise when we studied the behavior of axial loaded bars in Chapter 2. In this chapter we will revisit the one-dimensional axial loaded rod from our new perspective on the three-dimensional theory. We will see that the analysis we originally performed can be understood from the perspective of a kinematic assumption and a stress assumption on the full multi-dimensional theory. We will also look at the notion of thinness and its implications on functional variation; this will be done within the context of analyzing the behavior of thin-walled pressure vessels. This will be followed by a brief discussion of the validity of such procedures and important caveats which should always be kept in mind.

## 6.1 One-dimensional axially loaded rod revisited

How is the three-dimensional theory related to what we did earlier when looking at rods under the action of axial loads? The answer lies in making physical observations on the motion of three-dimensional bars under axial loads. The fundamental observation one makes when examining the deformation of axially loaded bars is that if one selects a plane orthogonal to the loading axis (i.e. a cross-sectional plane) before loading, then after loading the plane will still be a plane and will have merely displaced in the (axial) $x$-direction. This observation is known to engineers as the well-worn phrase: "Plane sections remain plane". The observation is amazingly robust, and even holds when the material properties are inhomogeneous[1] and the cross-sectional area is non-constant. Naturally, there are other aspects to the deformation such as thinning of the cross-section due to Poisson's effect, complexities of motion near points of loading and support, etc., but overall, the dominant feature of the motion is the displacement of the cross-sections in the $x$-direction.

[1] Inhomogeneous means not constant from point to point.

In terms of equations we are led to assume that

$$u = \hat{u}(x) \tag{6.1}$$

and that $v = w = 0$. The most important feature of this assumption is that $u$ has gone from a function of $x, y$ and $z$ to one that just depends on $x$. The assumptions on $v$ and $w$ are made since these displacements do not appear (feel) central to the dominant motion. This leads to only one fundamental strain-displacement relation:

$$\varepsilon_{xx}(x) = \frac{du}{dx}(x). \tag{6.2}$$

All the other strains are predicted to be zero according to the full three-dimensional strain-displacement relations.

To further the analysis we assume we are developing a theory for slender bodies so that the cross-sectional dimensions are small in comparison to the length. If we are not applying opposing loads in the $y$- and $z$-directions along the lateral surfaces then we know that the stresses associated with the $y$- and $z$-directions will be zero on the lateral surfaces. Since the body is thin we then assume that this holds throughout; this leads us to assume a uniaxial state of stress. The resulting strains are thus:

$$\boldsymbol{\varepsilon} = \begin{bmatrix} \frac{1}{E}\sigma_{xx} & 0 & 0 \\ 0 & -\frac{\nu}{E}\sigma_{xx} & 0 \\ 0 & 0 & -\frac{\nu}{E}\sigma_{xx} \end{bmatrix}. \tag{6.3}$$

Note that this assumption gives rise to transverse strains $\varepsilon_{yy}$ and $\varepsilon_{zz}$. This is at odds with the kinematic assumption which predicts zero transverse strains. This situation of incompatible assumptions is a common problem with analysis techniques in engineering mechanics. We are making assumptions to solve a complex problem, and in general it will be very hard to render everything compatible. The only way for us to decide whether or not we have done the correct thing is to either solve the complete three-dimensional problem or to compare our predictions with experiments. It turns out in this case that these opposing assumptions are "correct". One way of seeing that allowing for transverse strains is acceptable is to note that the contribution to the strain energy from the transverse motion is zero. Why? Because the transverse stresses are assumed zero. If one adopts an energy viewpoint of the behavior of the system, then allowing for these strains even though they are in contradiction to the kinematic assumption should not affect the final results, since there is no energy associated with them.

If we now take our stress assumption and look at the equilibrium equations we find that there is only one non-trivial equilibrium equation:

$$\frac{\partial \sigma_{xx}}{\partial x} + b_x = 0. \tag{6.4}$$

If we note that the total axial force on a cross-section is given by

$$R = \int_A \sigma_{xx} \, dA, \tag{6.5}$$

then we can also rewrite this equation as:

$$\frac{dR}{dx} + b = 0, \tag{6.6}$$

where $b = \int_A b_x \, dA$, whose units are force per unit length.

Our assumptions concerning the kinematics and the stresses thus lead us from a complex system of partial differential and algebraic equations in three dimensions to the following, much simpler, system of governing equations in one dimension for $u$, $\varepsilon_{xx}$, $\sigma_{xx}$, and $R$:

$$\varepsilon_{xx}(x) = \frac{du}{dx}(x) \tag{6.7}$$

$$\frac{dR}{dx}(x) + b(x) = 0 \tag{6.8}$$

$$\varepsilon_{xx}(x) = \frac{\sigma_{xx}(x,y,z)}{E(x,y,z)} \tag{6.9}$$

$$R(x) = \int_{A(x)} \sigma_{xx}(x,y,z) \, dA. \tag{6.10}$$

Note that these are the same equations which we used previously – with the exception of eqn (6.10), which is a generalization of $R = \sigma A$. Thus our one-dimensional rod formulation from earlier in the text can be interpreted as a tractable set of equations based upon full three-dimensional theory plus two fundamental assumptions – one kinematic (plane-sections-remain-plane) and one stress-based (uniaxial stress with respect to the $x$-direction). Additional information can now also be gleaned, such as information about the transverse strains.

**Example 6.1**

*Extension of a composite bar.* As an application of our new understanding of the theory of axially loaded bars, let us consider the composite bar shown in Fig. 6.1 and attempt to compute the end deflection of the bar assuming it is restrained at its far end.

*Solution*

If we construct an internal force diagram for the bar we see that equilibrium requires

$$R(x) = F, \tag{6.11}$$

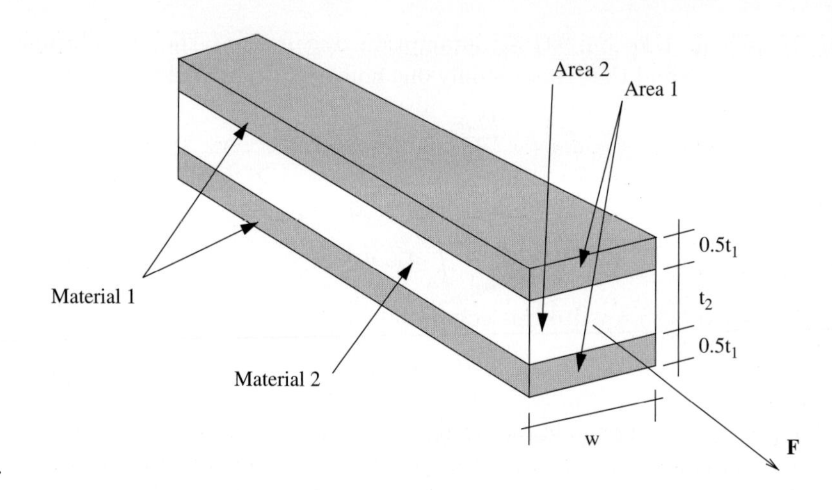


**Fig. 6.1** Axially loaded composite bar.

the applied end-load. The expression for the internal force tells us that

$$R(x) = \int_{A_1} \sigma \, dA + \int_{A_2} \sigma \, dA = t_1 w \sigma_1(x) + t_2 w \sigma_2(x). \tag{6.12}$$

If we now apply our constitutive law we find that

$$R(x) = t_1 w E_1 \varepsilon(x) + t_2 w E_2 \varepsilon(x). \tag{6.13}$$

Solving for the strain gives

$$\varepsilon(x) = \frac{R(x)}{t_1 w E_1 + t_2 w E_2}. \tag{6.14}$$

Integrating, finally gives us:

$$\Delta = u(L) - u(0) = \int_0^L \frac{du}{dx} \, dx = \frac{FL}{t_1 w E_1 + t_2 w E_2}. \tag{6.15}$$

**Remarks:**

(1) While the strains are constant on the cross-section, the stresses are not, because of the inhomogeneous moduli (in terms of $y$ and $z$). For example, if $\varepsilon(x) = \varepsilon_o$, then the stresses in material 1 are $E_1\varepsilon_o$, while those in material 2 would be $E_2\varepsilon_o$.

(2) The use of a kinematic assumption to simplify the strain field will be a central element of all the major problem classes we will treat in this book. Such assumptions generally emanate from physical observation, and their purpose is to reduce the number of unknowns in a problem and to reduce the space dimension of the problem being studied. In the remainder, such assumptions will always turn problems governed by partial differential equations into problems governed by ordinary differential equations.

(3) The use of a stress assumption will be a feature of some, but not all, problem classes with which we will work.

**Example 6.2**

*Plastic loading.* Consider the bar from Example 6.1 now using an elastic-perfectly plastic material model. Suppose the yield strain of material 2 is greater than the yield strain of material 1: $\varepsilon_{2Y} > \varepsilon_{1Y}$. (1) Assume the bar has been strained such that the strain is above the yield strain for material 1 but below that for material 2; i.e. that $\varepsilon_{2Y} > \varepsilon > \varepsilon_{1Y}$. Sketch the stress distribution. (2) Suppose $\varepsilon$ is increased to $\varepsilon = \varepsilon_{2Y}$. What is the load in the bar?

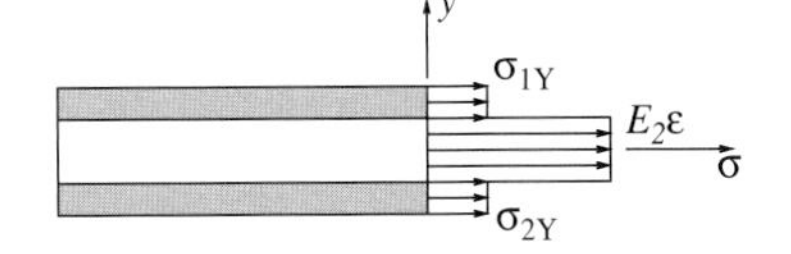


**Fig. 6.2** Stress distribution beyond yield.

*Solution*

For part (1) the stress in material 1 will be limited to $\sigma_{1Y}$, since it has been strained beyond yield. Material 2 is will still behave elastically and have a stress of $E_2\varepsilon$. The distribution is sketched in Fig. 6.2. For part (2) the stress in material 2 will now also be at yield. Thus

$$R(x) = \int_A \sigma\, dA = \int_{A_1} \sigma\, dA + \int_{A_2} \sigma\, dA = t_1 w \sigma_{1Y} + t_2 w \sigma_{2Y}. \quad (6.16)$$

**Remarks:**

(1) This load is sometimes termed the ultimate load for the bar, since it represents the maximum possible load one can apply.

## 6.2 Thinness

Many practical engineering structures can in some ways be considered thin. The meaning of thin in this context is that the changes in the coordinates in a particular direction are small when compared to the absolute values of the coordinates. As a particular example consider a function $f(x)$ where the $x$-coordinate can vary from $x = a$ to $x = b$; see Fig. 6.3. We will assume this situation qualifies as thin if $\frac{1}{2}|(a+b)/(b-a)| > 10$. The importance of this situation arises when we can additionally assume that the derivative $df/dx$ is small; i.e. $|df/dx| \ll 1$ for $x \in (a,b)$. In this situation we are justified in assuming that the function and its derivative are constant over the interval $(a,b)$. How we can take advantage of this is shown in the next few sections, where we analyze the behavior of thin-walled pressure vessels.

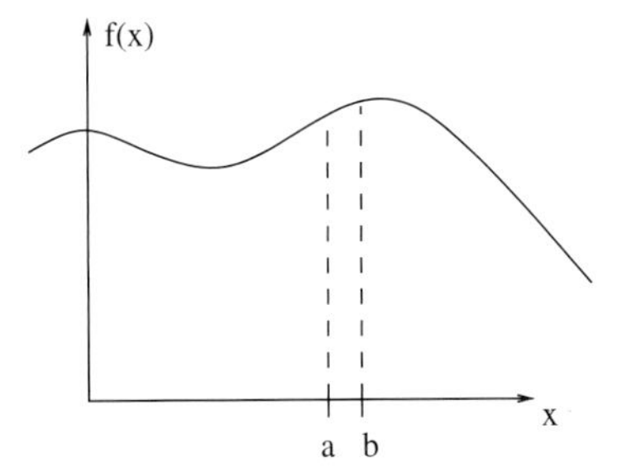


**Fig. 6.3** Graph of a generic function.

### 6.2.1 Cylindrical thin-walled pressure vessels

Figure 6.4 shows a thin-walled cylindrical pressure vessel under internal pressure $p$. The important dimension that we will assume to be thin in this situation is the radial one; i.e. we will require that the wall thickness be small in comparison to the radius:

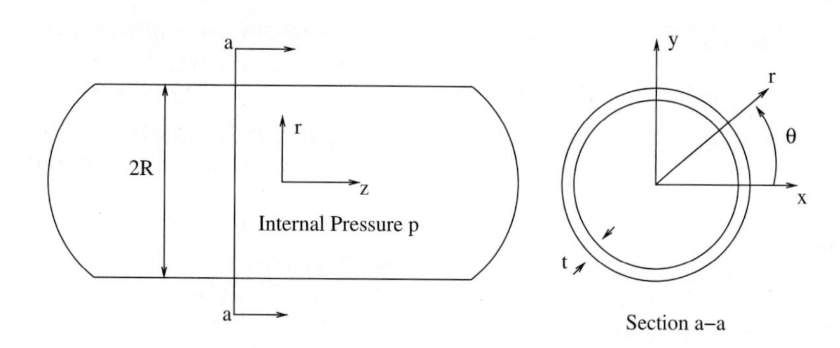


**Fig. 6.4** Cylindrical thin-walled pressure vessel.

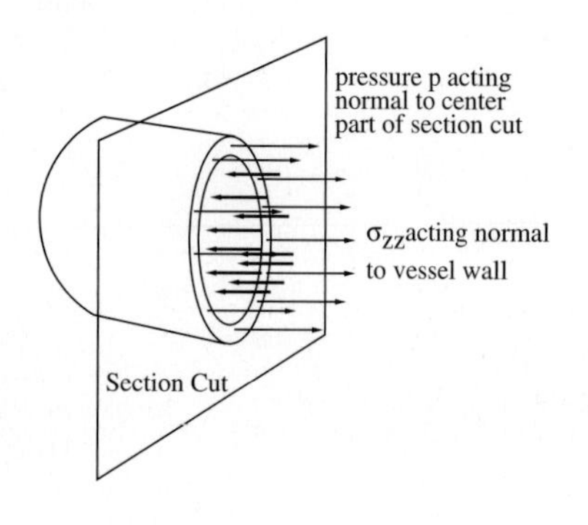


**Fig. 6.5** Cylindrical thin-walled pressure vessel with section cut whose normal is $\boldsymbol{e}_z$.

$$R/t > 10. \tag{6.17}$$

The primary implication of this is that we can assume the strains are constants as a function of $r$. In particular we will assume that the hoop strain, $\varepsilon_{\theta\theta}$, the axial strain, $\varepsilon_{zz}$, and the radial strain $\varepsilon_{rr}$ are all constants as a function of $r$. Note also that by symmetry of the system nothing can be a function of $\theta$; i.e. the problem is axis-symmetric.

Let us examine the use of this assumption on a cylinder made of an homogeneous linear elastic isotropic material. Since we are assuming that the strains are constant as a function of $r$, we also have that the stresses are constant as a function of $r$. Consider now a free-body diagram that is constructed by slicing the cylinder perpendicular to the $z$-axis with a section cut; see Fig. 6.5. If we sum the forces in the $z$-direction, then we have

$$\sum F_z = 0 = -p\pi R^2 + \sigma_{zz} 2\pi R t \tag{6.18}$$

$$\sigma_{zz} = \frac{pR}{2t}. \tag{6.19}$$

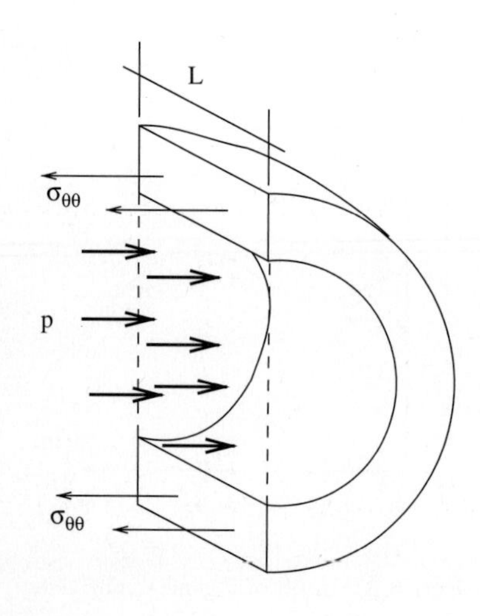


**Fig. 6.6** Cylindrical thin-walled pressure vessel with section cut whose normal is $\boldsymbol{e}_\theta$ at the top and $-\boldsymbol{e}_\theta$ at the bottom.

Thus our assumption allows us to determine the axial stress in terms of the applied pressure. The main thing we have taken advantage of here is that the integral of the stress over the material cross-section is simply the stress times the area, since the stress is a constant in terms $r$ and $\theta$.

If we now consider the free-body diagram shown in Fig. 6.6, we can determine the hoop stress by summing the forces in the $\theta$-direction:

$$\sum F_\theta = 0 = -p2RL + \sigma_{\theta\theta} 2Lt \tag{6.20}$$

$$\sigma_{\theta\theta} = \frac{pR}{t}. \tag{6.21}$$

The radial stresses are $-p$ on the inner radius and 0 on the outer radius. If we continue with our thinness assumption we need to pick a representative value for this stress. Since both the hoop and axial stresses are $p$ times a number that is at least 5 (for the axial) or 10 (for the hoop), we can see that the radial stress is small in comparison.

**Remarks:**

(1) The conventional assumption is $\sigma_{rr} = 0$. This is clearly questionable at $R/t = 10$, but rapidly becomes acceptable as R/t grows.

---

**Example 6.3**

*Circumference changes.* Suppose you are given a thin-walled cylindrical pressure vessel with welded end-caps, radius $R$, thickness $t$, modulus $E$ and Poisson ratio $\nu$. How much pressure is required to increase the circumference by 1%? How much pressure is required to decrease the thickness by 1%?

*Solution*

The state of stress in the vessel walls is given by $\sigma_{zz} = pR/2t$, $\sigma_{\theta\theta} = pR/t$, and $\sigma_{rr} \approx 0$. Changes in circumference will be related to hoop strains as $\Delta C = C\varepsilon_{\theta\theta}$, where $C$ is the circumference. Thus one needs a pressure such that

$$0.01 = \varepsilon_{\theta\theta} = \frac{\sigma_{\theta\theta}}{E} - \frac{\nu\sigma_{zz}}{E} \tag{6.22}$$

$$= \frac{pR}{tE} - \frac{\nu pR}{2tE} \tag{6.23}$$

$$p = \frac{0.02tE}{R(2-\nu)}. \tag{6.24}$$

Similarly, the change in wall thickness is related to $\varepsilon_{rr}$. Thus one needs

$$-0.01 = \varepsilon_{rr} = -\frac{\nu\sigma_{\theta\theta}}{E} - \frac{\nu\sigma_{zz}}{E} \tag{6.25}$$

$$= -\frac{3\nu pR}{2Et} \tag{6.26}$$

$$p = \frac{0.02tE}{3R\nu}. \tag{6.27}$$

---

## 6.2.2 Spherical thin-walled pressure vessels

Figure 6.7 shows a spherical pressure vessel with internal pressure $p$. The important dimension that we will assume to be thin in this situation is the radial one again; i.e. we will require that the wall thickness be small in comparison to the radius:

$$R/t > 10. \tag{6.28}$$

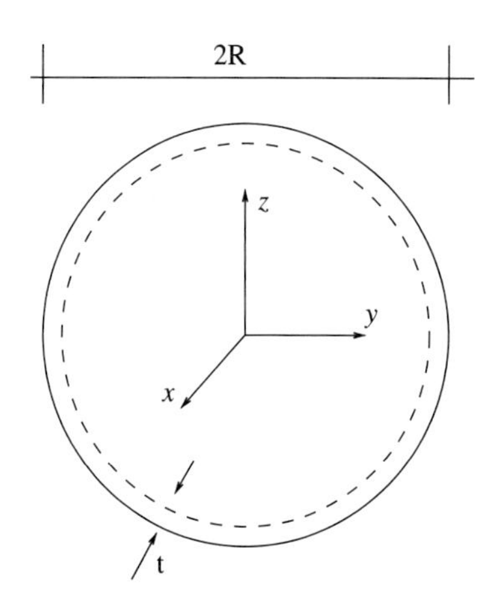

**Fig. 6.7** Spherical thin-walled pressure vessel.

The main implication of this is that we can assume the strains are constants as a function of $r$. In particular we will assume that the spherical hoop strain, $\varepsilon_{\theta\theta}$, the azimuthal strain, $\varepsilon_{\varphi\varphi}$, and the radial

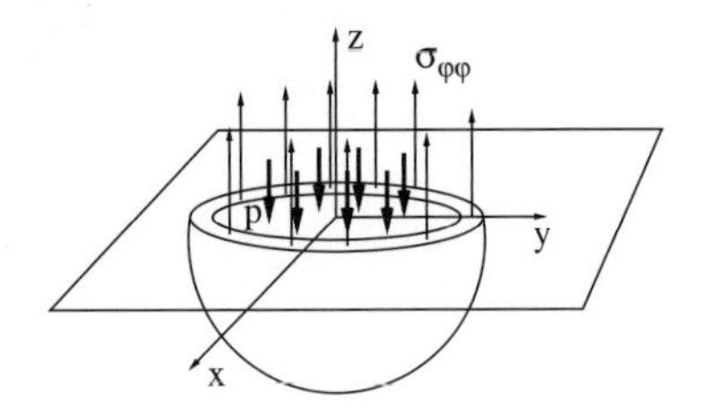


**Fig. 6.8** Spherical thin-walled pressure vessel with section cut whose normal is $-\boldsymbol{e}_{\varphi}$.

strain, $\varepsilon_{rr}$, are all constants as a function of $r$. Note also that by symmetry of the system nothing can be a function of $\theta$ or $\varphi$.

Let us assume that the sphere is made of an homogeneous linear elastic isotropic material. Since we have assumed that the strains are constant as a function of $r$, we also have that the stresses are constant as a function of $r$. Consider now a free-body diagram that is constructed by isolating the bottom half of the sphere with a section cut; see Fig. 6.8. If we sum the forces orthogonal to the section cut we have:

$$\sum F_n = 0 = p\pi R^2 - \sigma_{\varphi\varphi} 2\pi R t \tag{6.29}$$

$$\sigma_{\varphi\varphi} = \frac{pR}{2t}. \tag{6.30}$$

Thus our simple assumption allows us to determine the azimuthal stress in terms of the applied pressure. The main thing we have taken advantage of here is that the integral of the stress over the cross-section is the stress times the area, since the stress is a constant in terms of $r$ and $\theta$ (and $\varphi$).

If we now consider the free-body diagram shown in Fig. 6.9, then we can determine the spherical hoop stress by summing the forces orthogonal to the section cut:

$$\sum F_n = 0 = p\pi R^2 - \sigma_{\theta\theta} 2\pi R t \tag{6.31}$$

$$\sigma_{\theta\theta} = \frac{pR}{2t}. \tag{6.32}$$

This is also a result we could have also argued from symmetry.

The radial stresses are $-p$ on the inner radius and 0 on the outer radius. If we continue with our thinness assumption we need to choose a representative value for this stress. Since both the hoop and azimuthal stresses are $p$ times a number that is at least 5, we can see that the radial stress is small in comparison. The conventional assumption is that we assume $\sigma_{rr} \approx 0$.

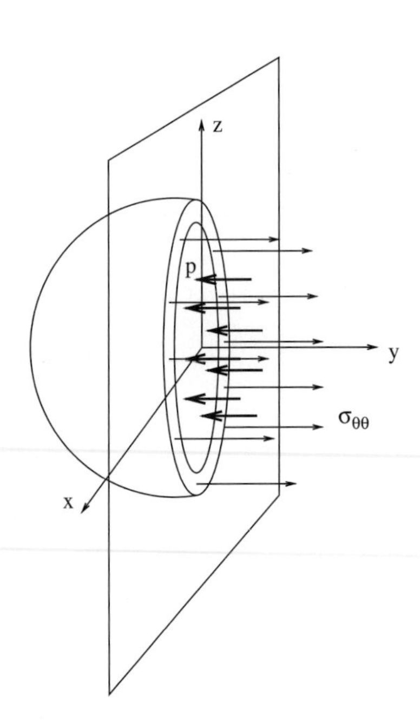


**Fig. 6.9** Spherical thin-walled pressure vessel with section cut whose normal is $\boldsymbol{e}_{\theta}$ in the front part and $-\boldsymbol{e}_{\theta}$ in the back part.

## 6.3 Saint-Venant's principle

The analysis in which we have engaged so far has always depended upon assumptions about the kinematics and also possibly the stresses. Before moving on to other types of load-bearing systems, it is worthwhile looking at the limitations of these assumptions. Firstly, the only real way to assess the limitations of such assumptions is to either solve the full set of governing equations or to perform extensive experimentation. Both of these topics are outside the scope of this text. Nonetheless, we can still have a qualitative discussion of the issue to gain some appreciation of the limitations.

If we look, for example, at a bar with an axial load, then our kinematic assumptions lead us to assume that the stresses are constant on the cross-section (for an homogeneous material); see Fig. 6.10. Clearly this

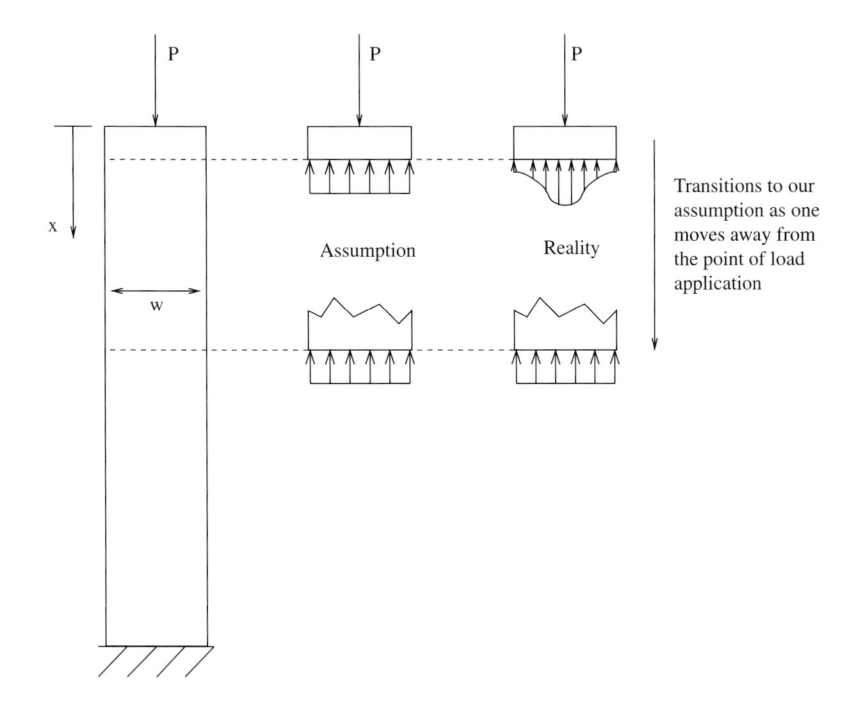


**Fig. 6.10** Bar with a point force.

cannot be true, for example, right at the end section unless the applied force is uniformly distributed to start with. If it is applied in any other way the stresses will have to transition from those consistent with the application of the load to a uniform distribution. Figure 6.10 qualitatively indicates the actual transition that takes place for a single point force.

The situation diagrammed in Fig. 6.10 is described by Saint-Venant's Principle:

The manner in which loads are applied only matters near the points of application.

Thus in Fig. 6.10, according to Saint-Venant's Principle, we can reasonably expect our basic computation of $\sigma_{xx} = R/A$ to be valid if we are sufficiently far from the application of the load, which in this case includes both the top and bottom ends of the bar. For slender bodies (such as those we deal with in this text), 'sufficiently far' is defined as one characteristic length. And characteristic length refers to some representative dimension of the cross-section of the body. As a quantitative example, consider that the bar in Fig. 6.10 is an homogeneous linear elastic bar, and that the load is applied by a single point force on the end. If we solve the complete equations for the two-dimensional theory, then we will find that the stresses one characteristic distance[2] from the end differ from those we would assume in our one-dimensional engineering mechanics solution by only 2.5%.

[2] For a two-dimensional bar the characteristic distance is the width, $w$.

Our theories, in addition to giving inaccurate results near supports and points of load application, give inaccurate results near rapid changes in cross-sectional area or modulus. For example, in a bar with a hole (Fig. 6.11) we would predict a stress $\sigma_{xx} = P/t(w - d)$ at the section with the hole. However, the hole represents a rapid change in area and leads, in reality, to a stress distribution more like that shown in

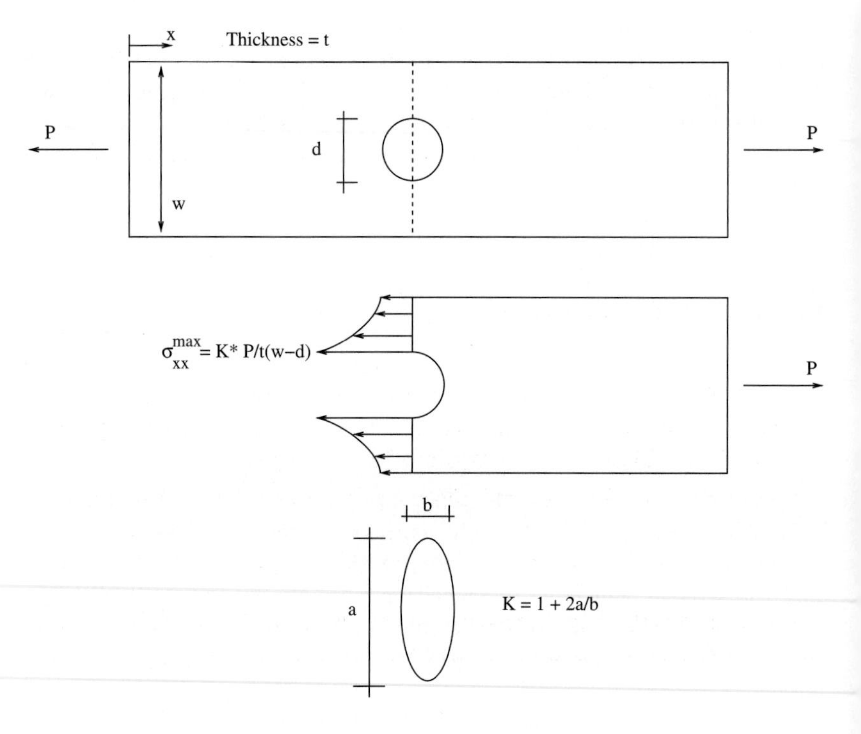


**Fig. 6.11** Bar with a hole showing a stress concentration.

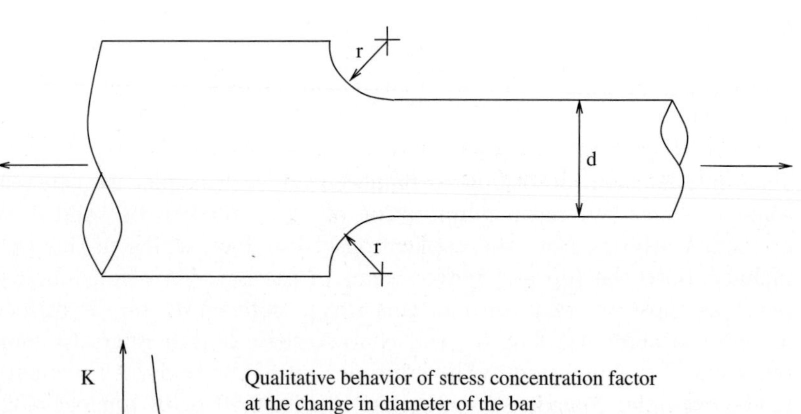


**Fig. 6.12** Stepped bar that will display a stress concentration at the fillet.

the middle part of Fig. 6.11. For such situations one defines a quantity called the stress concentration factor, $K$, to characterize the ratio of the maximum true stress to that computed from the engineering mechanics solution. In the case of an elliptic hole with semi-radii $a/2$ and $b/2$ the stress concentration factor is estimated (from the governing partial differential equations) to be $K = 1 + 2a/b$. This estimate is valid as long as $a$ is small in comparison to $w$; see Fig. 6.11. Figure 6.12 shows an example of another situation where one needs to be mindful of stress concentrations. The evaluation of stress concentration factors is most readily accomplished by numerical solution of the governing partial differential equations. Alternatively, for common situations, tables and graphs of stress concentrations are available in standard reference books.

## Chapter summary

- The fundamental kinematic assumption for axially loaded bars is that *plane sections remain plane*. This assumption together with the three-dimensional theory implies $\varepsilon = du/dx$.
- For axially loaded bars the general relation for the internal force is

$$R = \int_A \sigma \, dA$$

- In a thin-walled cylindrical pressure vessel ($R/t > 10$), the strains are a constant through the wall thickness and

$$\sigma_{\theta\theta} = \frac{pR}{t}$$
$$\sigma_{zz} = \frac{pR}{2t}$$
$$\sigma_{rr} \approx 0$$

- In a thin-walled spherical pressure vessel ($R/t > 10$), the strains are a constant through the wall thickness and

$$\sigma_{\theta\theta} = \frac{pR}{2t}$$
$$\sigma_{\varphi\varphi} = \frac{pR}{2t}$$
$$\sigma_{rr} \approx 0$$

- Saint-Venant's principle states:

  The manner in which loads are applied only matters near the points of application.

- Stress concentrations arise near rapid changes in dimensions.

# Exercises

(6.1) Consider a composite rod with steel core and aluminum casing under the action of arbitrary axial loads. If at a particular cross-section the stress in the steel is $\sigma_{xx} = 100$ MPa, what is $\sigma_{xx}$ in the aluminum at the same cross-section? What is the strain $\varepsilon_{xx}$ at this same cross-section?

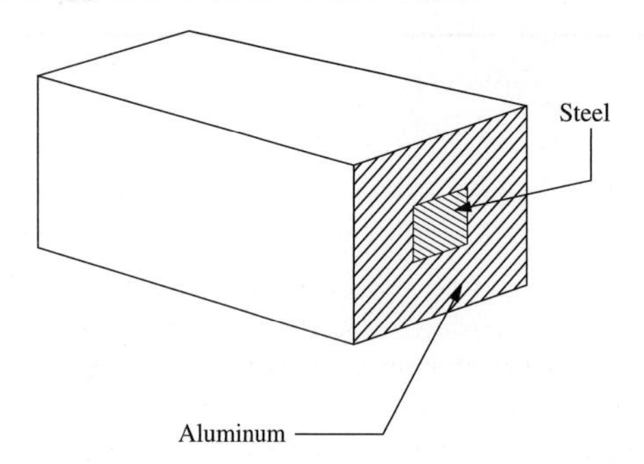


(6.2) For an elastic circular bar with homogeneous Young's modulus we know that the normal stress is given in terms of the axial force as $\sigma = R/A$. Find the corresponding formula for a bar that has been functionally graded in the radial direction so that the Young's modulus can be expressed as

$$E(r) = B + Cr,$$

where $B$ and $C$ are given material constants. Assume the bar has a radius $a$. (Hint: your answer should be equivalent to $\sigma_{xx}(r, z) = R(z)(B + Cr)/[2\pi(Ba^2/2 + Ca^3/3)]$, where $z$ is the axial coordinate.)

(6.3) The functionally graded round bar shown below is subjected to an axial force $F$. The Young's modulus is given as a function of radial position as $E(r) = E_o + \hat{E}r^2$, where $E_o$ and $\hat{E}$ and known material constants. By assuming that planar sections remain planar, find the axial displacement of the bar $u(x)$.

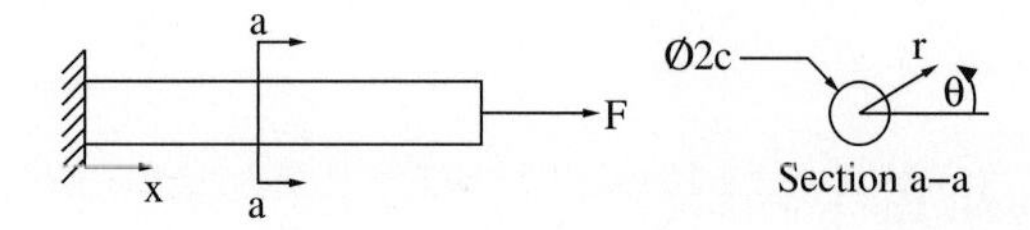


(6.4) The round slender composite bar, following is deformed by an end-load. The strain in the center of the rod is measured by neutron scattering to be $\varepsilon_{\text{center}}$. And $\varepsilon_{\text{center}}$ is found to lie between the yield strains for material 1 (inner material) and material 2 (outer material). How much force is being applied to the rod? Express your answer in terms of the material properties and the cross-sectional geometry. Assume the Young's modulus, $E$, for material 1 and 2 is the same. The radius of the inner rod is $R_1$ and the radius of the outer rod is $R_2$.

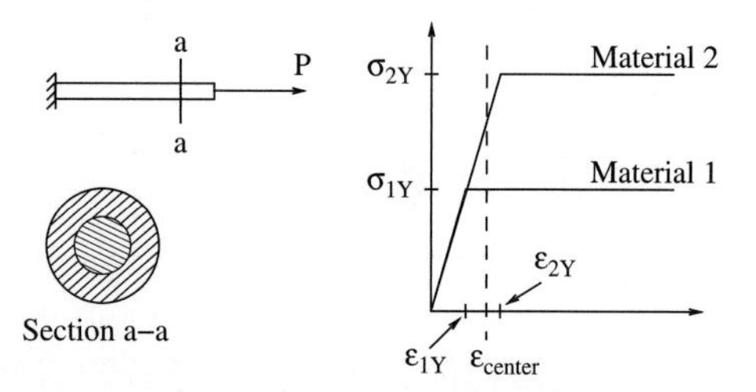


(6.5) In 1960 NASA launched its first communications satellite, ECHO. ECHO was a spherical Mylar balloon with a microwave reflective coating. Mylar is a registered DUPONT trademark for a thermoplastic composed of biaxially oriented polyethylene terephalate (BOPET); PET is the same material used in plastic soda bottles. Mylar sheet yields at 4 ksi and breaks at 27 ksi. ECHO was 100 feet in diameter and 0.0005 inches thick. What differential pressure would have caused ECHO to yield?

(6.6) Consider a thin-walled cylindrical pressure vessel of radius $R$, wall thickness $t$, and length $L$. Determine the changes in diameter and length of the vessel due to an internal pressure $p$. Assume homogeneous linear elastic material.

(6.7) A long cylindrical pressure vessel of length $L$, radius $R$, and thickness $t$ is subject simultaneously to an internal pressure $p$ and an axial force $F$. Find the relationship between $F$ and $p$ such that the axial stress and hoop stress are equal in the main part of the vessel.

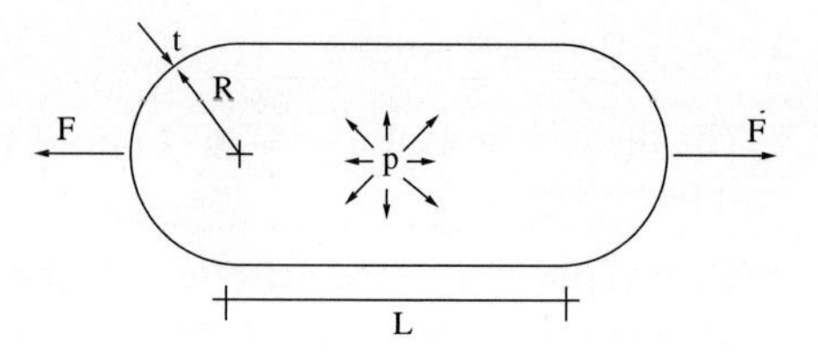

(6.8) Determine $\varepsilon_{\theta\theta}$, $\varepsilon_{\varphi\varphi}$, and $\varepsilon_{rr}$ for a thin-walled spherical pressure vessel with internal pressure $p$. Assume homogeneous linear elastic material.

(6.9) How much pressure is required to thin the walls of a thin-walled spherical pressure vessel by 1%? Assume homogeneous elastic properties.

(6.10) You are to design a cylindrical pressure vessel with spherical end caps as shown below. The caps and the main part of the vessel are to have the same thickness (which you need to determine). The radius of the vessel and the length are specified to be $R$ and $L$, respectively. The internal pressure is specified to be $p$. Design the vessel against yield using the following multi-axial criteria in the cylindrical portion of the vessel

$$\sqrt{(\sigma_r - \sigma_\theta)^2 + (\sigma_\theta - \sigma_z)^2 + (\sigma_z - \sigma_r)^2} \leq \sqrt{2}\sigma_Y$$

and

$$\sqrt{(\sigma_r - \sigma_\theta)^2 + (\sigma_\theta - \sigma_\varphi)^2 + (\sigma_\varphi - \sigma_r)^2} \leq \sqrt{2}\sigma_Y$$

in the spherical portion of the vessel. Allow for a safety factor of $SF$ against yield. In choosing the wall thickness, choose a material that minimizes the weight of the vessel. Choose your material from Table 6.1.

**Table 6.1** Material selection list.

| Material | E (GPa) | $\sigma_Y$ (MPa) | Density (kg/m$^3$) |
|---|---|---|---|
| Al | 70 | 470 | 2700 |
| Mild Steel | 200 | 260 | 7850 |
| HS Steel | 208 | 1300 | 7850 |
| HTS graphite-5208 epoxy | 183 | 1250 | 1550 |

What is the best material and what is the required wall thickness?

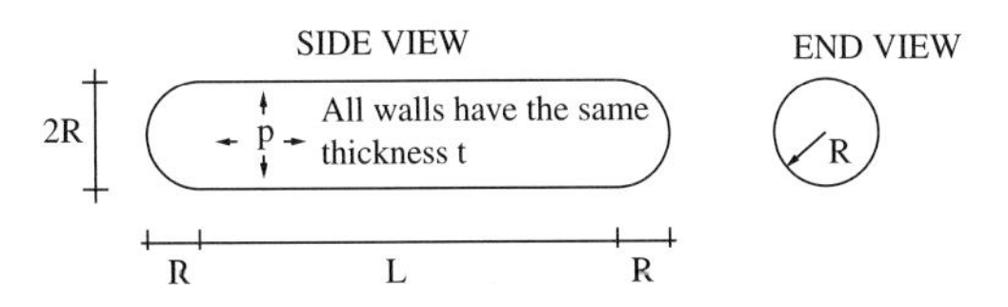


(6.11) A pressure vessel of radius $R$, length $L$, and wall thickness $t$ is constructed with a weld line at angle $\beta$. The dimensions are such that $t \ll R$ and $L \gg R$. The vessel is pressurized to a pressure $p$. What is the shear stress on the weld? Assume that the pressure vessel is closed on the ends.

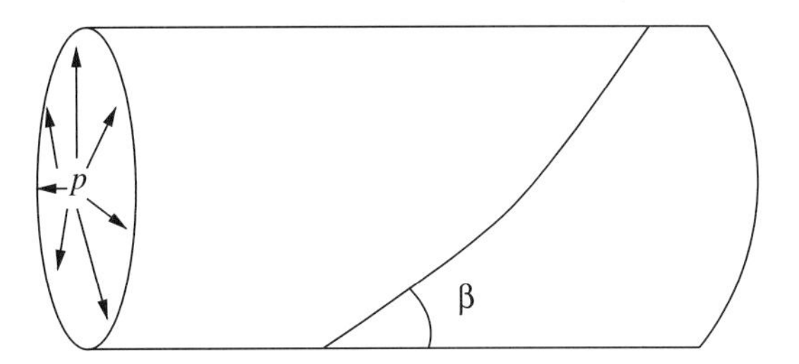


(6.12) A linear elastic tube with radius $(R - \delta)$ is to be shrink-fit on a rigid shaft of radius $R$. The tube is first heated so as to expand its inner radius to at least $R$. It is then slipped over the rigid shaft, and is allowed to cool. Assuming that the shaft is well lubricated, find a relation for the contact pressure between the tube and the shaft. Your final answer should be given in terms the isotropic elastic material constants, misfit $\delta$, thickness $t$, and rigid shaft radius $R$.

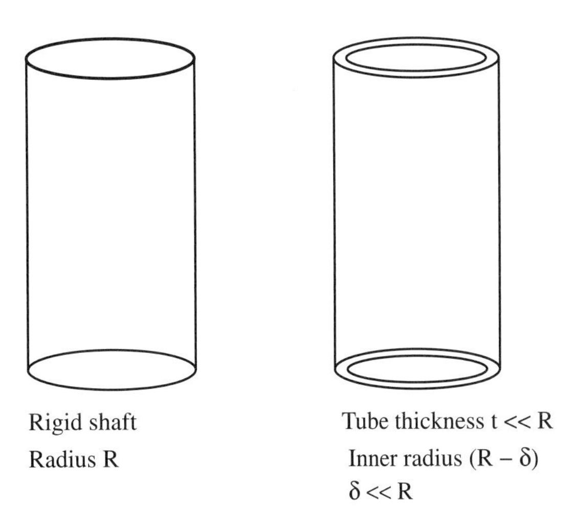


(6.13) Consider the following cross section of a composite cylindrical pressure vessel. Assume that the radius is much greater than the wall thicknesses. Assume that the strains are constants as a function of $r$. Thus, in particular, $\varepsilon_{zz}^{(1)} = \varepsilon_{zz}^{(2)}$ and

$\varepsilon_{\theta\theta}^{(1)} = \varepsilon_{\theta\theta}^{(2)}$. Also assume that the radial stresses are zero. Show that the hoop and axial stresses in the two materials are governed by the solution of the following linear equations.

$$\begin{pmatrix} p \\ p \\ 0 \\ 0 \end{pmatrix} = \begin{bmatrix} \frac{2t_1}{R} & \frac{2t_2}{R} & 0 & 0 \\ 0 & 0 & \frac{t_1}{R} & \frac{t_2}{R} \\ \frac{1}{E_1} & -\frac{1}{E_2} & -\frac{\nu_1}{E_1} & \frac{\nu_2}{E_2} \\ -\frac{\nu_1}{E_1} & \frac{\nu_2}{E_2} & \frac{1}{E_1} & -\frac{1}{E_2} \end{bmatrix} \begin{pmatrix} \sigma_{zz}^{(1)} \\ \sigma_{zz}^{(2)} \\ \sigma_{\theta\theta}^{(1)} \\ \sigma_{\theta\theta}^{(2)} \end{pmatrix}$$

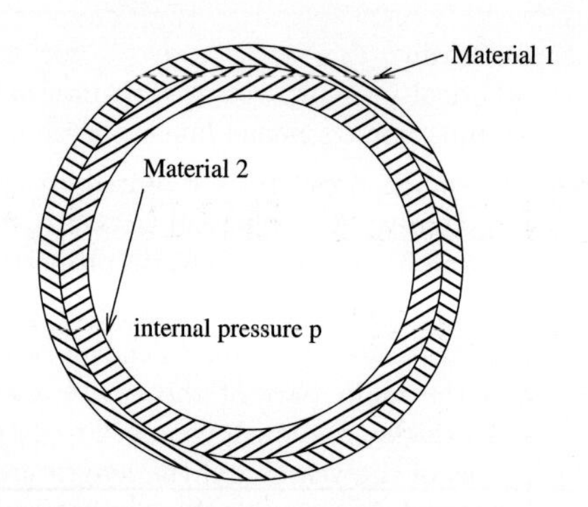


Hint: The first two relations are related to force equilibrium and the second two equations are related to kinematic considerations.

# Circular and Thin-Wall Torsion

The application of torsional loads to slender bars can be analyzed very effectively utilizing the techniques developed in Chapter 6. In particular, by making an appropriate kinematic assumptions we will be able to nicely analyze the behavior of twisted circular bars. For the case of bars with more general cross-sections the situation is a little more complex. But if we include the added requirement that the cross-section have the geometry of a thin-walled tube, then we will be able to take advantage of thinness to drive an effective analytical method for arbitrary thin-walled cross-sections. The subject of torsion of solid non-circular bars is left for more advanced courses where the solution of partial differential equations can be more comfortably tackled. Figures 7.1 and 7.2 shows some example situations where one finds structural members under torsional loads.

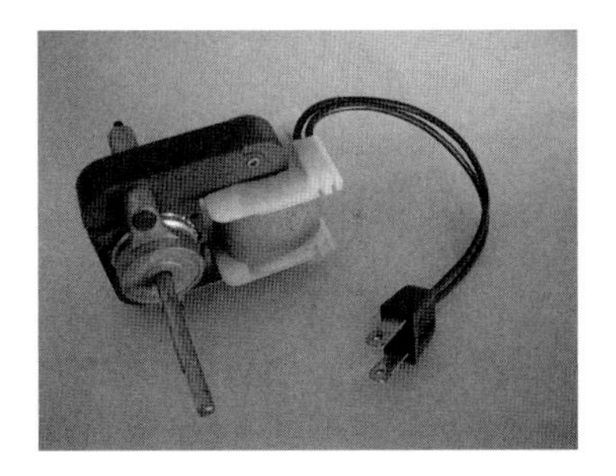

**Fig. 7.1** An electric motor's shaft is loaded in torsion when in operation.

## 7.1 Circular bars: Kinematic assumption

The fundamental observation associated with the motion of circular bars in torsion is that the bar cross-sections remain planar and rigidly rotate under load. This observation is predominantly true over a very wide range of loadings, elastic and plastic behavior, and material inhomogeneities (with circular symmetry). Figure 7.3a shows a solid circular bar that is clamped at the base. At the top is a lever arm that allows one to apply a torque to the bar. Figure 7.3b shows a close-up of the square grid that has been painted on the surface of the bar. After the application of a torque to the bar the grid distorts as shown in Fig. 7.3c. Note that the horizontal lines remain planar and that there is a progressive rotation that is increasing at a constant rate as one moves up the bar. In the special case of torsion of an infinitely long bar, one can show via symmetry arguments that these observations are exact.

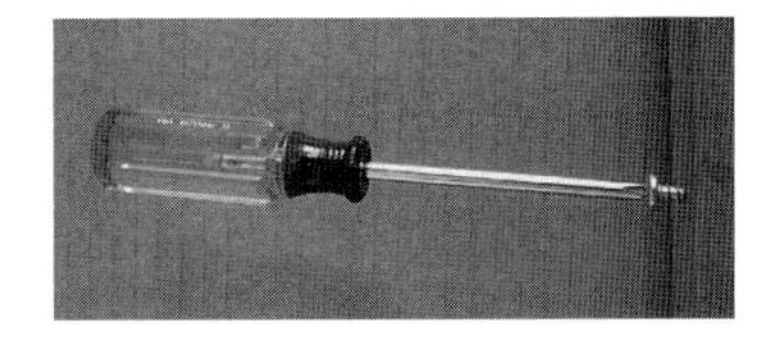

**Fig. 7.2** Both the screwdriver shaft and the screw are loaded in torsion when driving the screw into the wood member.

These physical observations can be turned into a usable kinematic relation by considering the geometric construction in Fig. 7.4, showing a circular bar under the action of two applied torques. First we slice the torsion bar at two elevations $z$ and $z + \Delta z$. Due to the applied torque the bottom of the piece rotates an amount $\phi(z)$ and the top an amount $\phi(z + \Delta z)$. In the theory of torsion, the rotation of the section $\phi$ is similar to the displacement $u$ in the theory of axial extension.

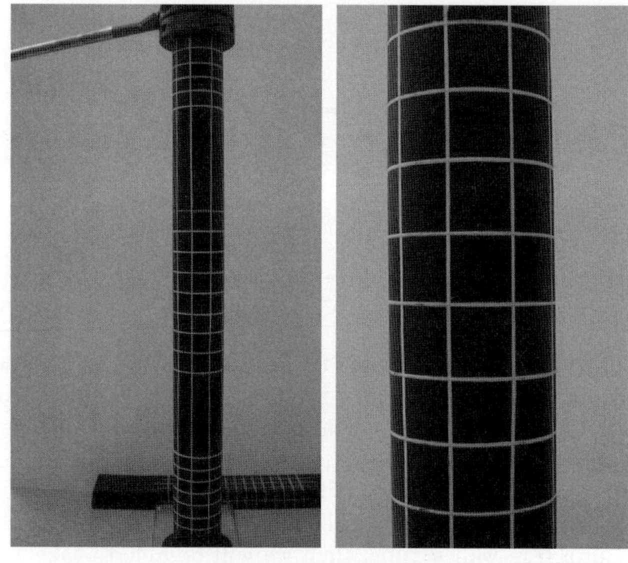

(a) Overall system. (b) Close-up of the square grid ruled on the outer surface.

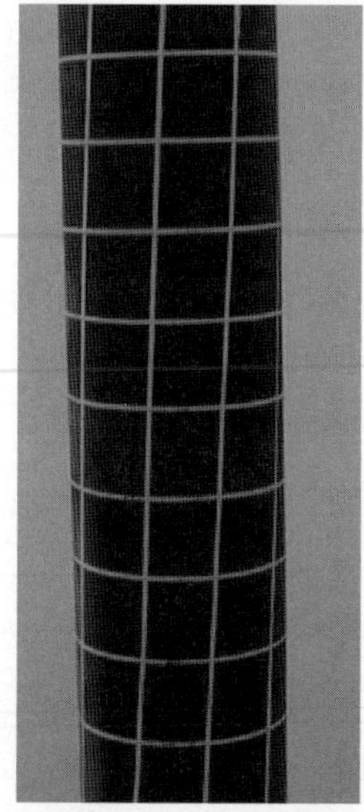

(c) Close-up of the square grid after the application of a torque to the bar.

**Fig. 7.3** Torsion experiment used to understand kinematics of torsion.

Before applying any torque to the bar we imagine that markers $a$, $b$, $c$, and $d$ have been placed on a core of radius $r$. After the application of the torque the markers will move as shown to locations $a'$, $b'$, $c'$, and $d'$. The physical arclength of displacement of markers $c$ and $d$ will be $r\phi(z)$ and that of $a$ and $b$ will be $r\phi(z + \Delta z)$. If we unwrap the picture we can better see the main characteristics of the deformation; see Fig. 7.5. It is

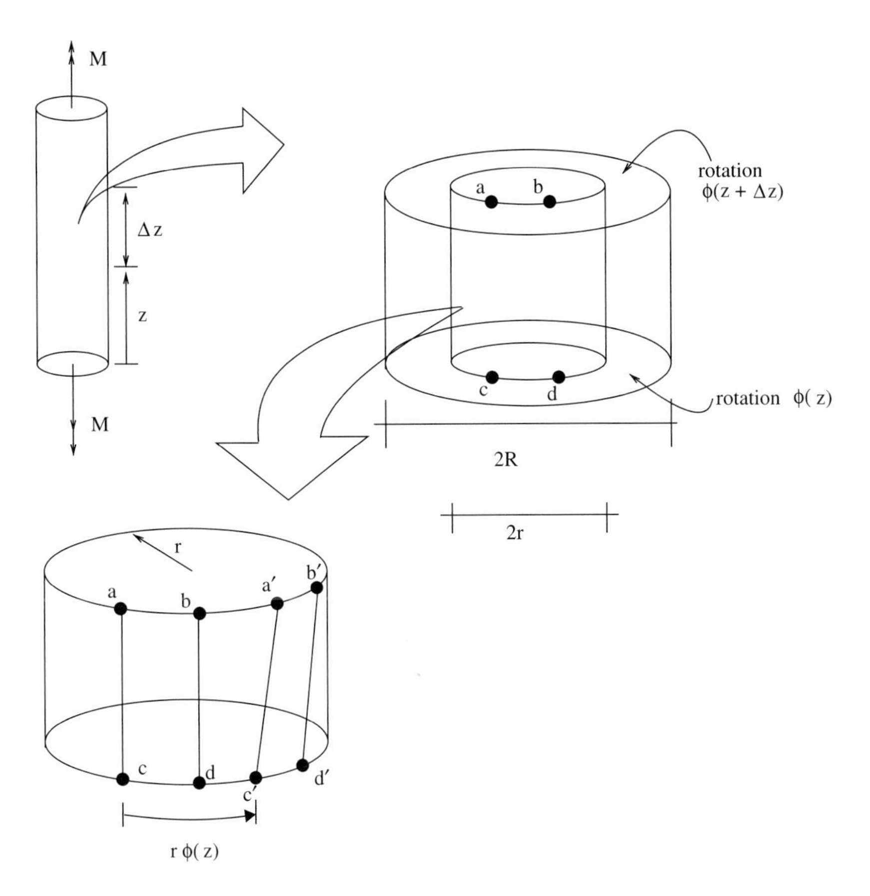


**Fig. 7.4** Kinematic assumption of circular torsion.

clear from this figure that the state of deformation is shear. Using our nomenclature from Chapter 4, we would characterize this deformation by the shear strain $\gamma_{z\theta}$ – change in angle in the "$z$-$\theta$ plane". Note that the "plane" in question is really a surface, and we adopt the convention that a surface is defined by the two basis vector which are tangent to the surface.

The expression for the average value of $\gamma_{z\theta}$ in the piece can be derived by applying the definition of shear strain from Chapter 4 to the construction in Fig. 7.5:

$$\gamma_{z\theta} = \frac{r\phi(z + \Delta z) - r\phi(z)}{\Delta z}. \tag{7.1}$$

In eqn (7.1) we have assumed that the angle change is small so that the arctangent of the right-hand side is equal to itself. To arrive at a fully pointwise measure of shear strain in the bar we need to take the limit as $\Delta z$ goes to zero. This then gives a pointwise measure of shear strain in the bar as a function of radial position $r$ and vertical location $z$.

$$\gamma_{z\theta}(r, z) = \lim_{\Delta z \to 0} \frac{r\phi(z + \Delta z) - r\phi(z)}{\Delta z} = r\frac{d\phi}{dz}. \tag{7.2}$$

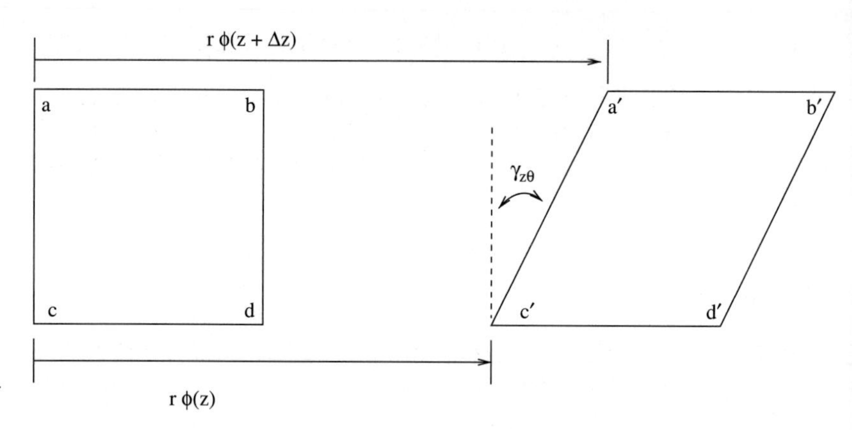


**Fig. 7.5** $z$-$\theta$ plane unwrapped to highlight kinematics.

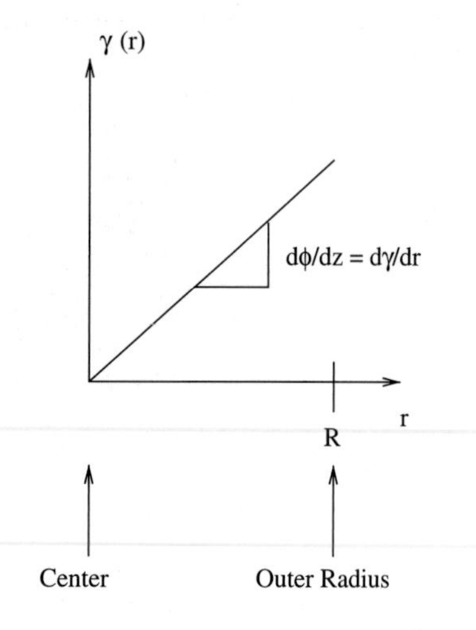


**Fig. 7.6** Shear strain distribution on a cross-section.

**Remarks:**

(1) Equation (7.2) is the circular torsion counterpart to $\varepsilon = du/dx$ from Chapter 2.

(2) Observe that $\gamma_{z\theta}$ is a function of $r$ and $z$ but is not a function of $\theta$. The problem is axis-symmetric.

(3) The derivation just given is completely independent of material response or the manner of application of the loads. It depends only upon the validity of the assumption (which in principle comes from a history of experimental observation).

(4) At any given cross-section the shear strain distribution is a linear function of radial position; see Fig. 7.6. It is zero at the center and takes its maximal value on the outer radius of the bar.

(5) Mathematically we could have expressed our kinematic assumption as $u_\theta = r\phi(z)$, $u_r = 0$, $u_z = 0$. Writing it this way allows us to use eqn (4.24) to show that $\gamma_{z\theta}$ is the only non-zero strain in the bar. Because of this we will from now on simply refer to $\gamma_{z\theta}$ as $\gamma$ – i.e. we will drop the subscripts when considering circular bars under torsion.

## 7.2 Circular bars: Equilibrium

Just as with the bar under axial forces we need to consider the torsional equilibrium of the bar. Paralleling the development in Section 2.2.2 we will look at a differential construction. Figure 7.7 shows a bar in torsion under the action of an end torque $M$ and a distributed body torque $t(z)$. $t(z)$ has units of torque per unit length and is similar to $b(x)$ which had units of force per unit length from Chapter 2. As shown, we will make two section cuts (one at $z$ and one at $z + \Delta z$) and then apply global equilibrium to the isolated segment. Summing the moments in the $z$-direction we find:

$$\sum M_z = 0 = T(z + \Delta z) + t(z)\Delta z - T(z), \tag{7.3}$$

where $T(z)$ represents the internal torque about the $z$-axis on the section cut. If we divide through by $\Delta z$ and take the limit as $\Delta z \to 0$, we will arrive at a pointwise expression of moment equilibrium about the $z$-axis:

$$\frac{dT}{dz} + t = 0. \tag{7.4}$$

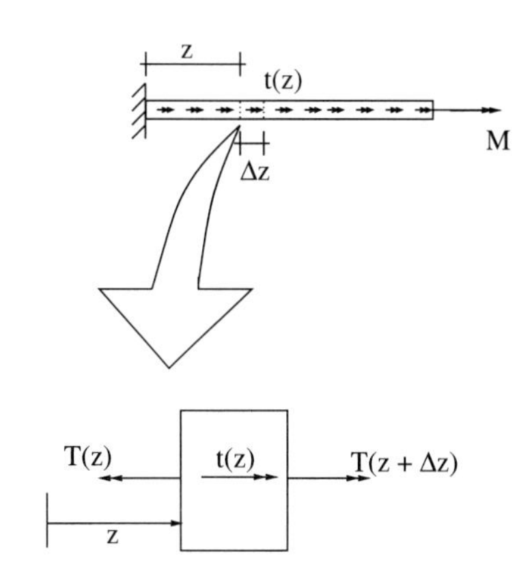


**Fig. 7.7** Construction for the derivation of the torsional equilibrium equation.

**Remarks:**

(1) Equation (7.4) is the torsion counterpart to $\frac{dR}{dx} + b = 0$ from Chapter 2; see Section 2.2.2.

### 7.2.1 Internal torque–stress relation

In our development of a theory to describe the behavior of bars under axial forces we profitably related the internal forces $R(x)$ to the stresses on the cross-section through the relation $R(x) = \int_A \sigma_{xx}\, dA$. For a bar in torsion there is a similar relation which we will derive through a direct construction. Figure 7.8 shows a cross-sectional slice at an arbitrary location along the $z$-axis. On each differential patch, $dA$, there is a force:

$$\boldsymbol{F} = F_z \boldsymbol{e}_z + F_r \boldsymbol{e}_r + F_\theta \boldsymbol{e}_\theta, \tag{7.5}$$

where

$$F_z = \sigma_{zz} dA \tag{7.6}$$

$$F_r = \sigma_{zr} dA \tag{7.7}$$

$$F_\theta = \sigma_{z\theta} dA. \tag{7.8}$$

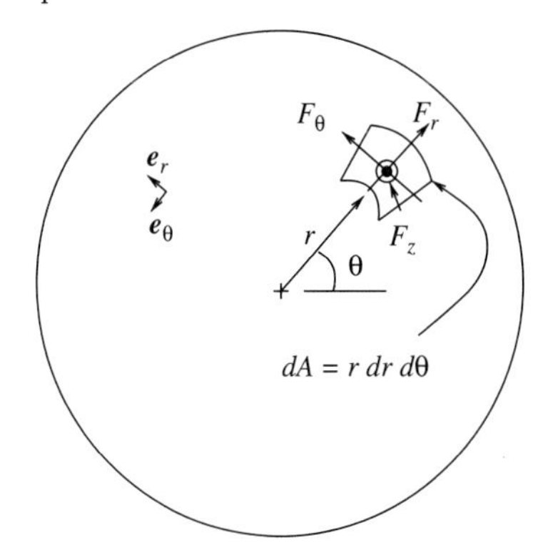


**Fig. 7.8** Stress contributions to forces on a differential patch on a torsion bar cross-section.

The $z$-component of the moment of this force about the origin will give the contribution of the stresses on this patch to the overall torque on the cross-section. Adding it over the whole cross-section will give the desired relation. Noting that the contribution from $F_r$ passes through the origin and that $F_z$ is parallel to the axis of interest, we see that the only contributions from the stresses to the moment about the $z$-axis come through $F_\theta \boldsymbol{e}_\theta$. The lever arm to the patch of material is simply $r$, and adding up we find that

$$T(z) = \int_{A(z)} \sigma_{z\theta} r\, dA = \int_{A(z)} \sigma_{z\theta} r^2 dr d\theta\,. \tag{7.9}$$

**Remarks:**

(1) Equation (7.9) gives the desired expression connecting stresses on a cross-section to internal torque.

(2) An alternative method of arriving at this relation comes from an application of the general expression for internal moments, eqn (3.27). To use eqn (3.27), one needs to take the dot product of

both sides with $\boldsymbol{e}_z$ to obtain the $z$-component, and further note that the cross-section normal $\boldsymbol{n} = \boldsymbol{e}_z$.

(3) As there is only one stress component contributing to the internal torque, we will by convention refer to the shear stress as $\tau$ when there is no chance of confusion.

## 7.3 Circular bars: Elastic response

The kinematic relation, $\gamma = r d\phi/dz$, the equilibrium relation, $dT/dz + t = 0$, and the resultant definition, $T = \int_A r\tau dA$, provide three of the needed expressions for a complete theory of one-dimensional torsion of circular bars. To close the set of equations we need one other ingredient: a description of the material response – i.e. the constitutive relation.

In the case of isotropic linear elastic materials this is given by

$$\tau = G\gamma, \tag{7.10}$$

where $G$ is the shear modulus of the material.

### 7.3.1 Elastic examples

At this stage we have a completely defined theory appropriate for solving problems associated with torsional loads on circular bars. The governing system of equations is given by the fundamental kinematic relation[1]

[1] This involves the plane-sections-remain-plane assumption, and is only appropriate for bars with circular cross-sections.

$$\gamma = r\frac{d\phi}{dz}, \tag{7.11}$$

the equilibrium relation

$$\frac{dT}{dz} + t = 0, \tag{7.12}$$

the internal torque relation

$$T = \int_A r\tau dA, \tag{7.13}$$

and the constitutive relation

$$\tau = G\gamma. \tag{7.14}$$

In total we have four linear equations in four variables: $\phi, \gamma, T,$ and $\tau$. Below we consider some example applications.

---

**Example 7.1**

*Twist rate to torque relation.* Consider a linear elastic homogeneous circular bar and find the relation between the twist rate and the internal torque.

*Solution*

This problem asks us to connect kinematics to resultants. Start with the expression for the definition of the resultant

$$T = \int_A r\tau dA. \tag{7.15}$$

Substitute for the shear stress in terms of the shear strain:

$$T = \int_A rG\gamma dA. \tag{7.16}$$

Substitute for the shear strain in terms of the twist rate:

$$T = \int_A r^2 G \frac{d\phi}{dz} dA. \tag{7.17}$$

Since the bar is homogeneous, $G$ is a constant and can be pulled out from under the integral sign. Also note that $d\phi/dz$ is only a function of $z$ so we have

$$T = G \underbrace{\left[\int_A r^2 dA\right]}_{J} \frac{d\phi}{dz} = GJ\frac{d\phi}{dz}. \tag{7.18}$$

**Remarks:**

(1) $J = \int_A r^2 dA$ is known as the polar moment of inertia. In the theory of torsion it occupies a position similar to that occupied by area in the theory of axial extension. For a solid circular section of radius $R$, $J = \pi R^4/2$. For a hollow tube, $J = \pi(R_o^4 - R_i^4)/2$, where $R_o$ is the outer radius and $R_i$ is the inner radius.

(2) In cases where the bar is inhomogeneous on the cross-section (i.e. $G$ is a function of $r$), $G$ cannot be removed from underneath the integral sign in eqn (7.17). In such cases one often defines an effective $(GJ)_{\text{eff}} = \int_A Gr^2 dA$. With this definition one can write $T = (GJ)_{\text{eff}} d\phi/dz$. This is a handy device for dealing with composite bars.

(3) In the case of inhomogeneous bars, $G$ is not permitted to be a function of $\theta$, because in such cases our kinematic assumption is known to be invalid.

**Example 7.2**

*Elastic stress distribution.* Consider the same linear elastic homogeneous circular bar as in the last example, and find the relation between the stress on a cross-section and the internal torque on the cross-section.

*Solution*
Starting from the result of the last example we have that

$$\frac{d\phi}{dz} = T/GJ. \tag{7.19}$$

Thus $\gamma = rd\phi/dz = rT/GJ$ in this special case. Substituting into the constitutive relation gives:

$$\tau(r,z) = G\gamma = \frac{T(z)r}{J}. \tag{7.20}$$

**Remarks:**

(1) Note that the shear stress varies linearly on the cross-section as long as the cross-section is homogeneous. The maximum value occurs at the outer radius.

(2) If the bar were not homogeneous on the cross-section, one can use $(GJ)_{\text{eff}}$. In this case one finds $\tau(r,z) = rG(r)T(z)/(GJ)_{\text{eff}}$, and the maximum stress occurs wherever the term $rG(r)$ is maximum (not necessarily at the outer radius).

---

**Example 7.3**

*Stresses in a composite bar.* Consider an elastic composite bar as shown in Fig. 7.9a. If the applied twist rate is $10^{-4}$ rad/mm, what is the maximum stress in the bar, and where does it occur?

*Solution*
The stresses are given as

$$\tau(r) = G(r)\gamma(r), \tag{7.21}$$

where $\gamma(r) = 10^{-4}r$ and

$$G(r) = \begin{cases} 100 & r < 20 \text{ mm} \\ 50 & r > 20 \text{ mm} \end{cases} \quad \text{GPa}. \tag{7.22}$$

Thus,

$$\tau(r) = \begin{cases} 10r & r < 20 \text{ mm} \\ 5r & r > 20 \text{ mm} \end{cases} \quad \text{MPa}. \tag{7.23}$$

The maximum happens at $r = 20$ mm just inside the inner material with a value of 200 MPa; see Fig. 7.9b, which shows the shear strain and shear stress distributions.

---

**Example 7.4**

*Power transmission.* Structural elements carrying torques often occur in power transmission systems. Consider the 1 − kW motor shown in

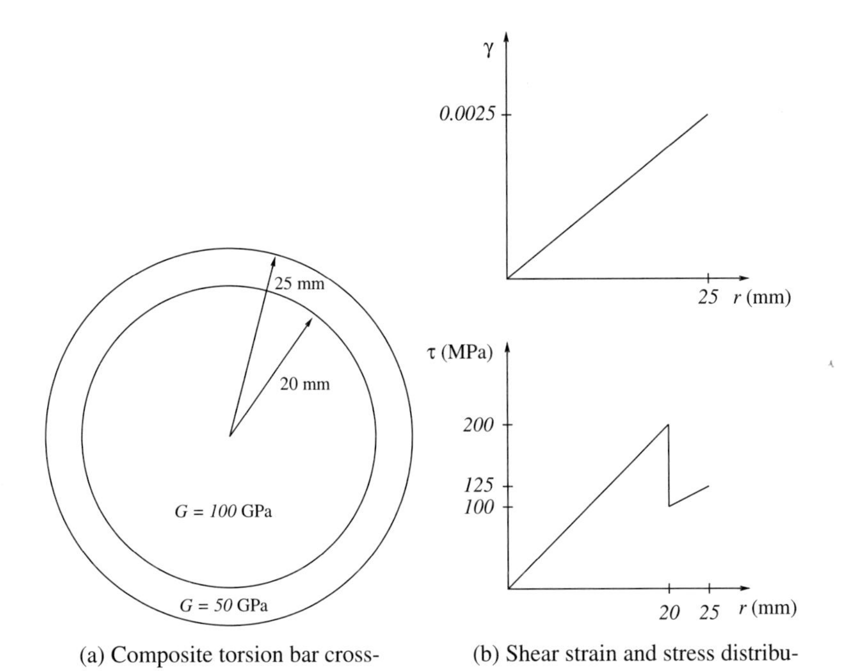


**Fig. 7.9** Example of a composite torsion bar.

Fig. 7.10. It is steadily spinning at 5,000 rpm, and is attached to an elastic shaft which is connected to two belts that deliver energy to two pieces of equipment in the amount of 700 W and 300 W, respectively. Determine (1) the internal torque distribution in the shaft, and (2) the shaft diameter between the two equipment loads such that the maximum shear stress along the entire length of the shaft is a constant. For part (2), express the answer relative to the diameter of the shaft between the motor and the first load.

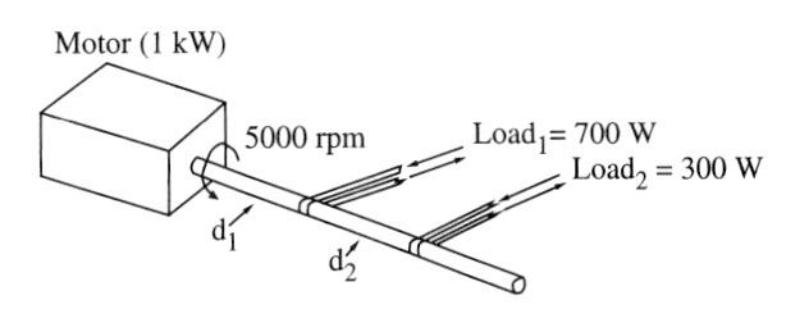


**Fig. 7.10** 1 − kW motor driving two belt loads at 5,000 rpm.

*Solution*

First we need to determine how the power ratings of the motor and the loads are related to applied torques. From elementary physics we recall that work input is force times displacement or torque times rotation depending upon the method of application of a load. Thus power input is either force times velocity or torque times angular velocity. Applying this concept to the motor tells us that

$$1,000 \text{ W} = T_M \,(5,000 \text{ rpm}) \left(\frac{2\pi \text{ rad}}{1 \text{ rev}}\right) \left(\frac{1 \text{ min}}{60 \text{ sec}}\right) \tag{7.24}$$

$$\Rightarrow T_M = \frac{6}{\pi} \text{ N} \cdot \text{m}, \tag{7.25}$$

where $T_M$ is the output torque of the motor at this speed and power level. The sense of this torque is in the same direction as the rotation.

At each of the loads we have a similar relation. The only question really is: What is the angular velocity at each of the loads? Since the shaft is in steady rotation, we immediately have that the angular velocity at each load point is the same as at the motor and in the same rotational direction. Thus each load imposes a torque on the shaft. For load 1, we have:

$$-700\ \mathrm{W} = T_1\,(5000\ \mathrm{rpm})\left(\tfrac{2\pi\ \mathrm{rad}}{1\ \mathrm{rev}}\right)\left(\tfrac{1\ \mathrm{min}}{60\ \mathrm{sec}}\right) \tag{7.26}$$

$$\Rightarrow T_1 = -\tfrac{42}{10\pi}\ \mathrm{N\cdot m}. \tag{7.27}$$

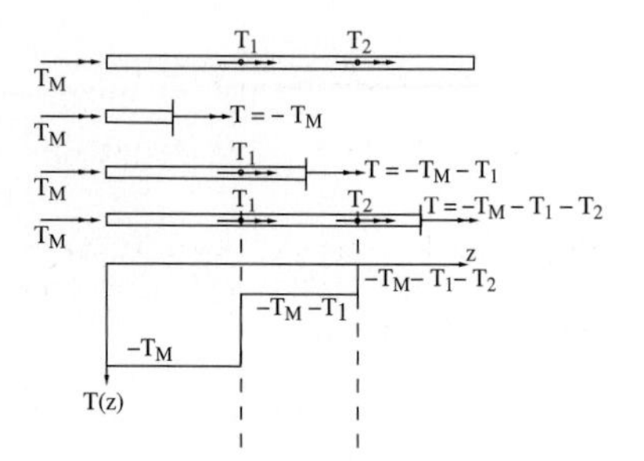


**Fig. 7.11** Internal torque diagram determined using section cuts.

Likewise, for load 2 it can be shown that

$$T_2 = -\frac{18}{10\pi}\ \mathrm{N\cdot m}. \tag{7.28}$$

Note that here, power is given with a negative sign because the power is being drawn off of the shaft. With these values determined we can make a free-body diagram as shown in Fig. 7.11. Making section cuts as shown allows us to determine the internal torque diagram for part (1) of the question.

For part (2) of the question, first determine the maximum shear stress as a function of $z$. We note that the maximum values will occur on the outer radius and be given by $\tau_{\max}(z) = T(z)r_{\max}/J(z)$. Since $\tau_{\max}$ is to be a constant, we have the requirement that

$$\frac{T_M r_1}{J_1} = \frac{(T_M + T_1)r_2}{J_2}. \tag{7.29}$$

Substituting in for the polar moment of inertia gives the answer to part (2) as:

$$d_2 = d_1\left(\frac{T_M + T_1}{T_M}\right)^{1/3} \tag{7.30}$$

$$d_2 = d_1(0.3)^{1/3} \approx 0.7d_1. \tag{7.31}$$

**Remarks:**

(1) Notice that even though the internal torque between the two belts is one-third the internal torque between the motor and the first load, we cannot reduce the size of the shaft in the same proportion; i.e. we can not let $d_2 = 0.3d_1$. Rather, we must have $d_2 \approx 0.7d_1$. The reason this occurs is two-fold: (1) the stresses are not constant on the cross-section (they are linear) and (2) the area measure that appears in the theory is non-linear in the cross-sectional dimension (it is quartic).

(2) Notice that on the last section of the shaft the internal torque is computed to be zero. This make sense, since the end of the shaft is free of applied load.

### Example 7.5

*Remote switch.* In hazardous environments equipment is often turned on and off using remote activation. One very simple such configuration is shown in Fig. 7.12. Here we have a linear elastic homogeneous wire of radius $c$ inside a sleeve (just like a brake cable housing on a bicycle). One end of the wire is attached to a rotary switch on some equipment which requires a torque $T_d$ to actuate it. The other end of the wire is located in a safe area where the operator twists it to activate the equipment. Assume that the sleeve provides a constant frictional resistance to rotation of the wire of an amount $t_f$ in units of torque per unit length. If the maximum allowed stress in the wire is $\tau_Y$, how long can the actuation wire be?

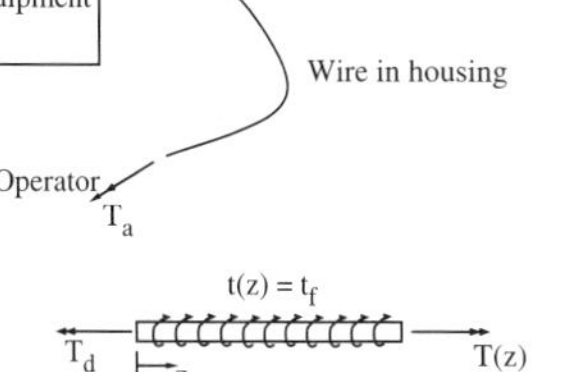


**Fig. 7.12** Equipment operated via a remote switch using a cable in a frictional cable housing.

*Solution*
First determine the internal torque in the wire using a free-body diagram as shown. Summing the moments in the $z$-direction gives

$$T(z) = T_d + t_f z. \tag{7.32}$$

Since the wire is homogeneous we have that

$$\tau_{\max}(z) = \frac{c(T_d + t_f z)}{\pi c^4/2} \leq \tau_Y. \tag{7.33}$$

This implies that

$$z \leq \left( \frac{\tau_Y \pi c^3}{2} - T_d \right) / t_f. \tag{7.34}$$

**Remarks:**

(1) To get a feel for numbers, let $c = 1.5$ mm, $t_f = 0.45$ N · mm/mm, $T_d = 100$ N · mm, and $\tau_Y = 100$ N/mm$^2$, then $z_{\max} = 0.96$ m; i.e. the maximum wire length is about one meter for these properties.

### Example 7.6

*Torsional stiffness of an end-loaded bar.* Consider an homogeneous end-loaded circular torsion bar of length $L$, polar moment of inertia $J$, and shear modulus $G$. Find its torsional stiffness.

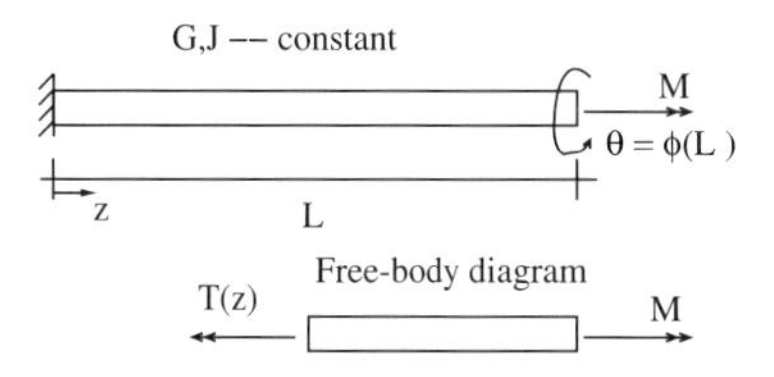


**Fig. 7.13** End-loaded torsion bar.

*Solution*
Torsional stiffness is defined as applied end-torque divided by net twist ($k = M/\theta$); see Fig. 7.13. The net rotation can be found by integrating the relation for the twist rate:

$$\theta = \phi(L) - \phi(0) = \int_0^L \frac{d\phi}{dz}\, dz. \tag{7.35}$$

Since the problem is homogeneous linear elastic we have

$$\theta = \phi(L) - \phi(0) = \int_0^L \frac{T}{GJ}\, dz. \tag{7.36}$$

Note that by equilibrium $T(z) = M$ a constant. Thus the entire integrand is a constant. Further, there is no rotation at $z = 0$; so we have that

$$k = \frac{M}{\theta} = \frac{M}{\phi(L)} = \frac{GJ}{L}. \tag{7.37}$$

**Remarks:**

(1) Compare this to the similar result for an end-loaded bar with axial forces, $k = AE/L$.

---

**Example 7.7**

*Rotation of bar with a point torque.* Consider the circular linear elastic homogeneous bar shown in Fig. 7.14a. Determine the rotation field, $\phi(z)$, the stress field, $\tau(r, z)$, the strain field, $\gamma(r, z)$, and the internal torque field, $T(z)$.

*Solution*

Begin first with a free-body diagram as shown in Fig. 7.14b. Making section cuts as indicated we see that the internal torque field is given by:

$$T(z) = \begin{cases} -M & z < a \\ 0 & z > a. \end{cases} \tag{7.38}$$

Since the bar is linear elastic homogeneous we know that $\tau = Tr/J$; thus

$$\tau(r, z) = \begin{cases} -Mr/J & z < a \\ 0 & z > a. \end{cases} \tag{7.39}$$

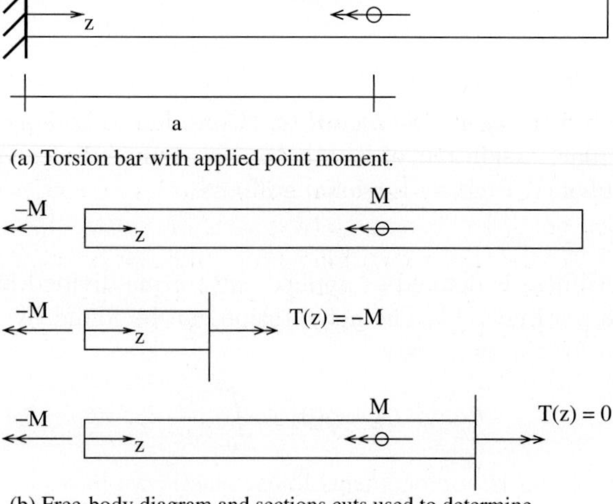


**Fig. 7.14** Analysis of torsion bar with an applied point torque using the method of section cuts.

The strains are connected to the stresses via the constitutive relation $\gamma = \tau/G$ which gives:

$$\gamma(r,z) = \begin{cases} -Mr/GJ & z < a \\ 0 & z > a. \end{cases} \tag{7.40}$$

The rotation field can now be found by integrating the fundamental kinematic relation

$$\phi(z) - \underbrace{\phi(0)}_{0} = \int_0^z \frac{d\phi}{dz}\, dz = \int_0^z \frac{\gamma}{r}\, dz = \begin{cases} -Mz/GJ & z < a \\ -Ma/GJ & z \geq a. \end{cases} \tag{7.41}$$

**Remarks:**

(1) Notice that the resulting rotation of the bar is negative. This simply means that the rotation is in the opposite sense as that defined by a right-hand rule about the $z$-axis.

(2) Notice that in this problem we were able to determine the internal torque field simply from statics. This problem was statically determinate. There was only one unknown support torque, and one equilibrium equation.

**Example 7.8**

*Indeterminate system.* Figure 7.15 shows a linear elastic homogeneous torsion bar with a distributed torque. Make a graph of the internal torque field (i.e. a torque diagram) and a graph of the rotation field.

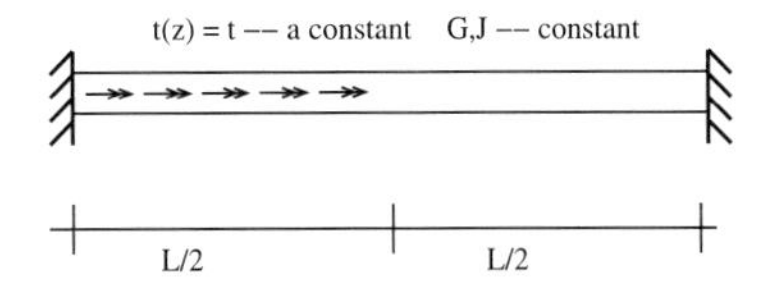


**Fig. 7.15** Indeterminate torsion bar with loads and supports.

*Solution*

First make a free-body diagram of the bar as shown in Fig. 7.16. From this diagram it is clear that the problem is statically indeterminate. There are two unknown support torques. Following the techniques developed in Chapter 2, let us assume that $T_o$ is known. At the end we will eliminate $T_o$ using the known kinematic information of the problem. From the free-body diagram, we can make section cuts progressively along the length of the bar (in this case only two are needed) to show that

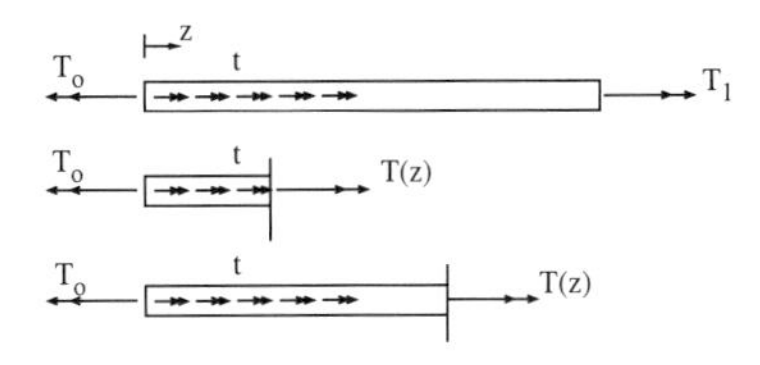


**Fig. 7.16** Free-body diagram and section cuts used in finding internal torque field.

$$T(z) = \begin{cases} T_o - tz & z < L/2 \\ T_o - t\frac{L}{2} & z \geq L/2. \end{cases} \tag{7.42}$$

Since the bar is homogeneous as in the last example we have that

$$\tau(r,z) = \frac{Tr}{J} = \begin{cases} \frac{(T_o - tz)r}{J} & z < L/2 \\ \frac{(T_o - tL/2)r}{J} & z \geq L/2, \end{cases} \tag{7.43}$$

$$\gamma(r,z) = \frac{\tau}{G} = \begin{cases} \frac{(T_o - tz)r}{GJ} & z < L/2 \\ \frac{(T_o - tL/2)r}{GJ} & z \geq L/2, \end{cases} \tag{7.44}$$

and that

$$\frac{d\phi}{dz}(z) = \frac{\gamma}{r} = \begin{cases} \frac{(T_o - tz)}{GJ} & z < L/2 \\ \frac{(T_o - tL/2)}{GJ} & z \geq L/2. \end{cases} \tag{7.45}$$

To eliminate the unknown $T_o$, we can integrate the twist rate from 0 to $L$ and apply the known kinematic information that there is no rotation at either end.

$$\underbrace{\phi(L)}_{=0} - \underbrace{\phi(0)}_{=0} = \int_0^L \frac{d\phi}{dz}\, dz \tag{7.46}$$

$$0 = \int_0^{L/2} \frac{(T_o - tz)}{GJ}\, dz + \int_{L/2}^{L} \frac{(T_o - tL/2)}{GJ}\, dz \tag{7.47}$$

$$0 = T_o L - \frac{3}{8} t L^2. \tag{7.48}$$

Thus $T_o = \frac{3}{8}tL$. If we now plug back into eqn (7.42) we find that

$$T(z) = \begin{cases} \frac{3}{8}tL - tz & z < L/2 \\ -\frac{1}{8}tL & z \geq L/2. \end{cases} \tag{7.49}$$

To determine the rotation field we need to integrate eqn (7.45):

$$\phi(z) - \underbrace{\phi(0)}_{=0} = \int_0^z \frac{d\phi}{dz}\, dz = \begin{cases} \frac{tLz}{2GJ}\left(\frac{3}{4} - \frac{z}{L}\right) & z < L/2 \\ \frac{tL^2}{8GJ}\left(1 - \frac{z}{L}\right) & z \geq L/2. \end{cases} \tag{7.50}$$

Both relations are plotted in Fig. 7.17.

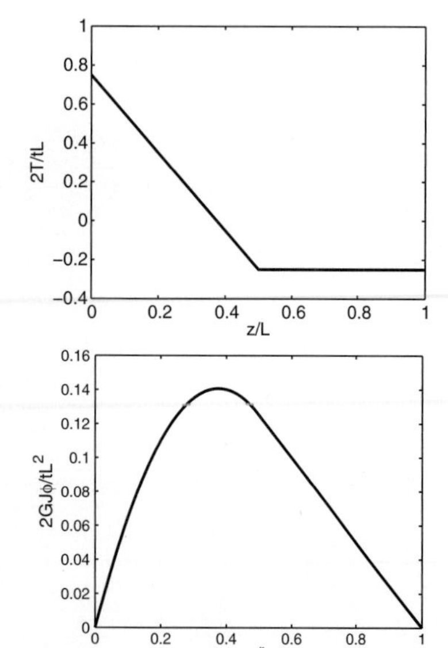


**Fig. 7.17** Torque diagram (top); rotation field (bottom).

**Remarks:**

(1) Observe that the final plots are given in non-dimensional form. This is useful for generating plots that can be used even when system parameters are varied.

(2) Note that the maximum rotation field in the bar does not occur in the middle. It occurs at $z = \frac{3}{8}L$; this point is easily located by finding where $d\phi/dz$ equals zero which is the same point where the internal torque is zero. The rotation at this point is $\phi(\frac{3}{8}L) = 9tL^2/128GJ$.

(3) Notice that when the internal torque field is a constant the rotation field is linear, and when the internal torque field is linear the

rotation field is quadratic. This is a general relation between polynomial orders in such problems; note here that $d\phi/dz = T/GJ$.

### 7.3.2 Differential equation approach

When we studied bars loaded by axial forces we solved many such problems using a methodology identical to the one just presented for torsion of bars. In a similar way we also have a method of solving torsion problems using second-order differential equations. One advantage of this scheme is that determinate and indeterminate problems are handled via a single-solution methodology.

We proceed in a manner similar to that in Chapter 2. Our first aim is to combine eqns (7.11)–(7.14) into a single equation. We start by substituting eqn (7.11) into eqn (7.14). Next, this result is inserted into eqn (7.13) to yield the intermediate result $T = GJd\phi/dz$. Finally, this result is substituted in the remaining equation, eqn (7.12), to give

$$\frac{d}{dz}\left(G(z)J(z)\frac{d\phi}{dz}\right) + t(z) = 0. \tag{7.51}$$

**Remarks:**

(1) This equation is a second-order ordinary differential equation for the rotation field. In the case where $G(z)J(z)$ is a constant it is a second-order ordinary differential equation with constant coefficients.

(2) To solve such an equation one needs two boundary conditions. Normally, we encounter two types of boundary condition: rotation boundary conditions and torque boundary conditions. Rotation boundary conditions will simply be the specification of the rotation at either end of the bar. Torque boundary conditions will involve the specification of the rate of change of rotation at either end of the bar. In particular, we will exploit the relation $T = GJd\phi/dz$.

(3) In the case that the system is radially inhomogeneous, then $GJ$ is replaced by $(GJ)_{\text{eff}}$, which can possibly be a function of $z$.

(4) Once one has solved eqn (7.51) for $\phi(z)$, all other quantities of interest can be determined by substitution into eqns (7.11)–(7.12).

As applications consider the prior examples.

**Example 7.9**

*Torsional stiffness of an end loaded bar revisited* For this example we will just set up the problem. What is needed is a specification of the distributed load and the conditions at the two ends (the boundary conditions). For this example we have

$$t(z) = 0 \tag{7.52}$$

$$\phi(0) = 0 \tag{7.53}$$

$$GJ\frac{d\phi}{dz}(L) = M; \tag{7.54}$$

i.e. no applied distributed torques, fixed rotation at $z = 0$, and an imposed torque $M$ at $z = L$.

---

**Example 7.10**

*Rotation of bar with a point torque revisited.* For this example we will also just set up the problem. Here we have

$$t(z) = -M\delta(z - a) \tag{7.55}$$

$$\phi(0) = 0 \tag{7.56}$$

$$GJ\frac{d\phi}{dz}(L) = 0. \tag{7.57}$$

Just as with axial loads, point loads (torque in this case) are properly represented by Dirac delta functions. There is no rotation at $z = 0$, and at $z = L$ we have a free end which is properly represented by a zero torque condition.

---

**Example 7.11**

*Indeterminate system revisited.* Here we have

$$t(z) = t - tH(z - \frac{L}{2}) \tag{7.58}$$

$$\phi(0) = 0 \tag{7.59}$$

$$\phi(L) = 0. \tag{7.60}$$

Recall that $H(\cdot)$ is the Heaviside step function. For this example let us continue and actually complete the solution by integrating eqn (7.51) twice:

$$\frac{d}{dz}\left(GJ\frac{d\phi}{dz}\right) = tH(z - \frac{L}{2}) - t \tag{7.61}$$

$$GJ\frac{d\phi}{dz} = t\langle z - \frac{L}{2}\rangle - tz + C_1 \tag{7.62}$$

$$GJ\phi = t\frac{1}{2}\langle z - \frac{L}{2}\rangle^2 - t\frac{1}{2}z^2 + C_1 z + C_2. \tag{7.63}$$

We can eliminate the integration constants using the boundary conditions. First,

$$\phi(0) = 0 \quad \Rightarrow \quad t\frac{1}{2}\underbrace{\langle 0 - \frac{L}{2}\rangle^2}_{=0} - t\frac{1}{2}0^2 + C_1 0 + C_2 \tag{7.64}$$

$$\Rightarrow \quad C_2 = 0.$$

Second,

$$\phi(L) = 0 \quad \Rightarrow \quad t\frac{1}{2}\langle L - \frac{L}{2}\rangle^2 - t\frac{1}{2}L^2 + C_1 L = 0 \tag{7.65}$$

$$\Rightarrow C_1 = \frac{3}{8}tL.$$

The final result is:

$$T(z) = GJ\frac{d\phi}{dz} = t\langle z - \frac{L}{2}\rangle - tz + \frac{3}{8}tL \tag{7.66}$$

and

$$\phi(z) = \frac{1}{GJ}\left[t\frac{1}{2}\langle z - \frac{L}{2}\rangle^2 - t\frac{1}{2}z^2 + \frac{3}{8}tLz\right]. \tag{7.67}$$

**Remarks:**

(1) On the surface this solution looks different from our previous one. However, this is only for cosmetic reasons, since here we have used Macaulay brackets to represent our solution. A plot of this solution against the prior one easily shows that they are the same.

---

**Example 7.12**

*Linearly varying distributed torque.* Consider the elastic homogeneous bar shown in Fig. 7.18, where $k$ is a constant with units of torque per unit length squared. Find the rotation field.

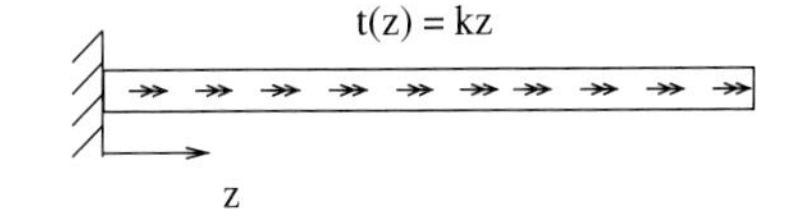


**Fig. 7.18** Torsion bar with a linearly varying load.

*Solution*

First identify the expression for the applied distributed torque and the boundary conditions. Here we see that $t(z) = kz$. The boundary conditions are a fixed end $\phi(0) = 0$ and a free end $GJ\phi'(L) = 0$.[2] Using eqn (7.51), by integration we have:

$$GJ\phi'' + kz = 0 \tag{7.68}$$

$$\phi'' = -kz/GJ \tag{7.69}$$

$$\phi' = -kz^2/2GJ + C \tag{7.70}$$

$$\phi = -kz^3/6GJ + Cz + D. \tag{7.71}$$

Now apply the boundary conditions to show that

[2] For convenience, we will often use a prime $(\cdot)'$ to denote the derivative of a function with respect to its argument. In this case the prime indicates differentiation with respect to $z$.

$$\phi(0) = 0 \tag{7.72}$$
$$\Rightarrow D = 0$$

and

$$\phi'(L) = 0 \tag{7.73}$$
$$\Rightarrow C = kL^2/2GJ.$$

Thus our final answer is:

$$\phi(z) = \frac{zkL^2}{2GJ}\left[1 - \frac{1}{3}\frac{z^2}{L^2}\right]. \tag{7.74}$$

---

**Example 7.13**

*Indeterminate bar with linearly varying distributed torque.* Suppose in the prior example the right-end was not free but rather built-in. Find the rotation field.

*Solution*

The procedure is the same as above except that in this case the boundary condition of the right-end is now $\phi(L) = 0$. From above we have that

$$\phi = -kz^3/6GJ + Cz + D. \tag{7.75}$$

Now apply the boundary conditions to show:

$$\phi(0) = 0 \tag{7.76}$$
$$\Rightarrow D = 0$$

and

$$\phi(L) = 0 \tag{7.77}$$
$$\Rightarrow C = kL^2/6GJ.$$

Thus our final answer is:

$$\phi(z) = \frac{zkL^2}{6GJ}\left[1 - \frac{z^2}{L^2}\right]. \tag{7.78}$$

**Remarks:**

(1) Notice that in this example and the prior one, some pains have been taken to express the final answer in a form different from simply inserting $C$ and $D$ back into, say, eqn (7.75). The manner in which this has been achieved involved two principles: (1) the answer should readily make evident the value of the function at the ends, and (2) the answer should be expressed in as non-dimensional a fashion as possible. This practice make the result easier to use and it also serves as a way to quickly spot errors.

---

## 7.4 Energy methods

In Chapter 2 we introduced the important concept of conservation-of-energy methods for elastic bars with axial loads. A similar notion exists for bars in torsion. First recall that any work we do on the system is stored (elastically) in the system:

$$W_{\text{in}} = W_{\text{stored}}. \tag{7.79}$$

From elementary physics we know that work equals force times distance, which is the same as torque times rotation angle. If we twist an elastic rod (not necessarily linear elastic) and measure the angle of rotation, then we can make a plot of torque versus rotation as shown in Fig. 7.19. The work that we have done on the rod is the area under the curve:

$$W_{\text{in}} = \int_0^{\theta_f} M(\theta)\,\mathrm{d}\theta. \tag{7.80}$$

If the material is linear elastic then the response will be linear and we find $W_{\text{in}} = \frac{1}{2}M_f\theta_f$.

The energy stored in the material can be found from the general expression for the strain energy density eqn (5.17). This expression is in terms of Cartesian coordinates, which is not convenient. However, by the discussion in Section 5.4, it is easily converted for use with cylindrical coordinates. In our case all the terms are zero except $\gamma_{z\theta}$ and $\tau_{z\theta}$.[3] This then gives the stored energy density in a linear elastic bar in torsion as $w = \frac{1}{2}\tau\gamma = \frac{1}{2}G\gamma^2 = \frac{1}{2}\tau^2/G$. If one integrates this over the volume of the material, then one will come to an expression for the energy stored in the material:

$$W_{\text{stored}} = \int_V \frac{1}{2}\tau\gamma\,\mathrm{d}V. \tag{7.81}$$

[3] We will drop the subscripts for convenience, as we have done throughout this chapter.

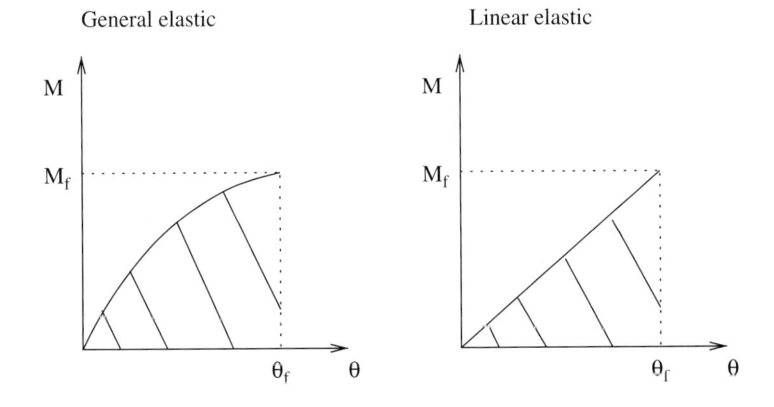


**Fig. 7.19** Torque-rotation response curves for elastic materials.

Thus we come to the final result (by conservation of energy) that in the linear elastic case:

$$\frac{1}{2}M_f\theta_f = \int_V \frac{1}{2}\tau\gamma\,\mathrm{d}V. \tag{7.82}$$

**Example 7.14**

*Rotation of an end-loaded rod by conservation of energy.* Find the rotation of an end-loaded bar. Assume $J$ and $G$ to be constants.

*Solution*
Starting from eqn (7.82) we have

$$\frac{1}{2}M_f\theta_f = \int_V \frac{1}{2}\tau\gamma\,\mathrm{d}V \tag{7.83}$$

$$= \int_0^L \int_A \frac{1}{2}\tau\gamma\,\mathrm{d}A\,\mathrm{d}z \tag{7.84}$$

$$= \int_0^L \int_A \frac{1}{2}\frac{1}{G}\tau^2\,\mathrm{d}A\,\mathrm{d}z \tag{7.85}$$

$$= \int_0^L \int_A \frac{1}{2}\frac{1}{G}\left(\frac{Tr}{J}\right)^2\,\mathrm{d}A\,\mathrm{d}z \tag{7.86}$$

$$= \int_0^L \frac{1}{2}\frac{T^2}{GJ}\,\mathrm{d}z \tag{7.87}$$

$$= \frac{1}{2}\frac{M_f^2}{GJ}\int_0^L \mathrm{d}z \tag{7.88}$$

$$= \frac{1}{2}\frac{M_f^2 L}{GJ}. \tag{7.89}$$

If we now cancel $\frac{1}{2}M_f$ from both sides we find $\theta_f = M_f L/GJ$ – a result we had from before. Thus we see that by using conservation of energy it is possible to determine rotations in elastic systems.

## 7.5 Torsional failure: Brittle materials

For circular bars in torsion there are two basic failure mechanisms depending upon the bar's material. In this section we will examine briefly what happens when a brittle bar fails in torsion, and in the next section, what happens when a ductile bar fails in torsion. Recall, first, that brittle materials fail by cracking (fracturing). The characteristic feature of such failures is that the fracture surface appears orthogonal to the direction of maximum (tensile) normal stress.

So how can the twisting of a such a bar cause brittle failure? After all, we have just computed that there are only shear stresses present when twisting a bar.

The answer is intimately connected to the discussion in Section 3.1.1, where we saw that depending upon the orientation of a section cut one can compute different amounts of shear and normal stress at the same location. Thus, for the section cuts we have chosen so far in this chapter we have only seen shear stresses. However, if we reorient our section cuts we will be able to see that the bar also carries normal stresses.

To proceed, let us consider an equilibrium construction. Shown in Fig. 7.20 is a bar with applied torque. Let us consider cutting out a thin polar wedge-shaped piece of material from the outer surface of the bar. The piece is thin in the $r$-direction. On the top and bottom surfaces we have stresses $\tau = TR/J$ as shown. By moment equilibrium, we have the same stresses on the sides (recall the discussion of Section 3.2.3). By a thinness assumption in the $r$-direction we can assume that these stresses are constant on all four sides; see Fig. 7.21.

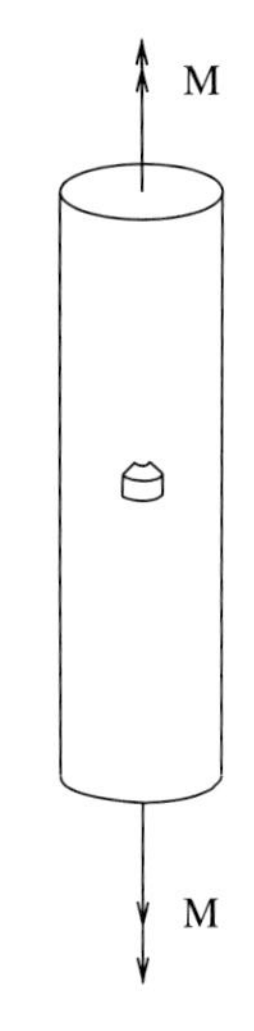


**Fig. 7.20** Construction for finding the stresses on an inclined plane in a torsion bar.

Let us now determine the normal stress on a plane inclined $\alpha$ radians with respect to the $z$-axis. Figure 7.22 shows our piece with a section cut with normal $\boldsymbol{e}_1$. The forces on the piece are given by the constant stress values times the areas over which they act: $\sigma$ is the unknown normal stress we are trying to determine. If we sum the forces in the 1-direction we find:

$$\sigma \Delta r l \sec(\alpha) = \tau \Delta r l \sin(\alpha) + \tau \Delta r l \tan(\alpha) \cos(\alpha) \tag{7.90}$$

$$\Rightarrow \sigma = 2\tau \sin(\alpha) \cos(\alpha). \tag{7.91}$$

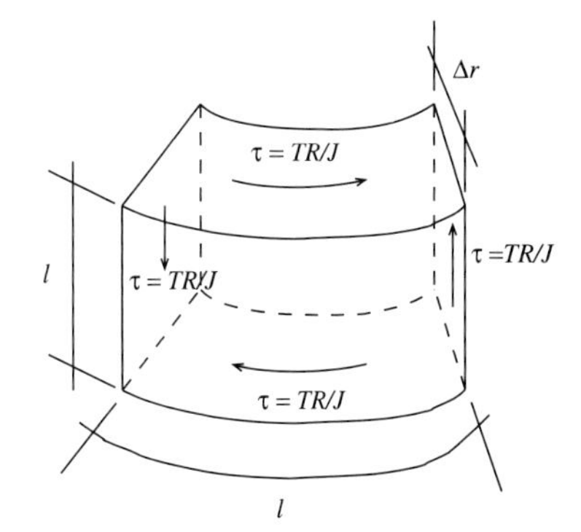


**Fig. 7.21** Small section from outer surface.

If we maximize this function with respect to $\alpha$ we find that $\alpha = \pi/4$ and that the maximum normal stress is given by $\sigma = \tau$. Thus we see, by an equilibrium argument, that the maximum normal stress $\sigma = TR/J$ on a plane inclined $\pi/4$ radians to the axis of the bar. Thus if one twists a brittle bar one would expect a fracture to appear that is aligned $\pi/4$ radians to the bar axis. Figure 7.23 shows a piece of chalk that has been so twisted. The failure orientation at all points on the surface of the bar is $\pi/4$ radians to the bar axis; the net result is a $\pi/4$ radian helical failure.

### Remarks:

(1) One may be concerned with the fact that the piece we have cut out of the bar is curved. Thus, while the stress values may be constant their direction is not, and this would invalidate the idea of simply multiplying stresses by area to get forces. After-all $\tau = \sigma_{z\theta}$ and $\boldsymbol{e}_\theta$ changes direction from point to point. If one performs a more rigorous analysis by integrating the stress fields over the surfaces of the piece, followed by shrinking the size of the piece to a point, then one recovers exactly the same result as we have found. Thus our result is correct even if our method of analysis was not 100% rigorous.

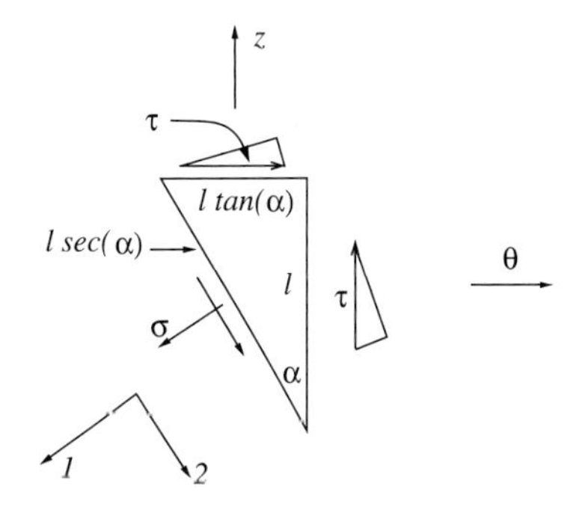


**Fig. 7.22** View down $r$-axis.

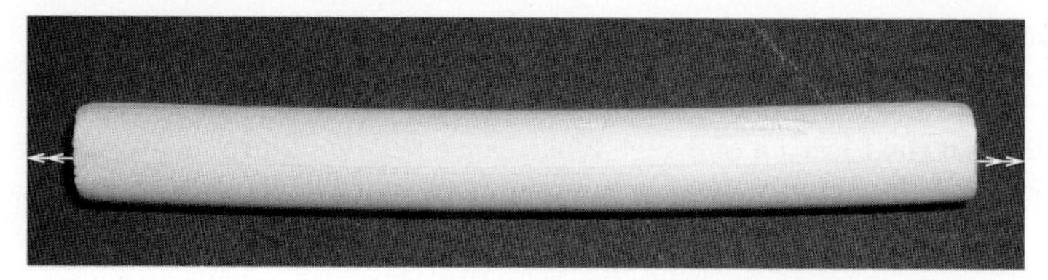

(a) Chalk before applied torque exceeds critical value.

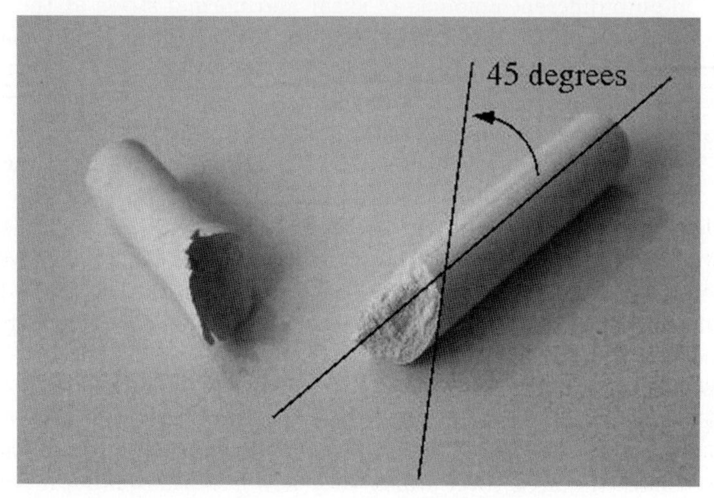


(b) Chalk after applied torque exceeds critical value.

**Fig. 7.23** Demonstration of the fracture of a brittle solid due to the application of torques.

## 7.6 Torsional failure: Ductile materials

In ductile materials, torsional failure follows a completely different path. In brittle failure, when the normal stress on the $\pi/4$ radian inclined plane reaches the fracture strength of the material, the bar fails suddenly and the load-carrying capacity of the bar is lost immediately. In ductile failure, failure proceeds in a much more gradual and less dramatic fashion. Thus sometimes one defines torsional failure as simply reaching the yield torque, $T_Y$, the torque at which yielding starts, and sometimes one defines it as reaching the ultimate torque, $T_u \neq T_Y$, the torque associated with complete yielding of the cross-section.

### 7.6.1 Twist-rate at and beyond yield

First recall that when the bar is elastic the stress and strain distributions are linear, as shown in Fig. 7.24. The connection between the strain distribution and the stress distribution is through the stress–strain relation for shear, $\tau = G\gamma$. This is valid as long as the stresses stay below the yield stress, $\tau_Y$, or equivalently the strains stay below the yield strain, $\gamma_Y = \tau_Y/G$. If we increase the twist rate in the bar beyond the elastic limit, some things change but some do not. The shear strains stay linear

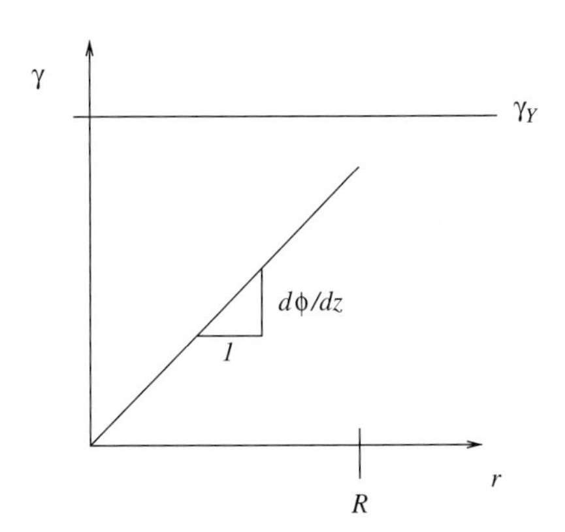


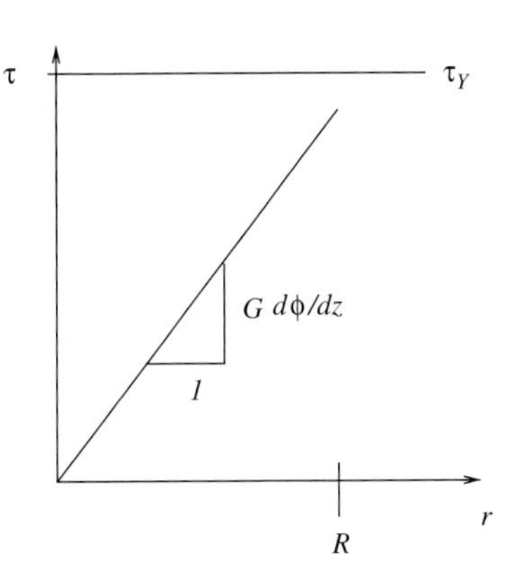


**Fig. 7.24** Shear strain and stress distribution when the response is fully elastic.

on the cross-section because we continue to assume that our kinematic assumption holds. However, the shear stresses are no longer linear on the cross-section, because $\tau \neq G\gamma$ at all points on the cross-section. As we progressively increase the twist rate the shear strain begins to exceed the yield strain on the outside of the bar. As the twist rate is further increased the radial location where the shear strain is equal to the yield strain moves inwards, with the material inside this radius still elastic and the material outside this radius plastically deformed. The radial location is known as the elastic–plastic interface. Figure 7.25 shows three different cases, each with an increasing amount of twist rate. Case 1 is elastic and cases 2 and 3 go beyond yield.

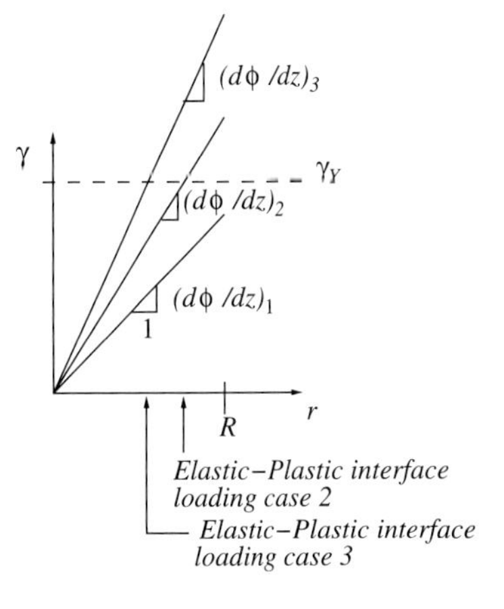


**Fig. 7.25** Shear strain distribution for three progressively increasing twist rates: $(d\phi/dz)_1 < (d\phi/dz)_2 < (d\phi/dz)_3$. States 2 and 3 represent twist rates beyond yield.

The relation between an applied twist rate and the location of the elastic–plastic interface is easy to determine, since we continue to assume the validity of our fundamental kinematic assumption, $\gamma = r d\phi/dz$. At the interface we know that $\gamma = \gamma_Y$; thus

$$\frac{d\phi}{dz} = \frac{\gamma_Y}{r_p}, \tag{7.92}$$

where $r_p$ is the symbol we use for the interface location. For a given yield stress, we can also write

$$\frac{d\phi}{dz} = \frac{\tau_Y}{G r_p}. \tag{7.93}$$

When $r_p = R$, i.e. the initiation of yield, we denote the applied twist rate as $(d\phi/dz)_Y = \gamma_Y/R = \tau_Y/(GR)$ – the twist rate at initial yield.

### 7.6.2 Stresses beyond yield

If we assume the constitutive relation is elastic–perfectly plastic (as shown in Fig. 7.26), then we can determine the stress distribution for a given twist rate beyond yield. Note that given a twist rate the shear strains are known, $\gamma = r d\phi/dz$. Thus for each radial location $r$ we can "look up" the corresponding stress from the stress–strain diagram which gives

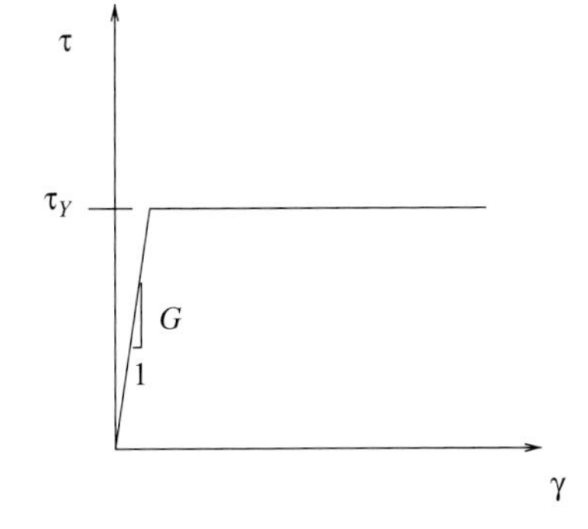


**Fig. 7.26** Elastic–perfectly plastic response curve.

$$\tau = \begin{cases} \tau_Y \frac{r}{r_p} & r < r_p \\ \tau_Y & r \geq r_p, \end{cases} \tag{7.94}$$

where $r_p = \gamma_Y/(d\phi/dz)$. For the strain distributions shown in Fig. 7.25, the resulting stress distributions are given in Fig. 7.27.

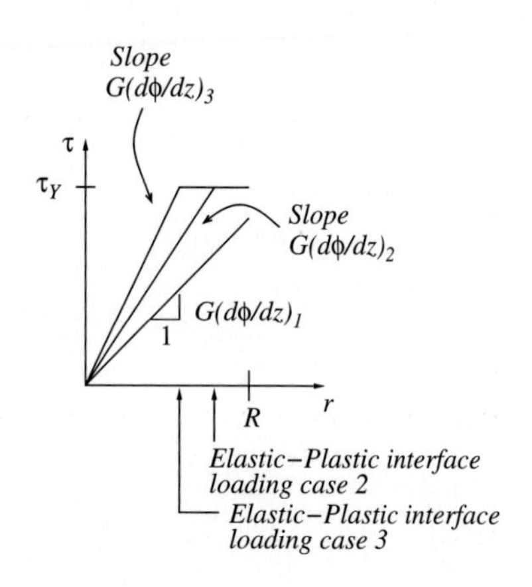


**Fig. 7.27** Stress response for an elastic–perfectly plastic material for the twist rates in Fig. 7.25.

### 7.6.3 Torque beyond yield

To determine the torque beyond yield for a given twist rate, we need to apply the general relation between torque and shear stresses, eqn (7.13). Note that eqn (7.13) holds independent of constitutive response. Inserting eqn (7.94) into eqn (7.13), gives

$$T = \int_A r\tau \, dA = \int_0^{2\pi} \int_0^R r\tau \, r dr d\theta \tag{7.95}$$

$$= 2\pi \left[ \int_0^{r_p} \tau_Y \frac{r^3}{r_p} \, dr + \int_{r_p}^R \tau_y r^2 \, dr \right] \tag{7.96}$$

$$= \tau_Y \left[ \frac{2\pi R^3}{3} - \frac{\pi r_p^3}{6} \right]. \tag{7.97}$$

This is also conveniently written as:

$$T = T_u \left[ 1 - \frac{1}{4} \left( \frac{r_p}{R} \right)^3 \right], \tag{7.98}$$

or

$$T = T_u \left[ 1 - \frac{1}{4} \left( \frac{(d\phi/dz)_Y}{d\phi/dz} \right)^3 \right], \tag{7.99}$$

where $T_u = \tau_Y 2\pi R^3/3$.

**Remarks:**

(1) In the final result we have used the symbol $T_u$. This is known as the ultimate torque that the bar can carry, and it occurs when the entire cross-section of the bar is fully yielded; i.e. $\tau = \tau_Y$ everywhere. This is the same as saying $r_p = 0$, for the solid bar.

(2) The torque at initial yield is given as $T_Y = \tau_Y \pi R^3/2$. This occurs when $r_p = R$. Notice that $T_Y = 3T_u/4$. Thus after yield initiates one can continue to increase the torque on the bar another 33%.

(3) This last remark can be better appreciated by making a plot of torque versus applied twist rate as shown in Fig. 7.28. Before yield the response is linear. After yield the response is non-linear due to the plastic yielding. Note that the ultimate torque can never be reached, as it represents an asymptote that is only approached in the limit as the twist rate goes to infinity.

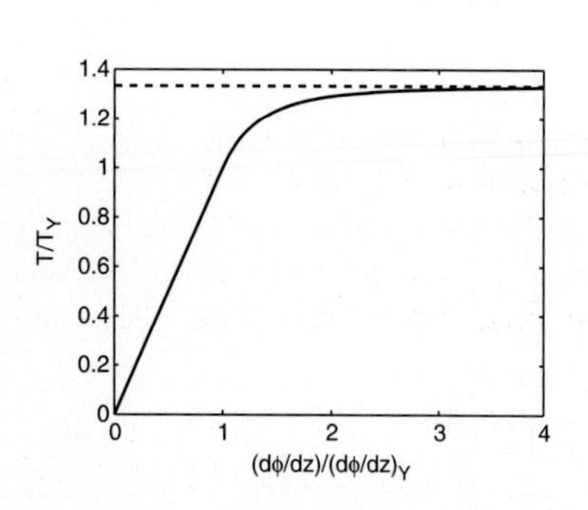


**Fig. 7.28** Torque versus twist rate for an homogeneous solid circular bar made from an elastic–perfectly plastic material.

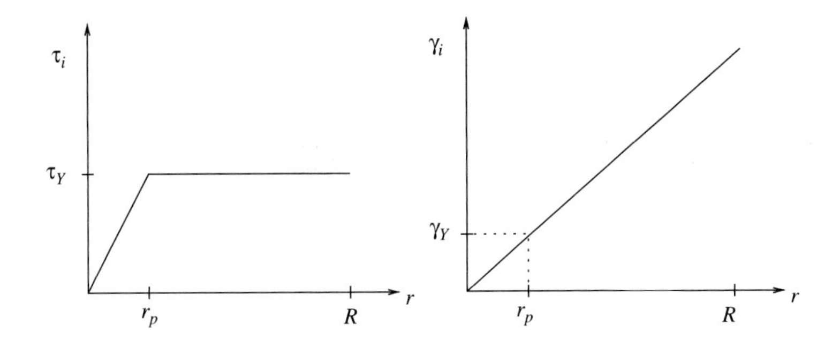


**Fig. 7.29** Initial stress and strain fields.

### 7.6.4 Unloading after yield

In the preceding sub-sections we have considered the case of yielding a bar in torsion through the progressive increase in applied twist rate. Let us now consider what happens when one releases a circular elastic–perfectly plastic bar that has been twisted beyond yield. Assume the bar has been twisted so that the elastic–plastic interface is now located at a radius $r_p$. We will call this the initial state and denote quantities at this state by a subscript $i$. Figure 7.29 plots of the initial stress and strain in the bar as a function of radial position, $r$. The value of the initial twist rate is

$$\left(\frac{d\varphi}{dz}\right)_i = \frac{\gamma_Y}{r_p} = \frac{\tau_Y}{Gr_p} \tag{7.100}$$

and the value of the initial torque is given by

$$T_i = \int_A r\tau \, dA = \tau_Y \left[\frac{2\pi R^3}{3} - \frac{\pi r_p^3}{6}\right]. \tag{7.101}$$

When we let go of the bar the state changes. We will call the new state the final state and denote quantities at this state by a subscript $f$. There are two observations we can immediately make about the final state:

(1) $T_f = 0$ since we have let go of the bar, and
(2) $\gamma_f = r(d\varphi/dz)_f$ holds since the strain is assumed linear, independent of material behavior from our physical observations. Note that we do not yet know what $(d\varphi/dz)_f$ equals.

Using observation (2) we can define the change in the strain as a function of radial position as $\Delta\gamma(r) = \gamma_f(r) - \gamma_i(r)$; i.e.

$$\Delta\gamma = r\left(\left(\frac{d\varphi}{dz}\right)_f - \left(\frac{d\varphi}{dz}\right)_i\right). \tag{7.102}$$

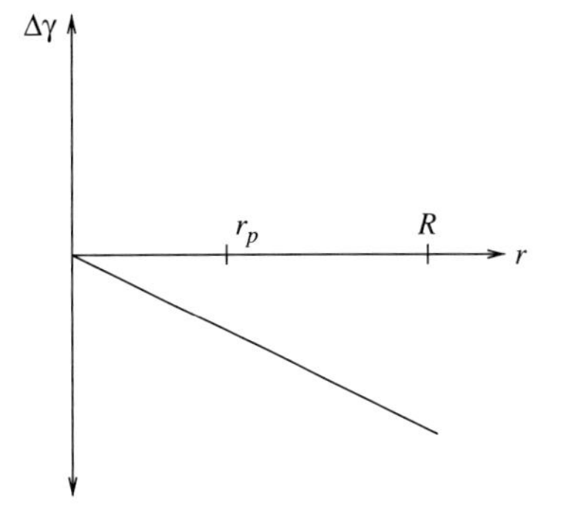


**Fig. 7.30** Increment of shear strain from initial to final configuration.

Figure 7.30 qualitatively shows this relation. Note that in Fig. 7.30 we have assumed that $(d\varphi/dz)_f < (d\varphi/dz)_i$; i.e. we have assumed that the bar springs back when released.

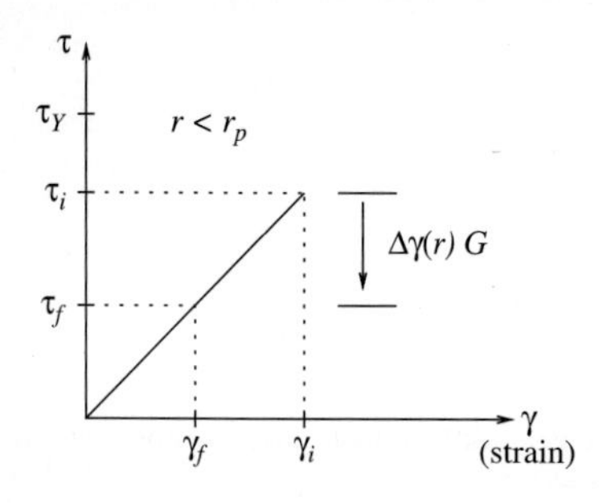


**Fig. 7.31** Stress–strain behavior for points $0 < r < r_p$.

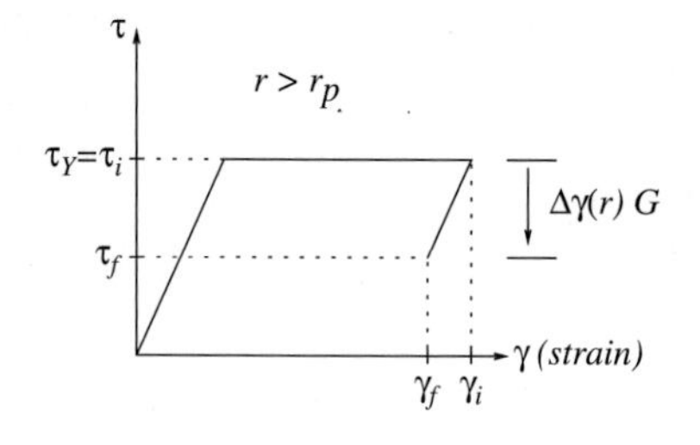


**Fig. 7.32** Stress–strain behavior for points $r_p < r < R$.

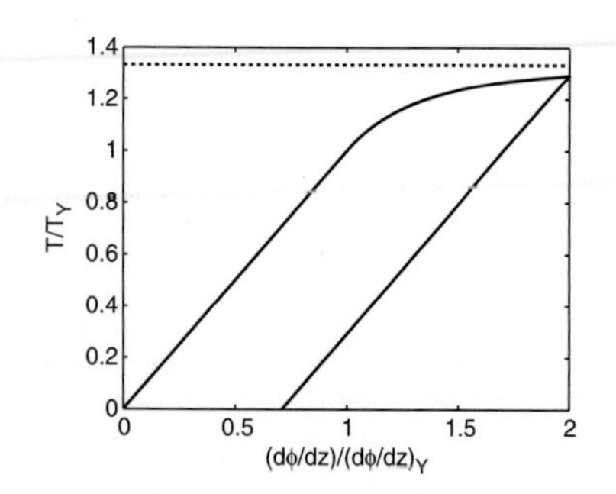


**Fig. 7.33** Torque versus twist rate response with unloading after an initial twist of $(d\phi/dz)_i = 2(d\phi/dz)_Y$. The initial torque $T = (31/24)T_Y$ and the final twist rate is $(d\phi/dz)_f = (17/24)(d\phi/dz)_Y$.

Consider now what happens to the stresses. For a point located between 0 and $r_p$, the stress–strain behavior appears as shown in Fig. 7.31. For a point located between $r_p$ and $R$ the stress–strain behavior looks as shown in Fig. 7.32. Thus we can write an expression for the final stress state as

$$\tau_f(r) = \begin{cases} \frac{\tau_Y}{r_p} r + \Delta\gamma(r)G & \text{for } 0 < r < r_p \\ \tau_Y + \Delta\gamma(r)G & \text{for } r_p < r < R. \end{cases} \tag{7.103}$$

The expression for $\Delta\gamma$ is still in terms of the unknown final twist rate, $(d\varphi/dz)_f$. To determine this we use the other bit of information that we know about that final state; viz. $T_f = 0$. If we use eqn (7.103) and the general relation between stress and torque, eqn (7.13), then

$$0 = T_f = \int_A r\tau_f \, dA \tag{7.104}$$

$$= 2\pi \left\{ \int_0^{r_p} r^2 [\frac{\tau_Y}{r_p} r + \Delta\gamma(r)G] \, dr \right.$$

$$\left. + \int_{r_p}^{R} r^2 [\tau_Y + \Delta\gamma(r)G] \, dr \right\} \tag{7.105}$$

$$= \tau_Y \left[ \frac{R^3}{3} - \frac{r_p^3}{12} \right] + \frac{R^4}{4} \left[ \left( \frac{d\varphi}{dz} \right)_f - \left( \frac{d\varphi}{dz} \right)_i \right] G. \tag{7.106}$$

This equation can be solved for the unknown twist rate; doing so gives

$$\left( \frac{d\varphi}{dz} \right)_f = \left( \frac{d\varphi}{dz} \right)_i - \frac{4\tau_Y}{3GR} \left[ 1 - \frac{1}{4} \left( \frac{r_p}{R} \right)^3 \right] \tag{7.107}$$

$$= \left( \frac{d\varphi}{dz} \right)_i - \frac{T_i}{GJ}. \tag{7.108}$$

**Remarks:**

(1) Figure 7.33 shows the complete load/unload curve for the torque versus twist rate. Here we have assumed that the initial twist rate is $(d\phi/dz)_i = 2(d\phi/dz)_Y$. As just computed the unloading process is completely elastic and thus appears as a straight line. Beware, however, as unloading is not completely elastic in all cases. The present result holds only for the case of a solid bar with homogeneous elastic–perfectly plastic material properties. In particular, for composite bars reverse yielding can take place during unloading, and this must be checked for.

(2) Having determined $(d\phi/dz)_f$ we can substitute back into the previous expressions to get the final strain and stress states in the bar. Since $(d\phi/dz)_f \neq (d\phi/dz)_i$, we will have a residual stress and strain field in the bar; i.e. even though there is no torque on the bar there are non-zero stresses and strains in it.

(3) If the bar were to be retorqued it would remain elastic until the applied torque reached $T_i$. Thus by such a procedure the effective yield torque of the bar has been increased in the direction of initial twisting. Such procedures are often used, for example, in the manufacture of stabilizing bars for vehicle suspensions.

---

**Example 7.15**

*Plastically twisted bar.* Consider a solid 24–inch long metal rod with a 1–inch radius. Determine how much torque is required to yield the bar halfway through ($r_p = 0.5$ inch). Also compute the amount of residual twist and stress in the bar if it is subsequently released. Assume $G = 12 \times 10^6$ psi and that the material is elastic–perfectly plastic with yield stress $\tau_Y = 24$ ksi.

*Solution*

Starting from eqn (7.97) we find that the required torque is given by

$$T_i = 24000\left[\frac{2\pi 1^3}{3} - \frac{\pi 0.5^3}{6}\right] = 48.7 \times 10^3 \text{ in} \cdot \text{lbf}. \tag{7.109}$$

Since the bar is solid and elastic–perfectly plastic, we can utilize the results of the last section. First we have that:

$$\left(\frac{d\phi}{dz}\right)_i = \frac{24000}{12 \times 10^6 \cdot 0.5} = 4 \times 10^{-3} \text{ rad/in.} \tag{7.110}$$

So the final twist rate is given by:

$$\begin{aligned}\left(\frac{d\phi}{dz}\right)_f = \left(\frac{d\phi}{dz}\right)_i - \frac{T_i}{GJ} &= 4 \times 10^{-3} - \frac{48.7 \times 10^3}{12 \times 10^6 \cdot \pi 1^4/2} \\ &= 1.4 \times 10^{-3} \text{ rad/in.}\end{aligned} \tag{7.111}$$

Knowing this value, we can now use eqn (7.103) to find the residual stress field as:

$$\tau_f(r) = \begin{cases} 24000\frac{r}{0.5} - \frac{48.7\times 10^3 r}{\pi 1^4/2} & r < 0.5 \\ 24000 - \frac{48.7\times 10^3 r}{\pi 1^4/2} & r \geq 0.5. \end{cases} \tag{7.112}$$

A plot of this stress field is shown in Fig. 7.34 along with the initially applied stress field.

**Remarks:**

(1) The initial twist rate is roughly 2.8°/ft which says that the differential rotation between the ends of the bar is initially 5.6°.

(2) After the load is released the twist rate is approximately 1°/ft; so there is a differential rotation between the ends of 2° even after release.

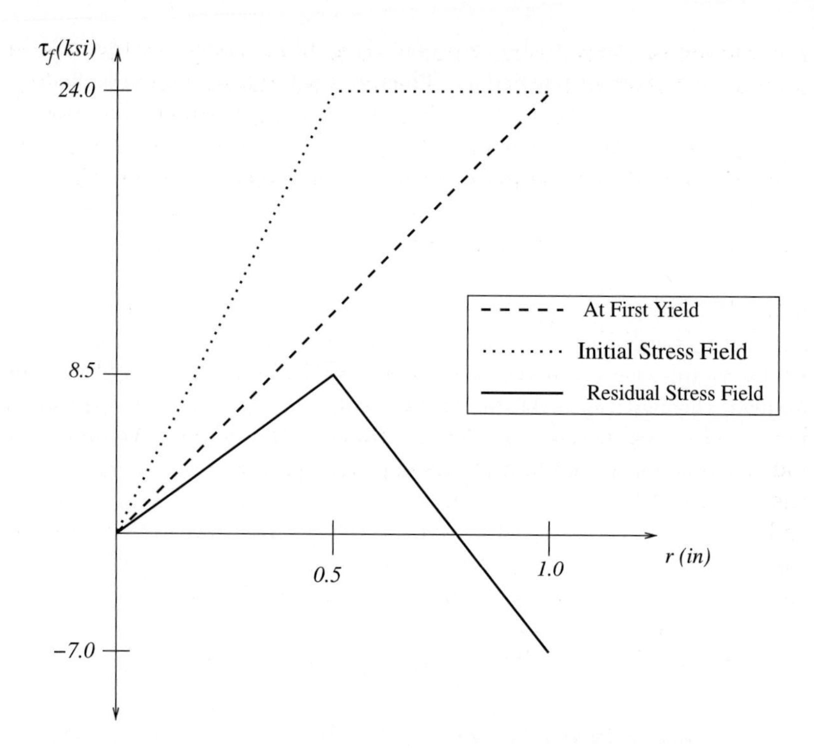


**Fig. 7.34** Stress distribution on cross-section during and after loading past the elastic limit.

(3) The residual shear stress distribution displays both positive and negative stresses. The inner parts of the bar have a shear stress which has the same sense (sign) as the initially applied stress; however, the outer parts of the bar have a stress of opposite sense. This negative stress is needed so that the integral of $r\tau_f$ over the cross-section equals zero; i.e. the final torque needs to be zero.

## 7.7 Thin-walled tubes

So far we have concentrated on the torsion of circular bars. While this is a very important class of structural members that experience torsion, it is far from comprehensive. The torsion of non-circular cross-sectional shapes is a fair bit more complex than what we have seen. This occurs because our fundamental kinematic observation that plane sections remain plane and only rotate is no longer true. Very basic observation of the torsion of non-circular cross-sections shows quite clearly that the cross-sections warp in addition to rotating. Warping refers to displacements in the $z$-direction. Thus the observed motion consists of rigid rotation of the cross-section coupled with out-of-plane displacements. The pattern of the out-of-plane displacements, warping, is also seen to be quite complex. The net result is that in general the solution of the problem of non-circular torsion requires the explicit solution of the governing partial differential equations. Notwithstanding, there is one case where we can effectively deal with non-circular torsion without solving partial differential equations. This case arises when we make the added assumption of thinness; i.e. here we assume that the member in question is hollow with arbitrary cross-sectional shape but where the tube walls are assumed thin.

### 7.7.1 Equilibrium

Equilibrium in a thin-walled tube under torsion is governed by the same expression as we used for the circular bar, viz. eqn (7.12). This follows because the expression was derived solely using equilibrium concepts independent of the kinematic assumption. To connect the stresses on the section to the internal torque we can also use the same relations as before, viz. eqn (7.13). However, because of the thinness assumption we can actually reduce this expression to an explicit algebraic relation that gives the stress for a given torque, and vice versa. In contrast to the circular bar case, we can do this without employing a kinematic assumption. This occurs because we will be using a thinness assumption.

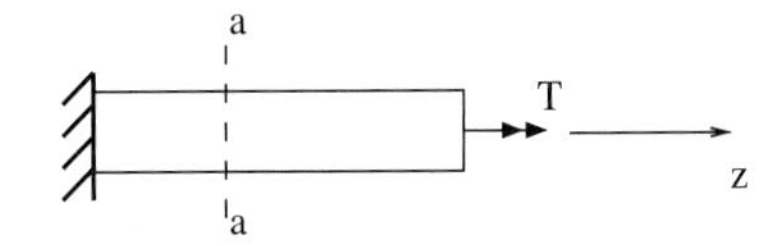


**Fig. 7.35** Thin-walled tube under load.

### 7.7.2 Shear flow

Consider the tube shown in Fig. 7.35. It is subjected to a torque, $T$, at one end and is fixed at the other. Points on the tube cross-section will be described either by the angle $\alpha$ which is measured counter-clockwise from the $x$-axis, or by arclength $s$, also measured counter-clockwise from the $x$-axis; see Fig. 7.36. The origin of the $x$-$y$ coordinate system is taken at the center of twist; the point about which the section rigidly rotates. Knowledge of this location is not needed in general, and its determination is beyond the scope of this text as it involves the solution of partial differential equations. The geometry of the tube is fully specified when one is given the two functions $r(\cdot)$ and $t(\cdot)$, where $r(\alpha)$ denotes the distance from the center of twist to the tube wall at angle $\alpha$, and $t(\alpha)$ denotes the wall thickness at angle $\alpha$. The wall thickness is measured perpendicular to the tube wall and not along the radial ray. The assumption of thinness amounts to assuming that

$$\frac{t(\alpha)}{r(\alpha)} < 10 \tag{7.113}$$

for all $\alpha$.

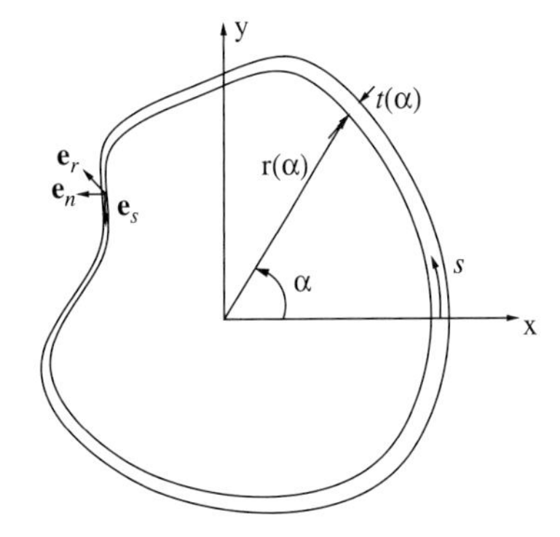


**Fig. 7.36** Section *a-a* from Fig. 7.35; coordinate definitions for thin-walled tube geometry.

To support the applied torque there must be a shear stress on the cross-section. Further, because we are assuming the walls to be thin we can assume that the shear stress at each location $\alpha$ is constant across the thickness. We also have that the shear stress at each location $\alpha$ must be tangential to the wall. If the stresses were not tangential to the wall, then moment equilibrium expressed in terms of stresses would require shear stresses on the surface of the tube in contradiction to the manner

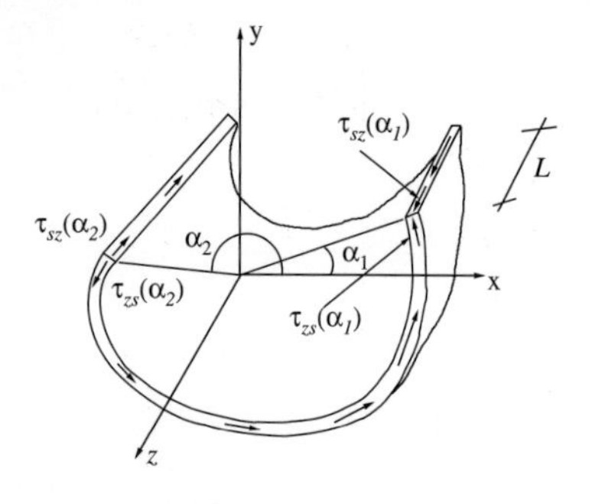


**Fig. 7.37** Free-body diagram for finding shear flow in a thin-walled tube.

in which the load is being applied. Thus the appropriate designation for this stress would be $\tau_{zs}(\alpha)$ – stress on the $z$-face in the $s$-direction (arclength direction).

As a step in determining the value of the shear stress in the tube for a given torque let us consider the free-body diagram shown in Fig. 7.37. By moment equilibrium we have that the shear stress $\tau_{sz}(\alpha) = \tau_{zs}(\alpha)$; thus we can express the shear stresses on the longitudinal slices at $\alpha_1$ and $\alpha_2$ by the corresponding shear stress on the cross-section. Since these are the only stresses that will appear in the analysis, we will drop the subscripts for convenience. If we now sum the forces on our free-body diagram in the $z$-direction we find that

$$\sum F_z = \tau(\alpha_1)t(\alpha_1)L - \tau(\alpha_2)t(\alpha_2)L = 0 \tag{7.114}$$

$$\Rightarrow \tau(\alpha_1)t(\alpha_1) = \tau(\alpha_2)t(\alpha_2). \tag{7.115}$$

Since the locations of the longitudinal cuts were arbitrary we have that

$$\tau(\alpha)t(\alpha) = q, \tag{7.116}$$

a constant for all $\alpha$. $q$ is known as the shear flow. The fact that it is a constant on the cross-section is a consequence of force equilibrium in the $z$-direction.

### 7.7.3 Internal torque–stress relation

Now that we have a characterization of the shear stress on the cross-section let us try to relate it to the applied torque. If we examine a small piece of the tube wall of length $ds$ we see that there is a force on this piece equal to $\tau(s)t(s)\boldsymbol{e}_s\, ds$; see Fig. 7.38. This force will generate a torque about the center of twist of amount $(r\boldsymbol{e}_r) \times (\tau(s)t(s)\boldsymbol{e}_s)\, ds$. We can now sum the contributions to the torque over the whole cross-section to find that:

$$T\boldsymbol{e}_z = \oint (r\boldsymbol{e}_r) \times (\tau(s)t(s)\boldsymbol{e}_s)\, ds \tag{7.117}$$

$$= q \oint r\boldsymbol{e}_r \times \boldsymbol{e}_s\, ds \tag{7.118}$$

$$= q \oint r\boldsymbol{e}_r \cdot \boldsymbol{e}_n\, ds\boldsymbol{e}_z \tag{7.119}$$

$$= q \oint (x\boldsymbol{e}_x + y\boldsymbol{e}_y) \cdot \boldsymbol{e}_n\, ds\boldsymbol{e}_z \tag{7.120}$$

$$= q \int_{A_e} \mathrm{div}[x\boldsymbol{e}_x + y\boldsymbol{e}_y]\, dA\boldsymbol{e}_z \tag{7.121}$$

$$= q \int_{A_e} 2\, dA\boldsymbol{e}_z \tag{7.122}$$

$$= q2A_e\boldsymbol{e}_z \tag{7.123}$$

where $A_e$ represents the area enclosed by the cross-section. Thus we see that

$$q = \frac{T}{2A_e} \tag{7.124}$$

and that

$$\tau(\alpha) = \frac{T}{2A_e t(\alpha)}. \tag{7.125}$$

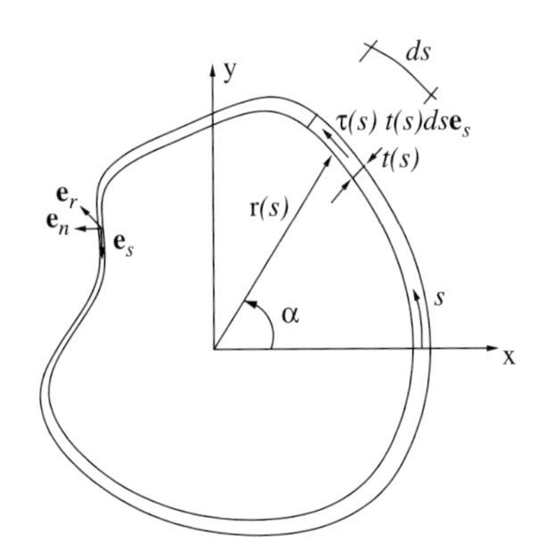


**Fig. 7.38** Construction for finding the torque-stress relation in a thin-walled tube.

**Remarks:**

(1) In the above, $\boldsymbol{e}_s$ is the unit vector tangent to the tube, and $\boldsymbol{e}_n$ is the unit vector normal to the tube. Note that $\boldsymbol{e}_r \neq \boldsymbol{e}_n$ in general.

(2) In going from eqn (7.118) to eqn (7.119) we have exploited the fact that we can decompose any vector $\boldsymbol{v}$ in the plane as $\boldsymbol{v} = (\boldsymbol{v} \cdot \boldsymbol{e}_s)\boldsymbol{e}_s + (\boldsymbol{v} \cdot \boldsymbol{e}_n)\boldsymbol{e}_n$, and additionally that $\boldsymbol{e}_n \times \boldsymbol{e}_s = \boldsymbol{e}_z$.

(3) In going from eqn (7.119) to eqn (7.120) we have used the fact that $r\boldsymbol{e}_r = x\boldsymbol{e}_x + y\boldsymbol{e}_y$.

(4) In going from eqn (7.120) to eqn (7.121) we have employed the divergence theorem.

(5) In going from eqn (7.121) to eqn (7.122) we used the fact that $\text{div}[x\boldsymbol{e}_x + y\boldsymbol{e}_y] = \partial x/\partial x + \partial y/\partial y = 1 + 1 = 2$.

### 7.7.4 Kinematics of thin-walled tubes

The preceding relations provide us with the equation for equilibrium in the tube in terms of the internal torque, $dT/dz + t = 0$, and the relation between internal torque and shear stress, $\tau = T/(2A_e t)$. The only thing missing is a characterization of the kinematics of the tube. We shall pose the question here in terms of finding the connection between the twist rate of the tube and the internal torque, and will restrict ourselves to the case of linear elastic homogeneous materials. Our approach will be an energetic one. Thus, in contrast to the theory of circular bars we will not derive a general kinematic relation that is valid independent of equilibrium and constitutive response.

If we consider an end-loaded thin-walled tube, the work input to the tube is given by:

$$W_{\text{in}} = \frac{1}{2} M\theta, \tag{7.126}$$

where $M$ is the applied torque and $\theta = \phi(L)$ is the end-rotation. The twist rate for this type of loading is constant, and thus $d\phi/dz = \theta/L$. The stored energy density in the tube is given by $w = (1/2)\tau^2/G$ since there is only one stress in the tube and the total stored energy in the tube can be expressed as

$$W_{\text{stored}} = \int_V \frac{1}{2}\frac{\tau^2}{G}\, dV. \tag{7.127}$$

The integration volume can be written as $dV = dtdsdz$. Since the stresses are constant across the thickness due to the thinness assumption, we can explicitly integrate out the thickness to give

$$W_{\text{stored}} = \int_V \frac{1}{2}\frac{\tau^2}{G}\,dV \tag{7.128}$$

$$= \oint \int_0^L \frac{1}{2}\frac{\tau^2}{G} t(s)\,dzds \tag{7.129}$$

$$= \oint \int_0^L \frac{T^2}{8A_e^2 tG}\,dzds \tag{7.130}$$

$$= \oint L\frac{T^2}{8A_e^2 tG}\,ds \tag{7.131}$$

$$= L\frac{T^2}{8A_e^2 G}\oint \frac{1}{t}\,ds. \tag{7.132}$$

Setting this equal to the work input and noting that $T = M$, we find that

$$\frac{\theta}{L} = \frac{M}{4A_e^2 G}\oint \frac{1}{t}\,ds. \tag{7.133}$$

**Remarks:**

(1) By analogy to the relation $d\phi/dz = T/(GJ)_{\text{eff}}$ in the circular bar case, we identify $(GJ)_{\text{eff}} = 4A_e^2 G/\oint(1/t)\,ds$ for a thin-walled tube.

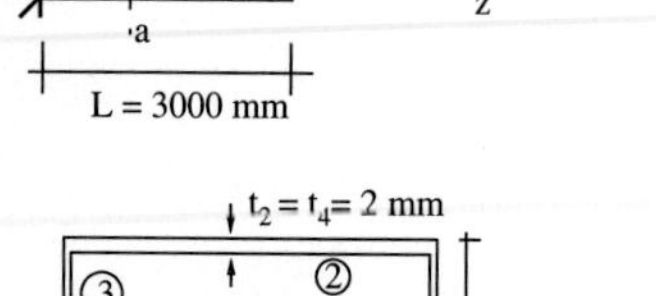


**Fig. 7.39** Torsion of a thin-walled box tube.

### Example 7.16

*Stiffness of a thin-walled tube.* Consider the thin-walled tube shown in Fig. 7.39. Compute the tube's torsional stiffness and determine the maximum permissible applied torque before yielding. Assume $G = 10$ GPa and $\tau_Y = 100$ MPa.

*Solution*

The torsional stiffness is given by $k_T = M/\theta$ for an end-loaded tube. Thus we have, using eqn (7.133):

$$k_T = \frac{4A_e^2 G}{L\oint \frac{1}{t}\,ds}. \tag{7.134}$$

The line integral is most easily computed by breaking up the integration to sections of constant thickness.

$$k_T = \frac{4A_e^2 G}{L\sum_{i-1}^{4}\frac{b_i}{l_i}}. \tag{7.135}$$

The yield torque will be given by the torque value at which any part of the cross-section starts to yield. Since $\tau = T/2A_e t$, this will occur at the thinnest section. Hence

$$T_Y = 2A_e t_{\min} \tau_Y . \tag{7.136}$$

Using the given dimensions and material properties we find

$$k_T = \frac{4 \cdot (200 \cdot 300)^2 \cdot 10000}{3000 \left[2\frac{200}{1} + 2\frac{300}{2}\right]} = 68.6 \text{ kN} \cdot \text{m/rad} \tag{7.137}$$

and

$$T_Y = 2 \cdot (200 \cdot 300)^2 \cdot 1 \cdot 100 = 12.0 \text{ kN} \cdot \text{m}. \tag{7.138}$$

## Chapter summary

- General torsion bars
  - Equilibrium: $dT/dz + t = 0$
  - Resultant-stress relation: $T = \int_A r\tau \, dA$
- Cylindrical bars
  - Kinematic assumption: $\gamma = r d\phi/dz$
- Homogeneous cylindrical linear elastic bars
  - Constitutive relation: $\tau = G\gamma$
  - Stress–torque: $\tau = Tr/J$
  - Polar moment of inertia: $J = \int_A r^2 \, dA$
  - Twist rate: $d\phi/dz = T/GJ$
  - Differential equation: $(GJ\phi')' + t = 0$
  - Boundary conditions: fixed and forced

$$\phi = 0, \qquad GJ\frac{d\phi}{dz} = M$$

- Energy: elastic

$$W_{\text{in}} = W_{\text{stored}}$$

  For linear elastic systems the strain energy density is $w = \frac{1}{2}\tau\gamma$ and the work input is $\frac{1}{2}M\theta$.
- Solid elastic–perfectly plastic cylindrical bars
  - Torque at yield: $T_Y = \tau_Y \pi R^3/2$
  - Ultimate torque: $T_u = \tau_Y 2\pi R^3/3$
  - Torque general: $T = T_u \left(1 - (r_p/R)^3/4\right)$
  - Twist rate: $d\phi/dz = \gamma_Y/r_p$

- Thin-walled tubes
  - Shear flow: $q = \tau(\alpha)t(\alpha) = T/2A_e$
  - Shear stress: $\tau = q/t = T/(2A_e t)$
  - Effective stiffness:

$$(GJ)_{\text{eff}} = 4A_e^2 G \left(\oint \frac{1}{t}\, ds\right)^{-1}$$

## Exercises

(7.1) Show that the state of strain in a slender circular prismatic bar under torsion can be expressed as $\gamma_{z\theta} = r\frac{d\varphi}{dz}$. Assume that planar cross-sections remain plane after deformation and radial lines remain straight after deformation. Assume small deformations, and *clearly* state the reasoning for *all* steps.

(7.2) Complete the missing entries in Table 7.1.

**Table 7.1** Analogy table for Exercise 7.2.

| | Axial loads | Circular torsion |
|---|---|---|
| Kinematic relation | $\varepsilon = \frac{du}{dx}$ | |
| Resultant definition | | $T = \int_A r\tau \, dA$ |
| Equilibrium | $\frac{dR}{dx} + b = 0$ | |
| Elastic Law | | $\tau = G\gamma$ |

(7.3) State which global equilibrium principle is associated with each of the following equations. Be specific about the direction.

$$\frac{dR}{dx} + b(x) = 0$$

$$\frac{dT}{dz} + t(z) = 0$$

$$\sigma_{xz} = \sigma_{zx}$$

(7.4) Each of the equations in Table 7.2 corresponds to a form of global equilibrium: $\sum F_x = 0$, $\sum F_y = 0$, $\sum F_z = 0$, $\sum M_x = 0$, $\sum M_y = 0$, or $\sum M_z = 0$. For each equation, identify the corresponding global form.

**Table 7.2** Equilibrium equations for Exercise 7.4.

| Given equation |
|---|
| $\sigma_{xy} = \sigma_{yx}$ |
| $dT/dz + t = 0$ |
| $\partial\sigma_{xz}/\partial x + \partial\sigma_{yz}/\partial y + \partial\sigma_{zz}/\partial z + b_z = 0$ |
| $dR/dx + b = 0$ |
| $\sigma_{zx} = \sigma_{xz}$ |
| $\partial\sigma_{xy}/\partial x + \partial\sigma_{yy}/\partial y + \partial\sigma_{zy}/\partial z + b_y = 0$ |
| $\sigma_{yz} = \sigma_{zy}$ |
| $\partial\sigma_{xx}/\partial x + \partial\sigma_{yx}/\partial y + \partial\sigma_{zx}/\partial z + b_x = 0$ |

(7.5) An elastic solid stepped cylindrical shaft is shown below with shear modulus $G = 50$ MPa. It is subject to a set of torques as shown from a set of gears (not shown). Make a torque diagram, a max shear strain diagram, and a max shear stress diagram for the shaft. What is the maximum shear stress in the shaft, and between which two gears does it occur?

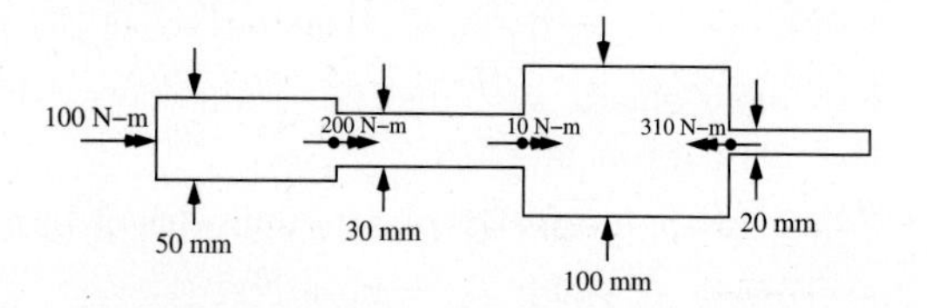


(7.6) A solid cylindrical shaft of radius $r_1$ has a hole bored out of it of radius $r_2$ so that it is now a hollow cylinder. Derive a formula for the percent reduction in torsional strength induced by the boring operation. Note that torsional strength in this context is defined as the torque at which the shaft first starts to yield.

(7.7) A 12-mm diameter steel shaft is used to deliver rotary power at 1,000 rpm. The differential twist from one end to the other is measured to be $5^{\circ}$. Assuming the shaft is 300 mm long, how much power is being transmitted through the shaft? Assume $G = 80$ GPa.

(7.8) Determine the torsional stiffness of the solid circular cone shown below.

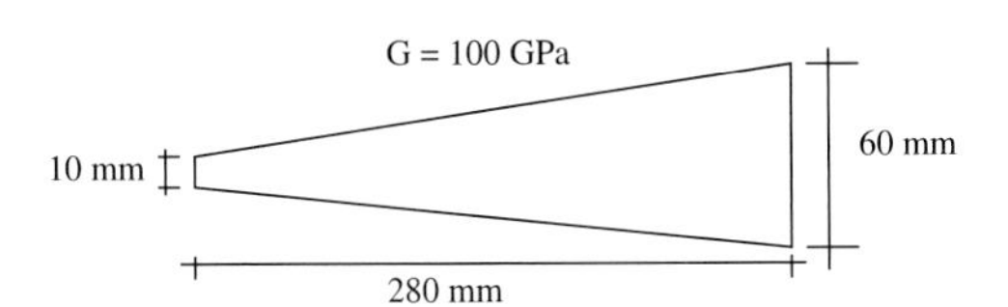


(7.9) (a) Determine the support reactions for the stepped circular shaft shown below. (b) Plot the angle-of-twist diagram for the shaft. The applied torques are $T_1 = 200$ lbf · in, $T_2 = 100$ lbf · in, $T_3 = 500$ lbf · in. The shaft diameters are $d_1 = 3$ in and $d_2 = 2$ in. Let $E = 30 \times 10^6$ psi and $\nu = 0.3$.

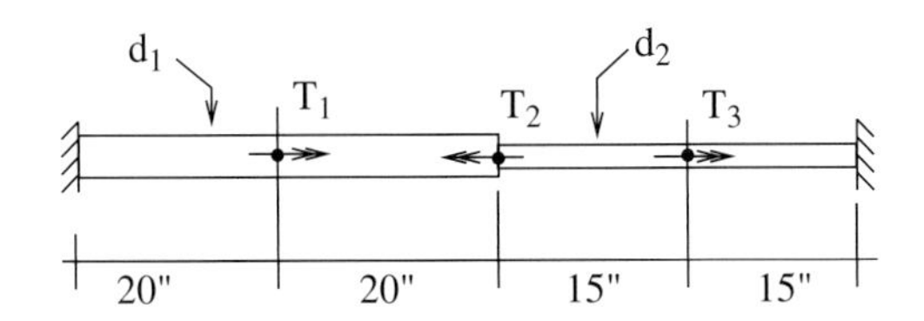


(7.10) What is the second-order ordinary differential equation that governs the rotation of an elastic circular bar in torsion?

(7.11) Using eqn (3.27), derive eqn (7.9).

(7.12) Prove that $\frac{dT}{dz} + t(z) = 0$ for a bar in torsion where $T$ is the internal torque and $t$ is the distributed torque; further, use this result to show that $GJ\frac{d^2\phi}{dz^2} + t(z) = 0$ for an homogeneous linear elastic circular bar, where $\phi$ is the section rotation. Explain each step by a concise and complete sentence.

(7.13) Consider a composite circular bar of length $L$ with cross-section as shown in the following (the materials are bonded together). Find the torsional stiffness assuming both materials are linear elastic.

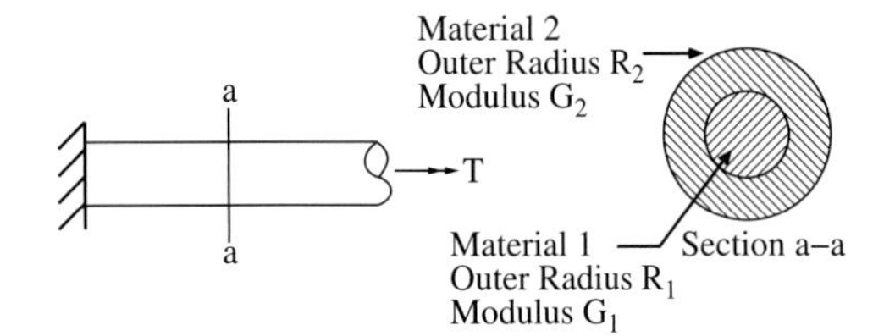


(7.14) Consider an elastic circular torsion bar of radius $R$ where the shear modulus is a function of position $G(r) = A + Br$. Assume that $A > 0$ and $B < 0$ are given constants. For a given twist rate $d\phi/dz$ at a given location $z$, find the radius $r$ at which the shear stress will be maximum?

(7.15) Consider a circular bar of length $L = 300$ mm and diameter $d = 5$ mm. The bar is built-in at $z = 0$ and is subjected to a distributed torque $t(z) = 50z$, where the constant (50) has units of N/mm. At $z = 300$ mm the bar is free. Find the rotation in the bar as a function of $z$. Assume $G$ to be a constant.

(7.16) Resolve Exercise 7.15 with an additional built-in end at $z = 300$ mm.

(7.17) Consider a circular bar which is built-in at both ends and loaded by a constant distributed torque. Derive a formula that gives the amount of load needed to induce a rotation $\hat{\theta}$ at the mid-point of the bar. Assume $GJ$ is a constant.

(7.18) Consider a circular bar which is built-in at both ends and loaded by a linear distributed torque, $t(z) = t_o\, z$. By solving the governing second-order ordinary differential equation find a relation that gives the amount of load needed to induce yield in the bar. Assume $GJ$ is a constant and $\tau_Y$ given.

(7.19) A motor shaft is driven with a 10-kW motor at 500 rpm. Every 0.5 meters the shaft drives a 2-kW load. Assume $G = 100$ GPa and that the maximum allowable shear stress is 50 MPa. If the shaft is solid, what diameter is required between each load? For this sizing, plot the twist rate and maximum shear stress in the shaft.

(7.20) Consider a $\beta$-kW motor that drives a round elastic shaft at $\omega$ rad/s with two loads of power $\mathcal{P}_1 = \alpha\beta$ and $\mathcal{P}_2 = (1-\alpha)\beta$, where $0 < \alpha < 1$. If the maximum allowable shear stress in the shaft is the yield stress $\tau_Y$ and we desire a safety factor SF against yield, how large must the shaft be between the two loads. Assume that the load $\mathcal{P}_1$ is closest to the motor.

(7.21) An elastic circular bar is fixed at one end and attached to a rubber grommet at the other end. The grommet functions as a torsional spring with spring constant $k$. If a concentrated torque of magnitude $T_a$ is applied in the center of the bar, what is the rotation at the end of the bar, $\phi(L)$? Assume a constant shear modulus $G$ and polar moment of inertia $J$.

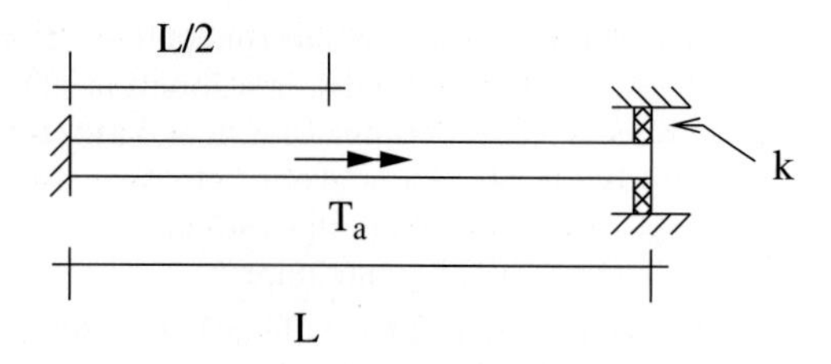


(7.22) For the system shown: (1) state the relevant boundary conditions in terms of the kinematic variables and (2) give an appropriate expression for the distributed load acting on the system.

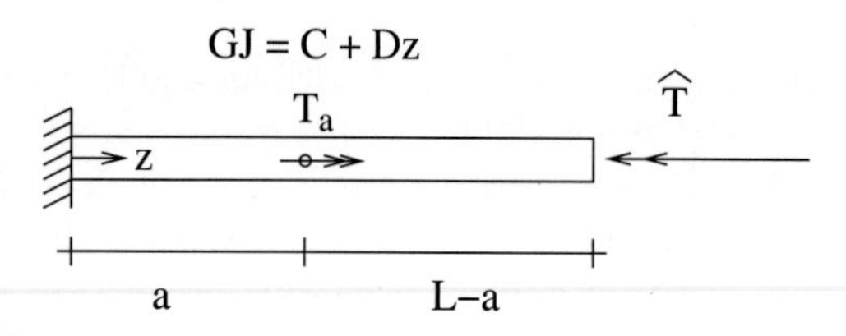


(7.23) An elastic solid circular bar of length $L$ with polar moment of inertia $J$ and shear modulus $G$ is built-in at both ends and subject to a system of distributed torques:

$$t(z) = \begin{cases} 0 & z < d \\ c & z \geq d \end{cases}$$

Determine the support torques $T(0)$ and $T(L)$ at the two ends of the bar.

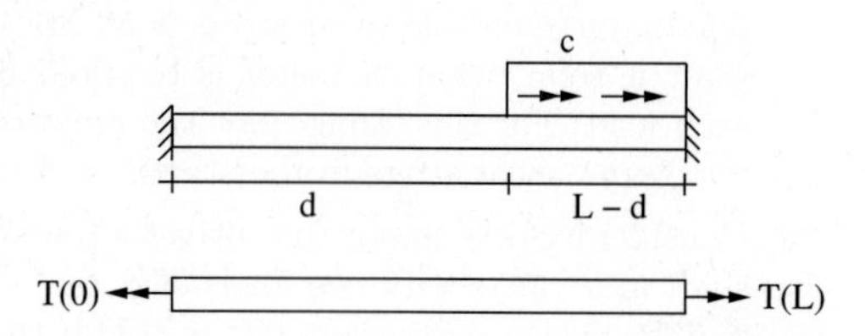


(7.24) A circular rod of length $L$ in made of a *non-linear elastic* material with constitutive relation $\tau = C\gamma^n$, where $C$ and $n$ are given material constants. Find an expression for the torsional stiffness of the bar, $k = T/\theta$. Note that your answer will not be independent of $T$ as it is in the linear case.

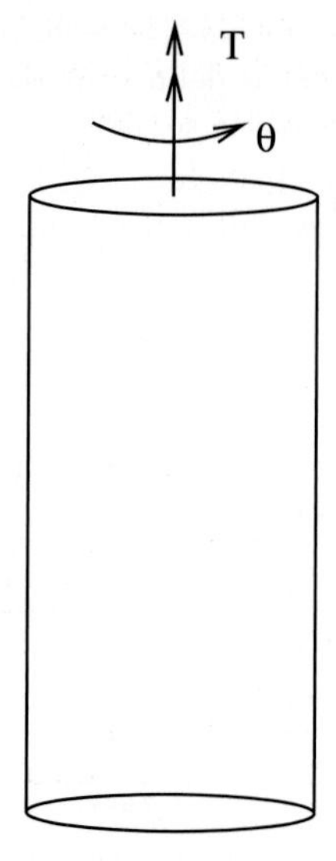


(7.25) The circular bar shown below is made of a *non-linear elastic* material which is governed by the constitutive relation $\tau = B\text{sign}(\gamma)|\gamma|^n$, where $B > 0$ and $n$ are material constants. The bar is subjected to an end torque, $T_a > 0$, and a constant distributed load, $t_o > 0$. Find $\phi(z)$.

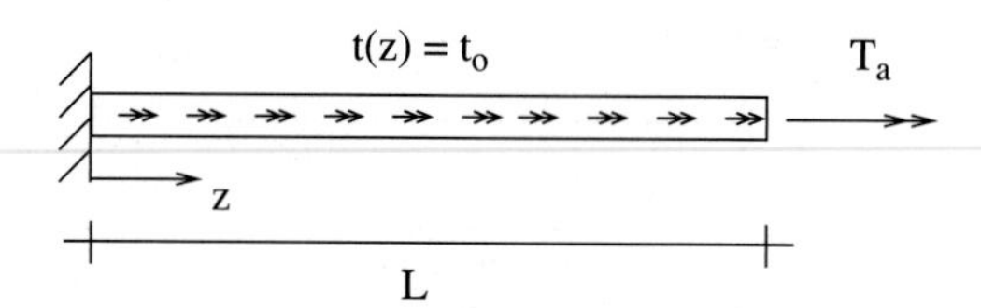


(7.26) Consider a solid round elastic bar with constant shear modulus, $G$, and cross-sectional area, $A$. The bar is built-in at both ends and subject to a spatially varying distributed torsional load

$$t(x) = p\sin(\frac{2\pi}{L}x),$$

where $p$ is a constant with units of torque per unit length. Determine the location and magnitude of the maximum internal torque in the bar.

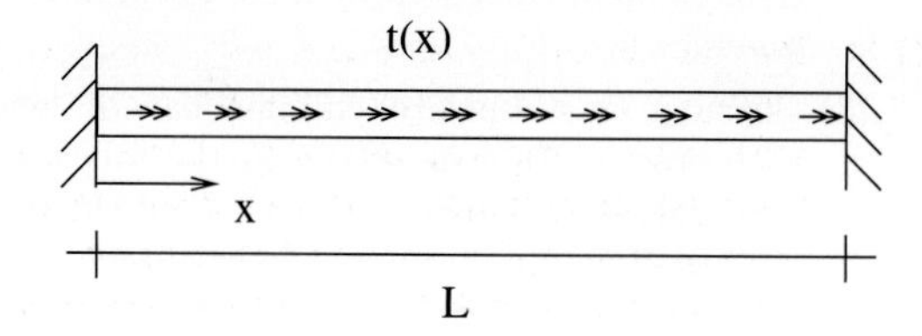

(7.27) A tapered circular torsion bar is built-in at both ends and subjected along its length to a constant distributed torque of 10 N-cm/cm: (1) Determine and plot the internal torque diagram. (2) Determine the twist rate as a function of axial coordinate. Assume the shear modulus $G$ is a given constant and that the outer radius as a function of axial coordinate $z$ is given by $R(z) = (1 \text{ cm}) \exp(-z/(4 \text{ cm}))$.

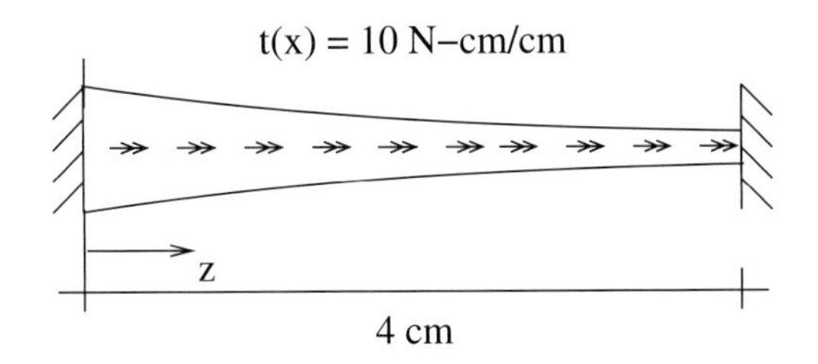


(7.28) A *non-linear elastic* solid circular bar, as shown, is subjected to a torque of 100 N-mm. The constitutive relation for the material is given as

$$\tau = B\text{sign}(\gamma)|\gamma|^n,$$

where $B = 60$ MPa and $n = \frac{1}{2}$. What is the maximum shear stress in the bar?

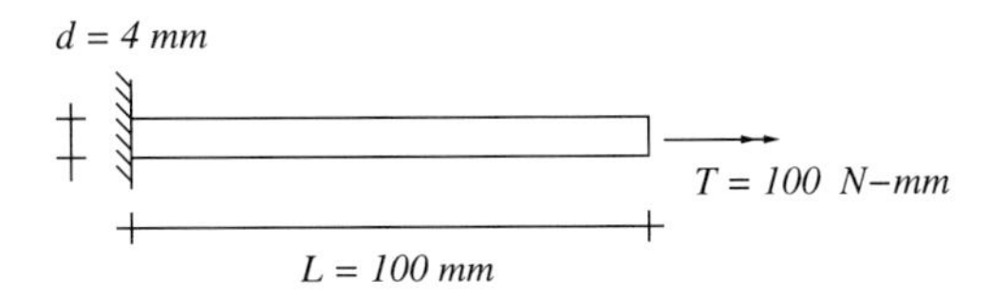


(7.29) The stepped round bar shown below is subjected to two opposing torques. Find the support torque at the right-end.

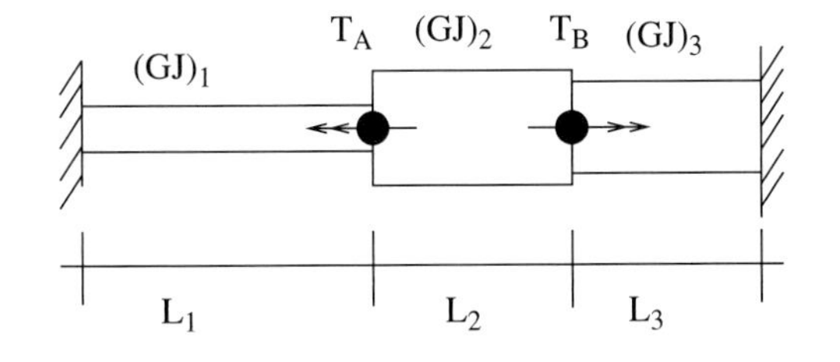


(7.30) For an elastic circular bar with constant shear modulus we know that the shear stress is given in terms of the torque as $\tau = Tr/J$. Find the corresponding formula for a circular bar that has been functionally graded in the radial direction so that the shear modulus can be expressed as

$$G(r) = A + Br,$$

where $A$ and $B$ are given constants.

(7.31) A cylindrical torsion bar is made by bonding a solid bar made of material 1 inside a tube made of material 2. Material 1 has a yield stress in shear $\tau_{Y1} = C/4$ and material 2 has a yield stress in shear $\tau_{Y2} = C$, where $C$ is a given constant. Both materials have the same shear modulus $G$ and can be considered to be elastic–perfectly plastic. What twist rate $\frac{d\phi}{dz}$ is required to yield all of material 2? Accurately plot the shear stress $\tau$ as a function of $r$ (radius) at this state of twist.

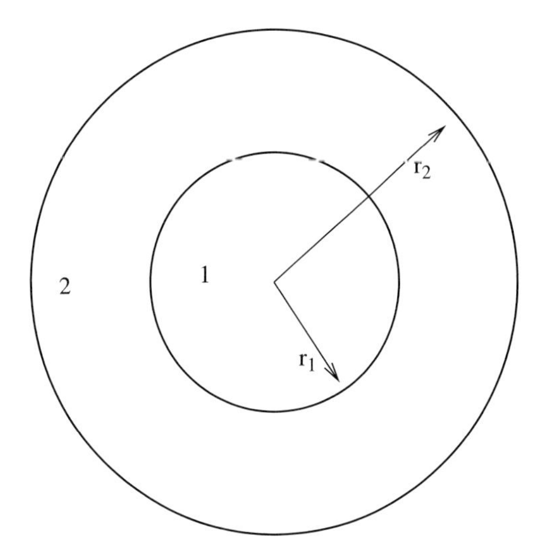


(7.32) A solid shaft, with a diameter of 20 mm and 1,000 mm in length, is twisted by end-torques such that it yields to a point where there is an elastic core of 16 mm. (a) How much torque is being applied to the shaft? (b) Assume the applied load is released. Plot the residual stress distribution. (c) What is the residual twist rate in the shaft? For all parts, assume $G = 50$ kN/mm$^2$, $\nu = 0.1$, and $\tau_Y = 50$ N/mm$^2$.

(7.33) If the shaft in Exercise 7.32 was twisted so that the differential end-rotation was 1.0 radians and then released, how much differential end-rotation would remain?

(7.34) Consider a hollow circular tube under torsional loads with inner radius $a$ and outer radius $b$. The tube is made of an elastic–perfectly plastic material with shear modulus $G$ and shear yield stress $\tau_Y$. What is the ultimate torque capacity of the tube?

(7.35) A thin tube of high-strength steel alloy is shrink-fitted to a solid circular rod of low-carbon steel. The low-carbon steel rod has a diameter of 12 mm and the high-strength steel alloy tube has a thickness of 2 mm. An end-torque is applied to the composite shaft and the outer surface stress is determined to be 500 N/mm$^2$. What is the value of the applied torque? Assume both steels have a shear modulus $G = 120\text{kN/mm}^2$. The low-carbon steel has a yield stress in shear of 100 N/mm$^2$; the high-strength alloy has a yield stress in shear of 600 N/mm$^2$.

(7.36) For a solid circular bar in pure torsion, derive the relation between the applied torque and the radius of the elastic–plastic interface. Assume the material is elastic–perfectly plastic. Express your answer in terms of the applied torque $T$, bar radius $R$, elastic-plastic interface radius $r_p$, and shear yield stress $\tau_Y$. For each step, write one or two complete sentences describing the step.

(7.37) Consider a solid elastic–perfectly plastic circular bar of length $L$ and radius $c$. What amount of imposed rotation at the end is required to yield a layer of material of thickness $0.1c$? Give your answer in terms of $L$, $G$, $\tau_Y$, and $c$.

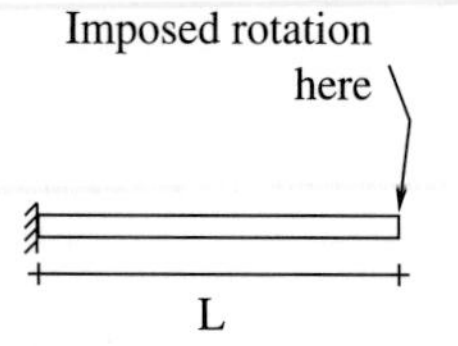


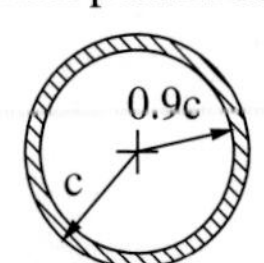


(7.38) A composite torsion bar of length $L$ is to carry a *given* ultimate torque $T_u$ and be of *minimum weight.* The bar is stipulated to have an aluminum core of *given* radius $R$ and a bonded *thin* jacket of *unknown* thickness $t$. Select the material for the jacket from Table 7.3 that will minimize the weight of the bar. To simplify the analysis, you should assume that $t \ll R$. (Hint: First determine an expression for the *given* ultimate torque $T_u$ in terms of the thickness of the jacket and the other parameters in the exercise; next, solve for the thickness; then determine the mass of the bar in terms of the jacket thickness etc; note that $T_u > 2\pi R^3 \tau_Y^{Al}/3$.)

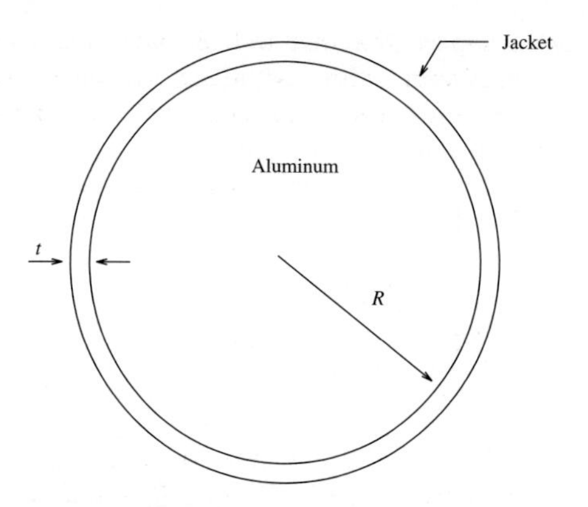


**Table 7.3** Material selection list.

| Material | G (GPa) | $\tau_Y$ (MPa) | Density (kg/m$^3$) |
|---|---|---|---|
| Al | 28.5 | 275 | 2700 |
| Mild steel | 80 | 154 | 7850 |
| HS Steel | 80 | 770 | 7850 |
| HTS graphite-5208 epoxy | 61 | 740 | 1550 |
| Alumina ($Al_2O_3$) | 161 | 9350 | 4000 |

(7.39) A solid circular shaft of radius $c$ is made of a material with the following stress–strain relation in shear:

$$\tau = \begin{cases} G\gamma, & \gamma < \gamma_Y \\ H\gamma + (G - H)\gamma_Y, & \gamma \geq \gamma_Y \end{cases}$$

Find the relation between the torque $T$ and the twist rate $\phi' = d\phi/dz$. Express your result in terms of $G$, $c$, $\phi'$, $H$, and $\gamma_Y$.

(7.40) Shown below is the cross-section of an elastic–perfectly plastic thin-walled tube. The dimensions $a$, $b$, and $d$ are all much greater than the wall thicknesses. Assume $t_1 < t_2 < t_3 < t_4$. Find the yield torque for the tube in terms of the given dimensions and the shear yield stress $\tau_Y$.

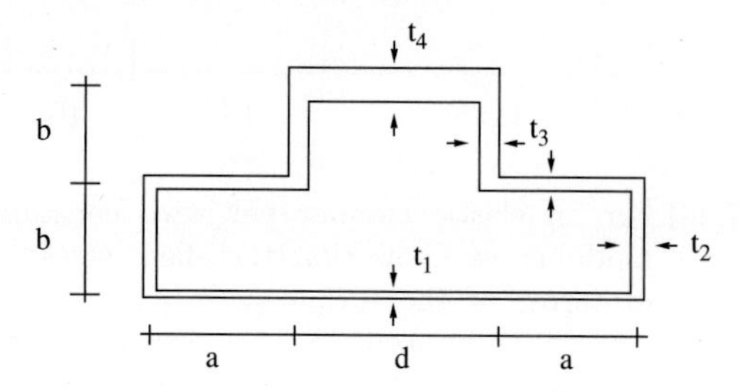

(7.41) Consider the thin-walled tube shown. Find the maximum shear stress, $\tau_{\max}$, in terms of the applied torque, $T$, and the geometry. Assume $t_1 < t_2$

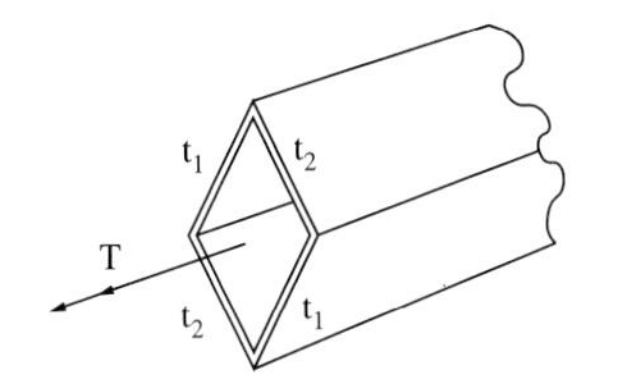


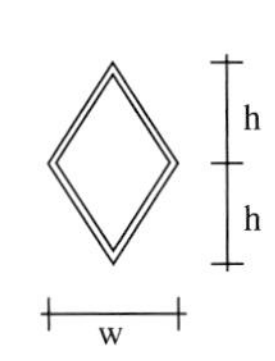


(7.42) Consider a thin-walled tube with cross-sectional geometry as shown below. Assuming that all wall thicknesses are small in comparison to the side lengths, determine the torsional stiffness per unit length of the tube.

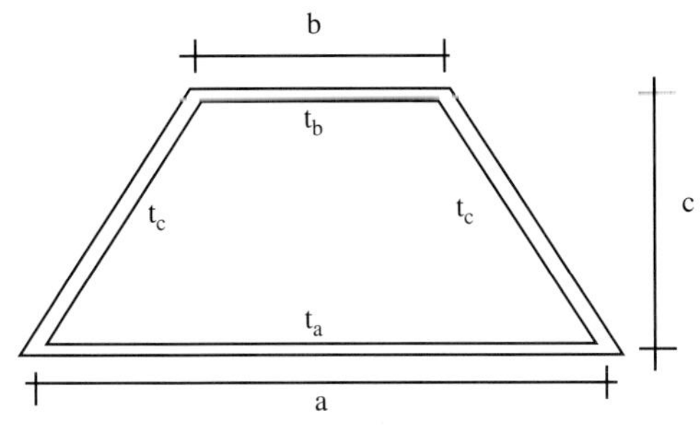

Thin-walled Tube Cross–section

(7.43) A hollow tube is fabricated by joining a rectangular channel section of uniform thickness $t_2$ to a semicircular channel of uniform thickness $t_1$. Both channels are made of the same material which has a yield stress in shear of $\tau_Y$. The dimensions are such that $t_1 < t_2$ and both thicknesses are much less than $h$ and $w$. Determine the torque at which the section will yield.

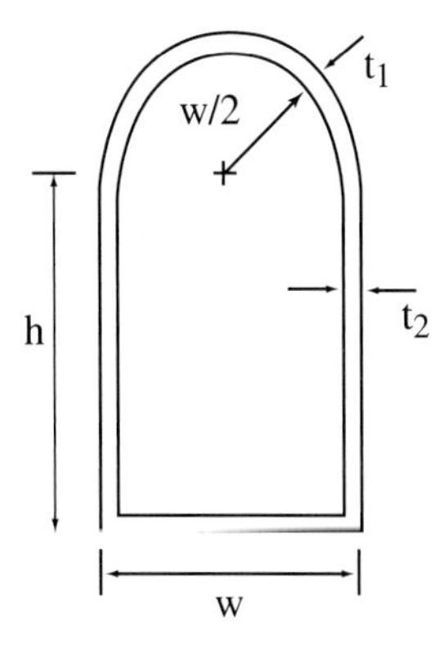


(7.44) What is the relation between shear stress and torque in a thin-walled tube?

(7.45) Consider the section shown in Exercise 7.43. If the channel has a length $L$ and shear modulus $G$, derive an expression for the torsional stiffness of the tube.

# 8 Bending of Beams

**Fig. 8.1** The support arm for this traffic light is being bent by its own weight and the weight of the two traffic lights.

So far we have addressed the analysis of slender bodies under the action of forces and moments (torques) aligned with the axial direction. In this chapter we will complete our basic development on the mechanics of slender bodies by looking at the case of applied moments and forces that are orthogonal to the axis of the body. Such loadings are known as bending loads. In many mechanical systems one can easily identify structural elements that act by resisting bending loads: vehicle axles, bridge decks, beams in buildings, flag poles, bones in animals, etc. To construct a viable theory for systems with these types of loads we will proceed in a manner similar to what we did for torsion. We will start with a kinematic assumption and then proceed to derive consequences of this assumption. To make the introduction to the bending behavior of beams tractable we will initially restrict ourselves to the case of symmetric bending of prismatic beams about a single axis. In the last sections of this chapter will briefly take up the topic of multi-axis bending and skew bending as well as the effects of plasticity.

## 8.1 Symmetric bending: Kinematics

Consider a prismatic slender body that has a cross-section with a vertical line of symmetry as shown in Fig. 8.2. For our analyses we will by convention take the $x$-axis as the axis of the body, the $y$-axis as the vertical axis, and the $z$-axis as their right-hand complement. The distance from the lower chord of the beam to the $x$-axis is as yet unspecified. If the body has a length to depth ratio of over 10, after the application of bending loads, one observes the following kinematic response:

(1) Plane cross-sections remain plane. Their motion can be fully characterized by a single displacement field and a rotation field.

(2) Normals remain normal; i.e. if one scribes two orthogonal lines (in the $x$-and $y$-directions) on the beam before the application of loads, then after the loads are applied the two lines remain orthogonal.

(3) The rotation of the cross-sections occurs about the same line for all cross-sections; i.e. the center of rotation for every cross-section is located at the same distance from the bottom of the beam.

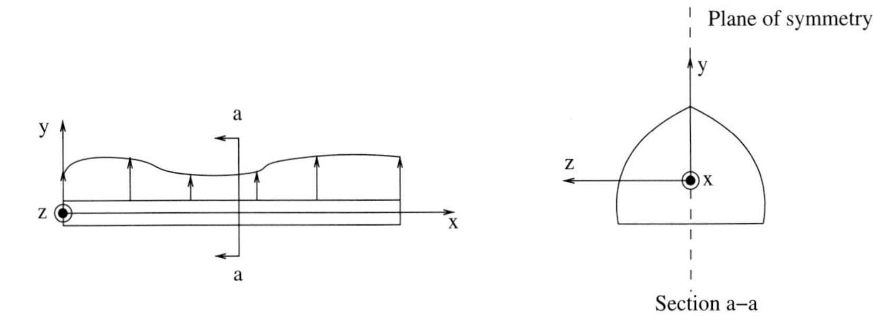


**Fig. 8.2** Beam with transverse loads and a cross-section with a vertical line of symmetry.

This situation is shown schematically in Fig. 8.3. The function $v(x)$ characterizes the displacement of the beam, and the function $\theta(x)$ characterizes the rotation of the cross-sections relative to the vertical.

**Remarks:**

(1) By this characterization the vertical motion of every point on a given cross-section is the same as long as one assumes small displacements and rotations.

(2) Because normals remain normals we have that $\tan(\theta) = dv/dx$. If we further assume small rotations, then we have that $\tan(\theta) \approx \theta$ and thus $\theta = dv/dx$.

(3) It is useful to also note that the curvature, $\kappa$, of any curve is given by $d\theta/ds$, where $\theta$ is the orientation of the curve's tangent to the horizontal and $s$ is the arc-length coordinate of the curve. If we assume small deformations and rotations, then we have for the beam that its curvature is approximately $\kappa = d\theta/dx$. Note, it is only a function of $x$ and is a constant for each cross-section. We also note that our assumptions imply that $\kappa = d\theta/dx = d^2v/dx^2$.

(4) Beams that obey these kinematic observations are known as Bernoulli–Euler beams.

These observations are very robust as long as the beam has a length to depth ratio of at least 10. They hold for both elastic and plastic

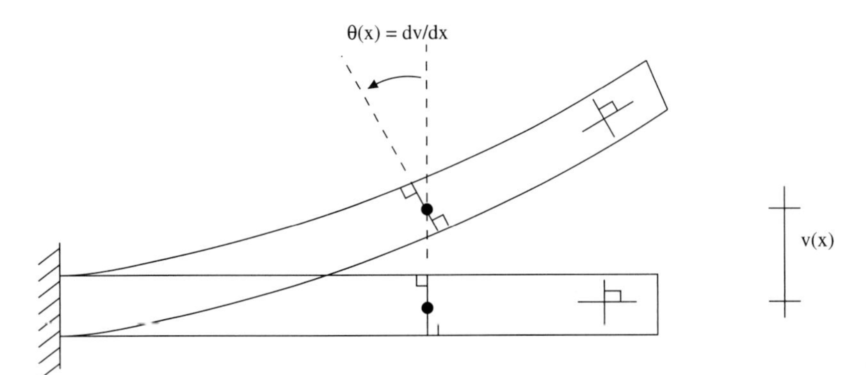


**Fig. 8.3** Beam kinematics.

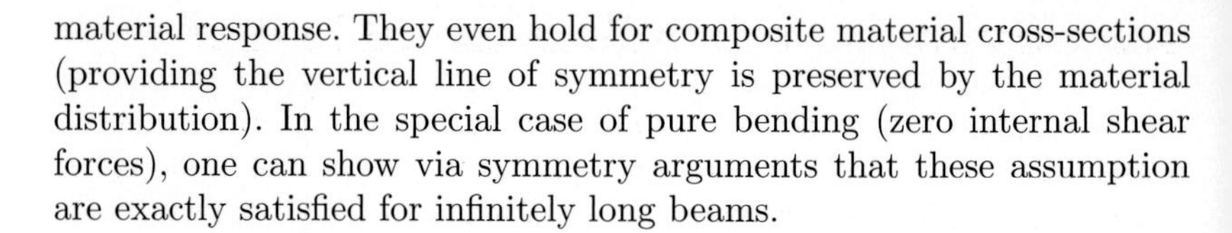

material response. They even hold for composite material cross-sections (providing the vertical line of symmetry is preserved by the material distribution). In the special case of pure bending (zero internal shear forces), one can show via symmetry arguments that these assumption are exactly satisfied for infinitely long beams.

*Strains in bending*

The essential motion of bending occurs in the $x$-$y$ plane. In this plane there are three components of the strain tensor that need to be determined. If we assume a plane stress conditions in the $y$-direction then we can safely ignore having to determine $\varepsilon_{yy}$, for the same reasons as discussed in Chapter 6. Because the normals remain normal there are no angles changes in the $x$-$y$ plane. Thus, $\gamma_{xy} = 0$. The remaining strain to be determined in the plane is $\varepsilon_{xx}$.

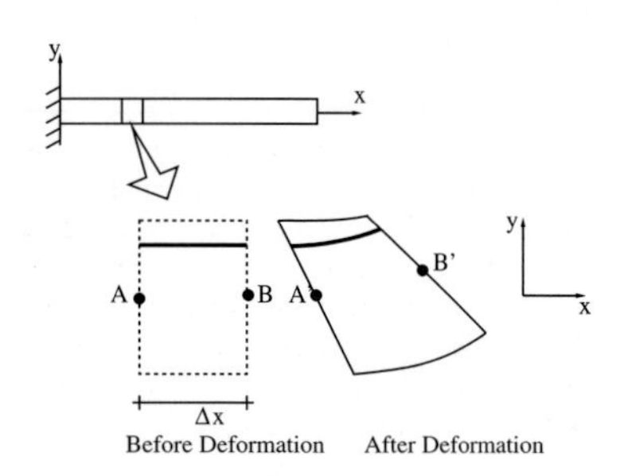


**Fig. 8.4** Bending strain construction.

To determine $\varepsilon_{xx}$ we can use a differential construction. Figure 8.4 shows a segment of length $\Delta x$ cut from the middle of a bent beam. The deformed shape of the differential element is shown shifted to the right for clarity. For convenience let us locate the $x$-axis along the line of centers of rotation. If we consider a strip of material at elevation $y$ from the $x$-axis, then it originally has a length $\Delta x$. After the segment deforms, it takes on the new length

$$\Delta x - y\theta(x + \Delta x) - [-y\theta(x)]. \tag{8.1}$$

In this expression we have assumed that the rotation is small. Thus the change in length of the strip is

$$\Delta x - y\theta(x + \Delta x) + y\theta(x) - \Delta x. \tag{8.2}$$

Dividing by the original length and taking the limit as $\Delta x \to 0$, we find

$$\varepsilon_{xx} = \lim_{\Delta x \to 0} \frac{-y\theta(x + \Delta x) + y\theta(x)}{\Delta x} = -y\frac{d\theta}{dx} = -y\frac{d^2 v}{dx^2} = -y\kappa. \tag{8.3}$$

**Remarks:**

(1) Equation (8.3) is the fundamental kinematic relation for the bending of beams. It is a direct consequence of our kinematic assumptions. As long as these hold, the strains will be given by eqn (8.3).

(2) The place of eqn (8.3) in the theory of beams is analogous to $\gamma = r d\phi/dz$ in the theory of torsion of circular bars.

(3) $\varepsilon_{xx}$ is usually called the bending strain. When there is no chance of confusion, we will omit the subscripts.

(4) Just as in torsion, our kinematic assumptions lead to a situation where the (bending) strains are linear on the cross-section; see Fig. 8.5.

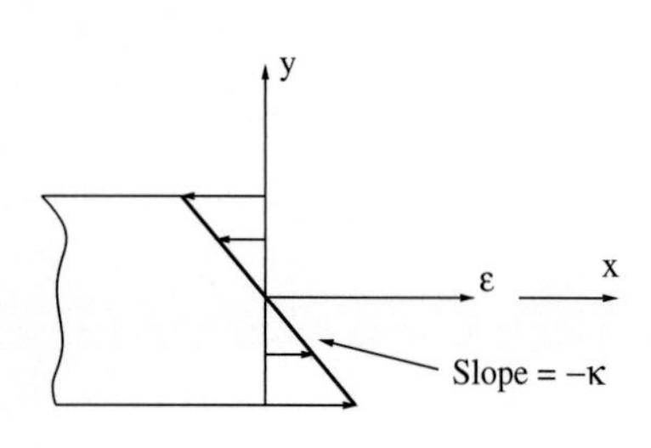


**Fig. 8.5** Bending strain distribution.

(5) The loci of points where the bending strains are zero is usually called the neutral axis, which may or may not lie in the center of the cross-section.

## 8.2 Symmetric bending: Equilibrium

Paralleling our developments of equilibrium for bars with axial forces and rods in torsion, we will consider the equilibrium of beams in differential form. Figure 8.6 shows a generically loaded beam. Let us consider a differential element of the beam as shown in Fig. 8.7. The segment is subject to a distributed load $q(x)$ which has units of force per unit length. On the two section cuts which define the segment, there are internal axial forces $R(x)$, shear forces $V(x)$, and bending moments $M(x)$. The axial force, shear force, and bending moments are drawn in the assumed positive sense.[1]

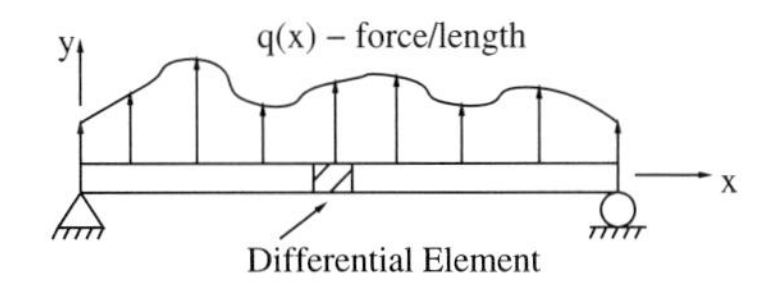


**Fig. 8.6** Beam with distributed load.

Let us first consider force equilibrium in the $x$-direction. This trivially tells us that $R(x)$ equals a constant. If we assume that there are no applied axial loads, then we have that

$$R(x) = 0. \tag{8.4}$$

If there are applied axial loaded, then $R(x) \neq 0$. The case $R(x) = 0$, is the one we will deal with the most.

If we now consider force equilibrium in the $y$-direction, then summing the forces gives us

$$\begin{aligned} 0 = \sum F_y &= -V(x) + V(x + \Delta x) + \int_x^{x+\Delta x} q(s)\, ds \\ &\approx -V(x) + V(x + \Delta x) + \Delta x q(x). \end{aligned} \tag{8.5}$$

We can divide through by $\Delta x$ and take the limit as $\Delta x \to 0$ to arrive at

$$\frac{dV}{dx} + q(x) = 0. \tag{8.6}$$

Equation (8.6) is the statement of force equilibrium in the vertical direction in differential form for a beam.

The third equilibrium equation in the plane is moment equilibrium about the $z$-axis. Let us sum moments about the point A shown in Fig. 8.7. This gives

$$\begin{aligned} 0 = \sum M_z &= M(x + \Delta x) - M(x) + V(x)\Delta x \\ &\quad - \int_x^{x+\Delta x} q(s)(x + \Delta x - s)\, ds \\ &\approx M(x + \Delta x) - M(x) + V(x)\Delta x - q(x)\Delta x^2/2. \end{aligned} \tag{8.7}$$

If we divide through by $\Delta x$ and take the limit $\Delta x \to 0$, then we find the differential expression for moment equilibrium for the beam:

$$\frac{dM}{dx} + V = 0. \tag{8.8}$$

[1] In many books the reverse sign convention is used for the shear forces. There is nothing wrong with using a different convention for positive shear forces. However, the convention we adopt here is consistent with the sign conventions we have used so far. When dealing with beam problems one should always make clear the assumed sign conventions.

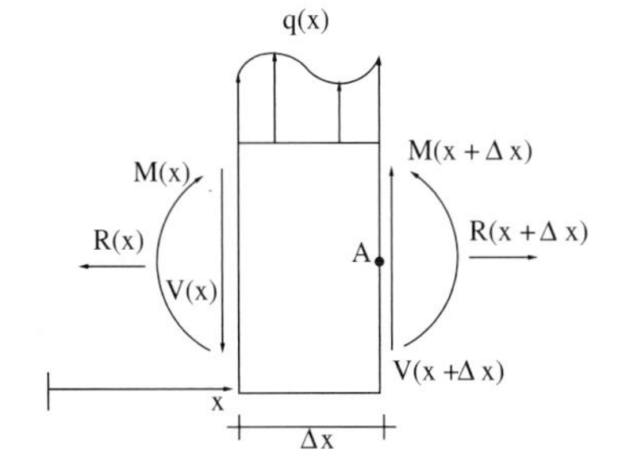


**Fig. 8.7** Differential element for deriving differential equilibrium equations.

**Remarks:**

(1) Equations (8.4), (8.6), and (8.8) provide the governing equations for equilibrium of internal forces in a beam. They play the same role as $dR/dx + b = 0$ did in bars with axial forces and $dT/dz + t = 0$ in rods with torsional loads.

(2) These equilibrium equations are independent of material response.

(3) It should be noted that the slope of the (internal) shear force diagram will always be equal to $-q(x)$ and the the slope of the (internal) moment diagram will always be equal to $-V(x)$. For example, if the distributed load is a constant, then the shear force will be linear in $x$, and the moment will be quadratic in $x$.

(4) An alternative way of expressing the equilibrium relations is to write them in integral form:

$$V(x) = V(0) - \int_0^x q(x)\, dx \tag{8.9}$$

$$M(x) = M(0) - \int_0^x V(x)\, dx. \tag{8.10}$$

This form is sometimes useful in practical problems. Note that $V(0)$ and $M(0)$ will need to be found from the boundary reactions of the particular problem.

(5) If the distributed load has point loads then $q(x)$ should be represented by a Dirac delta function, just as we did with point torques in torsion problems and point forces in axial load problems. If we have point moments in a beam we can represent them by a distribution of load over a small region $\zeta$ just as we did when we defined the Dirac delta function. Figure 8.8 shows one such possible distribution of load. The distribution $g_\zeta(x)$ has zero net force and a unit moment about the center point. Note also that $g_\zeta(x) = \frac{d}{dx} f_\zeta(x)$, where $f_\zeta(x)$ was the function used to define the Dirac delta function in Fig. 2.21. The idealization of a point moment comes in the limit $\zeta \to 0$, which implies that point moments (of unit magnitude) can be represented by $\delta'(x)$. Note the integration rule

$$\int \delta'(x)\, dx = \delta(x) + C. \tag{8.11}$$

It should also be observed that the sense of the moment is clockwise.

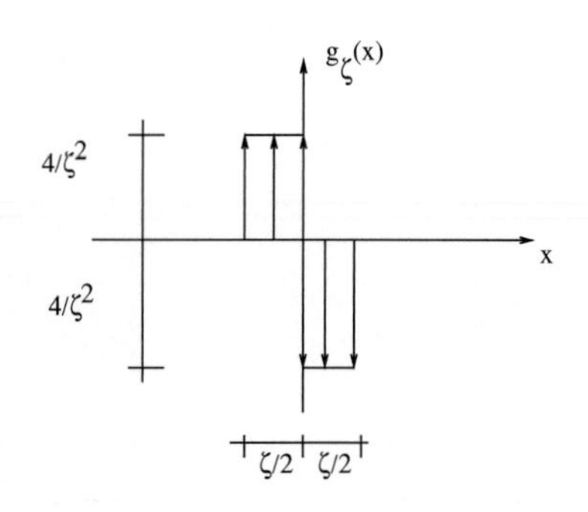


**Fig. 8.8** Doublet function construction.

### 8.2.1 Internal resultant definitions

The internal forces/moments (the internal resultants) arise from a distribution of stresses on the cross-section. The traction on the cross-section is given by Cauchy's law, eqn (3.31). Noting that the normal to our section cut is $\boldsymbol{e}_x$, Fig. 8.9 shows that the relevant two-dimensional traction components are given by:

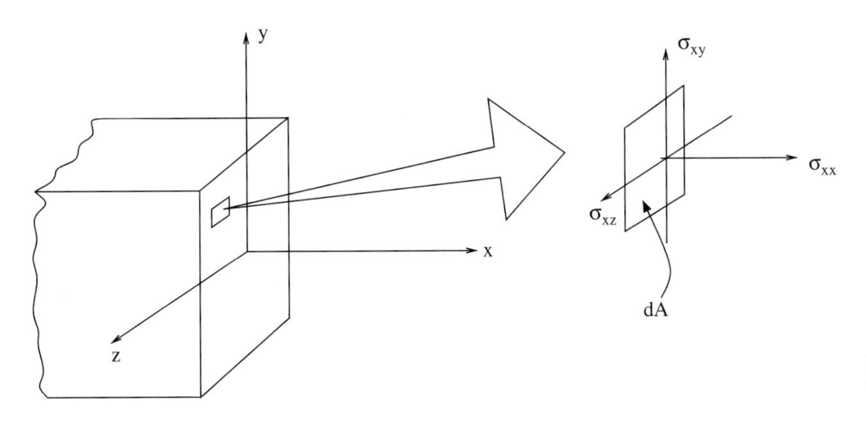


**Fig. 8.9** Traction components on beam cross-section.

$$\boldsymbol{\sigma}^T \boldsymbol{e}_x = \begin{pmatrix} \sigma_{xx} \\ \sigma_{xy} \end{pmatrix}. \tag{8.12}$$

The force normal to the section is given as it was when we treated axial force problems by

$$R = \int_A \sigma_{xx} \, dA. \tag{8.13}$$

The shear force will be given by integrating the shear stress contributions over the cross-section:

$$V = \int_A \sigma_{xy} \, dA, \tag{8.14}$$

where we have taken care to account for our assumed sign convention concerning the shear force. The moment about the $z$-axis will arise from the normal stresses $\sigma_{xx}$. The lever arm that the normal stresses act through is of length $y$. Taking care of the sense of the induced moment we have that

$$M = \int_A -y\sigma_{xx} \, dA. \tag{8.15}$$

**Remarks:**

(1) Equations (8.13)–(8.15) are analogous to $R = \int_A \sigma_{xx} \, dA$ in the axial force problems and $T = \int_A r\sigma_{z\theta} \, dA$ in torsion problems.

(2) These relations are independent of material response.

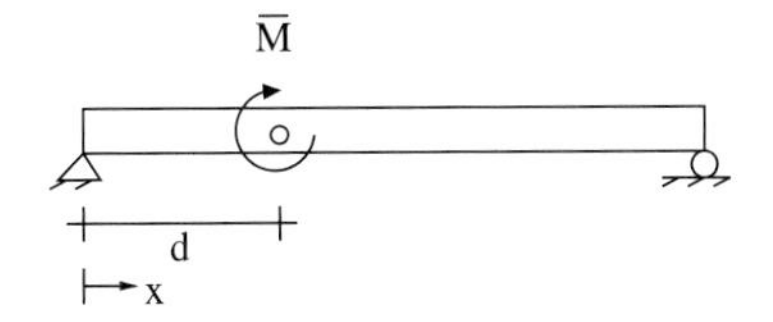


**Fig. 8.10** Beam of length $L$ with an applied point moment at $x = d$.

**Example 8.1**

*Shear-moment diagram: Section cuts.* Consider the beam shown in Fig. 8.10. Determine the internal force and moment distributions using a section cut method.

*Solution*
Figure 8.11 shows a free-body diagram of the beam with the supports removed. Force equilibrium in the $x$-direction tells us that $R_1 = 0$ and force equilibrium in the $y$-direction tells us that $R_2 = R_3$. Taking moments about any point along the axis of the beam shows that $R_2 = \bar{M}/L$ and, hence, $R_3 = \bar{M}/L$.

Internal force and moment diagrams can now be constructed by making successive section cuts and applying force/moment balance for different values of $x$, as shown in Fig. 8.11. For axial force balance, only a single section cut is necessary, since there are no distributed axial forces. This gives

$$0 = \sum F_x = R(x) + R_1 \qquad \Rightarrow \qquad R(x) = 0. \tag{8.16}$$

For vertical force balance we can also make just a single arbitrary cut. This yields

$$0 = \sum F_y = -\frac{\bar{M}}{L} + V(x) \qquad \Rightarrow \qquad V(x) = \frac{\bar{M}}{L}. \tag{8.17}$$

For moment equilibrium we need to make two different cuts – i.e. one before the applied point moment and one after. This gives

$$0 = \sum M_z = M(x) + \frac{\bar{M}}{L}x \qquad \Rightarrow M(x) = -\frac{\bar{M}}{L}x \qquad x < d \tag{8.18}$$

$$0 = \sum M_z = M(x) + \frac{\bar{M}}{L}x - \bar{M} \quad \Rightarrow M(x) = \bar{M} - \frac{\bar{M}}{L}x \quad x > d. \tag{8.19}$$

The resulting graphs are shown in Fig. 8.11.

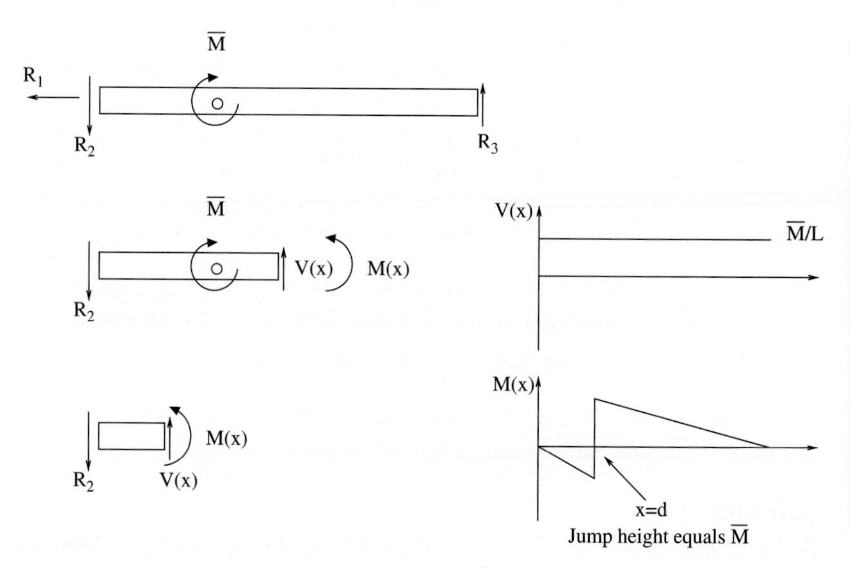


**Fig. 8.11** Free-body diagram for Example 8.1 and internal force diagrams.

**Example 8.2**

*Shear-moment diagram: Integration.* Resolve Example 8.1 using the differential expressions for equilibrium.

*Solution*

The distributed load is given by $q(x) = \bar{M}\delta'(x-d)$. Integrating the vertical force balance equation gives

$$V(x) - V(0) = \int_0^x \frac{dV}{dx} = -\int_0^x q(x)dx = -\bar{M}\delta(x-d). \tag{8.20}$$

Using the fact that $V(0) = \bar{M}/L$ from the computed support reactions, we have

$$V(x) = \frac{\bar{M}}{L} - \bar{M}\delta(x-d). \tag{8.21}$$

Integrating the moment equilibrium equation gives

$$M(x) - M(0) = \int_0^x \frac{dM}{dx} = -\int_0^x V(x)dx = -\frac{\bar{M}}{L}x + \bar{M}H(x-d). \tag{8.22}$$

Using the fact that $M(0) = 0$ from the computed support reactions, we have

$$M(x) = -\frac{\bar{M}}{L}x + \bar{M}H(x-d). \tag{8.23}$$

**Remarks:**

(1) This result is the same as that obtained in Example 8.1, except that the result for the shear force diagram has an extra term here (the delta function). This term did not appear when using the section cut method because it is only located at a single point – the point of the applied load. We do not make section cuts at applied point loads because we do not really know the exact distribution of the forces at such points. To make such cuts we would have to assume things which we probably do not know. Comparison of the two results shows that one needs to be cautious in interpreting results in the neighborhood of point moments (and forces).

**Example 8.3**

*Shear-moment diagram: Cantilever beam.* Determine the shear and moment diagrams for the cantilever beam shown in Fig. 8.12.

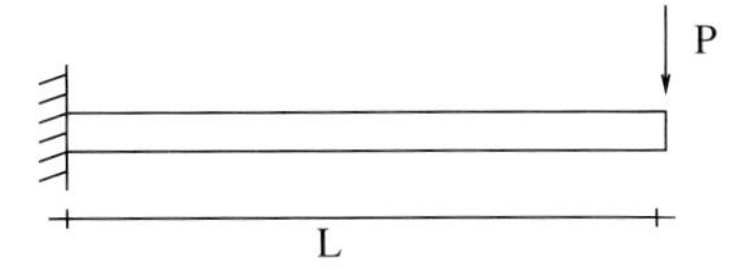


**Fig. 8.12** Cantilever beam for Example 8.3.

*Solution*

A free-body diagram, Fig. 8.13, shows that the support reactions at $x = 0$ are $V(0) = -P$ and $M(0) = -PL$. The distributed load $q(x) = 0$. Thus

$$V(x) - V(0) = \int_0^x \frac{dV}{dx} = \int_0^x 0 dx = 0$$
$$V(x) = -P \tag{8.24}$$

and

$$M(x) - M(0) = \int_0^x \frac{dM}{dx} = -\int_0^x V(x)dx = Px$$
$$M(x) = P(x - L). \tag{8.25}$$

The results are sketched in Fig. 8.13.

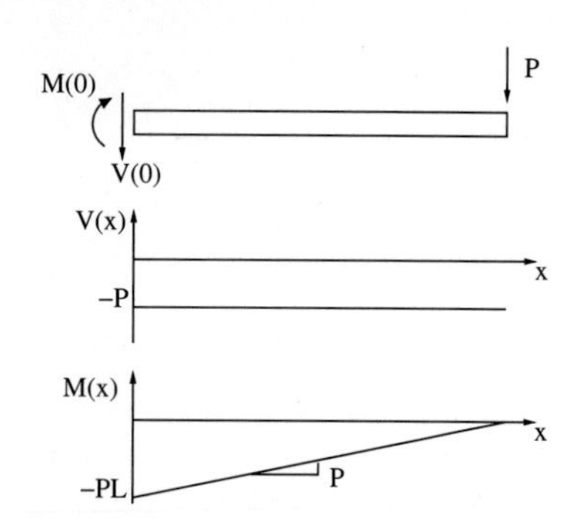


**Fig. 8.13** Free-body diagram and shear moment diagrams for the cantilever beam in Example 8.3.

## 8.3 Symmetric bending: Elastic response

In the last section we saw that it was possible to determine the internal force and moment distribution in a beam from the equilibrium equations. This was possible solely from equilibrium, because the problems that were posed were statically determinate. To be able to handle statically indeterminate problems, to be able to say anything about the stress distribution, the strain distribution, or the deflection and rotation of the beam, we will need to specify the constitutive relation for the beam. Let us start by considering linear elastic response, such that

$$\sigma = E\varepsilon. \tag{8.26}$$

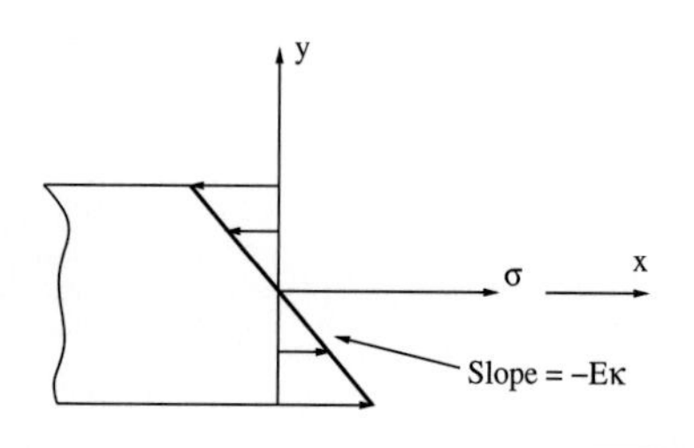


**Fig. 8.14** Stress distribution in elastic bending of an homogeneous material.

Given our kinematic assumptions we see that $\sigma = -E\kappa y$; i.e. the stress distribution is linear on the cross-section as shown in Fig. 8.14 (assuming that $E$ is a constant).[2]

[2] We have dropped the subscripts on the stress and strain, since it is clear we are discussing the bending stress and bending strain, $\sigma_{xx}$ and $\varepsilon_{xx}$.

### 8.3.1 Neutral axis

It is important to note that we have yet to determine the location of the neutral axis of the beam. We have stipulated that it corresponds to the loci of centers of rotation of the cross-sections, but we have not yet determined its location in terms of the other parameters of a beam bending problem. The location of the neutral axis of a beam is dependent upon the geometry of the beam cross-section, the distribution of material properties on the cross-section, and the type of material response elastic, plastic, etc. To determine its location we will use the fact that for a beam without axial loads, $R(x) = 0$. Thus $\int_A \sigma\, dA = 0$ on each cross-section. This tells us that the tensile forces on the section exactly balance the compressive forces on the section.

Using the stress–strain relation combined with the kinematic assumptions, we find

$$0 = R = \int_A -Ey\kappa\, dA. \tag{8.27}$$

Let us determine the location of the neutral axis relative to the bottom chord of the beam for a generic symmetric cross-section. As shown in Fig. 8.15, we introduce a second coordinate $y_b$ which measures distance from the bottom chord. The origin of the $y_{b-z}$ coordinate system is offset a distance $y_{na}$ from the origin of the $y$-$z$ coordinate system. $y_{na}$ is the quantity that we would like to determine. The relation between the vertical measures is given by $y = y_b - y_{na}$. Substituting into eqn (8.27) we find

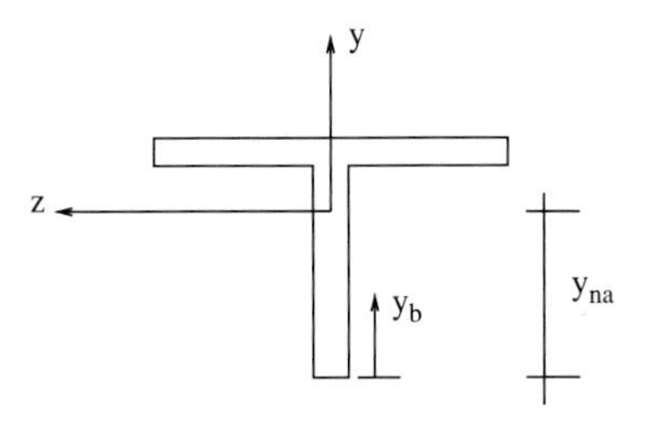


**Fig. 8.15** Generic symmetric cross-section for neutral axis construction.

$$0 = \int_A -E\kappa(y_b - y_{na})\, dA. \tag{8.28}$$

Because $\kappa$ is only a function of $x$, and $y_{na}$ is a constant, they can be pulled out from under the integral sign to yield

$$y_{na} = \frac{\int_A E y_b\, dA}{\int_A E\, dA}. \tag{8.29}$$

**Remarks:**

(1) When performing the integrations indicated in eqn (8.29), the area measure $dA$ should be taken as $dy_b dz$.

(2) In the special case where the beam cross-section is homogeneous (i.e. $E(y, z) = E$ a constant), one can pull the modulus out from under the integral signs and then they cancel. This leaves the special result for homogeneous cross-sections:

$$y_{na} = \frac{\int_A y_b\, dA}{\int_A dA}. \tag{8.30}$$

This is recognized to be the equation for the centroid of the cross-section. So in the homogeneous case, $y_{na} = y_c$. The cross-sections rotate about the line of centroids.

(3) The result shown in eqn (8.29) and the centroid result are restricted to the elastic case. For inelastic material behavior one needs to directly apply $R(x) = 0$ to determine the neutral axis location.

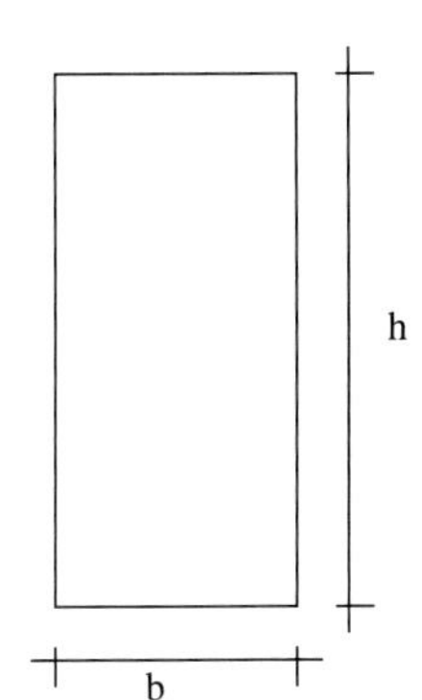


**Fig. 8.16** Rectangular cross-section.

### Example 8.4

*Centroid of a rectangular cross-section.* Consider the rectangular cross-section shown in Fig. 8.16. Find the location of the neutral axis, assuming the material is linear elastic homogeneous.

*Solution*

Apply eqn (8.29).

$$y_{na} = \frac{\int_A E y_b\, dA}{\int_A E\, dA} = \frac{\int_{-b/2}^{b/2} \int_0^h y_b\, dy_b dz}{bh} = \frac{b\frac{1}{2}h^2}{bh} = \frac{h}{2}. \tag{8.31}$$

This gives the distance as measured from the bottom of the beam, and it lies right in the center of the section, as was easily anticipated.

**Remarks:**

(1) This result shows that in a linear elastic homogeneous beam the bending strains and stress will be zero at the centroid of a cross-section. It also shows that the strains and stresses in such a beam will be maximum at the outer chords of the beam.

(2) For more general cross-sectional shapes, depending on the location of the centroid, the maximum stress and strain may occur at either the top or bottom chords or simultaneously at both the top and bottom.

### 8.3.2 Elastic examples: Symmetric bending stresses

At this stage we have a complete theory that can be utilized to solve a wide variety of beam bending problems. The governing system of equations is given by the fundamental kinematic relation (which depends upon our stated assumptions)

$$\varepsilon_{xx} = -y\kappa\,, \tag{8.32}$$

the equilibrium relations

$$R = 0\,, \tag{8.33}$$

$$\frac{dV}{dx} + q = 0\,, \tag{8.34}$$

$$\frac{dM}{dx} + V = 0\,, \tag{8.35}$$

the internal resultant relations

$$R = \int_A \sigma_{xx}\, dA\,, \tag{8.36}$$

$$V = \int_A \sigma_{xy}\, dA\,, \tag{8.37}$$

$$M = \int_A -y\sigma_{xx}\, dA\,, \tag{8.38}$$

and the constitutive relation

$$\sigma_{xx} = E\varepsilon_{xx}\,, \tag{8.39}$$

where $\kappa = d\theta/dx = d^2v/dx^2$.

**Example 8.5**

*Moment-curvature relation.* Consider a linear elastic homogeneous beam and find the relation between the moment and the curvature at a given cross-sectional location $x$.

*Solution*

This problem asks us to connect the kinematics of bending to the internal bending moment. Start with the expression for the bending moment

$$M = \int_A -y\sigma \, dA. \tag{8.40}$$

Substitute for the bending stress in terms of the bending strain:

$$M = \int_A y^2 E\kappa \, dA. \tag{8.41}$$

Since $E$ and $\kappa$ are not functions of $y$ and $z$. We find

$$M = E \underbrace{\int_A y^2 \, dA}_{I_z} \kappa = EI_z\kappa. \tag{8.42}$$

**Remarks:**

(1) $I_z = \int_A y^2 \, dA$ is known as the moment of inertia of the cross-section about the $z$-axis. It is sometimes also call the second area moment. In the theory of bending it occupies the same place that $J$, the polar moment of inertia, did in the theory of torsion. For a solid rectangular section, for example, $I_z = bh^3/12$, where $b$ is the beam width and $h$ the beam height/depth.

(2) In many situations one drops the subscript on $I_z$ and simply writes $I$. This can be done as long as no confusion could arise.

(3) The product $EI$ is often termed the bending stiffness.

(4) $M = EI\kappa$ is the bending counterpart to the relation $T = GJd\phi/dz$ from torsion.

(5) In cases where the bar is inhomogeneous on the cross-section (i.e. $E$ is a function of $y$ and/or $z$), $E$ cannot be removed from under the integral sign in eqn (8.41). In this situation one often defines an effective $(EI)_{\text{eff}} = \int_A Ey^2 \, dA$. With this definition one can write $M = (EI)_{\text{eff}}\kappa$. This is a useful device for dealing with composite cross-sections.

(6) Note that in the case of inhomogeneous material properties the dependency of $E$ on $z$ must be an even function; i.e. $E(z) = E(-z)$. This will insure that the vertical plane of symmetry we have been assuming is present.

**Example 8.6**

*Elastic stress distribution.* Consider the same elastic homogeneous beam as in the last example, and find the relation between the bending stress on the cross-section and the bending moment on the cross-section.

*Solution*
Starting from the result of the last example we have that

$$\kappa = \frac{M}{EI}. \tag{8.43}$$

We know $\kappa = -\varepsilon/y = -\sigma/(yE)$. Substituting in for $\kappa$, we find:

$$\sigma(x, y) = -\frac{M(x)y}{I}. \tag{8.44}$$

**Remarks:**

(1) This relation is the bending counterpart to $\tau = Tr/J$.

(2) If the beam cross-section is not homogeneous, then one can use $(EI)_{\text{eff}}$. In this case, one finds $\sigma(x, y, z) = -M(x)yE(y, z)/(EI)_{\text{eff}}$. The maximum bending stress on a given cross-section does not need to appear at the outer chords for such situations.

**Example 8.7**

*Maximum stress: T-beam.* Consider a linear elastic homogeneous T-beam with cross-sectional dimensions as shown in Fig. 8.17. For a given moment $M$ acting at some location $x$, determine the maximum bending stress (in absolute magnitude).

*Solution*
$|\sigma_{\text{max}}| = \frac{My_{\text{max}}}{I}$. We need to determine $y_{\text{max}}$ and $I$. Since the beam is homogeneous,

$$y_{na} = \frac{\int_A y\, dA}{\int_A dA}. \tag{8.45}$$

To make the computation of the integrals easier one can break up the domain of integration into a set of simpler shapes. In this case we can

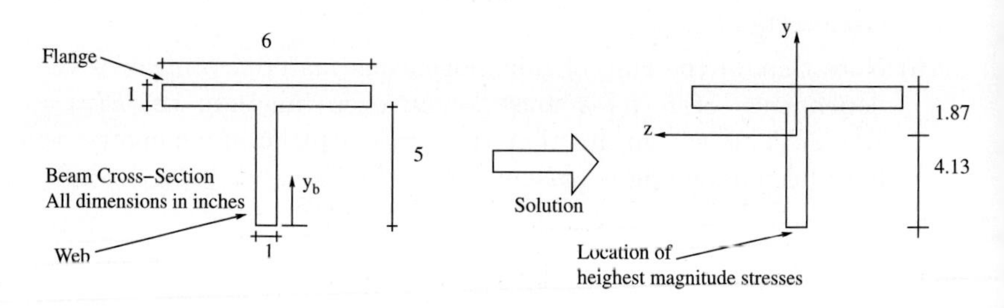


**Fig. 8.17** Cross-section of a T-beam.

break up the integral into an integral over the rectangular area of the flange and the rectangular area of the web. We can also simplify the process by noting that $\int_A y\,dA = \bar{y}A$, where $\bar{y}$ is the average value of $y$ over the area. Using these two devices,

$$y_{na} = \frac{A_{\text{flange}}\bar{y}_{\text{flange}} + A_{\text{web}}\bar{y}_{\text{web}}}{A} \tag{8.46}$$

$$= \frac{1 \times 6 \times 5.5 + 5 \times 1 \times 2.5}{11} = 4.13 \text{ in}, \tag{8.47}$$

where the distances have been measured from the bottom of the beam. $y_{\max} = \max\{y_{na}, 6 - y_{na}\} = 4.13$ in. This represents the distance to the chord furthest from the neutral axis.

To compute $I$, we need to compute it relative to the neutral axis. Again, it is convenient to break up the needed integral into an integral over the flange and one over the web. A second device that helps in the evaluation of the integrals is the *Parallel Axis Theorem.* This theorem (see Appendix D4) allows one to express the moment of inertia about a centroidal axis in terms of the moment of inertia about another (parallel) axis. Given the centroidal axis $z_c$ and a second parallel axis $z$ the moments of inertia of a given area $A$ about the two axes are related by the relation

$$I_z = I_{z_c} + Ad^2\,, \tag{8.48}$$

where $d$ is the distance separating the two axes. The expression for the moment of inertia is written as

$$I_z = \int_A y^2\,dA = \int_{A_{\text{web}}} y^2\,dA + \int_{A_{\text{flange}}} y^2\,dA. \tag{8.49}$$

Because we already know the moment of inertia of a rectangle about its own centroid, we can easily evaluate each term on the right-hand side using the parallel axis theorem:

$$I_z = \int_{A_{\text{web}}} y^2\,dA + \int_{A_{\text{flange}}} y^2\,dA \tag{8.50}$$

$$= I_{c,\text{web}} + A_{\text{web}}d^2_{\text{web}} + I_{c,\text{flange}} + A_{\text{flange}}d^2_{\text{flange}} \tag{8.51}$$

$$= \frac{1 \times 5^3}{12} + 5 \times (4.13 - 2.5)^2$$

$$+\frac{6 \times 1^3}{12} + 6 \times (4.13 - 5.5)^2 \tag{8.52}$$

$$= 35.5 \text{ in}^4. \tag{8.53}$$

Putting our results together, one finds

$$|\sigma_{\max}| \text{ (psi)} = \frac{M}{I/y_{\max}} = \frac{M}{S} = \frac{M \text{ (lbf} - \text{in)}}{8.59 \text{ in}^3}\,, \tag{8.54}$$

where $S = I/y_{\max}$ is known as the section modulus.

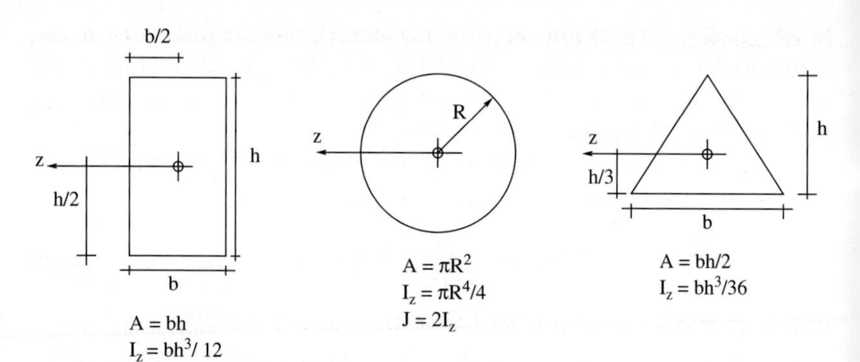


**Fig. 8.18** Centroid locations, areas, and moments of inertia of a few common shapes.

**Remarks:**

(1) Tables listing moments of inertia and section moduli for commonly manufactured beams' cross-sections are printed in all standard mechanical and civil engineering handbooks.

(2) Figure 8.18 provides some useful relations regarding the moments of inertia and centroid locations for a few basic shapes. For hollow sections one can use a similar procedure as used in this example, except that one subtracts values for missing areas.

## Example 8.8

*Bending of a composite beam.* Consider the beam cross-section shown in Fig. 8.19. Find: (1) the location of the neutral axis of the beam, (2) the moment-curvature relation for the beam, and (3) sketch the bending strain and stress distributions on the cross-section. Assume linear elasticity.

*Solution*

(1) Find the location of the neutral axis. From eqn (8.29) we have

$$y_{na} = \frac{\int_A Ey\,dA}{\int_A E\,dA} = \frac{\int_{A_w} E_w y\,dA + \int_{A_s} E_s y\,dA}{\int_{A_w} E_w\,dA + \int_{A_s} E_s\,dA} \tag{8.55}$$

$$= \frac{E_w A_w \bar{y}_w + E_s A_s \bar{y}_s}{E_w A_w + E_s A_s}. \tag{8.56}$$

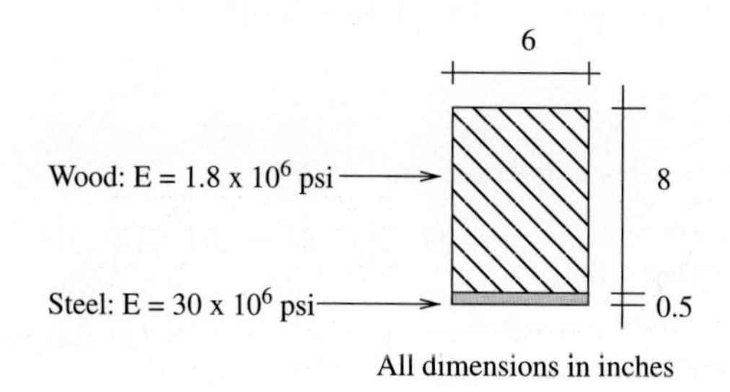


**Fig. 8.19** Composite beam cross-section made of wood and steel for Example 8.8.

Noting that $\bar{y}_s = 0.25$ in and $\bar{y}_w = 4.5$ in from the bottom of the beam, one finds

$$y_{na} = 2.3 \text{ in.} \tag{8.57}$$

(2) Since the beam has inhomogeneous material properties we can utilize an effective bending stiffness. In this case $M = (EI)_{\text{eff}}\kappa$, where

$$(EI)_{\text{eff}} = \int_A y^2 E(y)\, dA. \tag{8.58}$$

In this expression $y$ is measured from the neutral axis. Thus

$$(EI)_{\text{eff}} = b\left[\int_{-y_{na}}^{-y_{na}+0.5} y^2 E_s\, dy \int_{-y_{na}+0.5}^{8.5-y_{na}} y^2 E_w\, dy\right] \tag{8.59}$$

$$= bE_s \left.\frac{y^3}{3}\right|_{-y_{na}}^{-y_{na}+0.5} + bE_w \left.\frac{y^3}{3}\right|_{-y_{na}+0.5}^{8.5-y_{na}} \tag{8.60}$$

$$- 668 \times 10^6 \text{ lbf} - \text{in}^2. \tag{8.61}$$

Knowing this, the moment-curvature relation is fully defined.

(3) The strain distribution is linear, as shown in Fig. 8.20 (left). The stresses are given by multiplying by the modulus which is a discontinuous function of $y$. The result is shown in Fig. 8.20 (right).

**Remarks:**

(1) Note the low location of the neutral axis. The neutral axis is always "attracted" to the stiffest part of the cross-section.

---

**Example 8.9**

*Stiffness of an end-loaded cantilever.* Compute the stiffness of the linear elastic cantilever shown in Fig. 8.12. Assume homogeneous material properties and a general but symmetric cross-section.

*Solution*

The stiffness of the cantilever will be the ratio of the applied force $P$ to the downward deflection of the beam at the tip. From the results of Example 8.3 the internal shear force in the beam $V(x) = -P$ and

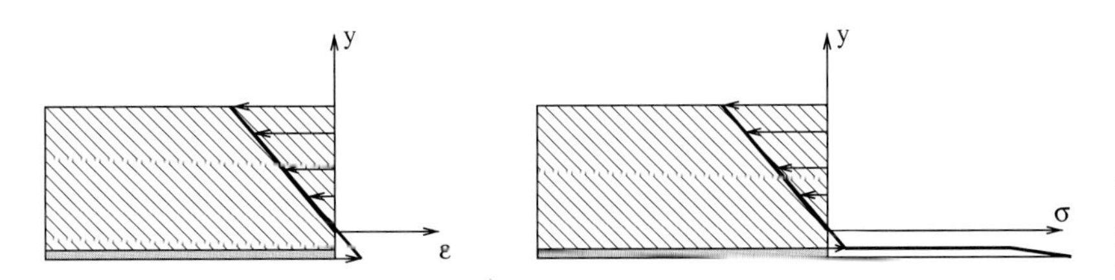


**Fig. 8.20** Sketch of strain and stress distributions for the composite beam of Example 8.8.

the internal bending moment $M(x) = P(x - L)$. Thus the curvature in the beam is given by $\kappa(x) = P(x - L)/EI$. The curvature can now be integrated to compute the rotation field:

$$\theta(x) - \underbrace{\theta(0)}_{=0} = \int_0^x \kappa \, dx = \frac{P}{EI}\left(\frac{x^2}{2} - Lx\right). \tag{8.62}$$

Integrating the rotation field gives

$$v(x) - \underbrace{v(0)}_{=0} = \int_0^x \theta(x) \, dx = \frac{Px^2}{2EI}\left(\frac{x}{3} - L\right). \tag{8.63}$$

Evaluating at the tip, we find $\Delta = v(L) = -PL^3/3EI$. The stiffness is thus $k = 3EI/L^3$.

**Remarks:**

(1) The appropriate boundary conditions at a built-in support are zero displacement and zero rotation.

## 8.4 Symmetric bending: Elastic deflections by differential equations

In Example 8.9 we were able to compute the deflection of the cantilever by first computing the bending moment in the beam and then relating it to the curvature, followed by double integration. This type of procedure will work only for statically determinate beams. For statically indeterminate beams we can apply either an assumed support reaction method such as was done in the earlier chapters or by employing a differential equation approach. Let us pursue this second option and restrict our attention to the linear elastic case where we will presume that $EI$ is known. To derive the governing differential equation for the deflection $v(x)$ of the beam we will proceed in a fashion similar to what we followed in Chapter 7 for torsion problems.

First, note that if we substitute the vertical force equilibrium equation into the moment equilibrium equation, then we have

$$\frac{d^2M}{dx^2} = q. \tag{8.64}$$

We can introduce the kinematics into this by using the moment-curvature expression to give

$$\frac{d^2}{dx^2}(EI\kappa) = q. \tag{8.65}$$

Using our small deformation approximation that $\kappa = d^2v/dx^2$, we come to a single differential equation for elastic beam deflections:

$$\frac{d^2}{dx^2}\left(EI\frac{d^2v}{dx^2}\right) = q. \tag{8.66}$$

**Remarks:**

(1) This equation embodies all the main equations of beam bending: equilibrium, kinematics, and constitutive response. It is a fourth-order ordinary differential equation.

(2) If $EI$ is not a function of $x$ (i.e. is a constant), then the governing differential equation simplifies to

$$EI\frac{d^4v}{dx^4} = q, \tag{8.67}$$

which is a fourth-order ordinary differential equation with constant coefficients.

(3) Equation (8.66) is the beam counterpart to eqn (7.51) for torsion bars.

(4) To solve eqn (8.66) we need four boundary conditions. Normally, one will encounter kinematic boundary conditions on $v$ and $\theta$ and force/moment boundary conditions on $V$ and $M$. As with the axial force and torque problems, the force/moment boundary conditions need to be converted to conditions on the derivatives of $v$. This is achieved by noting that $M = EId^2v/dx^2$ and $V = -EId^3v/dx^3$.

(5) In the case of inhomogeneous cross-sections one can replace $EI$ with $(EI)_{\text{eff}}$.

(6) For a given problem, once $v(x)$ has been determined, all other quantities of interest can be determined by differentiation and algebra.

**Example 8.10**

*End-loaded cantilever revisited.* Let us reconsider the end-loaded cantilever from Example 8.9 and compute the deflection $v(x)$.

*Solution*

The distributed load $q(x) = 0$. We can identify two boundary conditions at $x = 0$ of $v(0) = v'(0) = 0$ and two boundary conditions at $x = L$ of $EIv''(L) = 0$ and $EIv'''(L) = P$.[3] We can now proceed to integrate eqn (8.67) four times and then eliminate the constants of integration using the boundary conditions.

[3] As a short-hand we use a prime to indicate differentiation with respect to $x$.

$$EIv'''' = 0 \tag{8.68}$$

$$EIv''' = C_1 \tag{8.69}$$

$$EIv'' = C_1x + C_2 \tag{8.70}$$

$$EIv' = C_1x^2/2 + C_2x + C_3 \tag{8.71}$$

$$EIv = C_1x^3/6 + C_2x^2/2 + C_3x + C_4. \tag{8.72}$$

Applying the boundary conditions, the conditions at $x = 0$ give $C_3 = C_4 = 0$, the shear force condition at $x = L$ gives $C_1 = P$ and the moment condition gives $C_2 = -PL$. The final result is

$$v(x) = \frac{Px^2}{2EI}\left(\frac{x}{3} - L\right). \tag{8.73}$$

**Remarks:**

(1) This is the same result that we obtained in Example 8.9.

**Example 8.11**

*Cantilever with two point loads.* Find the deflection $v(x)$ of the beam shown in Fig. 8.21.

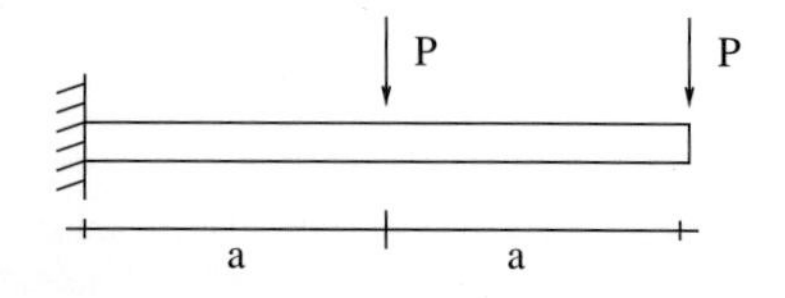


**Fig. 8.21** Cantilever with two point loads.

*Solution*

The distributed load for this beam is $q(x) = -P\delta(x - a)$. The boundary conditions are the same as in the last example: $v(0) = v'(0) = 0$, $EIv''(2a) = 0$, and $EIv'''(2a) = P$. We can now integrate eqn (8.66) four times.

$$EIv'''' = -P\delta(x - a) \tag{8.74}$$

$$EIv''' = -PH(x - a) + C_1 \tag{8.75}$$

$$EIv'' = -P\langle x - a\rangle + C_1x + C_2 \tag{8.76}$$

$$EIv' = -P\langle x - a\rangle^2/2 + C_1x^2/2 + C_2x + C_3 \tag{8.77}$$

$$EIv = -P\langle x - a\rangle^3/6 + C_1x^3/6 + C_2x^2/2 + C_3x + C_4. \tag{8.78}$$

Application of the conditions at $x = 0$, gives $C_3 = C_4 = 0$. The shear force condition at $x = 2a$, gives $C_1 = 2P$. The moment condition at $x = L$ gives $C_2 = -3Pa$. Substituting back in, we have

$$v(x) = \frac{1}{EI}\left[-\frac{P}{6}\langle x - a\rangle^3 + \frac{2Px^3}{6} - \frac{3Pax^2}{2}\right]. \tag{8.79}$$

**Example 8.12**

*Displacement based beam selection.* Consider a simply supported beam (Fig. 8.22) that is to carry a uniform load of $k$ (force per unit length) with a maximum allowed deflection of $\Delta_{\text{max}}$. Determine a formula for the required moment of inertia of the cross-section.

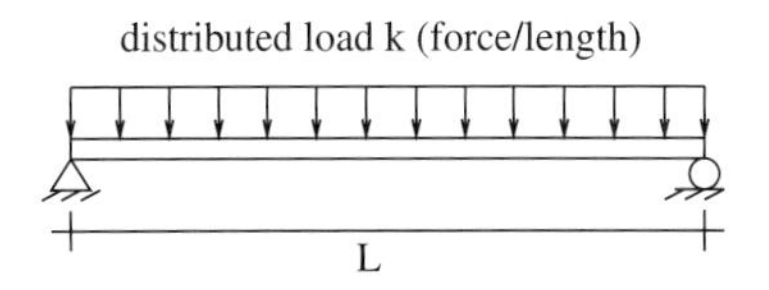


**Fig. 8.22** Simply supported beam with a uniform distributed load.

*Solution*
The distributed load for this beam is $q(x) = -k$. The boundary conditions for the pin supports are zero displacement and zero moment. Thus $v(0) = EIv''(0) = 0$ and $v(L) = EIv''(L) = 0$. Starting from eqn (8.66):

$$EIv'''' = -k \tag{8.80}$$

$$EIv''' = -kx + C_1 \tag{8.81}$$

$$EIv'' = -kx^2/2 + C_1x + C_2 \tag{8.82}$$

$$EIv' = -kx^3/6 + C_1x^2/2 + C_2x + C_3 \tag{8.83}$$

$$EIv = -kx^4/24 + C_1x^3/6 + C_2x^2/2 + C_3x + C_4. \tag{8.84}$$

The boundary conditions can be applied to find the values of the constants. The net result is

$$v(x) = \frac{kL^4}{24EI}\left[-\left(\frac{x}{L}\right)^4 + 2\left(\frac{x}{L}\right)^3 - \frac{x}{L}\right]. \tag{8.85}$$

By inspection (or by computing where the rotation $\theta$ is zero) we note that the deflection will be maximum at the center $x = L/2$. Evaluating, one finds

$$v(L/2) = -\frac{5kL^4}{384EI}. \tag{8.86}$$

The requirement is gives $\Delta_{\text{max}} \geq 5kL^4/(384EI)$. Thus we find

$$I \geq \frac{5kL^4}{384E\Delta_{\text{max}}}. \tag{8.87}$$

**Remarks:**

(1) To give a numerical sense to this result, consider a 40-ft span carrying a load $k = 100$ lbf/in, a displacement requirement $\Delta_{\text{max}} = 1$ in, and a steel material with modulus $E = 30 \times 10^6$ psi. Utilizing our result one finds the requirement of $I \geq 2304$ in$^4$. If one looks in a table of standard steel I-beams, then one finds that this requirement can be fulfilled by choosing a beam with a depth of around 20 inches.

---

**Example 8.13**

*Boundary conditions and distributed loads.* Figure 8.23 shows a number of beams with a variety of loadings and boundary conditions. For each beam state the distributed load function and the boundary conditions.

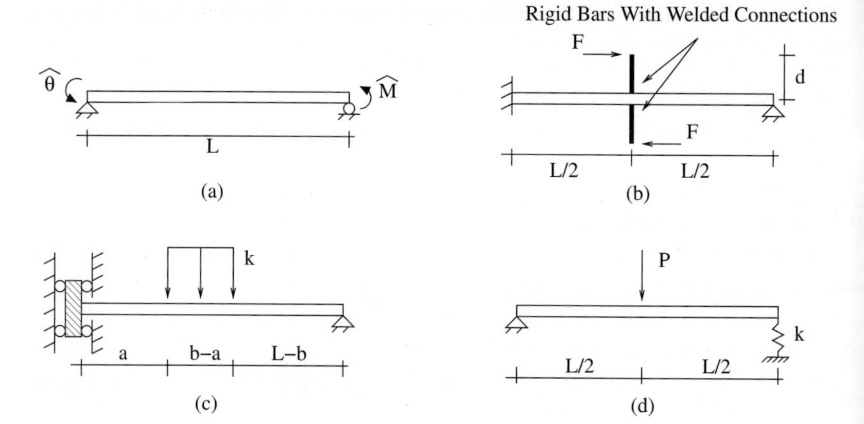


**Fig. 8.23** Beams for Example 8.13.

*Solution*

For beam (a) there is no distributed load, so, $q(x) = 0$. The pins prevent vertical displacement and allow free rotation. Thus one has $v(0) = 0$, $\theta(0) = v'(0) = \hat{\theta}$, $v(L) = 0$, and $M(L) = EIv''(L) = \hat{M}$.

For beam (b) the applied force delivers a point moment to the center of the beam. Thus $q(x) = 2Fd\delta'(x - L/2)$. The built-in end requires boundary conditions of $v(0) = v'(0) = 0$. The pinned end gives boundary conditions $v(L) = 0$ and $M(L) = EIv''(L) = 0$.

For beam (c) there is a distributed load that can be composed by superposing two step loadings as $q(x) = -kH(x - a) + kH(x - b)$. The pin end of the beam has zero deflection and moment, so $v(L) = 0$ and $M(0) = EIv''(L) = 0$. The left end of the beam is a slider. It travels freely in a slot, but cannot rotate. This implies $V(0) = -EIv'''(0) = 0$ and $\theta(0) = v'(0) = 0$.

For beam (d) the point load gives $q(x) = -P\delta(x - L/2)$ and the pin connection at the left gives $v(0) = 0$ and $M(0) = EIv''(0) = 0$. At the right the beam is supported by a linear spring through a pin. Thus the moment at the end is zero, $M(L) = EIv''(L) = 0$. We can use the behavior of the spring to determine the final boundary condition. The deflection of the end is related to the force at the end through the spring constant. Thus, $EIv'''(L) = kv(L)$.

## 8.5 Symmetric multi-axis bending

In the developments to this point we have restricted our attention to bending about the $z$-axis. In a more general setting one could have loads applied in a fashion to cause bending about both the $z$- and $y$-axes.

### 8.5.1 Symmetric multi-axis bending: Kinematics

If one observes the motion of a slender beam under multi-axis bending, then one sees a motion nearly identical to that observed for single-axis bending. In particular, one observes that a planar cross-section of a beam remains planar after deformation. The section will be seen to displace in both the $y$- and $z$-directions and to rotate about some axis. This leads naturally to the observation that the bending strain is a linear function of both $y$ and $z$. The general form for such a function is $\varepsilon_{xx} = A + By + Cz$. In the case of bending about just the $z$-axis, the case we have treated to this point, $C = 0$, $B = -\kappa_z$, and we selected the origin of the coordinate system such that $A = 0$. Note we have placed a subscript on the curvature to indicate that it represents curvature "about the $z$-axis". In the general case, using the additional assumption that normals remain normal one can show that $B = -\kappa_z(x) = -v''(x)$ and $C = \kappa_y(x) = -w''(x)$, where $w(x)$ is the deflection of the cross-section in the $z$ direction. Note that there has been a switch of sign; i.e., $\kappa_y \neq w''(x)$. We have done this so that later, positive bending moments (about the $y$-axis) produce positive curvature "about the $y$-axis". In what follows we will not necessarily assume that $A = 0$. We will consider a fixed placement of the coordinate axes and determine $A$ from axial equilibrium. Note that $A$ represents the normal strain at the origin of the coordinate system so we will call it $\varepsilon_o$. Thus the main kinematic assumption will be

$$\varepsilon_{xx} = \varepsilon_o - \kappa_z y + \kappa_y z. \tag{8.88}$$

### 8.5.2 Symmetric multi-axis bending: Equilibrium

Following a differential element equilibrium construction similar to what we followed for single axis bending it is not difficult to show that force balance in the three coordinate directions gives:

$$\frac{dR}{dx} + b = 0 \tag{8.89}$$

$$\frac{dV_y}{dx} + q_y = 0 \tag{8.90}$$

$$\frac{dV_z}{dx} + q_z = 0, \tag{8.91}$$

where we now permit distributed axial loads; see Fig. 8.24. If there are no axial loads then we recover the case of $R(x) = 0$ as we had in the prior sections. Subscripts have been added to indicate the directions of the internal forces, moments, and loads. If we take moment equilibrium about the $y$ and $z$ axes one finds

$$\frac{dM_z}{dx} + V_y = 0 \tag{8.92}$$

$$\frac{dM_y}{dx} - V_z = 0. \tag{8.93}$$

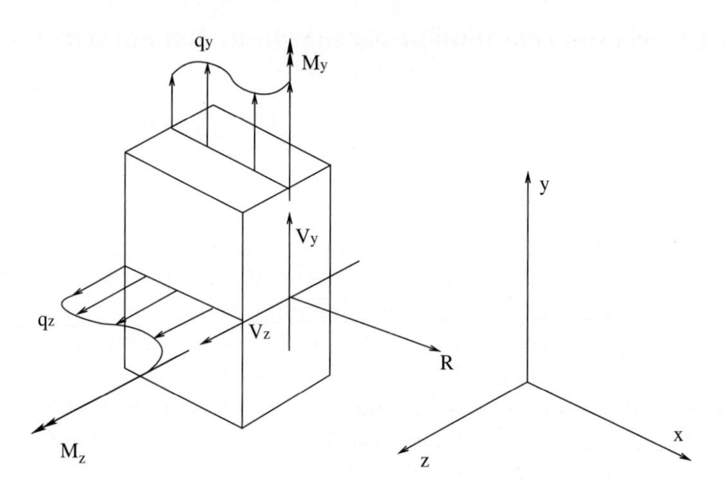


**Fig. 8.24** Differential element for multi-axial bending.

The internal forces and moments can be related to the stresses on the cross-section as

$$R = \int_A \sigma_{xx}\, dA \tag{8.94}$$

$$V_y = \int_A \sigma_{xy}\, dA \tag{8.95}$$

$$V_z = \int_A \sigma_{xz}\, dA. \tag{8.96}$$

It is likewise easy to show that the internal moments are given by

$$M_z = \int_A -y\sigma_{xx}\, dA \tag{8.97}$$

$$M_y = \int_A z\sigma_{xx}\, dA. \tag{8.98}$$

### 8.5.3 Symmetric multi-axis bending: Elastic

In the case of an elastic beam the necessary constitutive relation is $\sigma_{xx} = E\varepsilon_{xx}$. In order to gain a better understanding of multi-axis bending, let us look at an example problem.

---

**Example 8.14**

*Moment-curvature and stresses.* Consider a rectangular cross-section with width $b$ and depth $h$ under the action of an axial load $P$ and bending moments $M_z$ and $M_y$. Find a relation between the applied loads and the bending curvatures $\kappa_z$ and $\kappa_y$. Determine the stress distribution on the cross-section. Assume that the load $P$ is applied at the origin of the coordinate system which is located at the cross-section centroid. Assume linear elastic homogeneous material properties.

*Solution*
We can begin by replacing the bending stress in the resultant definitions by its expression in terms of bending strain. Further, we can use the kinematic assumption to yield:

$$P = \int_A E(\varepsilon_o - \kappa_z y + \kappa_y z)\, dA \tag{8.99}$$

$$= AE\varepsilon_o. \tag{8.100}$$

The integrals for the second and third terms drop out, since the integral of an odd function over an even interval will always be zero. If we proceed in the same manner with the moments we find:

$$M_z = \int_A -yE(\varepsilon_o - \kappa_z y + \kappa_y z)\, dA \tag{8.101}$$

$$= \int_A Ey^2 \kappa_z\, dA \tag{8.102}$$

$$= EI_z \kappa_z. \tag{8.103}$$

Note that the first and third integrals drop out again because the integrands are odd and the intervals of integration are even. The moment about the $y$-axis yields

$$M_y = \int_A zE(\varepsilon_o - \kappa_z y + \kappa_y z)\, dA \tag{8.104}$$

$$= \int_A Ez^2 \kappa_y\, dA \tag{8.105}$$

$$= EI_y \kappa_y, \tag{8.106}$$

where $I_y = \int_A z^2\, dA$ is the moment of inertia about the $y$-axis.

With these results we can solve for the coefficients in the strain distribution and insert into the constitutive relation to arrive at

$$\sigma_{xx} = \frac{P}{A} - \frac{M_z y}{I_z} + \frac{M_y z}{I_y}. \tag{8.107}$$

**Remarks:**

(1) Notice that the resulting stress state is a summation of the stresses caused by each individual internal force/moment on the cross-section. This is an example of the superposition principle; i.e. in linear problems solutions add.

(2) The symmetry of the cross-section and the placement of the coordinate origin played a crucial role in simplifying the analysis of this problem. By proper placement of the coordinate system origin we were able to eliminate a number of terms in the expressions.

(3) In a more general situation – for example, a non-symmetric cross-section – terms such as $Q_z = \int_A y\,dA$ and $I_{yz} = \int_A yz\,dA$ etc. do not necessarily equal zero. $Q_z$ is known as the first moment of the area about the $z$-axis and there is similarly a $Q_y$. $I_{yz} = I_{zy}$ is known as the cross-product of inertia. If the coordinate system is placed at the centroid of a cross-section, then one always has that $Q_z = Q_y = 0$. $I_{yz}$, however, may still be non-zero. This situation is often known as *skew-bending* and it couples moments about one axis to bending about another axis; for example, $M_z$ produces a $\kappa_y$. More specifically, we have that

$$\begin{pmatrix} M_y \\ M_z \end{pmatrix} = \begin{bmatrix} EI_y & -EI_{yz} \\ -EI_{yz} & EI_z \end{bmatrix} \begin{pmatrix} \kappa_y \\ \kappa_z \end{pmatrix}, \tag{8.108}$$

or the other way around, that,

$$\begin{pmatrix} \kappa_y \\ \kappa_z \end{pmatrix} = \frac{1}{E(I_y I_z - I_{yz}^2)} \begin{bmatrix} I_z & I_{yz} \\ I_{yz} & I_y \end{bmatrix} \begin{pmatrix} M_y \\ M_z \end{pmatrix}. \tag{8.109}$$

## 8.6 Shear stresses

In the preceding sections we have analyzed a number of different features of beam bending: bending stresses, bending strains, and bending deflections. Throughout this entire development we have ignored the issue of shear stresses in bending (which is sometimes referred to as direct shear). In this section we will examine the issue of shear stresses in beams. For the sake of simplicity, we will restrict our attention to linear elastic homogeneous beams.

To begin, let us recall our basic kinematic assumptions: plane sections remain plane and normals remain normal. The last of these two assumptions led us in Section 8.1 to the conclusion that $\gamma_{xy} = 0$. If we are assuming linear elastic behavior, then we are then naturally led to the conclusion that $\sigma_{xy} = G\gamma_{xy}$ will also be zero. This, however, seems to be in direct contradiction to the resultant relation $V = \int_A \sigma_{xy}\,dA$. If the shear stresses are zero, then so must the shear forces. The apparent paradox is circumvented by noting that if we assume that normals always remain normal, then we are in a way assuming that the material is infinitely stiff in shear; i.e. $G = \infty$. Thus the theory with which we are working actually gives $\sigma_{xy} = \infty \cdot 0$, with the product being finite. The Bernoulli–Euler theory is what is known as a non-shear deformable theory of beams. The reality is, of course, that there is a small amount of shear strain (and shear stress) in the beam. As a first-order estimate we can of course simply note that the average shear stress $\sigma_{xy,\text{avg}} = V/A$, and thus $\gamma_{xy,\text{avg}} = V/GA$. These estimates, however, can be substantially improved upon, and we will do so using an equilibrium argument.

### 8.6.1 Equilibrium construction for shear stresses

Let us begin with a qualitative discussion. Consider an I-beam that is loaded in a fashion to produce a non-zero internal shear force. If we consider a differential section cut from such a beam we will find that the stress distribution on the two sides of the section will differ as is shown in Fig. 8.25(a). The reason for this is that non-zero internal shear force leads to a moment variation (in $x$), and the stresses are directly proportional to the bending moments. If we consider making a horizontal section cut (see Fig. 8.25(b)) through our differential element at the neutral axis, then we clearly see that there will be a need for a shear stress $\sigma_{yx} \neq 0$ in order to balance the unequal stress distributions on either side of the differential element. This can be made more evident by subtracting an equal distribution of stresses from both sides, as shown in Fig. 8.25(c).

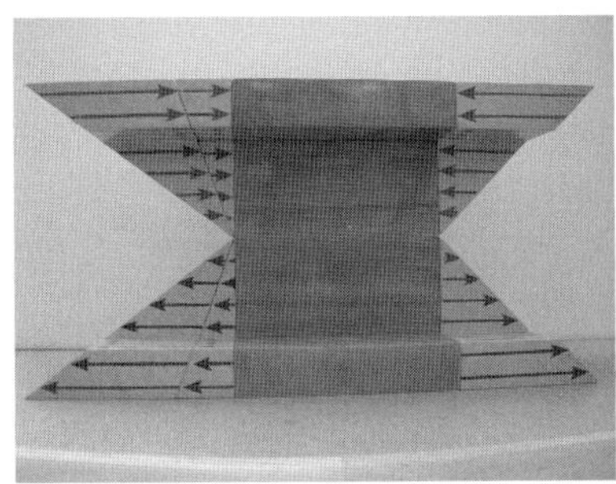

(a) Differential section of beam.

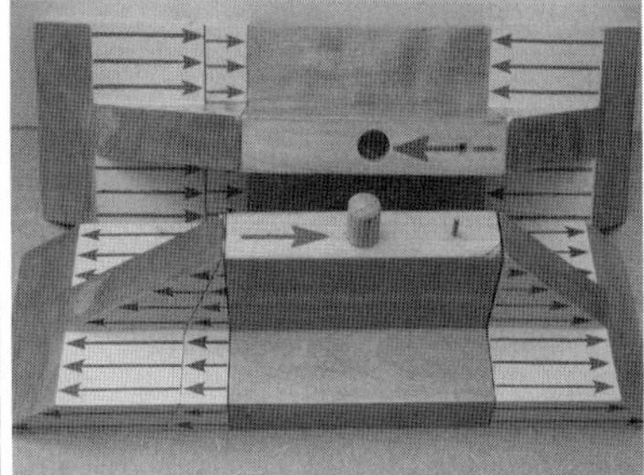

(b) Horizontal slice.

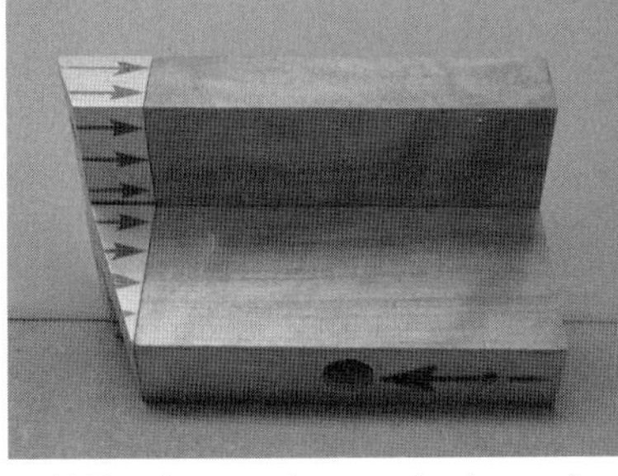

(c) Equal stresses(moments) subtracted.

(d) Stresses on orthogonal faces.

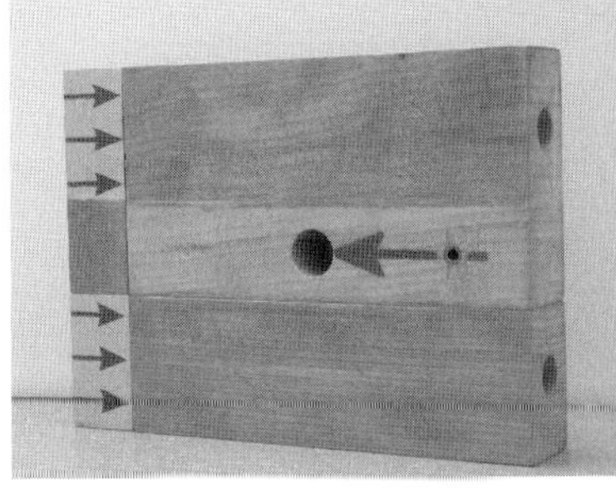

(e) Stresses depend on horizontal slice.

(f) $\sigma_{zx}$ is also possible.

**Fig. 8.25** Shear stress equilibrium construction.

If there is a shear stress $\sigma_{yx}$ then we know by moment equilibrium in terms of stresses that there will be an equal magnitude shear stress $\sigma_{xy}$; i.e. $\sigma_{xy} = \sigma_{yx}$ pointwise in a solid, always, as shown in Fig. 8.25(d). Note that if one changes the location of the horizontal slice, to say just under the flange, then the amount of shear stress will necessarily have to change as the amount of differential bending stress to be balanced will change; see Fig. 8.25(e). This implies that the shear stresses will be functions of $y$. The beam shown in Fig. 8.25 is an I-beam, and in such beams there is also the possibility for non-zero stresses in the $z$-direction. The existence of these stresses can be seen by adjusting our horizontal section cut to intersect the beam flange, as shown in Fig. 8.25(f).

Given this qualitative discussion, one can now formulate the expression for the shear stress $\sigma_{xy}$ in a beam. Assume a horizontal section cut through a differential element at a distance $y_1$ from the neutral axis, as shown in Fig. 8.26. The total force on the horizontal cut (for small $\Delta x$) will be $\sigma_{yx} b \Delta x$, where $b$ is the width of the beam at the location of the cut. This force arises in response to the difference of the forces generated by the bending stresses. If we define $A(y_1)$ to be the cross-sectional area above $y_1$, then by force equilibrium in the $x$-direction we have in the limit as $\Delta x \to 0$ that

$$\sigma_{yx} b \Delta x = -\int_{A(y_1)} \frac{M(x+\Delta x)y}{I}\, dA + \int_{A(y_1)} \frac{M(x)y}{I}\, dA \tag{8.110}$$

$$\sigma_{yx} b = -\int_{A(y_1)} \frac{M(x+\Delta x) - M(x)}{\Delta x} \frac{y}{I}\, dA \tag{8.111}$$

$$\sigma_{yx} b = \frac{V}{I} Q(y_1), \tag{8.112}$$

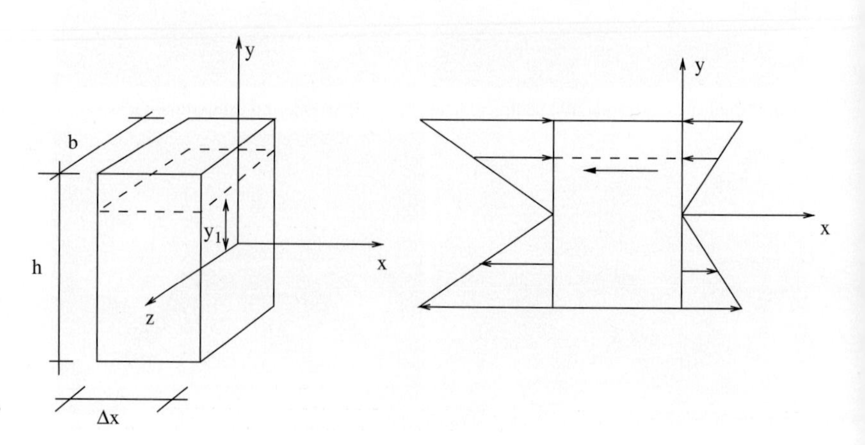


**Fig. 8.26** Section from a beam with a rectangular cross-section.

where $Q(y_1) = \int_{A(y_1)} y\,dA$ is the first moment of the area above $y = y_1$. Thus,

$$\sigma_{xy}(x, y) = \frac{V(x)Q(y)}{Ib}. \qquad (8.113)$$

**Remarks:**

(1) The product $\sigma_{xy}b$ is known as the shear flow, and is typically denoted by the letter $q$. Shear flow represents the force per unit length along the horizontal cut.

### Example 8.15

*Glued T-beam.* Consider the T-beam shown in Fig. 8.27. The beam is constructed by gluing two boards together. If the maximum allowable shear stress in the glue is $\tau_{\max}$, how much force can the beam support. Assume that $I$ and $y_{\text{glue}}$ are given, and that the coordinate axes have been aligned with the neutral axis.

*Solution*

The internal shear force in the beam $V(x) = P$. The shear stress in the glue is thus

$$\sigma_{yx}(y_{\text{glue}}) = \frac{PQ(y_{\text{glue}})}{bI}. \qquad (8.114)$$

The first moment of $A(y_{\text{glue}})$ is

$$Q(y_{\text{glue}}) = tw(y_{\text{glue}} + t/2) \qquad (8.115)$$

This gives the requirement that

$$P < \frac{\tau_{\max} bI}{tw(y_{\text{glue}} + t/2)}. \qquad (8.116)$$

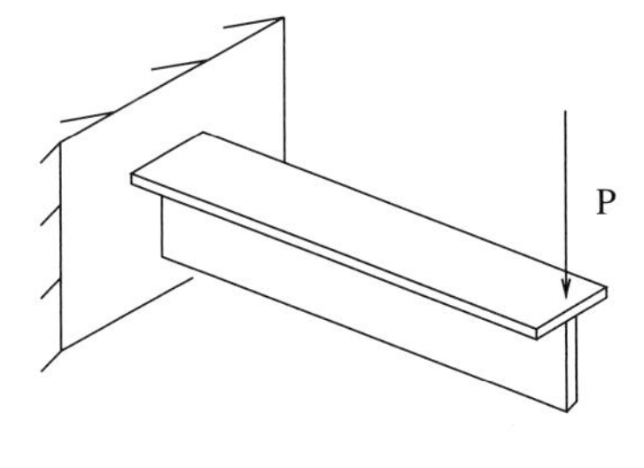


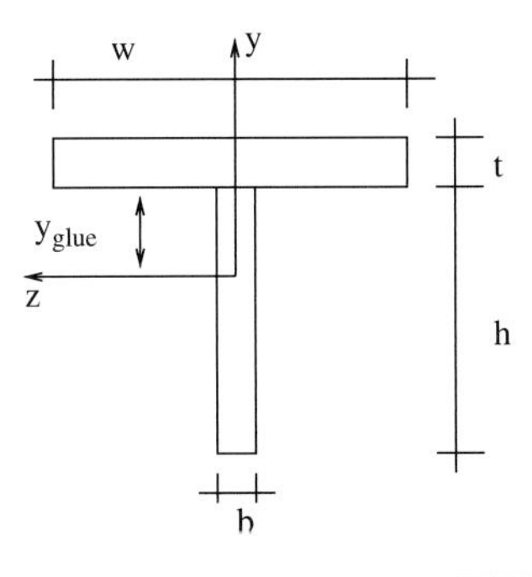


**Fig. 8.27** Cantilevered T-beam with glue joint.

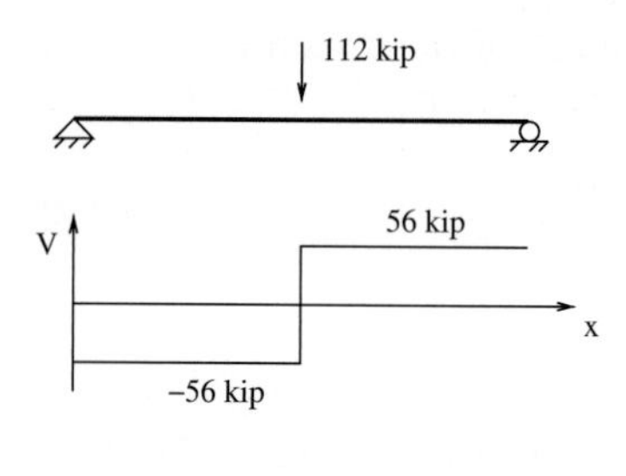


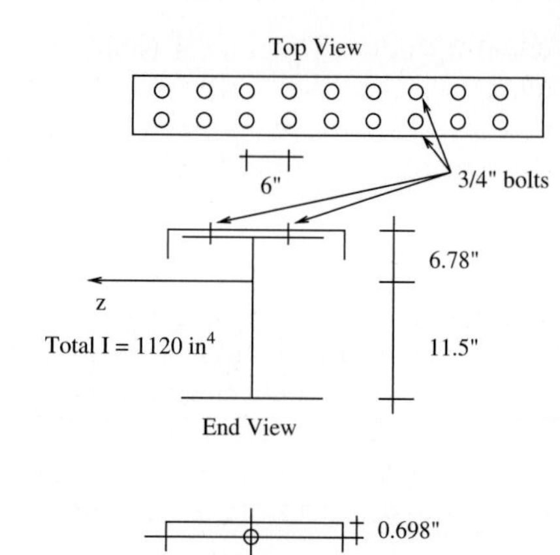


**Fig. 8.28** Built-up beam made from a channel section and I-beam.

### Example 8.16

*Shear stress distribution.* Compute the shear stress distribution in a beam with rectangular cross-section – height $h$ and width $b$.

*Solution*
From eqn (8.113) one has

$$\sigma_{xy}(y_1) = \frac{VQ(y_1)}{Ib} \tag{8.117}$$

$$= \frac{V}{Ib}\int_{A(y_1)} y\, dA \tag{8.118}$$

$$= \frac{V}{Ib}(h/2 - y_1)b(h/2 + y_1)/2 \tag{8.119}$$

$$= \frac{V}{2I}\left((h/2)^2 - y_1^2\right) \tag{8.120}$$

**Remarks:**

(1) The shear stress distribution is parabolic on the cross-section.
(2) The maximum value, $3V/2A$ occurs at the neutral axis and is 50% higher that the average shear stress values, $V/A$.

### Example 8.17

*Built-up beam.* Consider a simply supported beam with a central point load. The beam is constructed by bolting together an I-beam and a channel section. The bolting is done pairwise with a 6-inch spacing. What is the shear stress in the bolts? See Fig. 8.28 for all dimensions and properties.

*Solution*
Make a horizontal section cut that cuts only through the bolts as shown in Fig. 8.29 (note it need not be flat!). The shear flow with respect to this cut will be

$$q = \frac{56 \times 6.09 \times (6.78 - 0.698)}{1120} = 1.85 \text{ kip/in}. \tag{8.121}$$

The force carried per bolt pair will be

$$1.85 \frac{\text{kip}}{\text{in}} \times 6 \frac{\text{in}}{\text{bolt pair}} = 11.10 \frac{\text{kip}}{\text{bolt pair}}. \tag{8.122}$$

This gives 5.55 kip per bolt. Dividing by the area of the bolt, one finds a shear stress in the bolts of 12.6 ksi.

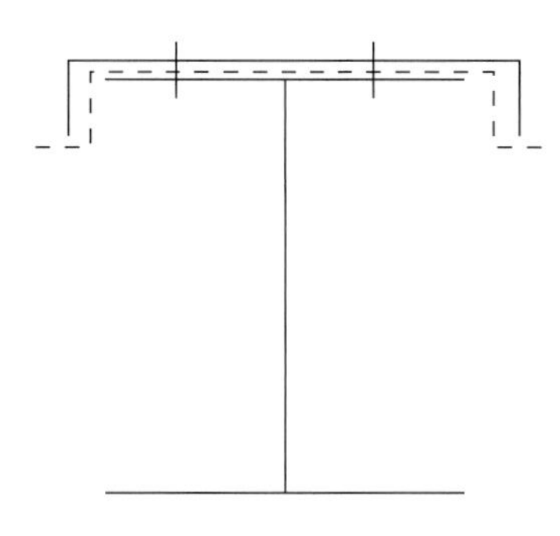

**Fig. 8.29** Horizontal cut just thorough the bolts.

### Example 8.18

$\sigma_{xz}$ *in flanges.* For a given shear force on the cross-section shown in Fig. 8.30, find the horizontal shear stresses in the upper flange.

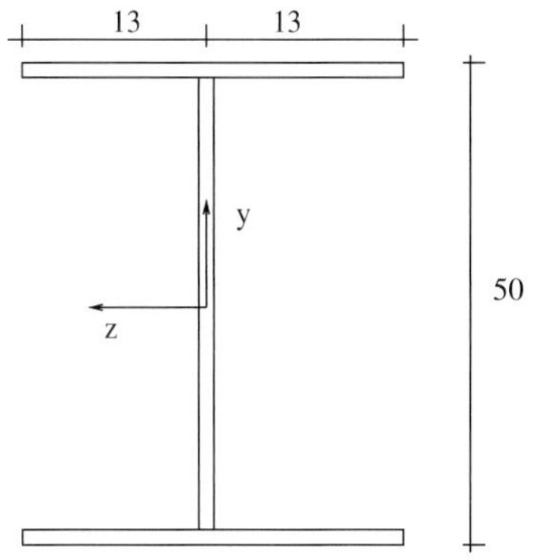

**Fig. 8.30** Thin-walled section. Flanges and webs have thickness $t = 1$.

*Solution*
At an arbitrary location $\alpha$ from the right end of the flange, make a section cut as shown in Fig. 8.31. The shear stress is then given via a modification to eqn (8.113) as

$$\sigma_{xz} = \frac{V}{It}(\alpha t)24.5 = \frac{V}{I}(\alpha)24.5. \tag{8.123}$$

If we do the same for the flange portion to the left we find

$$\sigma_{xz} = -\frac{V}{It}(\beta t)24.5 = -\frac{V}{I}(\beta)24.5. \tag{8.124}$$

**Remarks:**

(1) The horizontal shear stresses are seen to vary linearly in the flange.
(2) At the intersection with the web the value computed from the right and that from the left are not the same (in general). This is a deficiency in the theory, and highlights the fact that the stress state at the intersection of the flange and the web is rather complex. The solution computed is acceptable outside this region.
(3) If one were to compute the vertical shear stress distribution in the web, one would find a parabolic stress distribution just as we did with the rectangular cross-section beam. The only difference being that the value at the top of the web will not be zero but rather $(V/I) \times 24.5 \times 26$.
(4) The horizontal shear stresses in the lower flange are the same as in the upper flange, except that they are flipped in sign.

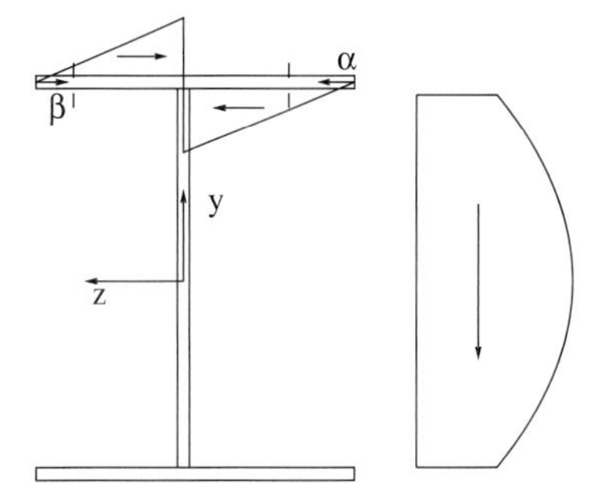


**Fig. 8.31** Shear stress distribution in a thin-walled section. Linear in the flanges and parabolic in the web.

### 8.6.2 Energy methods: Shear deformation of beams

If there are non-zero shear stresses then there are in reality non-zero shear strains. It is natural to ask if these shears have any impact upon the beam deflection computations we performed in Section 8.4. One way of approaching this question is to consider an energy method.

As a concrete example, consider a linear elastic cantilevered beam of length $L$ with a tip load $P$. Assume that the cross-section of the beam is rectangular. In this case, the work input, $W_{\text{in}} = \frac{1}{2}P\Delta$, where $\Delta$ is the tip deflection. The work input is stored as strain energy in the beam. If we account for bending stresses and direct shear stresses we find that the strain energy density $w = \frac{1}{2}\sigma_{xx}\varepsilon_{xx} + \frac{1}{2}\sigma_{xy}\gamma_{xy} = \frac{1}{2}\sigma_{xx}^2/E + \frac{1}{2}\sigma_{xy}^2/G$. Integrating over the volume of the beam gives the stored energy as:

$$W_{\text{stored}} = \int_0^L \int_A \frac{1}{2}\frac{\sigma_{xx}^2}{E}\, dAdx + \int_0^L \int_A \frac{1}{2}\frac{\sigma_{xy}^2}{G}\, dAdx. \tag{8.125}$$

Recalling that $\sigma_{xx} = -My/I = -Pxy/I$ and that $\sigma_{xy} = (P/2I)((h/2)^2 - y^2)$ one finds

$$W_{\text{stored}} = \frac{P^2L^3}{6EI} + \frac{3P^2L}{5AG}, \tag{8.126}$$

where the first term is from the bending stresses and the second from the shearing stresses. Setting this equal to the work input we arrive at

$$\Delta = \frac{PL^3}{3EI}\left[1 + \frac{3E}{10G}\left(\frac{h}{L}\right)^2\right], \tag{8.127}$$

where we have factored out the bending stress contribution (i.e. our solution obtained by ignoring shear stresses). Thus the second term in the square brackets represents the error one makes when ignoring shear effects on deflections. A typical situation (for a metal) is that $E/G \approx 2.5$. In this case one can see that for a beam that has $L/h > 10$, the error in tip deflection is already less that 0.75%. This result justifies our neglect of shear effects on deflection for cases where beams are "slender".

## 8.7 Plastic bending

In ductile beams it is possible to apply loads beyond those that cause initial yielding. This is a situation similar to that which occurs in ductile torsion. Plastic failure proceeds in a gradual fashion with yield initiating on the outer fibers and then progressing with increasing load towards the center of the beam. Sometimes one defines bending failure as simply reaching the yield moment, $M_Y$, the moment at which yielding starts, and sometimes one defines it as reaching the ultimate moment, $M_u \neq M_Y$, the moment associated with complete yielding of the cross-section. Throughout the process of plastic deformation of a beam, one can reasonably assume that our kinematic assumptions hold. In particular,

$\varepsilon = -\kappa y$. In order to simplify the discussion throughout this section we will assume that the beam is loaded in pure bending, $V(x) = 0$, so that we can simply discuss the system behavior by considering a single cross-section of the beam and so that we can ignore the complexity of dealing with yield under the simultaneous action of normal and shear stresses. Further, we will always assume linear elastic–perfectly plastic behavior.

### 8.7.1 Limit cases

The first limit case mentioned above is the case of initiation of yield. This capacity of a beam cross-section can be easily computed using the elastic analysis developed to this point. Using our moment-stress relation gives

$$M_Y = \frac{\sigma_Y I}{y_{\max}}, \tag{8.128}$$

where $\sigma_Y$ is the uniaxial yield stress and $y_{\max}$ is the maximal distance of a material point on the cross-section from the neutral axis. In cases where the yield stress differs in compression and tension, then one must pay careful attention to the signs as the yield moment will differ for positive and negative moments.

The second limit case occurs with full yield of the cross-section. In such a case

$$\sigma = \begin{cases} -\sigma_Y & y > 0 \\ +\sigma_Y & y < 0 \end{cases}, \tag{8.129}$$

where for simplicity we have assumed the same yield stress in tension and compression. The ultimate moment is then given by integrating this stress distribution (against $y$) over the cross-section. An added difficulty that arises in plastic bending of beams is that one must take care of the position of the neutral axis ($y = 0$). In particular it should be noted that in plastic bending the neutral axis (where the bending strains are zero) is not necessarily co-located with the neutral axis of the elastic case. In fact, for some beams the neutral axis moves continuously as the loads are increased (in the plastic range). To determine the location of the neutral axis we apply our basic technique, which says that at all times, $\int_A \sigma \, dA = 0$ (assuming no axial loads); i.e. the total axial tensile resultant must be equal to the total axial compressive resultant.

**Example 8.19**

*Limit moments: Rectangular beam.* Compute the yield and ultimate moments for a rectangular cross-section with width $b$ and height $h$. Assume an homogeneous material.

*Solution*
The yield moment is given by

$$M_Y = \frac{\sigma_Y bh^2/12}{h/2} = \sigma_Y \frac{bh^2}{6}. \tag{8.130}$$

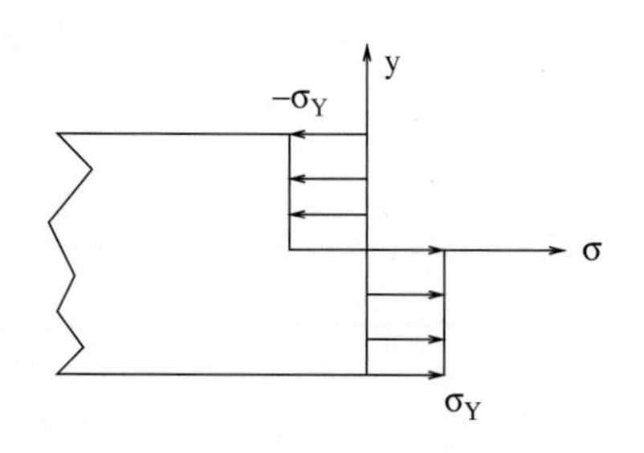


**Fig. 8.32** Ultimate stress distribution rectangular cross-section.

This is the moment at which yielding first starts. For the ultimate moment, one assumes that material points are either at the positive yield stress or at the negative yield stress. The transition location (the neutral axis) is found by requiring that $\int_A \sigma \, dA = 0$. In the homogeneous rectangular cross-section case, this simply means that the neutral axis is located in the middle (i.e. it does not move); see Fig. 8.32. The ultimate moment is then computed as

$$\begin{aligned} M_u = \int_A -y\sigma \, dA &= \int_{-b/2}^{b/2} \int_0^{h/2} y\sigma_Y \, dydz \\ &+ \int_{-b/2}^{b/2} \int_{-h/2}^{0} -y\sigma_Y \, dydz \qquad (8.131) \\ &= \frac{bh}{2}\sigma_Y \frac{h}{4} - \frac{bh}{2}\sigma_Y \frac{-h}{4} \qquad (8.132) \\ &= \frac{bh^2}{4}\sigma_Y = \frac{3}{2}M_Y. \qquad (8.133) \end{aligned}$$

**Remarks:**

(1) The ultimate moment capacity of a rectangular cross-section is 50% higher than the initial yield moment.

(2) The expressions for the yield moment and the ultimate moment are unique to the shape of the cross-section, as is the ratio between the two.

(3) For doubly symmetric cross-sections the neutral axis does not move as long as the yield stress in tension and compression are the same.

### Example 8.20

*Ultimate moment of a T-beam.* Consider a T-beam as shown in Fig. 8.33 (left). Find the ultimate moment.

*Solution*
If the tensile (axial) force on the cross-section is to equal the compressive (axial) force, then the area in compression must equal the area in tension. This will occur if the neutral axis is located at the intersection of the web and flange. The stress distribution is then as shown in Fig. 8.33 (right). The ultimate moment is then

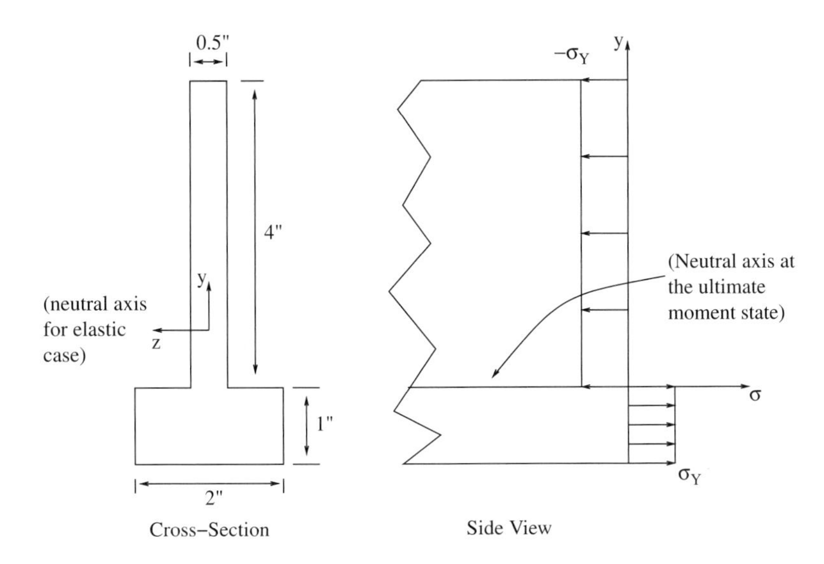


**Fig. 8.33** T-beam (left) with the ultimate moment stress state (right).

$$M_u \text{ (lbf} - \text{in)} = \int_A -y\sigma\, dA \tag{8.134}$$

$$= (2 \times 1) \times \sigma_Y \times \frac{1}{2} + (4 \times \frac{1}{2}) \times \sigma_Y \times 2 \tag{8.135}$$

$$= 5 \text{ (in}^3)\sigma_Y \text{ (psi)}. \tag{8.136}$$

**Remarks:**

(1) Notice that the neutral axis at the start of yield will be located 1.75 inches from the bottom of the cross-section. As yield progresses to the ultimate state the neutral axis translates downwards until it reaches 1 inch from the bottom. At all times the strain distribution remains linear and of the form $\varepsilon = -\kappa y$, where $y$ is always measured from the present location of the neutral axis.

### 8.7.2 Bending at and beyond yield: Rectangular cross-section

To understand the behavior of a plastically bent beam for loads between the yield moment and the ultimate moment, let as focus upon the behavior of the doubly symmetric rectangular cross-section with equal yield stresses in tension and compression. With these assumptions the neutral axis will not shift, and this greatly simplifies the analysis. To begin, recall that when a beam is elastic the stress and strain distributions are linear, as shown in Figs. 8.5 and 8.14. The connection between the strain distribution and the stress distribution is through the stress–strain relation, $\sigma = E\varepsilon$. This is valid as long as the stresses stay below the yield stress, $\sigma_Y$, or equivalently the strains stay below the yield strain, $\varepsilon_Y = \sigma_Y/E$. If we increase the curvature in the beam beyond the elastic limit, some things change but some do not. The bending strains stay linear on the cross-section because we continue to assume that our kinematic assumption holds. However, the bending stresses are no longer linear on the cross-section because $\sigma \neq E\varepsilon$ at all points on the cross-section. As we progressively increase the curvature the bending strain begins to exceed the yield strain on the outer fibers of the beam. As the curvature is further increased the vertical location where the bending strain is equal to the yield strain moves inwards (from the top and bottom), with the material inside these limits still elastic and the material outside these limits plastically deformed. This vertical location is known as the elastic-plastic interface. Figure 8.34 shows three different cases, each with an increasing amount of curvature. Case 1 is elastic, and cases 2 and 3 go beyond yield.

The relation between an applied curvature and the location of the elastic–plastic interface is easy to determine since we continue to assume the validity of our fundamental kinematic assumption, $\varepsilon = -y\kappa$. At the interface we know that $\varepsilon = \varepsilon_Y$; thus

$$\kappa = \frac{\varepsilon_Y}{y_p}, \tag{8.137}$$

where $y_p > 0$ is the symbol we use for the interface location. For a given yield stress, we can also write

$$\kappa = \frac{\sigma_Y}{Ey_p}. \tag{8.138}$$

When $y_p = y_{\max}$, i.e. the initiation of yield, we denote the applied curvature as $\kappa_Y = \varepsilon_Y/y_{\max} = \sigma_Y/(Ey_{\max})$ – the curvature at initial yield. Note that the $M_Y = EI\kappa_Y$.

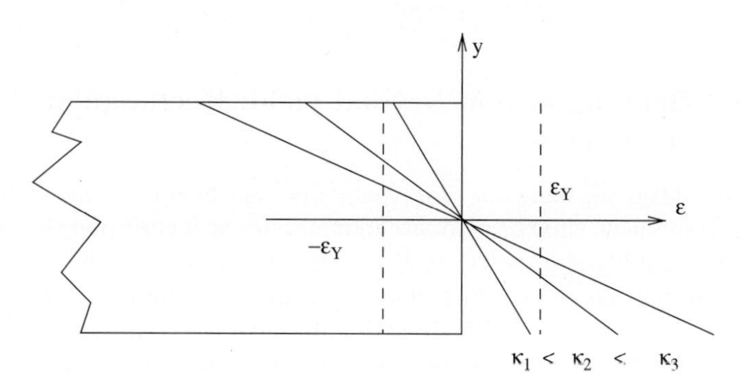


**Fig. 8.34** Bending strain distribution for three progressively increasing curvatures: $\kappa_1 < \kappa_2 < \kappa_3$. Cases 2 and 3 represent curvatures beyond yield.

### 8.7.3 Stresses beyond yield: Rectangular cross-section

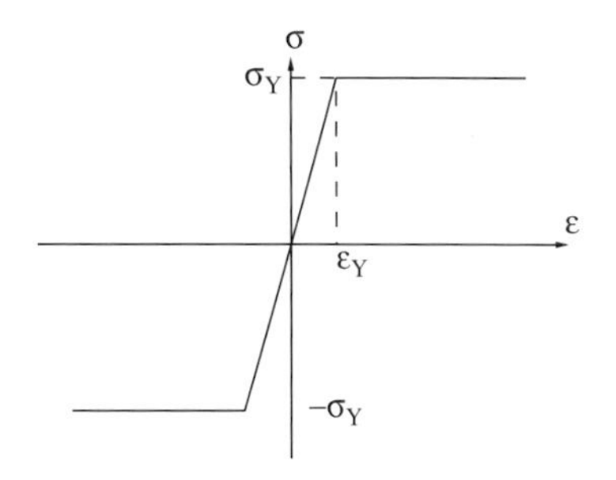


**Fig. 8.35** Elastic–perfectly plastic response curve.

If we assume that the constitutive relation is elastic–perfectly plastic (as shown in Fig. 8.35), then we can determine the stress distribution for a given curvature beyond yield. Note that given a curvature, the bending strains are known, $\varepsilon = -y\kappa$. Thus for each vertical location $y$ we can "look up" the corresponding stress from the stress–strain diagram, which gives

$$\sigma = \begin{cases} -\sigma_Y & y \geq y_p \\ -\sigma_Y \frac{y}{y_p} & |y| < y_p \\ \sigma_Y & y \leq -y_p \,, \end{cases} \tag{8.139}$$

where $y_p = \varepsilon_Y/\kappa$. For the strain distributions shown in Fig. 8.34, the resulting stress distributions are given in Fig. 8.36.

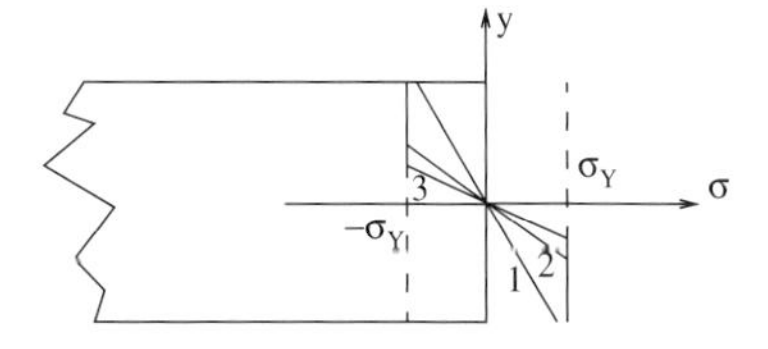


**Fig. 8.36** Stress response for an elastic–perfectly plastic material for the curvatures in Fig. 8.34.

### 8.7.4 Moment beyond yield: Rectangular cross-section

To determine the moment beyond yield for a given curvature, we need to apply the general relation between moment and bending stresses, eqn (8.15). Note that eqn (8.15) holds independent of constitutive response. Inserting eqn (8.139) into eqn (8.15), gives

$$M = \int_A -y\sigma \, dA \tag{8.140}$$

$$= -b\left\{\int_{-y_p}^{y_p} y\frac{-\sigma_Y y}{y_p}\, dy + \int_{y_p}^{h/2} y(-\sigma_Y)\, dy + \int_{-h/2}^{-y_p} y\sigma_Y \, dy\right\} \tag{8.141}$$

$$= b\left\{\frac{2\sigma_Y y_p^2}{3} + \sigma_Y\left(\frac{h^2}{4} - y_p^2\right)\right\} \tag{8.142}$$

$$= \sigma_Y \frac{bh^2}{4} - \frac{b\sigma_Y y_p^2}{3} \,. \tag{8.143}$$

This is also conveniently written as:

$$M = M_u\left[1 - \frac{1}{3}\left(\frac{y_p}{h/2}\right)^2\right], \tag{8.144}$$

or

$$M = M_u\left[1 - \frac{1}{3}\left(\frac{\kappa_Y}{\kappa}\right)^2\right] \tag{8.145}$$

where $M_u = \sigma_Y bh^2/4$.

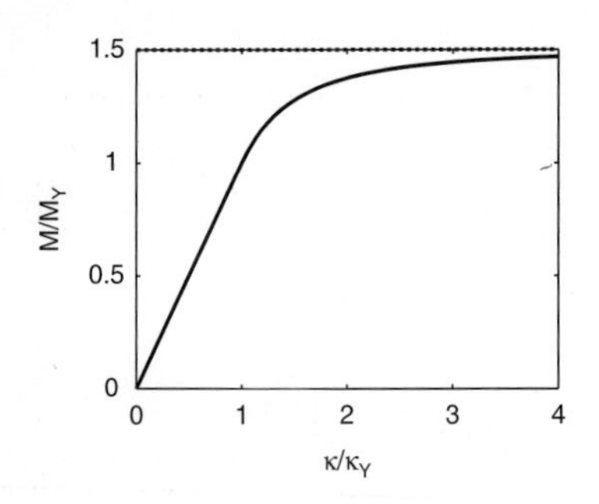


**Fig. 8.37** Moment versus curvature for an homogeneous rectangular cross-section made from an elastic–perfectly plastic material.

**Remarks:**

(1) A plot of moment versus applied curvature is shown in Fig. 8.37. Before yield the response is linear. After yield the response is non-linear due to the plastic yielding. Note that the ultimate moment can never be reached, as it represents an asymptote that is only approached in the limit as the curvature goes to infinity. This situation is in complete correspondence to what occurs in plastic torsion; see Fig. 7.28.

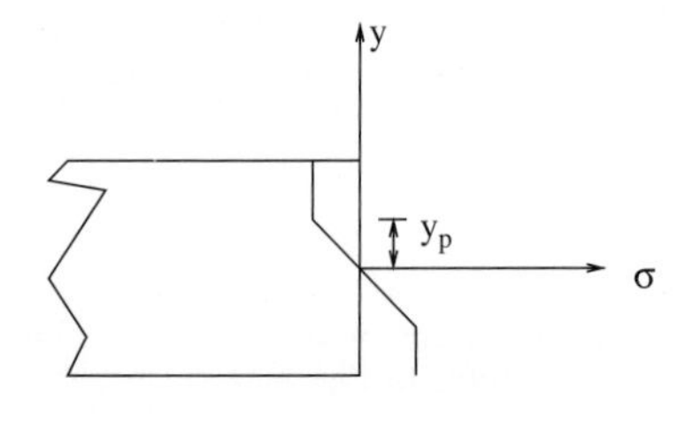


**Fig. 8.38** Initial stress field.

### 8.7.5 Unloading after yield: Rectangular cross-section

In the preceding sub-sections we have considered the case of yielding a beam in bending through the progressive increase in applied curvature. Let us now consider what happens when one releases the applied loads. The situation is quite similar to what we had for the analogous problem in torsion. Let us assume that the beam has been bent initially so that the elastic–plastic interface is located at $y_p$. The stress state will be as shown in Fig. 8.38.

The initially applied curvature, $\kappa_i$, will change upon release to a final curvature, $\kappa_f$, which is not necessarily zero.

$$\kappa_f - \kappa_i = \Delta\kappa. \tag{8.146}$$

This will induce a change in bending strain

$$\Delta\varepsilon = -y\Delta\kappa. \tag{8.147}$$

Following the same argument as in the torsion case, we note that the change in stress will be such that

$$\sigma_f = \sigma_i + E\Delta\varepsilon = \sigma_i - yE\Delta\kappa. \tag{8.148}$$

Since we know that the bar has been released, we know that the final moment is zero, $M_f = 0$. Using eqn (8.15), this tells us

$$0 = M_f = \int_A -y\sigma_f \, dA = M_i + EI\Delta\kappa\,, \tag{8.149}$$

where $M_i$ is the initially applied moment, $M_i = \int_A -y\sigma_i \, dA$. Thus we find that $\Delta\kappa = -M_i/EI$ and

$$\kappa_f = \kappa_i - \frac{M_i}{EI}. \tag{8.150}$$

The final stress state is then given as:

$$\sigma_f = \begin{cases} -\sigma_Y + \frac{M_i y}{I} & y > y_p \\ -\sigma_Y \frac{y}{y_p} + \frac{M_i y}{I} & |y| < y_p \\ \sigma_Y + \frac{M_i y}{I} & y < -y_p. \end{cases} \tag{8.151}$$

**Remarks:**

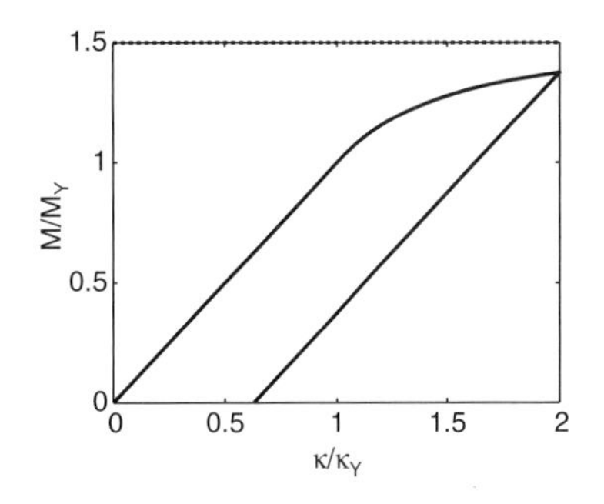


**Fig. 8.39** Moment versus curvature response with unloading after an initial curvature of $\kappa_i = 2\kappa_Y$. The initial moment $M = (33/24)M_Y$, and the final curvature is $\kappa_f = (15/24)\kappa_Y$.

(1) Figure 8.39 shows the complete load/unload curve for the moment versus curvature, where the initial curvature $\kappa_i = 2\kappa_Y$. As just computed, the unloading process is completely elastic and thus appears as a straight line. Beware, however, that unloading is not completely elastic in all cases. The present result only holds for the case of a rectangular cross-section with homogeneous elastic–perfectly plastic material properties. In particular, for composite beams reverse yielding can take place during unloading, and this must be checked for.

(2) Since $\Delta\kappa \neq -\kappa_i$, we have a residual stress and strain field in the beam after release; i.e. even though there is no net moment on the cross-section there are non-zero stresses and strains on it.

(3) If the cross-section were to be reloaded it would remain elastic until the applied moment reached $M_i$. Thus by such a procedure the effective yield moment of the bar has been increased in the direction of initial bending.

## Example 8.21

*Partial yield of a rectangular beam.* Consider a beam in pure bending with a rectangular cross-section with width $b = 3$ mm and height $h = 20$ mm. The Young's modulus is $E = 200$ GPa, and the yield stress in tension and compression is $\sigma_Y = 200$ MPa. (1) Find the yield moment and the ultimate moment. (2) Assume that the beam has been loaded in pure bending to the point where $y_p = 5$ mm. Find the applied moment and compute the residual curvature and stress field upon release of the load.

*Solution*

The yield moment is given as

$$M_Y = \frac{\sigma_Y I}{y_{\max}} = \frac{200 \times 3 \times 20^3/12}{10} = 40 \text{ kN} - \text{mm}. \tag{8.152}$$

The ultimate moment comes by integrating the ultimate stress state times minus $y$ over the cross-section; for the rectangular cross-section this will give

$$M_u = \frac{3}{2} M_Y = 60 \text{ kN} - \text{mm}. \tag{8.153}$$

For part (2) the initially applied moment is

$$M_i = \int_A -y\sigma_{xx}\, dA = M_u \left[1 - \left(\frac{y_p}{h/2}\right)^2\right] = 55 \text{ kN} - \text{mm}. \tag{8.154}$$

The initial curvature is

$$\kappa_i = \frac{\varepsilon_Y}{y_p} = \frac{200}{200 \times 10^3 \times 5} = 2 \times 10^{-4} \text{ 1/mm}. \tag{8.155}$$

The unloading curvature will be $\Delta\kappa = -M_i/EI = -1.38 \times 10^{-4}$ 1/mm, and thus the residual beam curvature will be

$$\kappa_f = \kappa_i + \Delta\kappa = 6.25 \times 10^{-5} \text{ 1/mm}. \tag{8.156}$$

The residual stress field (in MPa) will then be

$$\sigma_f = \begin{cases} -200 + 27.5y & y > 5 \\ -40y + 27.5y & |y| < 5\,, \\ 200 + 27.5y & y < -5 \end{cases} \tag{8.157}$$

where $y$ is measured in mm. A plot of the stress field is given in Fig. 8.40.

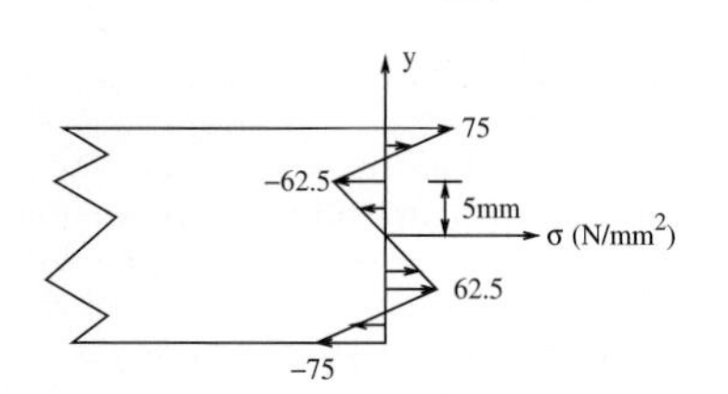


**Fig. 8.40** Residual stress field after unloading.

## Example 8.22

*Plastically deformed composite beam.* Shown in Fig. 8.41 is a composite beam cross-section. The core of the beam is made of high-strength steel, and the faces are made from an ordinary aluminum. The beam is placed in pure bending, and the strain at the top of the beam is measured to be $\varepsilon(20) = -7.5 \times 10^{-3}$. What moment is being applied to the beam? Find the residual stress distribution upon removal of the applied moment.

*Solution*

The strain at the top face exceeds the yield strain of the aluminum. Thus it has yielded at least partially. Since we know the strain at the top we can compute the curvature of the beam to be

$$\kappa = -\frac{\varepsilon(20)}{20} = \frac{7.5 \times 10^{-3}}{20} = 3.75 \times 10^{-4} \text{ 1/mm}. \tag{8.158}$$

This tells us that the strain at $y = 16$ mm is $\varepsilon(16) = -16\kappa = -6.0 \times 10^{-3}$. This is also above the yield strain for the aluminum. Thus the aluminum is fully yielded. This strain is, however, below the yield strain

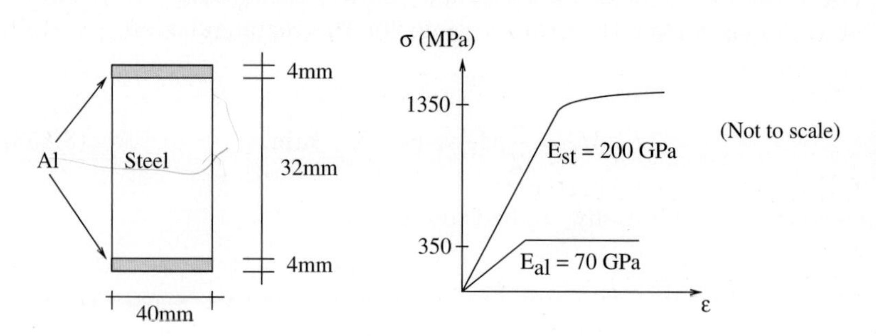


**Fig. 8.41** Composite beam made of a steel core and aluminum faces.

of the steel, $\varepsilon_Y^{st} = 6.75 \times 10^{-3}$. Thus the steel is completely elastic. The stress distribution is as sketched in Fig. 8.42. The applied moment is then given as

$$M = \int_A -y\sigma\, dA \tag{8.159}$$

$$= 2 \times 40 \left[ \int_0^{16} 200 \times 10^3 \times 3.75 \times 10^{-4} y^2\, dy \right.$$

$$\left. + \int_{16}^{20} 350y\, dy \right] \tag{8.160}$$

$$= 10.2 \text{ kN} - \text{m}. \tag{8.161}$$

Upon unload we will have $\kappa_f = \kappa_i + \Delta\kappa$, and thus $\varepsilon_f = \varepsilon_i - y\Delta\kappa$. Assuming elastic unloading the residual stresses will be

$$\sigma_f = \begin{cases} -\sigma_Y^{al} - E^{al}\Delta\kappa y & y > 16 \\ -E^{st}\kappa_i y - E^{st}\Delta\kappa y & |y| < 16 \\ \sigma_Y^{al} - E^{al}\Delta\kappa y & y < -16 \end{cases}. \tag{8.162}$$

Since the final moment is zero, we have

$$0 = \int_A -y\sigma_f\, dA \tag{8.163}$$

$$0 = 10.2\ (\text{kN} - \text{m}) + 2 \times 40 \left[ \int_0^{16} E^{st}\Delta\kappa y^2\, dy \right.$$

$$\left. + \int_{16}^{20} E^{al}\Delta\kappa y^2\, dy \right] \tag{8.164}$$

$$= 10.2 \times 10^6 + 80 \left[ 200 \times 10^3 \times \frac{16^3}{3} \right.$$

$$\left. + 70 \times 10^3 \times \frac{20^3 - 16^3}{3} \right] \Delta\kappa. \tag{8.165}$$

Solving for the unloading curvature gives $\Delta\kappa = -3.5 \times 10^{-4}$ 1/mm. The final curvature is then $\kappa_f = 0.25 \times 10^{-4}$ 1/mm. Note that $|\Delta\kappa \times 20 \times 70 \times 10^3| = 490$ MPa $< 2 \times 350$ MPa. Thus our assumption of elastic unloading is valid; i.e. the aluminum does not reverse yield during the unloading process. The final stress distribution is shown in Fig. 8.43.

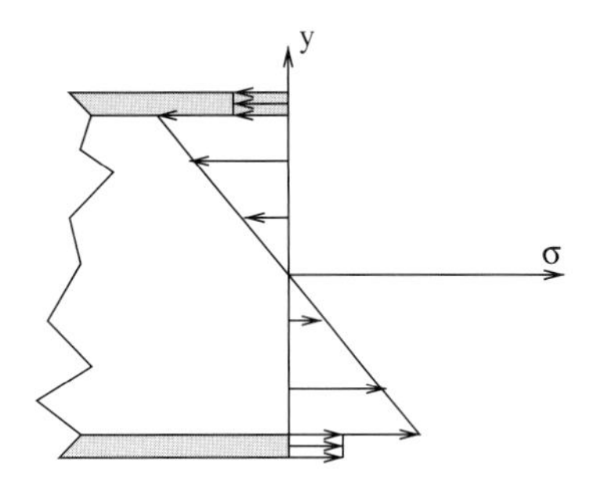


**Fig. 8.42** Initial stress distribution.

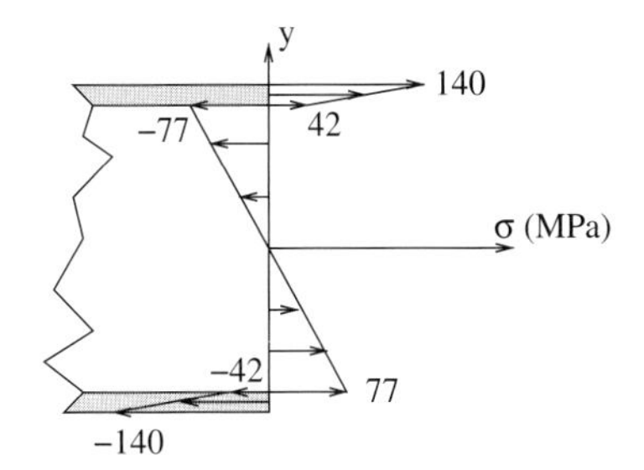


**Fig. 8.43** Final stress distribution.

## Chapter summary

- Bending of symmetric beams
  - Kinematic assumption: $\varepsilon_{xx} = -y\kappa$

- Equilibrium: $R = 0$, $dM/dx + V = 0$, $dV/dx + q = 0$
- Neutral Axis Condition: $R = 0$
- Resultant-stress relations: $R = \int_A \sigma_{xx}\, dA$, $M = \int -y\sigma_{xx}\, dA$, $V = \int_A \sigma_{xy}\, dA$
- Moment of inertia: $I = \int_A y^2\, dA$

• Bending of symmetric linear elastic beams

- Constitutive relation: $\sigma = E\varepsilon$
- Stress-moment: $\sigma = -My/I$
- Neutral axis: $y_{na} = \int_A Ey_b\, dA / \int_A E\, dA$
- Moment-curvature: $M = EI\kappa$
- Differential equation: $(EIv'')'' = q$

• Boundary conditions: fixed and forced

$$v,\ v' = 0\,, \qquad EIv'' = M,\ -EIv''' = V$$

• Shear in beams

- Shear flow: $q = VQ/I$
- First moment of area: $Q(y_1) = \int_{A(y_1)} y\, dA$
- Shear stress: $\tau = q/b$

• Energy: elastic

$$W_{\text{in}} = W_{\text{stored}}$$

For linear elastic systems the strain energy density is $w = \frac{1}{2}(\sigma_{xx}\varepsilon_{xx} + \sigma_{xy}\gamma_{xy})$ and the work input is $\frac{1}{2}M\theta$ or $\frac{1}{2}P\Delta$.

• Elastic–perfectly plastic bending

- Moment at yield: $M_Y = \sigma_Y I / y_{\max}$
- Ultimate moment: $M_u = \int_A \sigma\, dA$, where $\sigma = \pm\sigma_Y$
- Curvature: $\kappa = \varepsilon_Y / y_p$

# Exercises

(8.1) Derive the fundamental kinematic relation $\varepsilon = -y\kappa$ using appropriate differential arguments. Be specific and explain your steps with short phrases.

(8.2) The composite cross-section shown experiences a curvature $\kappa = 10^{-4}$ mm$^{-1}$ (bending about the $z$-axis). Plot the stress distribution on the cross-section, assuming linear elastic behavior.

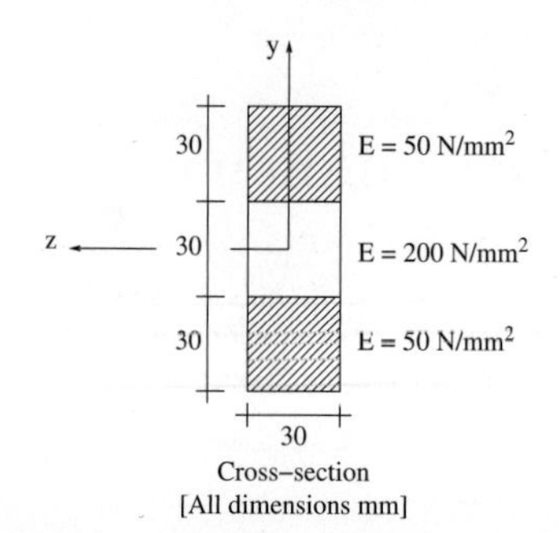

Cross–section
[All dimensions mm]

(8.3) Consider a slender rectangular beam with cross-sectional dimensions of $b$ and $h$. The material is non-linearly elastic with the following constitutive law $\sigma = C_1\varepsilon + C_2\varepsilon^3$, where $C_1$ and $C_2$ are given material constants. If the beam is in a state of pure bending with constant curvature $\kappa$, what will the bending moment $M$ in the beam be? Express your answer in terms of the given constants: $b$, $h$, $\kappa$, $C_1$ and $C_2$.

(8.4) Derive the differential equations of vertical and moment equilibrium for a beam. Explain each major step with a short concise phrase.

(8.5) For the following equations, state the physical principle that each represents in at most *one sentence*.

$$(a): \quad \frac{dM}{dx} + V = 0$$

$$(b): \quad \frac{dV}{dx} + q = 0$$

(8.6) For the beam shown, determine the shear and moment diagrams. (Note the hinge.)

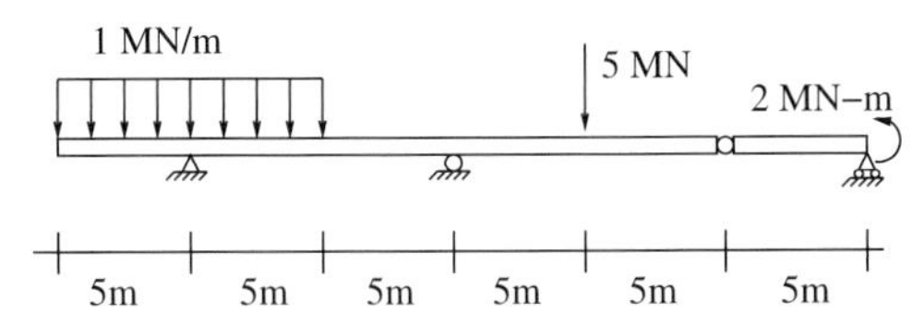


(8.7) Certain nickel–titanium alloys display stress-strain behavior as shown below; the behavior in compression is the same as in tension. The beam shown is in pure bending about the $z$-axis. The strain on the top face has been measured to be $\varepsilon_g$ with $\varepsilon_g > \varepsilon_b$.

(a) Accurately sketch the strain distribution.

(b) Accurately sketch the stress distribution.

Label all critical points and values on the graphs in terms of $\varepsilon_g$, $\varepsilon_a$, $\varepsilon_b$, $\sigma_p$, $h$, $b$, and $E$.

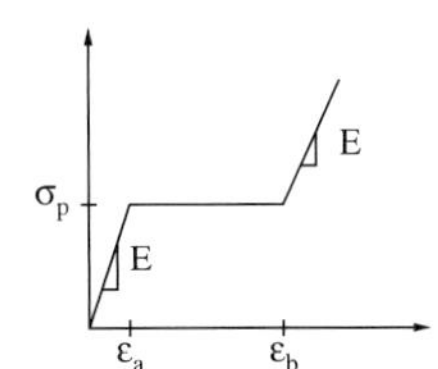


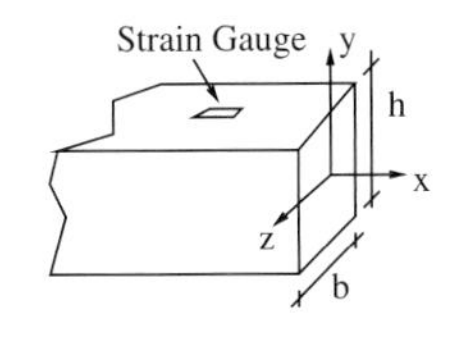


(8.8) A beam is constructed by joining a triangular prism with a rectangular prism. The cross-section is shown in the following. Determine the location of the neutral axis, assuming bending about a horizontal axis; assume homogeneous material properties. Find, also, the moment of inertia about this axis.

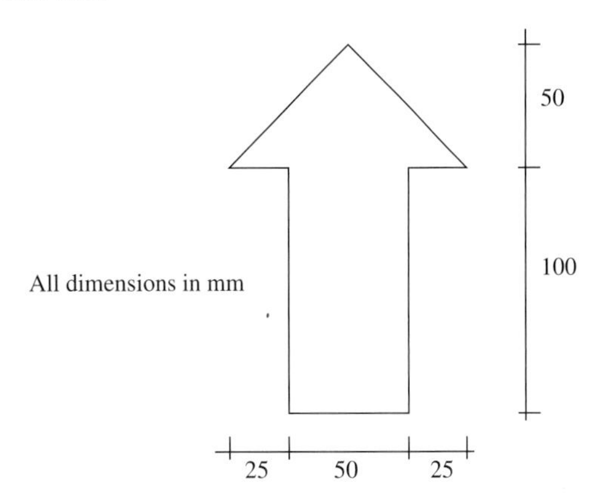


(8.9) Consider a beam with an elliptical cross-section. Assume the major axis is parallel to the $y$-axis and the minor axis is parallel to the $z$-axis. What is the moment of inertia of the cross-section about the $z$-axis? Define any needed quantities.

(8.10) Consider a generic doubly symmetric I-beam cross-section. Determine an expression for the moment of inertia. Define any dimensions needed.

(8.11) Find the location of the neutral axis and the moment of inertia about the neutral axis of a T-beam cross-section with flange width 400 mm, flange thickness 40 mm, web height 200 mm, and web thickness 20 mm. Assume homogeneous properties.

(8.12) Find the neutral axis location in Exercise 8.11 assuming that the Young's modulus of the flange is $E = 100$ MPa and the Young's modulus of the web is $E = 200$ MPa.

(8.13) Consider a beam cross-section composed of an elastomer square stock with side length 200 mm and modulus $E = 10$ MPa. To the lower side of the bar we have glued a 1-mm thick sheet of steel (width 200 mm) whose modulus is $E = 200$ GPa. Determine the effective bending stiffness of the cross-section $(EI)_{\text{eff}}$.

(8.14) The following beam has an inhomogeneous linear coefficient of thermal expansion that is a function of depth, $\alpha(y')$, where $y'$ is distance measured up from the bottom of the beam. Find the formula for the location of the neutral axis when the beam is subjected to a temperature change. Assume the beam is linear elastic with a constant Young's modulus.

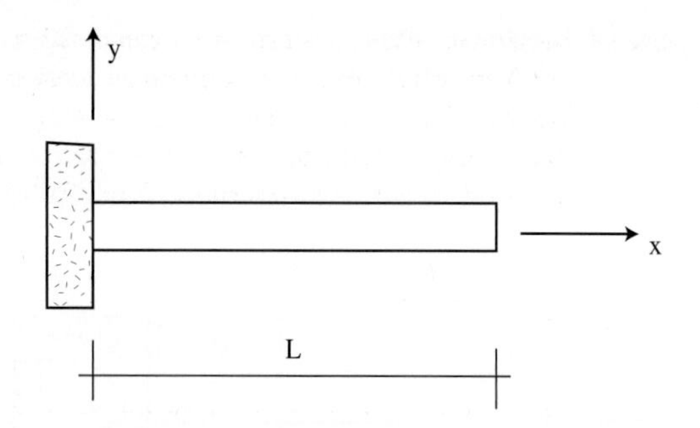


(8.15) A strain gauge has been placed on the cantilever beam shown at a distance $L/2$ from the support and at a distance $h/4$ from the top face of the beam. The beam is subjected to an unspecified load and the gauge gives a reading of $\varepsilon_g$ for the bending strain at this location. Find the curvature of the beam at $x = L/2$ in terms of this gauge value, the geometric dimensions, and the material properties. Note that the beam is composed of two different materials with differing elastic moduli $E_1$ and $E_2$.

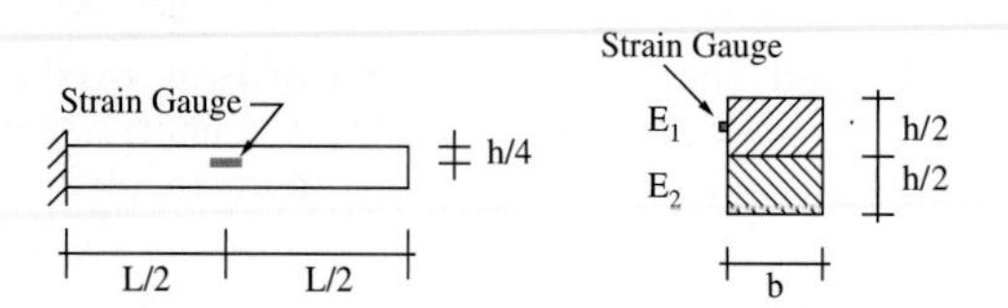


(8.16) An elastic beam with Young's modulus $E$ and moment of inertia $I$ is shown below. Find $v(x)$ (the displacement field) for the given load.

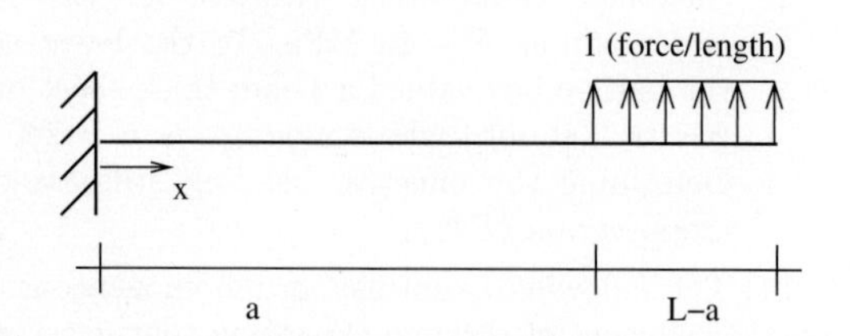


(8.17) In the statically indeterminate beam shown, find the reactions at the wall by integrating the differential equation for the deflection of the beam. Assume $EI$ is a constant.

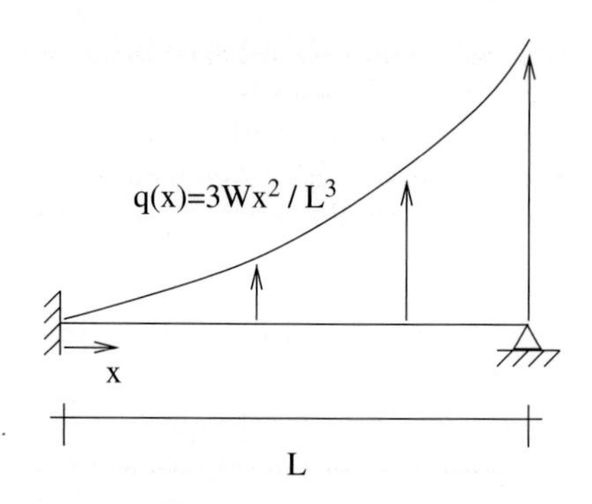


(8.18) For each of the four elastic structure (a)–(d), determine the appropriate boundary conditions (at both ends of the structure) and the distributed load function that would be needed to determine the deformation of the structural element via integration. *Express all boundary conditions in terms of* $E$, $G$, $I$, $J$, $A$, $v$, $u$, $\phi$ *and their derivatives as appropriate.*

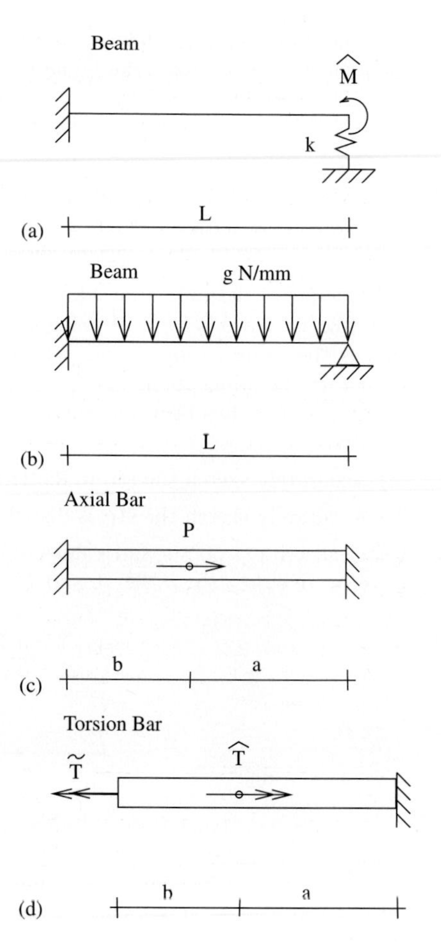

(8.19) For the beam shown below, with imposed deflection $\Delta$, find the deflection curve $v(x)$ and the location and magnitude of the maximum bending stress. Assume $EI$ is constant and the maximum distance from the neutral axis to the outer fibers of the beam is $c$.

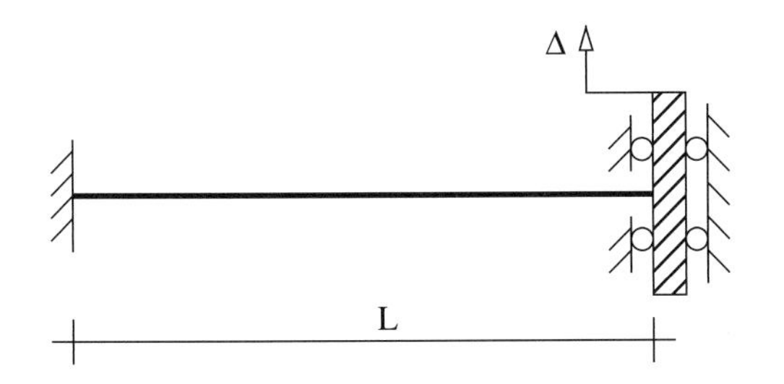


(8.20) For the system shown: (1) state the relevant boundary conditions in terms of the kinematic variables, and (2) give an appropriate expression for the distributed load acting on the system.

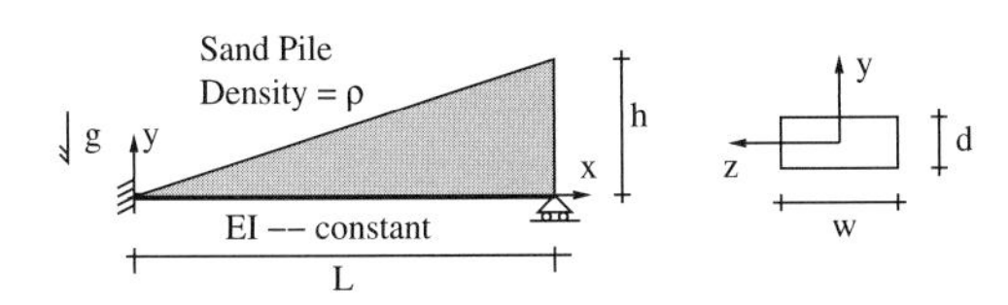


(8.21) Consider an elastic beam of length $L + a$ with constant Young's modulus, $E$, and cross-sectional area moment of inertia, $I$. The beam is subject to a point moment as shown. Determine the deflection of the beam as a function of $x$.

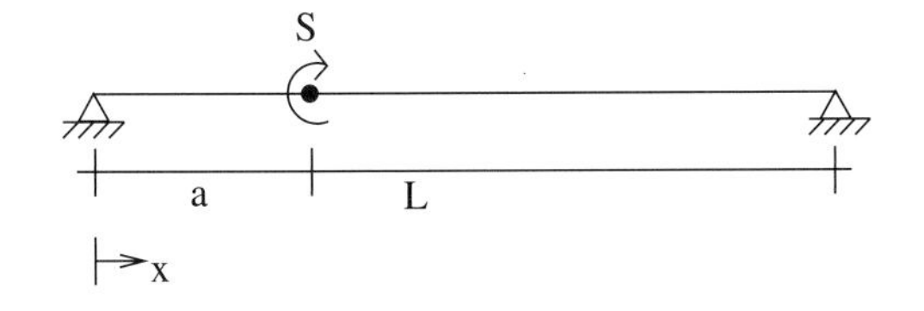


(8.22) Consider an elastic beam of length $L$ with constant Young's modulus, $E$, and cross-sectional area moment of inertia, $I$. The beam is subject to a point moment as shown. Determine the torsional stiffness at $x = a$; i.e. determine $k_T = S/\theta(a)$.

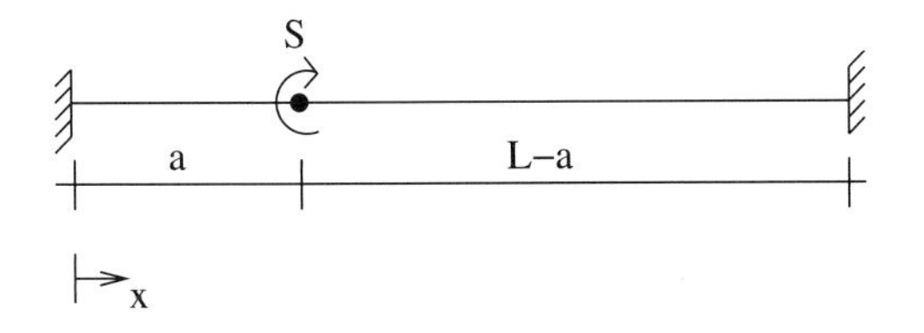


(8.23) Consider an elastic beam of length $L$ with a constant Young's modulus, $E$, and cross-sectional area moment of inertia, $I$. The beam is subject to a point force as shown. Determine the transverse stiffness at $x = b$; i.e. determine $k = P/v(b)$.

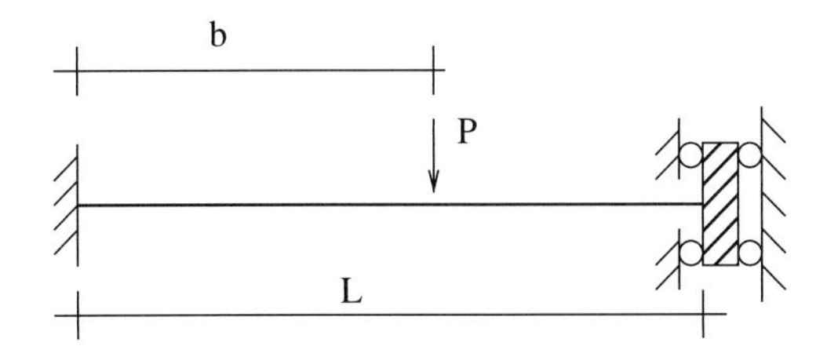


(8.24) For the four beams (a)–(d): state (a) the boundary conditions, and (b) the distributed load function $q(x)$. Assume all beams are of length $L$.

(a)

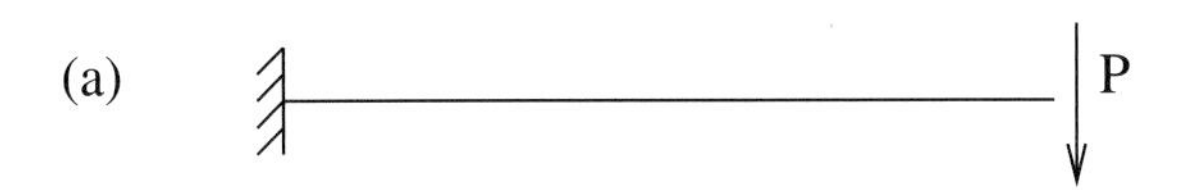


(b)

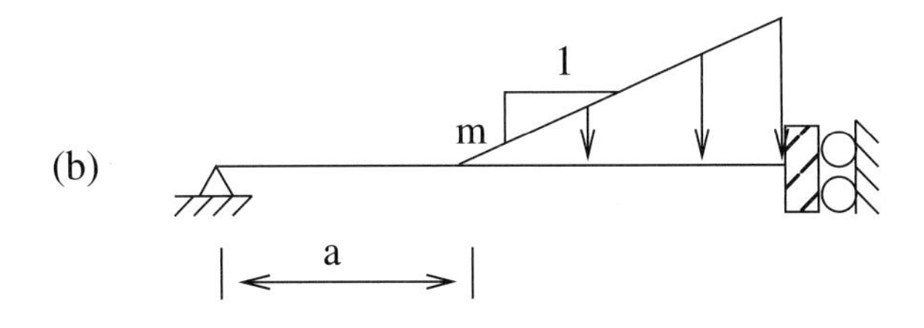


(c)

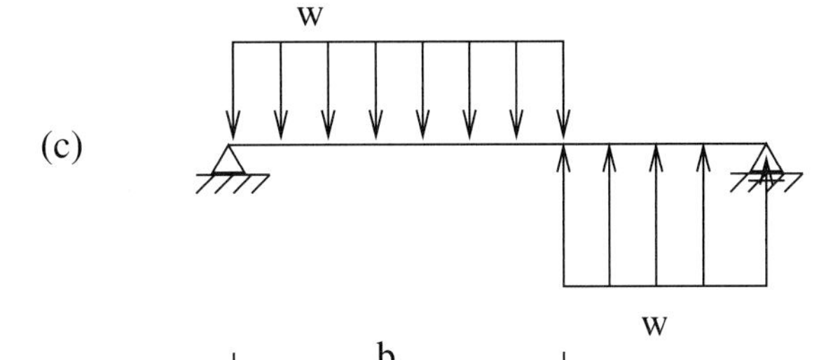


(d)

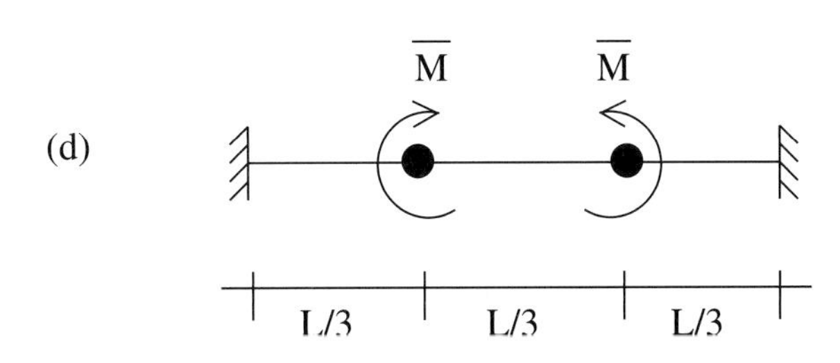

(8.25) For each of the four elastic beams (a)–(d) write down the appropriate boundary conditions and distributed load function that would be needed to determine the deflection of the beam via integration. *Express all boundary conditions in terms of $E$, $I$, $v$ and its derivatives.*

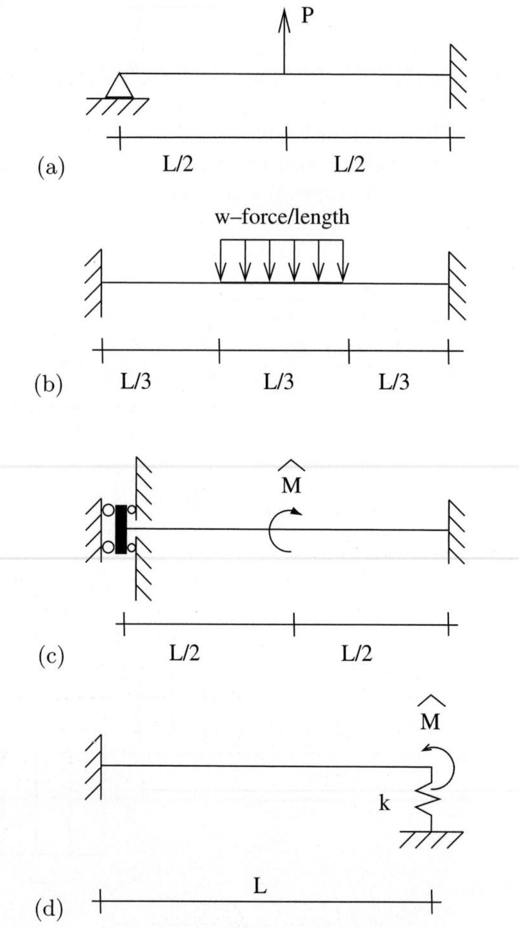


(8.26) For each of the four beams (a)–(d), write down the appropriate boundary conditions and distributed load expression, which are needed to compute each beam's deflection via the ODE $(EIv'')'' = q$. *Express all boundary conditions in terms of $E$, $I$, $v$ and its derivatives.*

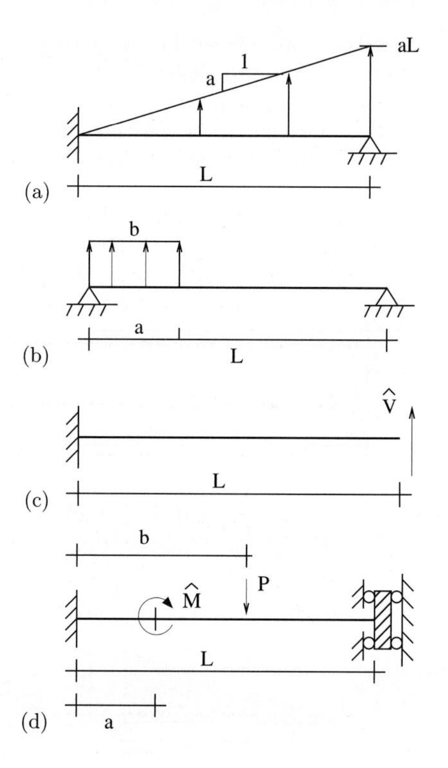


(8.27) The beam shown below is loaded by a point moment at $x = 2$; find the maximum internal moment. Assume consistent units.

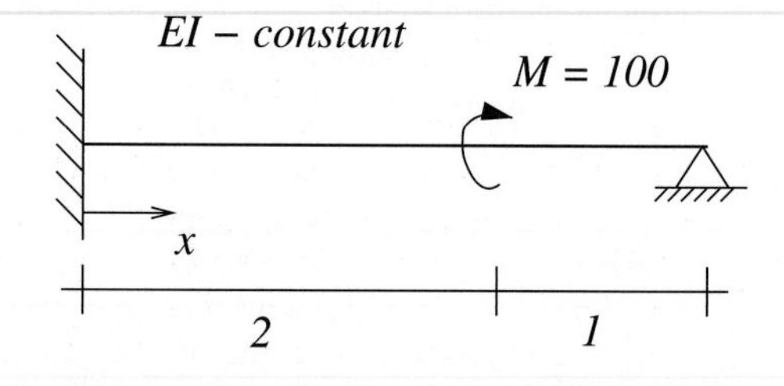


(8.28) Find the equation for the deflection of the beam shown. Assume a constant value for $EI$.

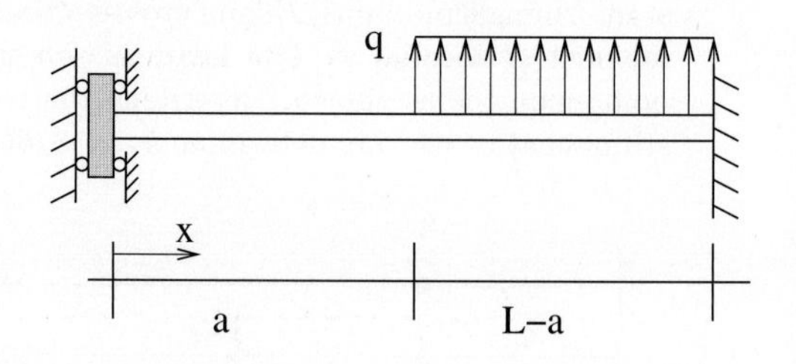


(8.29) For the following linear elastic beam, determine the deflection $v(x)$. Assume a constant value $E$ for the Young's modulus.

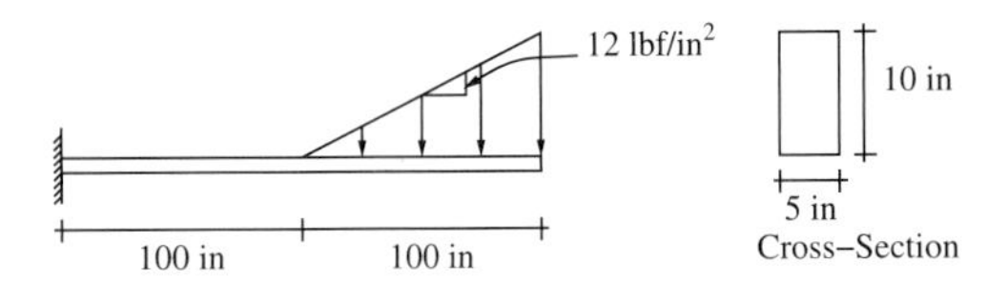


(8.30) Consider a simply supported beam with a transverse load $P$ applied at its mid-span. Determine the deflection of the beam from the governing ordinary differential equation.

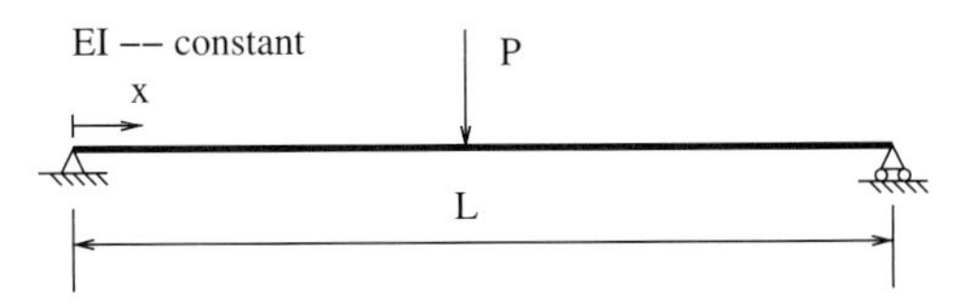


(8.31) Determine the deflection $v(x)$ for the beam shown below. Assume a constant $EI$. Hint: the intermediate roller support can be replaced by an unknown reaction plus an additional kinematic conditions.

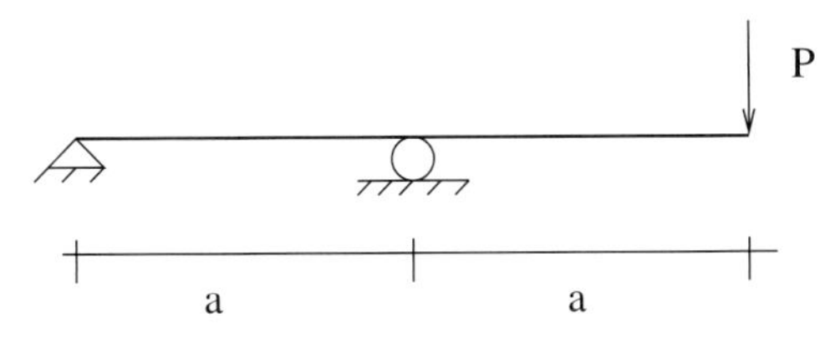


(8.32) Consider a beam supported by a distributed spring foundation (such as a railroad rail or grade-beam); such supports are know as *Winkler Foundations*. Assume the beam is 100 ft long with a Young's modulus of $E = 30 \times 10^6$ psi and a cross-sectional area moment of inertia $I = 77.4$ in$^4$. Assume a foundation stiffness of $k = 100$ lb/in$^2$ and determine the maximum positive and negative moments in the beam for a $30 \times 10^3$ lb load distributed over 3 in around the beam's center. Note that for this exercise the governing equation is given by $EIv(x)'''' = q(x)$, where $q(x) = q_{\text{applied}}(x) - kv(x)$. For boundary conditions, assume zero moment and shear at each end.

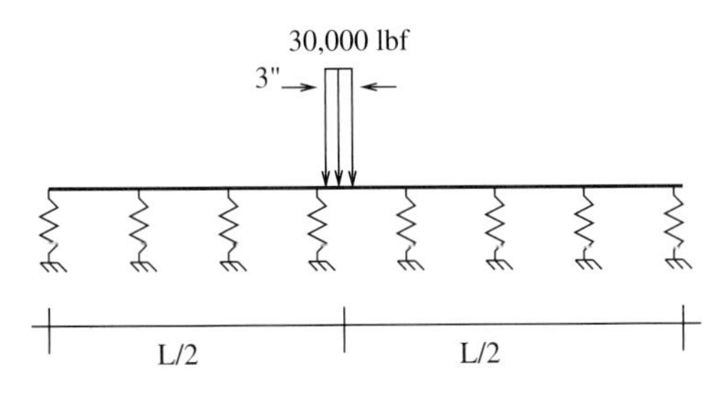


(8.33) Consider the beam in Exercise 8.32 and determine the maximum rotation (in absolute value) and the location at which it occurs.

(8.34) A device has been proposed as a momentum sensor for molecular beams. A molecular beam when it hits a solid object imparts a force, $P$, to the object. In our case the object is a pin-clamped beam. The force bends the beam and the bending is detected using a light source that reflects off the surface of a beam. The angle of reflection is measured on a circular screen at some distance from the beam. Determine the sensitivity of the device ($S/P$) assuming $L = 300$ mm, $a = 85$ mm, $b = 0.5L$, $E = 70$ GPa, and $R = 1.5$ m. The beam has a 4-mm by 0.9-mm cross-section, and bends about the weak axis.

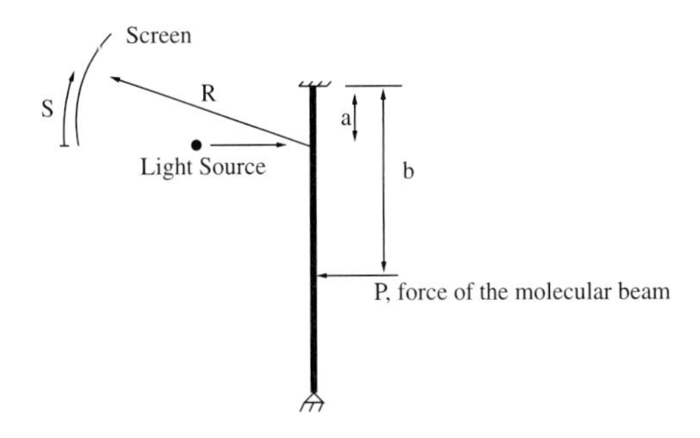


(8.35) A thermoelastic switch is made by bonding together two materials of differing linear thermal expansion coefficients. Upon a change in temperature the system will deflect up or down, depending upon the sign of the temperature change and the difference in thermal expansion coefficients. Assuming the materials have the same Young's modulus, find an expression for the deflection of the tip of the beam as a function of temperature change.

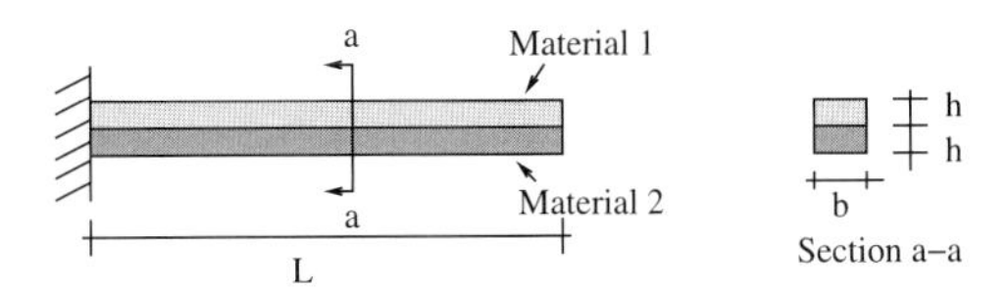


(8.36) Using a differential element argument, show that

$$\frac{dM_y}{dx} - V_z = 0.$$

Clearly explain all steps.

(8.37) Using a differential element argument show that

$$\frac{dV_z}{dx} + q_z = 0.$$

Clearly explain all steps.

(8.38) Consider the beam shown with an offset axial force $P$ and moment $M$. What is the minimum value of $P$ for which the normal stress $\sigma_{xx}$ is everywhere non-negative. Assume $L = 3$ ft, $h = 5$ in, $c = 4$ in, $a = 0.25$ in, $b = 1$ in, and $M = 100$ in – lbf,

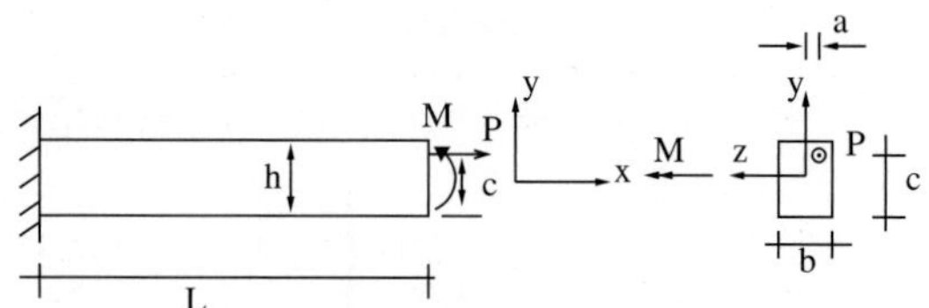


(8.39) Shown is a beam with a square cross-section ($a \times a$). The beam is built-in at one end and has a plate welded onto the other. There is an applied force $-P\boldsymbol{e}_y$ at the center of the plate. There is a second force $P\boldsymbol{e}_x$ applied to the plate. Determine the possible points of application of this second load so that the neutral axis is oriented at 45° with respect to the $z$-axis at mid-span.

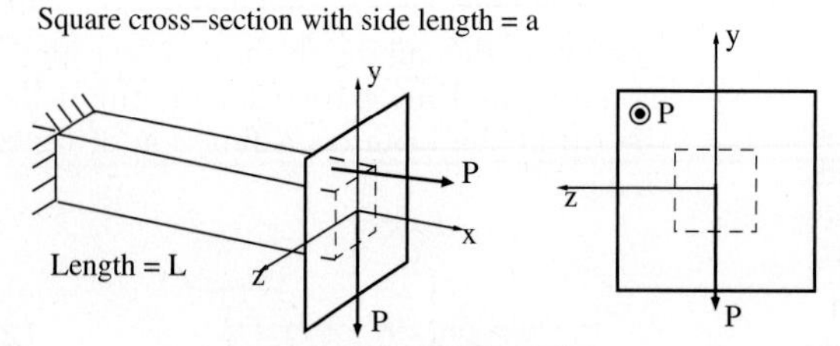


(8.40) Consider a spread footing as shown in the following. By treating the footing as a beam, show that one may misplace the applied load $P$ in the $z$ direction by an amount $\pm h/6$ from the centroid and the base of the footing will still be entirely in compression. A similar result holds in the $y$ direction, and the region on the top of the footing where "misplacement" is allowed is known as the *kern* of the footing (or column).

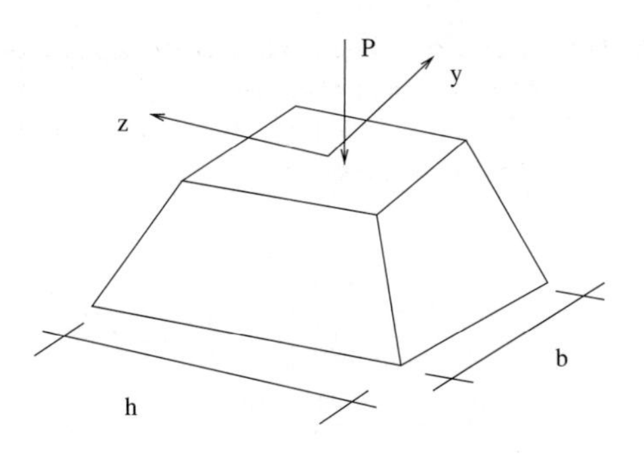


(8.41) Shown below is a 10-ft long cantilevered W21×62 I-beam. Consider a section cut made at $x = 5$ ft and determine the stress state at points $A$ and $B$ in terms of the load $P$. This is a standard wide-flange steel section with the following properties: area 18.3 in$^2$, depth 20.99 in, flange width 8.240 in, flange thickness 0.615 in, web thickness 0.400 in, moment of inertia about z-axis 1330 in$^4$, moment of inertia about y-axis 57.5 in$^4$.

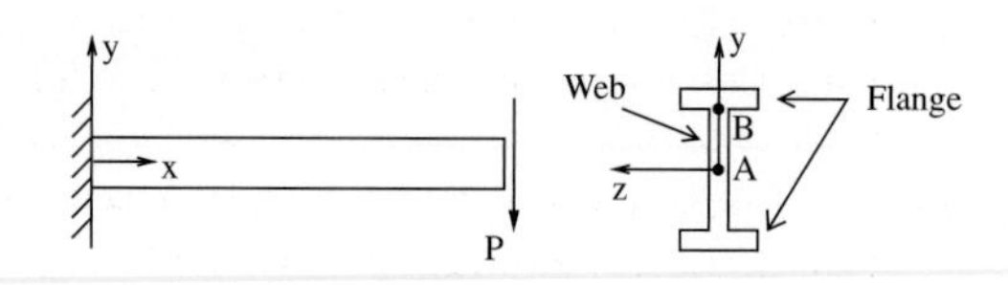


(8.42) What is the magnitude of the uniform load $q$ that the beam shown can carry? The beam is made of two wooden planks which are bolted together at a spacing of 150 mm. The allowable normal stress in the wood is 5 MPa, the allowable shear stress is 0.8 MPa, and the allowable shear force per bolt is 30 kN. Check all critical sections.

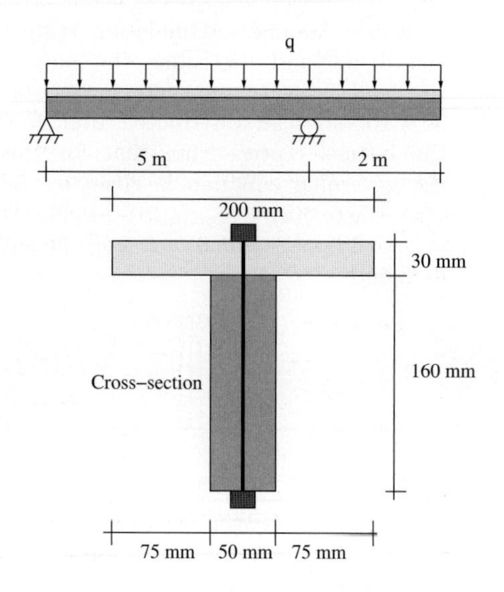

(8.43) Consider a T-beam with cross-section dimensions as shown. The section is subject to a shear force of 10 kN. Find the magnitude of the maximum shear stress on the cross-section.

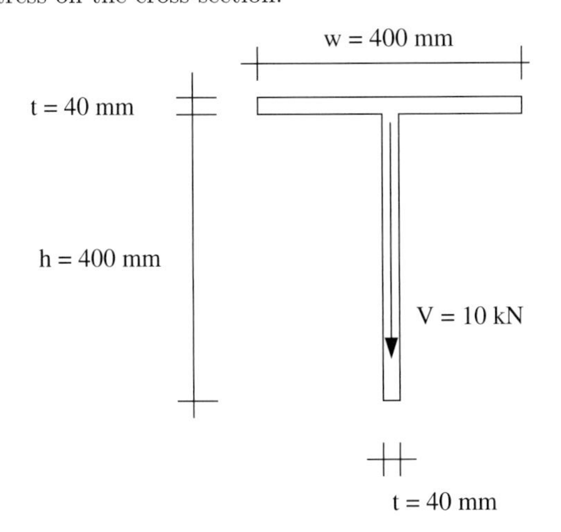


(8.44) Derive the formula $q = VQ/I$. Define all terms in your derivation, and clearly state the meaning of each step.

(8.45) The beam shown below is to be made by joining two pieces of wood together with uniformly spaced dowels. If the shear force capacity of each dowel is $F_d$, what is the maximum allowable spacing between the dowels?

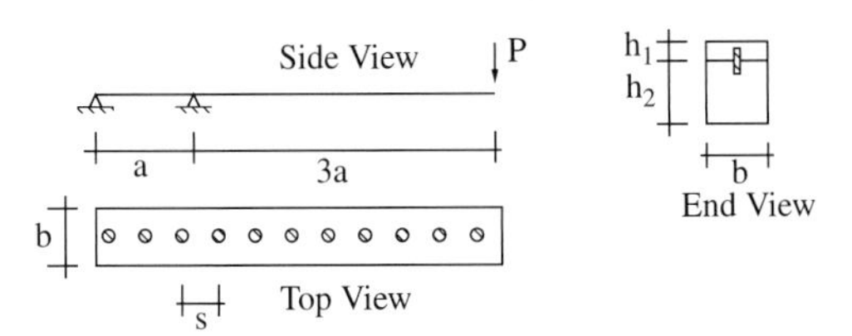


(8.46) A beam is made by gluing together two pieces of wood. The beam is to be loaded as shown. What is the required shear stress strength of the glue? Express your answer in terms of the geometric dimensions of the system and the constitutive properties.

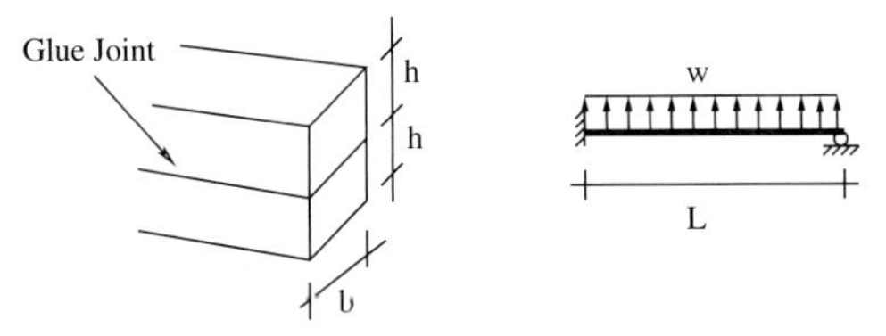


(8.47) Consider the 1,500-mm long simply-supported beam shown. The beam cross-section in an inverted T, which has been made by welding two plates together with a series of short strip welds. The welds are spaced apart every 60 mm and are 20 mm in length. The welds are able to hold 600 N per mm of weld. What load $P$ will cause weld failure?

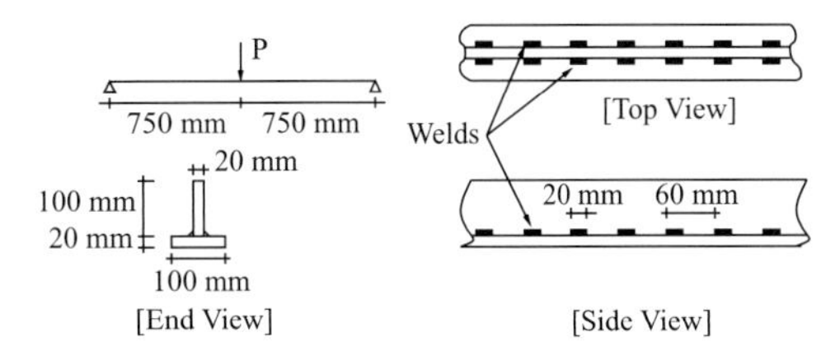


(8.48) The beam shown has a rectangular cross-section with a glue joint as indicated. Determine the required shear-stress strength of the glue to prevent shear failure in the joint. Assume $a \ll b$ and $a \ll h$.

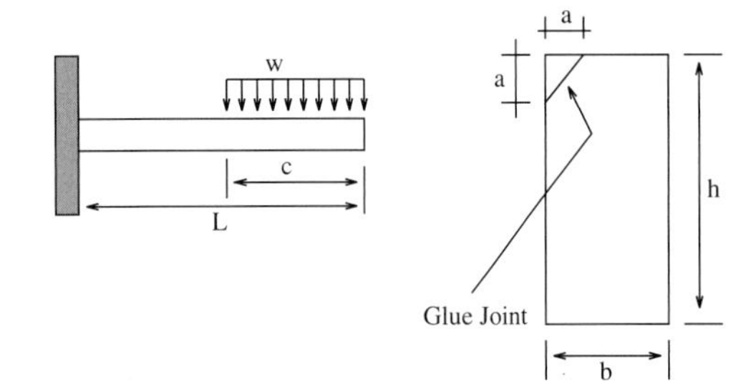


(8.49) Consider the beam in Exercise 8.41. Determine the length of the beam so that the tip deflection due to shear is equal to the tip deflection due to bending.

(8.50) A *slender* metal band with constant $EI$, as shown, is subjected to a force $P$. Find the horizontal deflection at the point where the load is applied using conservation of energy. Ignore axial and shear effects.

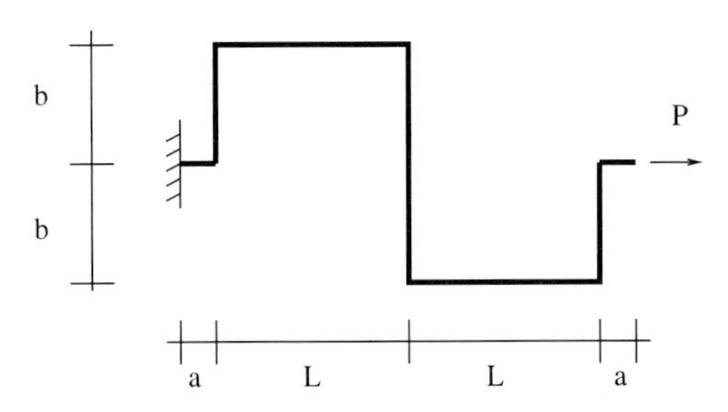


(8.51) For the composite cross-section shown, find the yield moment $M_Y$ and the ultimate moment $M_u$. Let $E_1 = 20{,}000$ (ksi), $E_2 = 40{,}000$ (ksi), $\sigma_{1Y} = 60$ (ksi), $\sigma_{2Y} = 50$ (ksi).

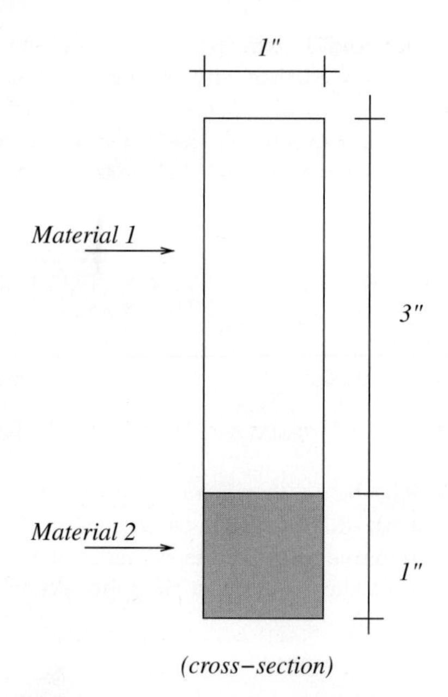


(8.52) Consider an elastic–perfectly plastic beam with a rectangular cross-section. The beam has an inhomogeneous yield stress

$$\sigma_Y(\widehat{y}) = 200 + 4\widehat{y},$$

where $\widehat{y}$ is measured from the bottom face of the beam. Find the ultimate moment capacity of the beam. Assume consistent units.

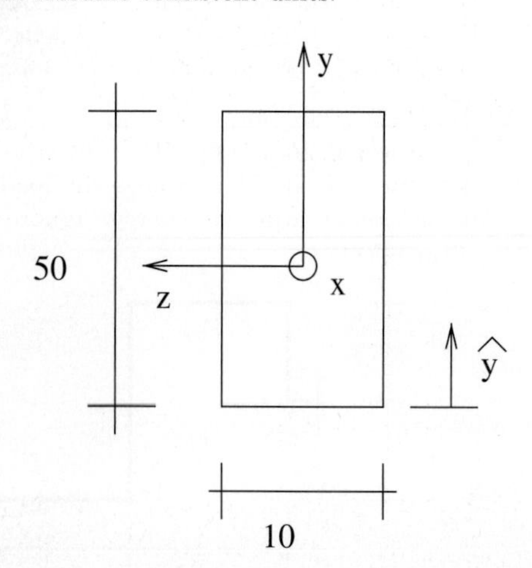


(8.53) Shown in the following is the cross-section of a composite beam. Both materials are elastic–perfectly plastic with the same Young's modulus. The cross-hatched material has a yield stress of 100 MPa, and the remaining material has a yield stress of 300 MPa. The beam has been bent so that at a given cross-section the cross-hatched material has just completely yielded. What is the applied moment at this cross-section?

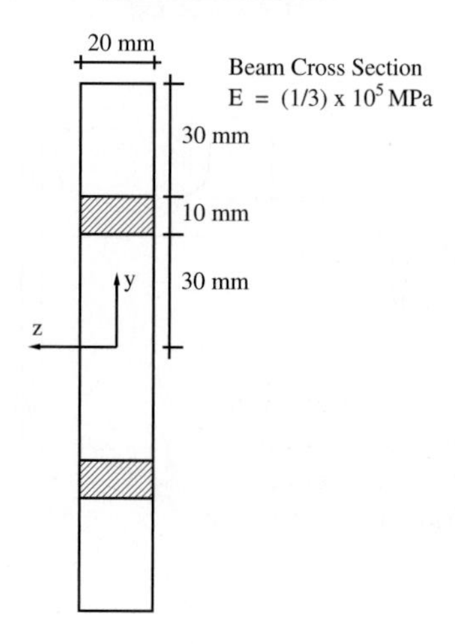


(8.54) Consider the I-beam cross-section shown below. Assume the material to be elastic–perfectly plastic with yield stress $\sigma_Y$ and Young's modulus $E$.

(a) What moment is required to completely yield the flanges with the web remaining elastic? Express your answer in terms of $\sigma_Y$ and the geometric parameters defining the cross-section.

(b) What will the curvature be at this state of load. Express your answer in terms of $\sigma_Y$, $E$, and the geometric parameters defining the cross-section.

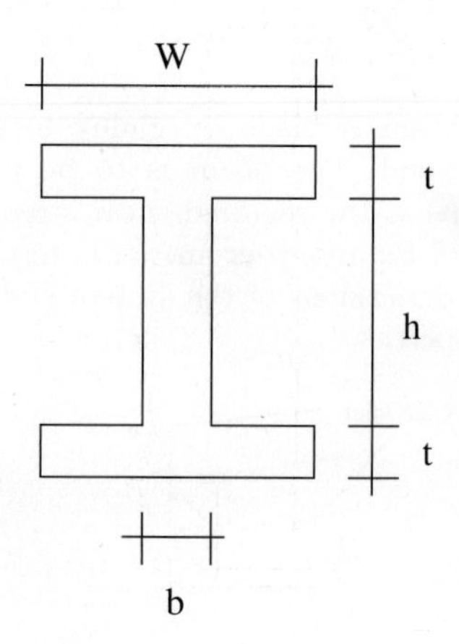


(8.55) A rectangular beam is made of an elastic–perfectly plastic material with differing properties in tension and compression as indicated. Find the ultimate moment, $M_u$, for bending about the $z$-axis. Express your answer in terms of $B$, $b$, $h$.

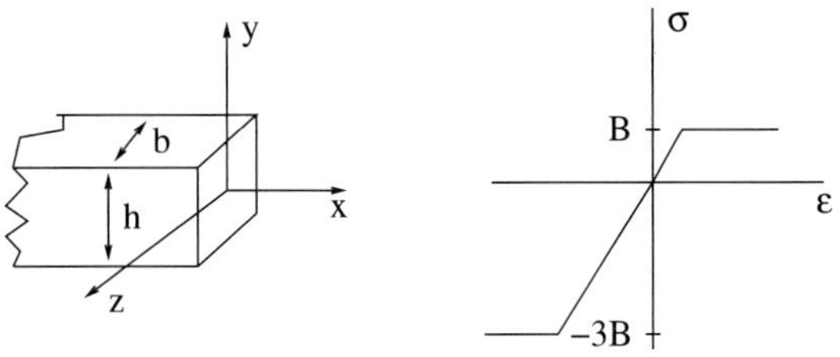


(8.56) Consider a rectangular beam of cross-sectional width $b$ and depth $h$. The beam has been surface hardened by rolling so that the yield stress varies quadratically with depth:

$$\sigma_Y(y) = \begin{cases} A + By^2 & \text{Tension} \\ -(A + By^2) & \text{Compression} \end{cases}$$

Assume $A > 0$ and $B > 0$. What is the ultimate moment for the beam? What curvature corresponds to the ultimate moment?

(8.57) Compute the ultimate moment for a beam with the cross-section shown below. Assume perfect plasticity with a yield stress $\sigma_Y = 100$ ksi; all dimensions are in inches.

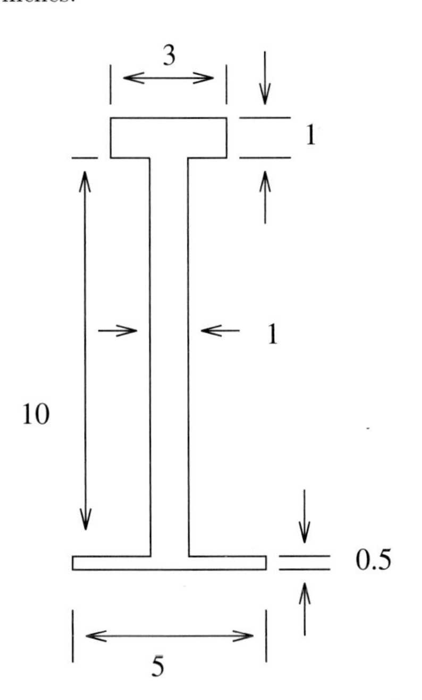


(8.58) An elastic–perfectly plastic beam with rectangular cross-section is shown below. It is loaded with forces $P = (1.4M_Y)/a$, where $M_Y$ is the *initial* yield moment. Find the deflection at the center. Suggested method: Find the extent of the plastic deformation; get the curvature; knowing the curvature, find the deflection.

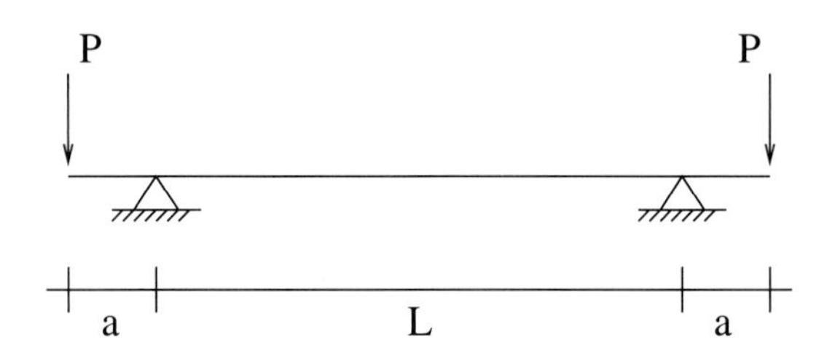


(8.59) The elastic–perfectly plastic beam shown below is loaded past the yield moment, $M_Y$. Find the deflection at mid-span.

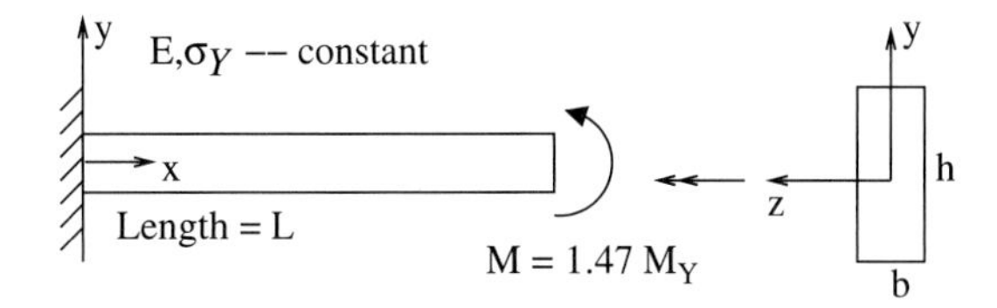


(8.60) The beam below has a rectangular cross-section and is elastic–perfectly plastic. The load is increased slowly from zero to a value of $P = 1.1\frac{M_Y}{a}$ and then slowly decreased again to zero. Determine the final deflection of the center of the beam at the end of the loading cycle.

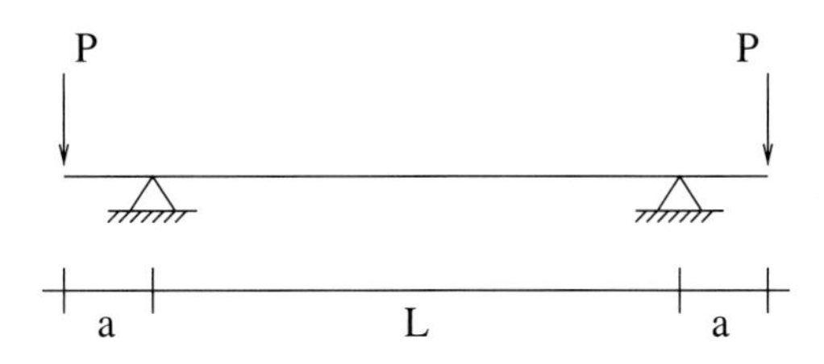


(8.61) Shape memory alloys are rapidly becoming very important materials in engineering applications. Above their austenite finish temperature they have the unusual stress–strain behavior as shown on the left in the figure. In pure bending the moment-curvature relation for a rectangular $b \times h$ beam made of such a material is as shown on the right in the figure. The response curve can be expressed as

$$M(\kappa) = \begin{cases} E_a I \kappa & \kappa < \kappa_s \\ \frac{3}{2} M_s [1 - \frac{1}{3}(\kappa_s/\kappa)^2] & \kappa_s < \kappa < \kappa_f \\ \underline{\qquad ?? \qquad} & \kappa > \kappa_f \end{cases}$$

where $\kappa_s = 2\varepsilon_s/h$, $\kappa_f = 2\varepsilon_f/h$, and $M_s = E_a I \kappa_s$. Determine the appropriate relation for the case where $\kappa > \kappa_f$.

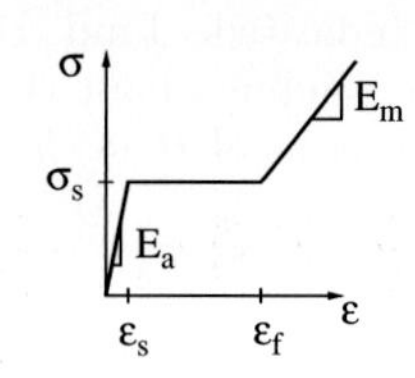


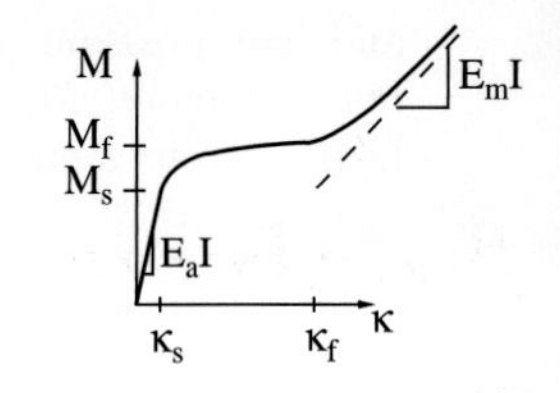

# Analysis of Multi-Axial Stress and Strain

# 9



We now have at our disposal the means to determine the stress state of a variety of common mechanical elements. If the element is under multiple types of loading and is elastic, then the total stress state can be computed using superposition. For each of these computations, a preferred coordinate system is used to determine the stresses. Sometimes the coordinate systems to do not coincide, and thus there is a need to transform stresses associated with one coordinate system to those of another. Further, as we saw in Chapter 7 when we analyzed brittle failure in torsion, there is sometimes a utility in expressing a result computed in one coordinate system in another. Both of these points brings up the need to develop some machinery for computing the transformations of stress (and strain). We will begin by first reviewing the transformation rules for vectors and then move on to look at tensors. Later in this chapter we will use our understanding to examine failure criteria for multi-axial states of stress.

## 9.1 Transformation of vectors

The central question in the transformation of vectors is to compute the components of a given vector in one coordinate system in terms of the components of the same vector in a second coordinate system when the relative orientation between the coordinate frames is known. Consider the vector $\boldsymbol{F}$ shown in Fig. 9.1. The components of the vector in the $x$-$y$ coordinate system are given by the orthogonal projections of $\boldsymbol{F}$ onto the coordinate axes. Thus

$$F_x = \boldsymbol{e}_x \cdot \boldsymbol{F} \tag{9.1}$$

$$F_y = \boldsymbol{e}_y \cdot \boldsymbol{F}, \tag{9.2}$$

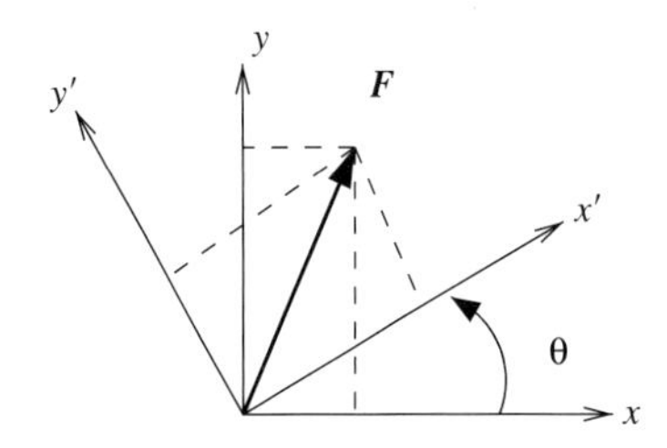


**Fig. 9.1** Relative orientation of two given coordinate systems.

where $\boldsymbol{e}_x$ and $\boldsymbol{e}_y$ are the unit vectors in the coordinate directions. Combined together we can write

$$\boldsymbol{F} = F_x\boldsymbol{e}_x + F_y\boldsymbol{e}_y. \tag{9.3}$$

Similarly, in the $x'$-$y'$ coordinate system we have

$$F_{x'} = \boldsymbol{e}_{x'} \cdot \boldsymbol{F} \tag{9.4}$$

$$F_{y'} = \boldsymbol{e}_{y'} \cdot \boldsymbol{F}, \tag{9.5}$$

where $\boldsymbol{e}_{x'}$ and $\boldsymbol{e}_{y'}$ are the unit vectors in the $x'$- and $y'$-coordinate directions. Combined together we can also write

$$\boldsymbol{F} = F_{x'}\boldsymbol{e}_{x'} + F_{y'}\boldsymbol{e}_{y'}. \tag{9.6}$$

Because both decompositions (eqns (9.3) and (9.6)) refer to the same physical vector we can assert the equality

$$F_x\boldsymbol{e}_x + F_y\boldsymbol{e}_y = F_{x'}\boldsymbol{e}_{x'} + F_{y'}\boldsymbol{e}_{y'}. \tag{9.7}$$

Equation (9.7) can be used to derive the transformation rules between coordinate systems. For example, if we want to find $F_{x'}$ in terms of $F_x$ and $F_y$ we can dot both sides of eqn (9.7) with $\boldsymbol{e}_{x'}$. This yields

$$F_x\boldsymbol{e}_{x'} \cdot \boldsymbol{e}_x + F_y\boldsymbol{e}_{x'} \cdot \boldsymbol{e}_y = F_{x'}. \tag{9.8}$$

Noting the relation between the coordinate axes this can also be written as

$$F_{x'} = F_x \cos(\theta) + F_y \sin(\theta). \tag{9.9}$$

Likewise one can show that

$$F_{y'} = -F_x \sin(\theta) + F_y \cos(\theta). \tag{9.10}$$

These are the desired transformation rules.

**Remarks:**

(1) The results can be collected in matrix-vector form so that

$$\begin{pmatrix} F_{x'} \\ F_{y'} \end{pmatrix} = \begin{bmatrix} \cos(\theta) & \sin(\theta) \\ -\sin(\theta) & \cos(\theta) \end{bmatrix} \begin{pmatrix} F_x \\ F_y \end{pmatrix}. \tag{9.11}$$

(2) The matrix in the above expression is recognized to be a rotation matrix (of rotation $\theta$); i.e. its transpose is its inverse. The elements of the rotation matrix are sometimes called the direction cosines as they represent the dot products between the unit vectors in the coordinate directions. These happen to be equal to the cosines of the angles between the coordinate axes.

(3) The analysis presented is for two-dimensional vectors. For three-dimensional vectors one has a similar result except that the rotation matrix is a three dimensional rotation matrix.

## 9.2 Transformation of stress

The objective of stress transformation rules is similar to the objective of vector transformation rules. For simplicity we will begin with the analysis of two-dimensional states of stress. Given the components of the stress tensor in the $x$-$y$ coordinate system we wish to find the components of the stress tensor in the $x'$-$y'$ coordinate system. To develop the transformation rules we can utilize the technique developed for vectors.

### 9.2.1 Traction vector method

Consider a point, P, in a body. At this point we will assume that we know the stress state relative to the $x$-$y$ coordinate system; i.e., we will assume that we know the stress tensor components $\sigma_{xx}$, $\sigma_{yy}$, and $\sigma_{xy}$. We wish to determine the components of the stress tensor relative to the $x'$-$y'$ coordinate system. Assuming the angle between the coordinate systems is $\theta$, this means that we wish to find the normal and tangential components of the traction vector on planes with normal vectors $\boldsymbol{e}_{x'}$ and $\boldsymbol{e}_{y'}$. Referring to Fig. 9.2, these vectors are given by

$$\boldsymbol{e}_{x'} = \cos(\theta)\boldsymbol{e}_x + \sin(\theta)\boldsymbol{e}_y, \tag{9.12}$$

$$\boldsymbol{e}_{y'} = -\sin(\theta)\boldsymbol{e}_x + \cos(\theta)\boldsymbol{e}_y. \tag{9.13}$$

Using Cauchy's Law we know that the traction vector on any plane with normal vector $\boldsymbol{n}$ is given by

$$\boldsymbol{t}(\boldsymbol{n}) = \boldsymbol{\sigma}^T\boldsymbol{n}. \tag{9.14}$$

Thus we know that on the plane with normal vector $\boldsymbol{e}_{x'}$, the traction vector is

$$\begin{aligned}\begin{pmatrix} t_x(\theta) \\ t_y(\theta) \end{pmatrix} &= \begin{bmatrix} \sigma_{xx} & \sigma_{yx} \\ \sigma_{xy} & \sigma_{yy} \end{bmatrix} \begin{pmatrix} \cos(\theta) \\ \sin(\theta) \end{pmatrix} \\ &= \begin{pmatrix} \cos(\theta)\sigma_{xx} + \sin(\theta)\sigma_{yx} \\ \cos(\theta)\sigma_{xy} + \sin(\theta)\sigma_{yy} \end{pmatrix}. \end{aligned} \tag{9.15}$$

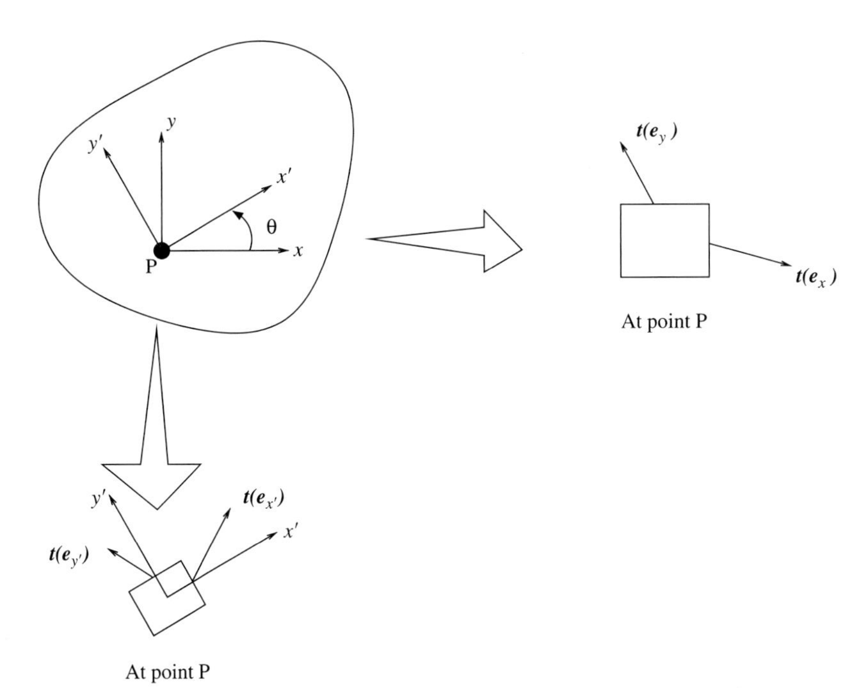


**Fig. 9.2** Geometry used for deriving stress transformation rules.

Notice that the traction vector can be written as $\boldsymbol{t} = t_x \boldsymbol{e}_x + t_y \boldsymbol{e}_y$, or equivalently as $\boldsymbol{t} = \sigma_{x'x'} \boldsymbol{e}_{x'} + \sigma_{x'y'} \boldsymbol{e}_{y'}$. Thus we take the dot product of eqn (9.15) with $\boldsymbol{e}_{x'}$ to find an expression for $\sigma_{x'x'}$. This gives

$$\begin{aligned}\sigma_{x'x'} &= +\sigma_{xx}\cos(\theta)\cos(\theta) + \sigma_{yx}\sin(\theta)\cos(\theta)\\ &\quad + \sigma_{xy}\cos(\theta)\sin(\theta) + \sigma_{yy}\sin(\theta)\sin(\theta).\end{aligned} \tag{9.16}$$

If we take the dot product of eqn (9.15) with $\boldsymbol{e}_{y'}$ then we find the transformation rule which gives $\sigma_{x'y'}$:

$$\begin{aligned}\sigma_{x'y'} &= -\sigma_{xx}\cos(\theta)\sin(\theta) - \sigma_{yx}\sin(\theta)\sin(\theta)\\ &\quad + \sigma_{xy}\cos(\theta)\cos(\theta) + \sigma_{yy}\sin(\theta)\cos(\theta).\end{aligned} \tag{9.17}$$

If we consider the plane with normal $\boldsymbol{e}_{y'}$, we can additionally show that

$$\begin{aligned}\sigma_{y'y'} &= +\sigma_{xx}\sin(\theta)\sin(\theta) - \sigma_{yx}\cos(\theta)\sin(\theta)\\ &\quad - \sigma_{xy}\sin(\theta)\cos(\theta) + \sigma_{yy}\cos(\theta)\cos(\theta).\end{aligned} \tag{9.18}$$

**Remarks:**

(1) The three transformation rules can be conveniently re-expressed in matrix form as:

$$\begin{aligned}\begin{bmatrix}\sigma_{x'x'} & \sigma_{x'y'}\\ \sigma_{y'x'} & \sigma_{y'y'}\end{bmatrix} &= \begin{bmatrix}\cos(\theta) & \sin(\theta)\\ -\sin(\theta) & \cos(\theta)\end{bmatrix}\begin{bmatrix}\sigma_{xx} & \sigma_{xy}\\ \sigma_{yx} & \sigma_{yy}\end{bmatrix}\\ &\qquad \begin{bmatrix}\cos(\theta) & -\sin(\theta)\\ \sin(\theta) & \cos(\theta)\end{bmatrix}.\end{aligned} \tag{9.19}$$

(2) The transformation rules are often written with aide of the double-angle formulae

$$\cos^2(\theta) = \frac{1}{2}\left[1 + \cos(2\theta)\right] \tag{9.20}$$

$$\sin^2(\theta) = \frac{1}{2}\left[1 - \cos(2\theta)\right] \tag{9.21}$$

$$\sin(\theta)\cos(\theta) = \frac{1}{2}\sin(2\theta). \tag{9.22}$$

This allows one to write the transformation rules in the following form:

$$\sigma_{x'x'} = \frac{\sigma_{xx} + \sigma_{yy}}{2} + \frac{\sigma_{xx} - \sigma_{yy}}{2}\cos(2\theta) + \sigma_{xy}\sin(2\theta) \tag{9.23}$$

$$\sigma_{y'y'} = \frac{\sigma_{xx} + \sigma_{yy}}{2} - \frac{\sigma_{xx} - \sigma_{yy}}{2}\cos(2\theta) - \sigma_{xy}\sin(2\theta) \tag{9.24}$$

$$\sigma_{x'y'} = -\frac{\sigma_{xx} - \sigma_{yy}}{2}\sin(2\theta) + \sigma_{xy}\cos(2\theta). \tag{9.25}$$

(3) The two-dimensional stress tensor possesses two *invariants*. The first is the trace of the tensor and the second is the determinant of the tensor. The trace and determinant are called invariants of the tensor because their numerical value does not depend upon the

coordinate system used to express the components of the tensor. In terms of the components this implies:

$$\mathrm{I}_\sigma = \mathrm{trace}[\boldsymbol{\sigma}] \quad = \sigma_{xx} + \sigma_{yy} \quad = \sigma_{x'x'} + \sigma_{y'y'} \tag{9.26}$$

$$\mathrm{II}_\sigma = \det[\boldsymbol{\sigma}] \quad = \sigma_{xx}\sigma_{yy} + \sigma_{xy}^2 \quad = \sigma_{x'x'}\sigma_{y'y'} + \sigma_{x'y'}^2. \tag{9.27}$$

The invariants are useful for double-checking numerical computations. They are also useful as they represent intrinsic properties of the state of stress independent of coordinate system.

(4) For three-dimensional stress transformations we have a result similar to the one presented except, that the rotation matrix is replaced by a three-dimensional rotation matrix. In three dimensions the invariants are defined as

$$\mathrm{I}_\sigma = \mathrm{trace}[\boldsymbol{\sigma}] \tag{9.28}$$

$$\mathrm{II}_\sigma = \frac{1}{2}[\mathrm{trace}(\boldsymbol{\sigma}^2) - (\mathrm{trace}(\boldsymbol{\sigma}))^2] \tag{9.29}$$

$$\mathrm{III}_\sigma = \det[\boldsymbol{\sigma}]. \tag{9.30}$$

**Example 9.1**

*Transformation of stress.* Consider a welded plate, as shown in Fig. 9.3, with applied loads such that it is in an homogeneous state of two-dimensional stress relative to the $x$-$y$ axes of

$$\begin{bmatrix} 100 & 50 \\ 50 & 20 \end{bmatrix}_{xy} \text{ksi.} \tag{9.31}$$

Find the state of stress in the plate relative to a set of axes aligned with the weld.

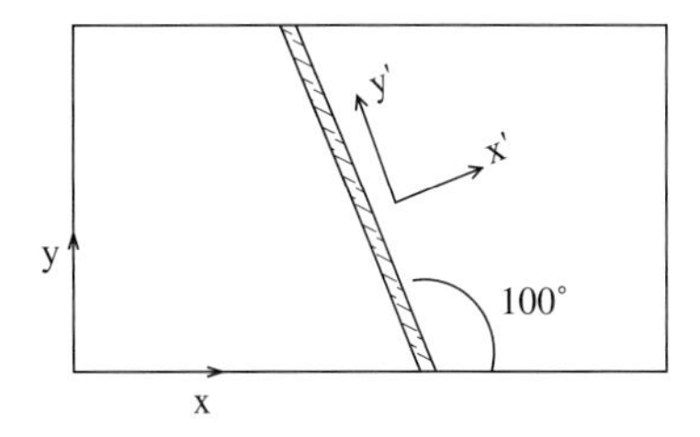


**Fig. 9.3** Uniformly stressed plate with a weld.

*Solution*

Select the $x'$ axis to be orthogonal to the weld line. Then we have that $\theta = 10 \times \pi/180$ rad. Inserting into the double angle transformation equations, we find

$$\sigma_{x'x'} = 60 + 40\cos(2\theta) + 50\sin(2\theta) \tag{9.32}$$

$$\sigma_{y'y'} = 60 - 40\cos(2\theta) - 50\sin(2\theta) \tag{9.33}$$

$$\sigma_{x'y'} = \quad -40\sin(2\theta) + 50\cos(2\theta), \tag{9.34}$$

which results in

$$\begin{bmatrix} 114.7 & 33.3 \\ 33.3 & 5.3 \end{bmatrix}_{x'y'} \text{ksi.} \tag{9.35}$$

**Remarks:**

(1) The matrix answer has been subscripted with the coordinate frame. This is to remind us that the components in the matrix are relative to the $x'$-$y'$ frame and not the $x$-$y$ coordinate frame.

Whenever expressing components of a tensor and confusion is possible, one should explicitly note the coordinate frame used.

(2) It is easily checked here that the invariants of the stress matrix did not change with coordinate frame. The trace is $\mathrm{I}_\sigma = 120$ ksi and the determinant is $\mathrm{II}_\sigma = -500\ \text{ksi}^2$, in both frames.

### 9.2.2 Maximum normal and shear stresses

The results of the previous section allow us to compute the normal stresses and shear stresses relative to an arbitrarily oriented coordinate system. Having this, it is natural to ask: what is the maximum value the normal stress can take, and what is the maximum value the shear stress can take?

To find the maximum normal stress we need to set the first derivative of eqn (9.23) to zero. This is the necessary condition for a maximum:

$$0 = \frac{d\sigma_{x'x'}}{d\theta} = -(\sigma_{xx} - \sigma_{yy}) \sin(2\theta) + 2\sigma_{xy} \cos(2\theta). \tag{9.36}$$

The solution to this equation is termed the principal angle, $\theta_p$, and is given by

$$\tan(2\theta_p) = \frac{2\sigma_{xy}}{\sigma_{xx} - \sigma_{yy}}. \tag{9.37}$$

Note that this equation shows that there are two solutions which differ from each other by $\pi/2$. One of these solutions corresponds to a maximum and the other to a minimum. If we evaluate the transformation equations for $\theta = \theta_p$, then we find

$$\sigma_{x'y'}(\theta_p) = 0 \tag{9.38}$$

$$\sigma_{x'x'}(\theta_p) = \frac{\sigma_{xx} + \sigma_{yy}}{2} + \sqrt{\left(\frac{\sigma_{xx} - \sigma_{yy}}{2}\right)^2 + \sigma_{xy}^2} \tag{9.39}$$

$$\sigma_{y'y'}(\theta_p) = \frac{\sigma_{xx} + \sigma_{yy}}{2} - \sqrt{\left(\frac{\sigma_{xx} - \sigma_{yy}}{2}\right)^2 + \sigma_{xy}^2}. \tag{9.40}$$

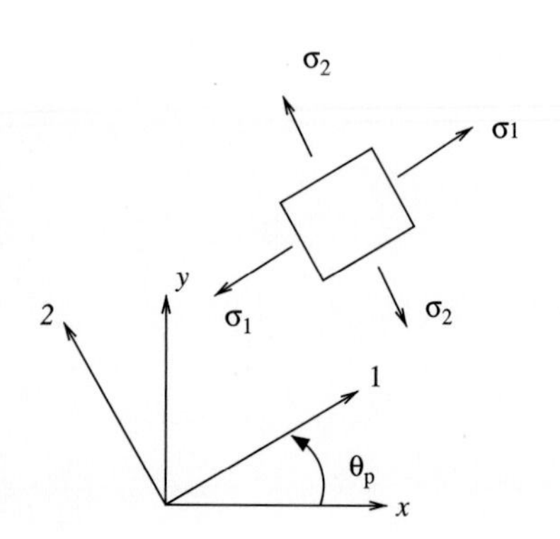

**Fig. 9.4** Stress in the principal frame.

**Remarks:**

(1) In the principal coordinate frame (principal axes) the shear stresses are zero. The two normal stresses in the principal frame are called the principal stresses. The larger represents the maximum normal stress and the smaller the minimum normal stresses. These are often denoted as $\sigma_1$ and $\sigma_2$, and by convention we order them as $\sigma_1 \geq \sigma_2$.

(2) Schematically, in the principal axes, we have the situation shown in Fig. 9.4.

(3) The principal axes are also often called the principal directions.

The orientation of the coordinate system in which the maximum shear stresses occur can be found in a similar manner:

$$0 = \frac{d\sigma_{x'y'}}{d\theta} = -(\sigma_{xx} - \sigma_{yy})\cos(2\theta) - 2\sigma_{xy}\sin(2\theta). \tag{9.41}$$

The solution to this equation gives the maximum shear angle, $\theta_s$, which is given by

$$\tan(2\theta_s) = -\frac{\sigma_{xx} - \sigma_{yy}}{2\sigma_{xy}}. \tag{9.42}$$

Just as above there are two solutions to this equation which differ from each other by $\pi/2$. One solution gives the maximum and the other the minimum. If we evaluate the stresses components in the maximum shear coordinate frame we find:

$$\sigma_{x'y'}(\theta_s) = \sqrt{\left(\frac{\sigma_{xx} - \sigma_{yy}}{2}\right)^2 + \sigma_{xy}^2} \tag{9.43}$$

$$\sigma_{x'x'}(\theta_s) = \frac{\sigma_{xx} + \sigma_{yy}}{2} \tag{9.44}$$

$$\sigma_{y'y'}(\theta_s) = \frac{\sigma_{xx} + \sigma_{yy}}{2}. \tag{9.45}$$

**Remarks:**

(1) The normal stresses in the maximum shear coordinate frame are equal to the average normal stress in the plane, $\sigma_m$.

(2) The maximum shear stress is often denoted by the symbol $\tau_{\max}$. Note that $\tau_{\max} = \frac{1}{2}(\sigma_1 - \sigma_2)$ in two dimensions.

(3) Schematically, in the maximum shear coordinate frame, we have the situation shown in Fig. 9.5.

(4) The maximum shear angle and the principal angle are related to each other by $\pi/4$:

$$\theta_p - \theta_s = \frac{\pi}{4}. \tag{9.46}$$

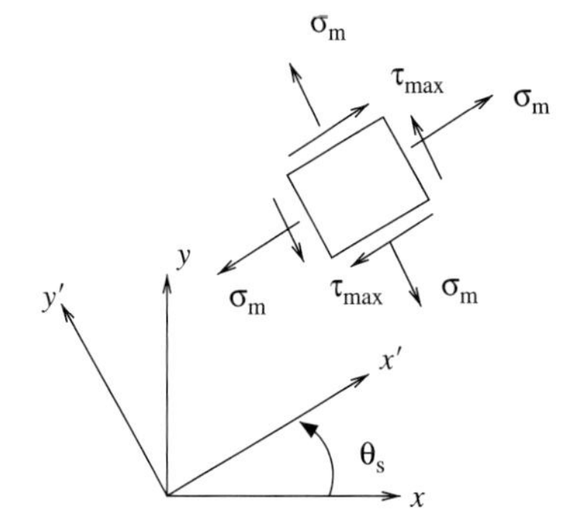

**Fig. 9.5** Stress in the maximum shear frame.

### 9.2.3 Eigenvalues and eigenvectors

If we evaluate the stress transformation eqn (9.19) for $\theta = \theta_p$, then we find that

$$\begin{bmatrix} \sigma_1 & 0 \\ 0 & \sigma_2 \end{bmatrix} = \begin{bmatrix} \cos(\theta_p) & \sin(\theta_p) \\ -\sin(\theta_p) & \cos(\theta_p) \end{bmatrix} \begin{bmatrix} \sigma_{xx} & \sigma_{xy} \\ \sigma_{yx} & \sigma_{yy} \end{bmatrix} \begin{bmatrix} \cos(\theta_p) & -\sin(\theta_p) \\ \sin(\theta_p) & \cos(\theta_p) \end{bmatrix}. \tag{9.47}$$

We can separate out the matrix products into two separate equations as:

$$\begin{bmatrix} \sigma_{xx} & \sigma_{xy} \\ \sigma_{yx} & \sigma_{yy} \end{bmatrix} \begin{pmatrix} \cos(\theta_p) \\ \sin(\theta_p) \end{pmatrix} = \sigma_1 \begin{pmatrix} \cos(\theta_p) \\ \sin(\theta_p) \end{pmatrix} \tag{9.48}$$

$$\begin{bmatrix} \sigma_{xx} & \sigma_{xy} \\ \sigma_{yx} & \sigma_{yy} \end{bmatrix} \begin{pmatrix} -\sin(\theta_p) \\ \cos(\theta_p) \end{pmatrix} = \sigma_2 \begin{pmatrix} -\sin(\theta_p) \\ \cos(\theta_p) \end{pmatrix}. \tag{9.49}$$

This reveals that the principal values are nothing more that the eigenvalues of the stress tensor and that the eigenvectors of the stress tensor correspond to the principal directions. The classical eigenvalue problem is usually written as

$$(\boldsymbol{\sigma} - \lambda \mathbf{1})\boldsymbol{n} = \mathbf{0}, \tag{9.50}$$

where $\lambda$ is the eigenvalue and $\boldsymbol{n}$ is the eigenvector. The condition for a non-trivial solution to these homogeneous equations is that $\det(\boldsymbol{\sigma} - \lambda\mathbf{1}) = 0$. This (in two dimensions) produces a quadratic polynomial in $\lambda$ (the characteristic polynomial):

$$-\lambda^2 + \mathrm{I}_\sigma \lambda - \mathrm{II}_\sigma = 0. \tag{9.51}$$

The two roots of this equation are the principal values.

**Remarks:**

(1) This observation about the connection between principal values and eigenvalues also holds true in three dimensions. In three dimensions the characteristic polynomial is given as

$$-\lambda^3 + \mathrm{I}_\sigma \lambda^2 - \mathrm{II}_\sigma \lambda + \mathrm{III}_\sigma = 0 \tag{9.52}$$

and there will be three principal values ($\sigma_1 \geq \sigma_2 \geq \sigma_3$) and three principal directions.

(2) While it may seem more complex to discuss the eigenvalues of a tensor, this is in practice the easiest way to compute the principal values of a general three-dimensional state of stress. This especially holds true due to efficient algorithms for numerically computing eigenvalues and eigenvectors.

(3) By the properties of symmetric tensors, we also have the result that the principal directions will always be orthogonal to each other – the eigenvectors of symmetric matrices can always be chosen to be orthogonal.

(4) The maximum shear stress is given by $\tau_{\max} = \frac{1}{2}(\sigma_1 - \sigma_2)$ in two dimensions and $\tau_{\max} = \frac{1}{2}(\sigma_1 - \sigma_3)$ in three dimensions.

---

**Example 9.2**

*Principal stresses.* Given a state of stress

$$\begin{bmatrix} 1.0 & 10.0 \\ 10.0 & 3.0 \end{bmatrix} \text{MPa} \tag{9.53}$$

find the principal values and principal directions.

*Solutions*
Compute the eigenvalues as the roots of the characteristic polynomial

$$\det \begin{bmatrix} 1.0-\lambda & 10.0 \\ 10.0 & 3.0-\lambda \end{bmatrix} = 0 \tag{9.54}$$

$$(1.0-\lambda)(3.0-\lambda) - 100.0 = 0. \tag{9.55}$$

Using the quadratic formula we find

$$\sigma_1 = 12.0 \text{ MPa} \tag{9.56}$$

$$\sigma_2 = -8.0 \text{ MPa}. \tag{9.57}$$

The principal directions are given by the eigenvectors which are found by solving the linear equations

$$\begin{bmatrix} 1.0-12.0 & 10.0 \\ 10.0 & 3.0-12.0 \end{bmatrix} \begin{pmatrix} n_x \\ n_y \end{pmatrix} = 0 \tag{9.58}$$

and

$$\begin{bmatrix} 1.0+8.0 & 10.0 \\ 10.0 & 3.0+8.0 \end{bmatrix} \begin{pmatrix} n_x \\ n_y \end{pmatrix} = 0. \tag{9.59}$$

The first principal direction is found to be

$$\begin{pmatrix} 0.6710 \\ 0.7415 \end{pmatrix} \tag{9.60}$$

and the second

$$\begin{pmatrix} -0.7415 \\ 0.6710 \end{pmatrix}. \tag{9.61}$$

To compute the principal angle we can always use the relation $\theta_p = \cos^{-1}(\boldsymbol{e}_x \cdot \boldsymbol{n}_1) = 0.8352$ rad, where $\boldsymbol{n}_1$ is the first principal direction.

### 9.2.4 Mohr's circle of stress

Mohr's circle is a graphical device that allows one to have a visual picture of all the normal and shear stress combinations which are possible by a change of coordinate basis. The device is usually applied to two-dimensional states of stress. It can, however, also be applied to three-dimensional states of stress when one of the principal stresses is known *a priori*.

Mohr's circle (of stress) is based upon the following writing of the transformation equations:

$$\begin{pmatrix} \sigma_{x'x'} \\ -\sigma_{x'y'} \end{pmatrix} = \begin{bmatrix} \cos(2\theta) & -\sin(2\theta) \\ \sin(2\theta) & \cos(2\theta) \end{bmatrix} \begin{pmatrix} \sigma_d \\ -\sigma_{xy} \end{pmatrix} + \begin{pmatrix} \sigma_m \\ 0 \end{pmatrix}, \tag{9.62}$$

where $\sigma_m = (\sigma_{xx} + \sigma_{yy})/2$ and $\sigma_d = (\sigma_{xx} - \sigma_{yy})/2$. These equations are observed to be the parametric equations for a circle centered at the point $(\sigma_m, 0)$ in the $\sigma_{x'x'}$–$\sigma_{x'y'}$ plane. The radius of the circle is $\sqrt{\sigma_d^2 + \sigma_{xy}^2}$.

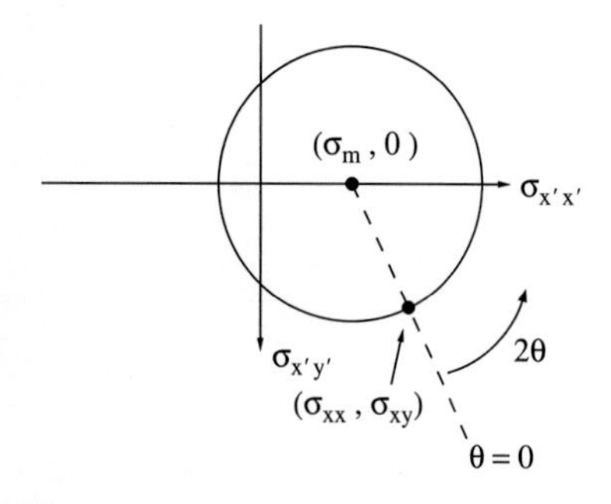


**Fig. 9.6** Mohr's circle.

**Remarks:**

(1) A plot of these equations is shown in Fig. 9.6. The loci of points shows all possible combinations of the normal and shear stresses on the plane with normal $\boldsymbol{e}_{x'}$.
(2) The intersections of the circle with the abscissa provides the principal values, and the angle from the dashed line to the abscissa gives twice the principal angle.
(3) The maximum shear stresses can be identified as the radius of the circle. The maximum shear angle is also identifiable as half the angle to the lowest point on the circle.
(4) Care should always be exercised in that rotations on the circle are double the physical rotation angles, and by convention we plot $\sigma_{x'y'}$ as positive downwards.[1]

[1] In some presentations, authors choose to plot $\sigma_{x'y'}$ as positive upwards. In this case, physical rotations occur in the direction opposite to the rotations on Mohr's circle.

## Example 9.3

*Mohr's circle.* Consider the two-dimensional state of stresses shown in Fig. 9.7. Using Mohr's circle find the principal values, the principal angle, the maximum shear, and the maximum shear angle. Sketch the principal stresses on a properly oriented element. Sketch the maximum shear state on a properly oriented element.

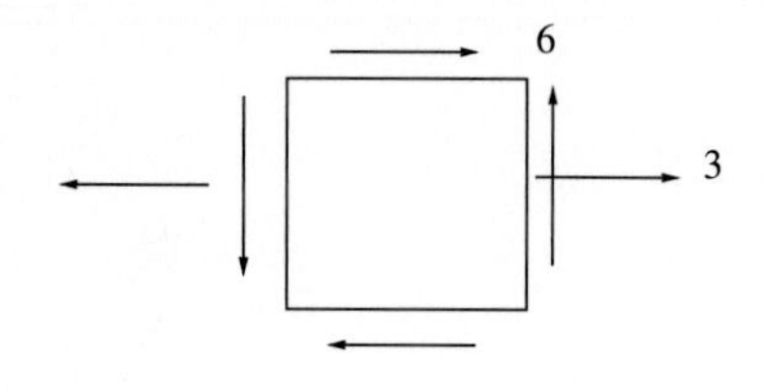


**Fig. 9.7** Stress state for Example 9.3.

*Solution*

The center of the Mohr's circle is located at $(1.5, 0)$ and the radius is $\sqrt{1.5^2 + 6^2} = 6.2$. Plotting the circle, we have the diagram shown in Fig. 9.8. From the diagram we can see that $\sigma_1 = 1.5 + 6.2 = 7.7$ and $\sigma_2 = 1.5 - 6.2 = -4.7$. The principal angle is seen to be $\theta_p = \frac{1}{2}\tan^{-1}(6/1.5) = 38.0^o$ (counter-clockwise). The maximum shear is seen to be $\tau_{\max} = 6.2$ and the maximum shear angle is seen to be $\theta_s = \frac{1}{2}[-2\theta_p + 90^o] = 7.0^o$ (clockwise). The corresponding states of stress are also sketched in Fig. 9.8.

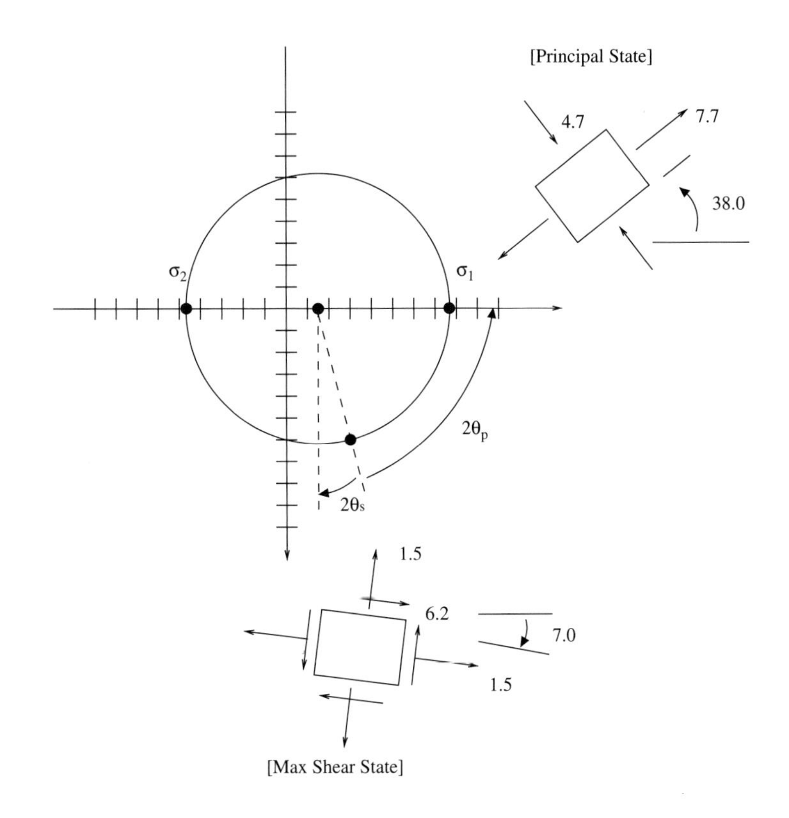


**Fig. 9.8** Mohr's circle for Example 9.3.

## Example 9.4

*Mohr's circle for torsion.* Consider the state of stress on the surface of an elastic circular bar of radius $R$, polar moment of inertia $J$, in torsion with an applied torque $T$:

$$\begin{bmatrix} 0 & \frac{TR}{J} \\ \frac{TR}{J} & 0 \end{bmatrix}_{\theta z}. \tag{9.63}$$

Draw the Mohr's circle and determine the principal stresses. Sketch them on a properly oriented element.

*Solution*

The center of the Mohr's circle is located at $(0, 0)$ and the radius is $\sqrt{0 + \frac{T^2R^2}{J^2}} = \frac{TR}{J}$. Plotting the circle, we have the diagram shown in Fig. 9.9. From the diagram we can see that $\sigma_1 = \frac{TR}{J}$ and $\sigma_2 = -\frac{TR}{J}$. The principal angle is seen to be $\theta_p = 45.0^{\mathrm{o}}$ (counter-clockwise). The principal state of stress is also sketched in Fig. 9.9.

**Remarks:**

(1) This analysis confirms our discussion in Section 7.5 on brittle failure in torsion, where we claimed that the maximum normal stress occurred on a plane angled at $45^{\mathrm{o}}$ to the bar axis.

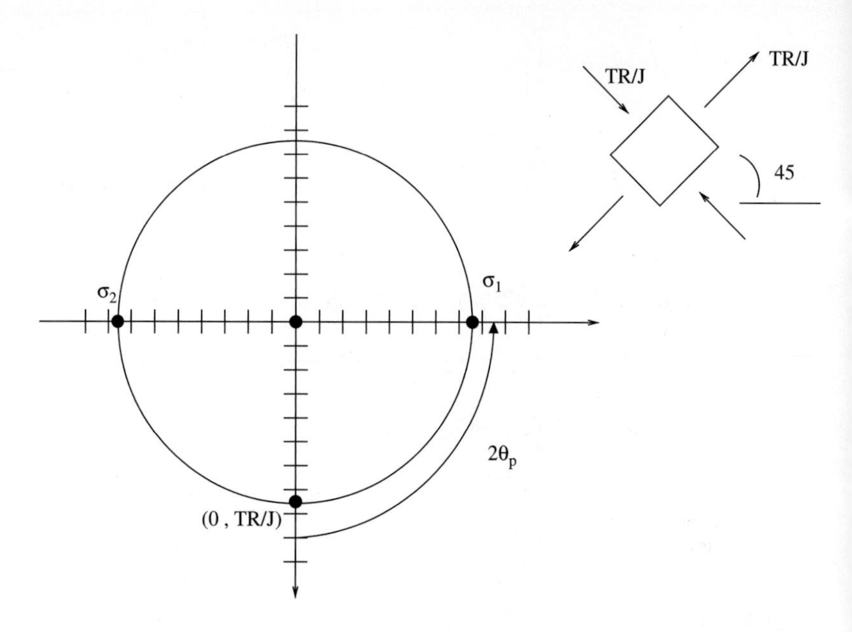


**Fig. 9.9** Mohr's circle for Example 9.4.

### 9.2.5 Three-dimensional Mohr's circles of stress

Mohr's circle as developed so far is a strictly two-dimensional device. We can, however, extend its use to three dimensions by noting that we have been implicitly assuming that $\sigma_{zz} = \sigma_{zx} = \sigma_{yz} = 0$. The key point to observe is that we have been assuming that there are no shear stresses on the plane normal to $\boldsymbol{e}_z$. So, we have been tacitly assuming that the $z$-direction is a principal direction with corresponding principal stress $\sigma_3 = 0$. Another way of stating this is that the Mohr's circle we have drawn corresponds to looking down the $z$-axis at the material in the $x$-$y$ plane. The generalization to this is that given a principal axis one can always draw a Mohr's circle for the state of stress in the plane normal to the the given axis. Since in three dimensions there are three principal axes, it is possible to draw three Mohr's circles for a general state of stress – one circle for each plane normal to a principal axis. This is sketched in Fig. 9.10.

**Remarks:**

(1) To construct the three Mohr's circles one must *a priori* know one of the principal values. The value is first plotted as a value on the abscissa. Then the two-dimensional Mohr's circle normal to this principal direction is drawn. The intersections with the abscissa can then be connected to create the three circles.

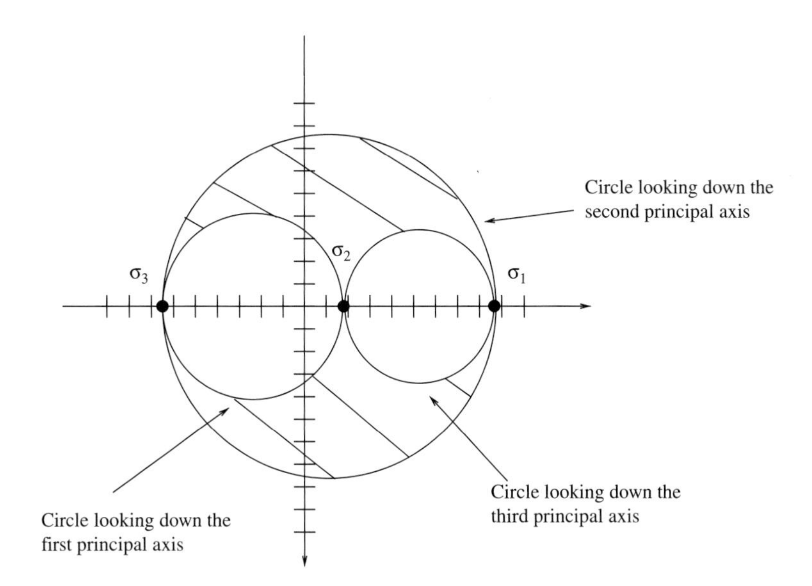


**Fig. 9.10** Three-dimensional Mohr's circles.

(2) Each individual circle has the same interpretation as we had before.

(3) The true maximum shear stress in three dimensions occurs on the circle with greatest radius.

(4) It can proved that all possible values for the normal stresses and shear stresses for any orientation of the coordinate axes is contained in the hatched region in Fig. 9.10.

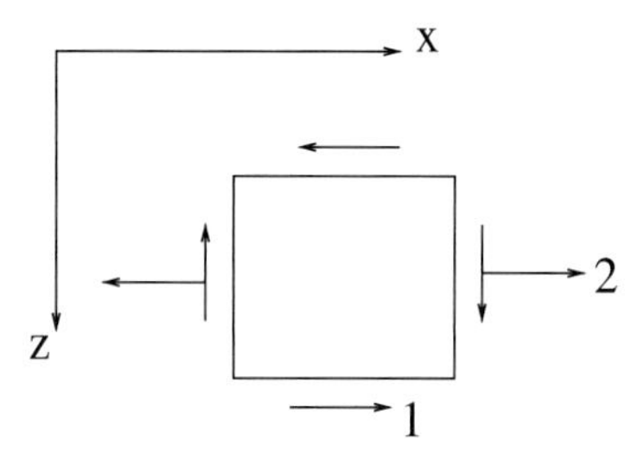


**Fig. 9.11** $z$-$x$ plane state of stress.

### Example 9.5

*Three-dimensional Mohr's circles.* Consider the state of stress:

$$\begin{bmatrix} 2 & 0 & 1 \\ 0 & -2 & 0 \\ 1 & 0 & 0 \end{bmatrix} \text{ ksi.} \tag{9.64}$$

Find the principal values and the maximum shear stress for this state of stress.

*Solution*

The $y$-axis is seen to be a principal axis since the plane with normal vector in the $y$-direction has no shear stresses on it ($\sigma_{yx} = \sigma_{yz} = 0$). Looking down the $y$-axis at the material in the $z$-$x$ plane, we have the state of stress shown in Fig. 9.11. The Mohr's circle for this state is shown in Fig. 9.12. If we add the principal value $-2$ ksi to the diagram we can then construct the three circles of stress as shown in Fig. 9.13.

**Remarks:**

(1) The principal stresses can be determined from the diagram to be $-2, 1 \pm \sqrt{2}$ ksi.

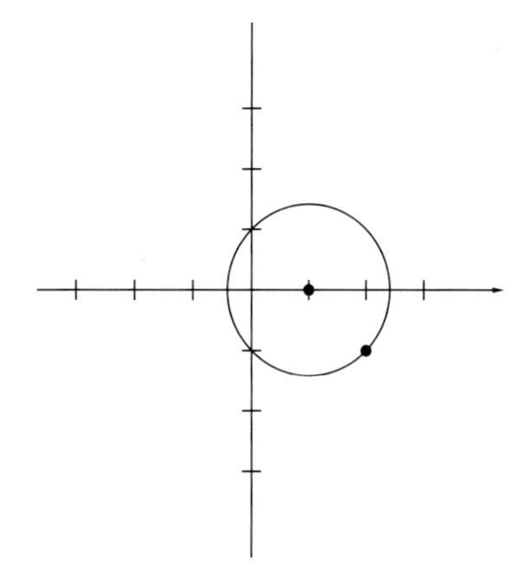

**Fig. 9.12** Mohr's circle in the $z$-$x$ plane.

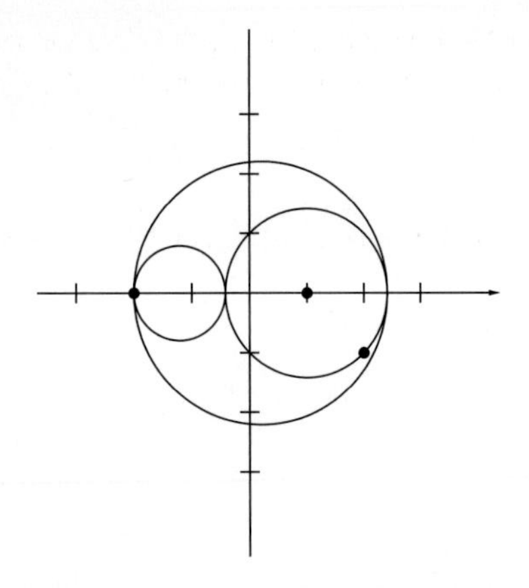

**Fig. 9.13** Three-dimensional Mohr's circles.

(2) The maximum shear stress is the radius of the largest circle, $(3 + \sqrt{2})/2$ ksi.

(3) This construction cannot be performed unless at least one of the principal values is known *a priori*. In the situation where no principal values are known one must use the eigenvalue method to compute the principal values.

## 9.3 Transformation of strains

In multi-dimensions, strains, like stresses, are represented by tensors. As such they also obey transformation rules upon change of coordinates. In fact, they obey the exact same transformation rules. To see this, let us derive the expression for the normal strain in the, say, $x'$-direction in terms of the strains expressed relative to an $x$-$y$ coordinate system. The setup will be identical to the one we used to discuss stress transformations; the angle between the $x$-axis and the $x'$-axis will be $\theta$.

By definition, the normal strain in the $x'$-direction is given by:

$$\varepsilon_{x'x'} = \frac{\partial u_{x'}}{\partial x'}. \tag{9.65}$$

Our goal is to expand the right-hand-side of eqn (9.65) in terms of the strain components in the $x$-$y$ coordinate basis – i.e. in terms of $\varepsilon_{xx}$, $\varepsilon_{yy}$, and $\varepsilon_{xy} = \gamma_{xy}/2$. To do this, we need expression for $u_{x'}$ in terms of $u_x$ and $u_y$. We also need an expression for $\frac{\partial}{\partial x'}$ in terms of $\frac{\partial}{\partial x}$ and $\frac{\partial}{\partial y}$. Both relations can be derived using the transformation rules for vectors from Section 9.1. In particular,

$$\begin{aligned} u_{x'} = \boldsymbol{e}_{x'} \cdot \boldsymbol{u} &= \boldsymbol{e}_{x'} \cdot (u_x \boldsymbol{e}_x + u_y \boldsymbol{e}_y) \\ &= u_x \cos(\theta) + u_y \sin(\theta). \end{aligned} \tag{9.66}$$

To transform the derivative, note that the chain rule tells us that:

$$\frac{\partial}{\partial x'} = \frac{\partial x}{\partial x'}\frac{\partial}{\partial x} + \frac{\partial y}{\partial x'}\frac{\partial}{\partial y}. \tag{9.67}$$

To compute the coefficients in this relations, note that the position vector transforms as

$$\begin{pmatrix} x' \\ y' \end{pmatrix} = \begin{bmatrix} \cos(\theta) & \sin(\theta) \\ -\sin(\theta) & \cos(\theta) \end{bmatrix} \begin{pmatrix} x \\ y \end{pmatrix} \tag{9.68}$$

and the inverse of this is

$$\begin{pmatrix} x \\ y \end{pmatrix} = \begin{bmatrix} \cos(\theta) & -\sin(\theta) \\ \sin(\theta) & \cos(\theta) \end{bmatrix} \begin{pmatrix} x' \\ y' \end{pmatrix}. \tag{9.69}$$

Thus, $\partial x/\partial x' = \cos(\theta)$ and $\partial y/\partial x' = \sin(\theta)$. Combining these results gives

$$\begin{aligned}
\varepsilon_{x'x'} &= \left(\cos(\theta)\frac{\partial}{\partial x} + \sin(\theta)\frac{\partial}{\partial y}\right)(u_x\cos(\theta) + u_y\sin(\theta)) \\
&= \cos^2(\theta)\frac{\partial u_x}{\partial x} + \cos(\theta)\sin(\theta)\left(\frac{\partial u_x}{\partial y} + \frac{\partial u_y}{\partial x}\right) + \sin^2(\theta)\frac{\partial u_y}{\partial y} \\
&= \cos^2(\theta)\varepsilon_{xx} + 2\cos(\theta)\sin(\theta)\varepsilon_{xy} + \sin^2(\theta)\varepsilon_{yy}. \qquad (9.70)
\end{aligned}$$

Similar expressions can be derived for $\varepsilon_{y'y'} = \partial u_{y'}/\partial y'$ and $\varepsilon_{x'y'} = \frac{1}{2}(\partial u_{x'}/\partial y' + \partial u_{y'}/\partial x')$. When combined, we find

$$\begin{bmatrix} \varepsilon_{x'x'} & \varepsilon_{x'y'} \\ \varepsilon_{y'x'} & \varepsilon_{y'y'} \end{bmatrix} = \begin{bmatrix} \cos(\theta) & \sin(\theta) \\ -\sin(\theta) & \cos(\theta) \end{bmatrix} \begin{bmatrix} \varepsilon_{xx} & \varepsilon_{xy} \\ \varepsilon_{yx} & \varepsilon_{yy} \end{bmatrix} \begin{bmatrix} \cos(\theta) & -\sin(\theta) \\ \sin(\theta) & \cos(\theta) \end{bmatrix}. \qquad (9.71)$$

**Remarks:**

(1) Equation (9.71) is identical in form to the transformation rule for stresses given in eqn (9.19).

(2) Just as for stresses, the transformation rules for strain are often expressed using the double angle formulae:

$$\varepsilon_{x'x'} = \frac{\varepsilon_{xx} + \varepsilon_{yy}}{2} + \frac{\varepsilon_{xx} - \varepsilon_{yy}}{2}\cos(2\theta) + \varepsilon_{xy}\sin(2\theta) \qquad (9.72)$$

$$\varepsilon_{y'y'} = \frac{\varepsilon_{xx} + \varepsilon_{yy}}{2} - \frac{\varepsilon_{xx} - \varepsilon_{yy}}{2}\cos(2\theta) - \varepsilon_{xy}\sin(2\theta) \qquad (9.73)$$

$$\varepsilon_{x'y'} = -\frac{\varepsilon_{xx} - \varepsilon_{yy}}{2}\sin(2\theta) + \varepsilon_{xy}\cos(2\theta). \qquad (9.74)$$

(3) For the same reason as with the stress tensor, the strain tensor also possesses two invariants in two dimensions. These are $\mathrm{I}_\varepsilon = \mathrm{trace}[\boldsymbol{\varepsilon}]$ and $\mathrm{II}_\varepsilon = \det[\boldsymbol{\varepsilon}]$ with associated characteristic polynomial:

$$-\lambda^2 + \mathrm{I}_\varepsilon\lambda - \mathrm{II}_\varepsilon = 0. \qquad (9.75)$$

In three dimensions $\mathrm{I}_\varepsilon = \mathrm{trace}[\boldsymbol{\varepsilon}]$, $\mathrm{II}_\varepsilon = \frac{1}{2}(\mathrm{trace}[\boldsymbol{\varepsilon}^2] - (\mathrm{trace}[\boldsymbol{\varepsilon}])^2)$, and $\mathrm{III}_\varepsilon = \det[\boldsymbol{\varepsilon}]$ with associated characteristic polynomial:

$$-\lambda^3 + \mathrm{I}_\varepsilon\lambda^2 - \mathrm{II}_\varepsilon\lambda + \mathrm{III}_\varepsilon = 0. \qquad (9.76)$$

### 9.3.1 Maximum normal and shear strains

As with stresses, we can speak of principal strains. In two dimensions, principal strains will represent the maximum and minimum normal strains in the plane. The mathematics of computing these values follows exactly our developments for stress. In this regard we can follow any of the approaches shown for stresses.

(1) We can simply maximize the expressions for the normal strain and the shear strain. For the normal strains, this will give

$$\varepsilon_1 = \varepsilon_{x'x'}(\theta_p) = \frac{\varepsilon_{xx} + \varepsilon_{yy}}{2} \tag{9.77}$$

$$+ \sqrt{\left(\frac{\varepsilon_{xx} - \varepsilon_{yy}}{2}\right)^2 + \varepsilon_{xy}^2} \tag{9.78}$$

$$\varepsilon_2 = \varepsilon_{y'y'}(\theta_p) = \frac{\varepsilon_{xx} + \varepsilon_{yy}}{2} \tag{9.79}$$

$$- \sqrt{\left(\frac{\varepsilon_{xx} - \varepsilon_{yy}}{2}\right)^2 + \varepsilon_{xy}^2}, \tag{9.80}$$

where $\tan(2\theta_p) = 2\varepsilon_{xy}/(\varepsilon_{xx} - \varepsilon_{yy})$. Again, like with stresses, the shear strain is zero in the principal frame. For the maximum shear strains we have:

$$\frac{1}{2}\gamma_{\text{max}} = \varepsilon_{x'y'}(\theta_s) = \sqrt{\left(\frac{\varepsilon_{xx} - \varepsilon_{yy}}{2}\right)^2 + \varepsilon_{xy}^2} \tag{9.81}$$

$$\varepsilon_{x'x'}(\theta_s) = \frac{\varepsilon_{xx} + \varepsilon_{yy}}{2} \tag{9.82}$$

$$\varepsilon_{y'y'}(\theta_s) = \frac{\varepsilon_{xx} + \varepsilon_{yy}}{2}, \tag{9.83}$$

where $\tan(2\theta_s) = -(\varepsilon_{xx} - \varepsilon_{yy})/2\varepsilon_{xy}$.

(2) We can also utilize an eigenvalue approach. In this case we can solve the polynomial $\det[\boldsymbol{\varepsilon} - \lambda\mathbf{1}] = 0$ for the eigenvalues of $\boldsymbol{\varepsilon}$. These will be the principal values. In two dimensions $\frac{1}{2}\gamma_{\text{max}} = (\varepsilon_1 - \varepsilon_2)/2$, and in three dimensions $\frac{1}{2}\gamma_{\text{max}} = (\varepsilon_1 - \varepsilon_3)/2$. The eigenvectors will point in the directions of the principal axes. This approach is required for general states of strain in three dimensions.

(3) We can apply a Mohr's circle idea. In this case the axes on the Mohr diagram will be $\varepsilon_{x'x'}$ on the abscissa and $\varepsilon_{x'y'}$ on the ordinate (pointing down). As with stresses, this approach is limited to two-dimensional planar cases and three-dimensional cases where one of the principal strains is already known.

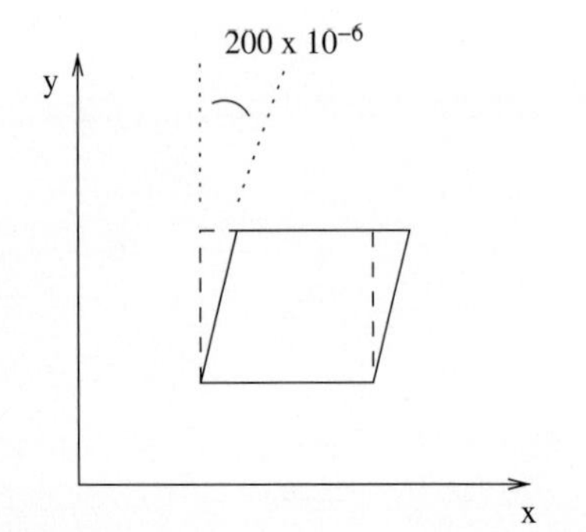


**Fig. 9.14** Sketch of deformation of a square aligned with the $x$-$y$ axes in Example 9.6.

### Example 9.6

*Mohr's circle for strain.* Consider a two-dimensional state of pure shear $\gamma_{xy} = 2\varepsilon_{xy} = 200 \times 10^{-6} = 200$ $\mu$strain and $\varepsilon_{xx} = \varepsilon_{yy} = 0$. Sketch the deformation associated with this strain state for an homogeneously strained square of material that is aligned with the $x$-$y$ coordinate axes. Compute the principal strains and sketch the deformation associated with this strain state for an homogeneously strained square of material which is aligned with the principal axes.

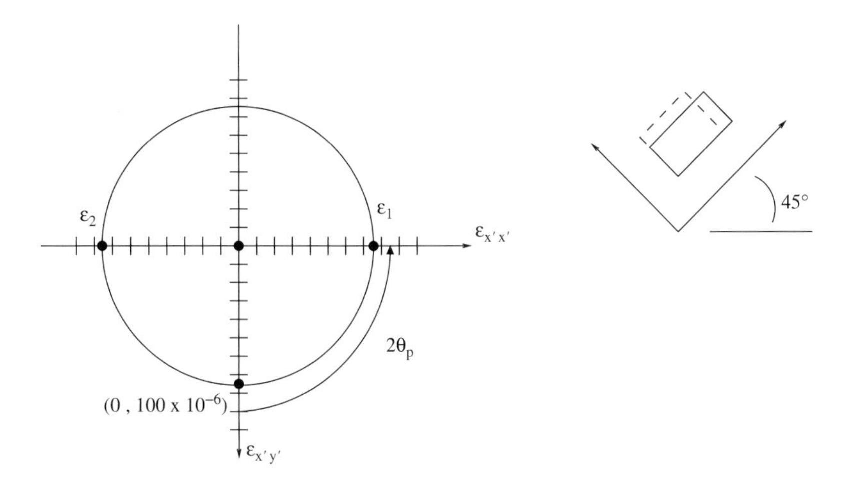


**Fig. 9.15** Mohr's circle of strain for Example 9.6.

*Solution*
For the first part of the question we need to consider a square of material that is aligned with the coordinate axes. The square undergoes no normal strains in the $x$- or $y$-directions but it does experience a change in angle of $200 \times 10^{-6}$ rad. This is sketched in Fig. 9.14. To compute the principal strains we plot Mohr's circle of strain. The center of the circle is the mean normal strain equal to zero. The radius of the circle is $\sqrt{\frac{\varepsilon_{xx}-\varepsilon_{yy}}{2} + \varepsilon_{xy}^2} = 100 \times 10^{-6}$. The circle is plotted in Fig. 9.15. The principal angle is seen to be 45° and in the principal frame the normal strains are $\pm 100 \times 10^{-6}$. The deformation state aligned with the principal frame is also sketched in Fig. 9.15. There is no angle change in the frame, and there is an elongational strain in the first principal direction and a contraction in the second.

## Example 9.7

*Strain rosette.* The measurement of strains is often performed using strain gauges – small strips of electrically conductive wire. They are glued to the surface of an object. When the object is strained the gauges are strained by the same amount. Strain in the gauge changes the electrical resistance of the wire. This can be detected electronically and converted to strain. The gauges are designed so that they only measure normal strains in a single direction. In order to measure a complete two-dimensional state of strain, several gauges are needed. Consider the layout of strain gauges shown in Fig. 9.16. Given the normal strains in gauges $a$, $b$, and $c$ determine the two-dimensional state of strain; i.e. find expressions for $\varepsilon_{xx}$, $\varepsilon_{yy}$, and $\varepsilon_{xy}$.

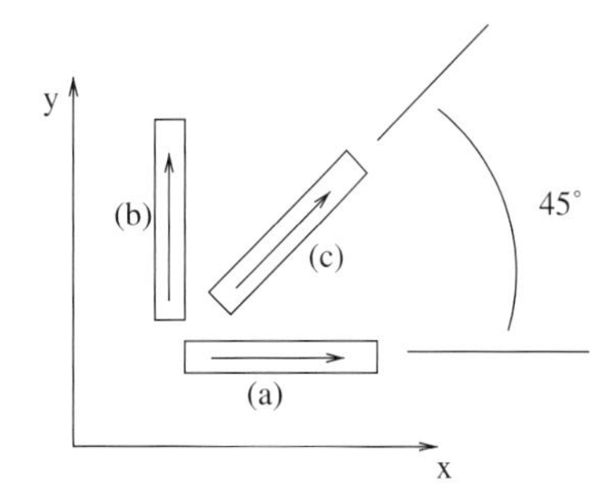


**Fig. 9.16** 0-45-90 strain-gauge rosette.

*Solution*
Gauges $a$ and $b$ immediately tell us that $\varepsilon_{xx} = \varepsilon_a$ and $\varepsilon_{yy} = \varepsilon_b$. Gauge $c$ tells us the normal strain in the 45° direction. Thus we have:

$$\varepsilon_c = \varepsilon_{x'x'}(\pi/4) = \frac{\varepsilon_a + \varepsilon_b}{2} + \frac{\varepsilon_a - \varepsilon_b}{2}\cos(\pi/2) + \varepsilon_{xy}\sin(\pi/2). \quad (9.84)$$

Solving this for $\varepsilon_{xy}$, gives

$$\varepsilon_{xy} = \varepsilon_c - \frac{\varepsilon_a + \varepsilon_b}{2}. \quad (9.85)$$

Assembling together we find

$$\begin{bmatrix} \varepsilon_a & \varepsilon_c - \frac{\varepsilon_a+\varepsilon_b}{2} \\ \varepsilon_c - \frac{\varepsilon_a+\varepsilon_b}{2} & \varepsilon_b \end{bmatrix}_{xy}. \quad (9.86)$$

---

**Example 9.8**

*Relation between elastic constants.* Use an energy argument to show that $G = E/[2(1+\nu)]$.

*Solution*

Let us consider a state of pure shear stress in two dimensions. The stress tensor is given by

$$\begin{bmatrix} 0 & c \\ c & 0 \end{bmatrix}_{xy}, \quad (9.87)$$

where $c$ is a constant representing the magnitude of the shear stress. This exact same state of stress can be represented in a coordinate system oriented at 45° to the $x$-$y$ axes. In this second coordinate system the state of stress appears as

$$\begin{bmatrix} c & 0 \\ 0 & -c \end{bmatrix}_{x'y'}, \quad (9.88)$$

where the $x'$- and $y'$-axes are rotated 45° relative to the $x$- and $y$-axes. Because both sets of stress components represent the same state of stress, one can claim that the associated strain energy density should be the same. In the original coordinate frame we find (using eqn (5.17)):

$$\begin{aligned} w &= \frac{1}{2}(\sigma_{xx}\varepsilon_{xx} + \sigma_{yy}\varepsilon_{yy} + \sigma_{zz}\varepsilon_{zz} + \sigma_{xy}\gamma_{xy} + \sigma_{yz}\gamma_{yz} + \sigma_{zx}\gamma_{zx}) \\ &= \frac{1}{2}(\sigma_{xy}\gamma_{xy}) \\ &= \frac{1}{2}(\sigma_{xy}^2/G) \\ &= \frac{1}{2}(c^2/G). \end{aligned} \quad (9.89)$$

In the rotated frame we have

$$\begin{aligned} w = \frac{1}{2}\big(&\sigma_{x'x'}\varepsilon_{x'x'} + \sigma_{y'y'}\varepsilon_{y'y'} + \sigma_{z'z'}\varepsilon_{z'z'} \\ &+ \sigma_{x'y'}\gamma_{x'y'} + \sigma_{y'z'}\gamma_{y'z'} + \sigma_{z'x'}\gamma_{z'x'}\big) \end{aligned}$$

$$
\begin{aligned}
&= \frac{1}{2}\left(\sigma_{x'x'}\varepsilon_{x'x'} + \sigma_{y'y'}\varepsilon_{y'y'}\right) \\
&= \frac{1}{2}\left(\sigma_{x'x'}\left[\frac{\sigma_{x'x'}}{E} - \frac{\nu}{E}\sigma_{y'y'}\right] + \sigma_{y'y'}\left[\frac{\sigma_{y'y'}}{E} - \frac{\nu}{E}\sigma_{x'x'}\right]\right) \\
&= \frac{1}{2}c^2\frac{2+2\nu}{E}. \qquad (9.90)
\end{aligned}
$$

If we equate the two results, which describe the exact same physical state, then we see that

$$G = \frac{E}{2(1+\nu)}. \qquad (9.91)$$

## 9.4 Multi-axial failure criteria

In the chapters up to this point we have analyzed systems dominated by a single component of the stress tensor. For this reason, we have been able to simply determine the load-carrying capacity of different systems by requiring that

$$|\sigma| \leq \sigma_Y \qquad (9.92)$$

for systems that are dominated by a single normal stress, or that

$$|\tau| \leq \tau_Y \qquad (9.93)$$

for systems that are dominated by a single shear stress $\tau$. When we have a multi-axial state of stress (more that one non-zero stress component), then we are faced with a new problem. Do we simply enforce eqns (9.92) and (9.93) component-by-component or should we do something else? As it turns out, component-by-component enforcement of eqns (9.92) and (9.93) does not comport with experimental experience. Further, it has the unfortunate side-effect of being coordinate system dependent. In other words, component by component enforcement of eqns (9.92) and (9.93) will not be able to predict material yielding/failure independent of the coordinate system used to express the stress components. What we would like is a criterion that evaluates some function of the stress tensor and outputs a value indicating yield/failure or not yield/failure. The function should be independent of coordinate system for it to be physically meaningful. Thus we would like something of the form

$$f(\boldsymbol{\sigma}) \leq \text{critial value}. \qquad (9.94)$$

For polycrystalline metallic materials at room temperature the two most common criteria of this form are Tresca's yield condition and the Henky–von Mises yield condition; these are both discussed next. Criteria for the multi-axial failure/yield of brittle materials and materials with different properties in tension and compression is left for more advanced texts.

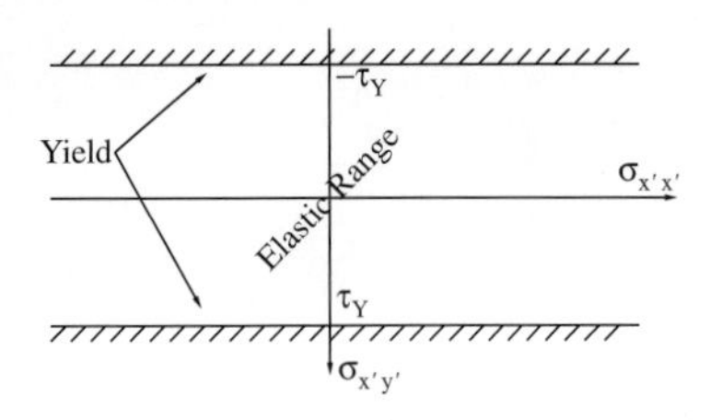


**Fig. 9.17** Admissible elastic range according to Tresca's condition.

### 9.4.1 Tresca's yield condition

Tresca's yield condition simply states that the maximum shear stress (in three dimensions) should be less that $\tau_Y$ the yield stress in shear. If $\tau_{\max} < \tau_Y$, then the stress state is elastic. Yield occurs where $\tau_{\max} = \tau_Y$. A Mohr diagram is one way to visualize Tresca's yield condition. As long as the three circles of stress remain between $\tau_Y$ and $-\tau_Y$ the state of stress is elastic; see Fig. 9.17.

**Remarks:**

(1) Tresca's condition meshes nicely with ideas from material science on dislocations. At a fundamental level, yield is associated with dislocation glide on slip planes, and this motion is driven by Schmidt resolved shear stress on the glide planes. In a polycrystalline material there are glide planes in all directions, and thus yield takes place when the shear stress on any plane reaches a critical value.

(2) Tresca's condition is often easy to use in hand computations. Simply figure out the three circles of stress and check that they lie within the admissible range.

(3) It has been experimentally determined that room-temperature yield in metals is independent of the hydrostatic (or volumetric) component of the stress state. The volumetric part of the stress state is defined to be

$$\boldsymbol{\sigma}_{\text{vol}} = p\mathbf{1}, \tag{9.95}$$

where $p = \frac{1}{3}\text{trace}[\boldsymbol{\sigma}] = \frac{1}{3}(\sigma_{xx} + \sigma_{yy} + \sigma_{zz})$ is the mean normal stress (otherwise known as the pressure). The remaining part of the stress tensor is known as the deviatoric part of the stress, and is usually denoted by the symbol $\boldsymbol{s}$. Thus

$$\boldsymbol{s} = \boldsymbol{\sigma} - \boldsymbol{\sigma}_{\text{vol}} \tag{9.96}$$

$$\begin{bmatrix} s_{xx} & s_{xy} & s_{xz} \\ s_{yx} & s_{yy} & s_{yz} \\ s_{zx} & s_{zy} & s_{zz} \end{bmatrix} = \begin{bmatrix} \sigma_{xx} & \sigma_{xy} & \sigma_{xz} \\ \sigma_{yx} & \sigma_{yy} & \sigma_{yz} \\ \sigma_{zx} & \sigma_{zy} & \sigma_{zz} \end{bmatrix} - p \begin{bmatrix} 1 & 0 & 0 \\ 0 & 1 & 0 \\ 0 & 0 & 1 \end{bmatrix}. \tag{9.97}$$

The statement that yield is independent of pressure implies that yield depends only on the deviatoric part of the stress. It is easy to see that Tresca's conditions satisfies this observation. Adding a hydrostatic component to any given stress state will only shift the corresponding three circles of stress to the right or left on the Mohr diagram. It will not affect the radius of the largest circle of stress.

(4) Tresca's condition can be written in the form of eqn (9.94) but it is not very convenient to use it in that format; so, we do not include it here.

**Example 9.9**

*Tresca's condition.* Consider the state of stress in Example 9.5, but scaled by the factor $k$.

$$k \begin{bmatrix} 2 & 0 & 1 \\ 0 & -2 & 0 \\ 1 & 0 & 0 \end{bmatrix} \text{ ksi} \tag{9.98}$$

If $\tau_Y = 20$ ksi, what value of $k$ causes yield?

*Solution*
Using the results from Example 9.5 we know that the radius of the largest circle of stress will be $k(3+\sqrt{2})/2$. Thus we have that a value of

$$k = \frac{40}{3+\sqrt{2}} \tag{9.99}$$

will initiate yield according to Tresca's condition.

**Remarks:**

(1) It is important to note that one needs to check the radius of the largest of the three circles of stress. It is not enough to check only the stresses in a single plane.

**Example 9.10**

*Calibration of $\tau_Y$.* The typical test for calibrating yield criteria is to use a uniaxial tension test to determine the yield stress $\sigma_Y$. What is the relation between $\sigma_Y$ and the $\tau_Y$?

*Solution*
In a uniaxial tension test at the point of yield the three-dimensional state of stress is

$$\begin{bmatrix} \sigma_Y & 0 & 0 \\ 0 & 0 & 0 \\ 0 & 0 & 0 \end{bmatrix}. \tag{9.100}$$

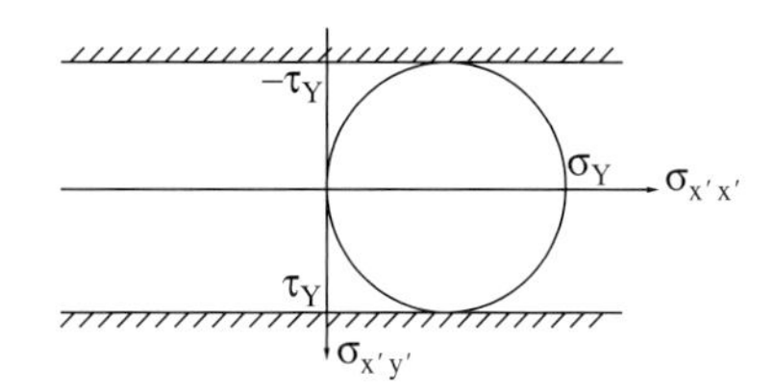


**Fig. 9.18** Uniaxial tension test on a Mohr diagram.

On a Mohr diagram, we have the situation sketched in Fig. 9.18. All three circle of stress are shown. One circle is centered at $(0,0)$ and has zero radius. The other two circles are centered at $(\sigma_Y/2, 0)$ and have radius $\sigma_Y/2$. Thus

$$\tau_Y = \frac{\sigma_Y}{2} \tag{9.101}$$

according to Tresca's condition.

### 9.4.2 Henky–von Mises condition

The Henky–von Mises condition for yield is based upon an energetic idea. Their idea was that yield in a multi-axial state of stress will occur when the strain energy density reaches a critical value. The critical value is found by computing the strain energy density in a uniaxial test specimen at the moment of initial yield. Because yield is generally observed to be independent of pressure, the Henky–von Mises condition involves only the part of the strain energy associated with the deviatoric stresses. The strain energy density is given by:

$$w = \frac{1}{2}(\sigma_{xx}\varepsilon_{xx} + \sigma_{yy}\varepsilon_{yy} + \sigma_{zz}\varepsilon_{zz} + \sigma_{xy}\gamma_{xy} + \sigma_{yz}\gamma_{yz} + \sigma_{zx}\gamma_{zx}). \tag{9.102}$$

If we substitute in for the strains using Hooke's Law and separate the pressure contributions from the deviatoric contributions, then we find that

$$\begin{aligned} w = & \underbrace{\frac{1+\nu}{2E}(s_{xx}^2 + s_{yy}^2 + s_{zz}^2 + 2s_{xy}^2 + 2s_{yz}^2 + 2s_{zx}^2)}_{w_{\text{dev}}} \\ & + \underbrace{\frac{3(1-2\nu)}{2E}p^2}_{w_{\text{vol}}}. \end{aligned} \tag{9.103}$$

The Henky–von Mises condition then provides that

$$w_{\text{dev}} \le w_{\text{dev}}^{1-\text{D}}, \tag{9.104}$$

where $w_{\text{dev}}^{1-\text{D}}$ is the calibration constant determined from a uniaxial test. In a uniaxial test at yield the stress is given as:

$$\begin{bmatrix} \sigma_Y & 0 & 0 \\ 0 & 0 & 0 \\ 0 & 0 & 0 \end{bmatrix}. \tag{9.105}$$

Thus the pressure is $p = \frac{1}{3}\sigma_Y$ and the deviatoric stress is given as

$$\begin{bmatrix} \frac{2}{3}\sigma_Y & 0 & 0 \\ 0 & -\frac{1}{3}\sigma_Y & 0 \\ 0 & 0 & -\frac{1}{3}\sigma_Y \end{bmatrix}. \tag{9.106}$$

This gives

$$w_{\text{dev}}^{1-\text{D}} = \frac{2(1+\nu)}{6E}\sigma_Y^2. \tag{9.107}$$

So we have as our yield criteria

$$(s_{xx}^2 + s_{yy}^2 + s_{zz}^2 + 2s_{xy}^2 + 2s_{yz}^2 + 2s_{zx}^2) \le \frac{2}{3}\sigma_Y^2. \tag{9.108}$$

This expression can be made a little more convenient by expressing it directly in terms of the stress components instead of the components of the deviatoric stress; this can be done using the definition of the stress deviator, eqn (9.97). The net result is given as

$$\frac{1}{2}\left((\sigma_{xx}-\sigma_{yy})^2+(\sigma_{yy}-\sigma_{zz})^2+(\sigma_{zz}-\sigma_{xx})^2\right) \\ +3(\sigma_{xy}^2+\sigma_{yz}^2+\sigma_{zx}^2)\leq\sigma_Y^2. \tag{9.109}$$

As long as the strict inequality is satisfied, the multi-axial stress state represents an elastic state of stress. When the equality is satisfied, yield starts.

**Remarks:**

(1) One advantage of the Henky–von Mises condition is that there is no need to determine Mohr's circles, principal stresses, or maximum shears. It works directly with the stress components. This point makes it especially easier to use in computer programs for automated stress analysis.

(2) Though not obvious, as written, the Henky–von Mises condition is independent of coordinate system.

(3) By construction, the Henky–von Mises condition is independent of pressure.

(4) Equation (9.109) is given in terms of an $x$-$y$-$z$ Cartesian coordinate system. It is also valid for any other orthonormal coordinate system. For example, for cylindrical/polar coordinates or for spherical coordinates under the substitutions $(x,y,z)\to(r,\theta,z)$ and $(x,y,z)\to(r,\varphi,\theta)$, respectively.

(5) The Henky–von Mises condition for yield is a model just as the Tresca condition is a model for yield. They are both models for the same phenomena and will differ slightly in their predictions. One way to appreciate their differences is to consider a state of plane stress in principal coordinates:

$$\begin{bmatrix}\sigma_1 & 0 & 0\\ 0 & \sigma_2 & 0\\ 0 & 0 & 0\end{bmatrix}. \tag{9.110}$$

Applying the Henky–von Mises condition to this state of stress we see that

$$(\sigma_1-\sigma_2)^2+\sigma_2^2+\sigma_1^2\leq 2\sigma_Y^2. \tag{9.111}$$

This expression tells us that in the $\sigma_1$-$\sigma_2$ plane that the set of elastic stress states is contained in an elliptical region. The application of Tresca's condition to this state of stress gives a hexagonal region of elastic states in the same plane. The situation is sketched in Fig. 9.19. As can be seen, both criteria are in reasonable agreement with each other. In terms of ability to accurately model data, both criteria provide decent accuracy for polycrystalline metals at room temperature.

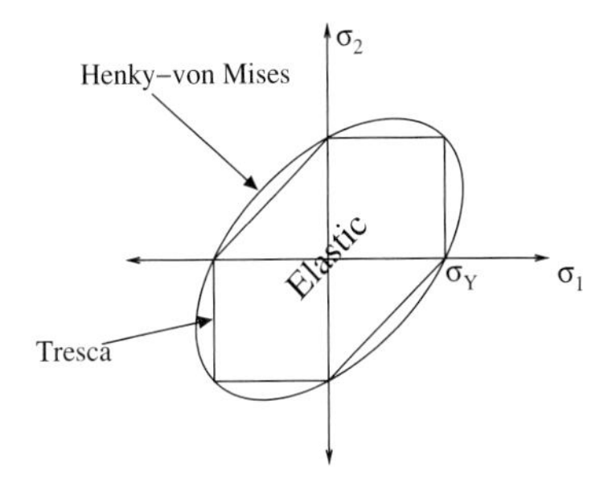


**Fig. 9.19** Comparison of Henky–von Mises condition with Tresca's criteria in plane stress.

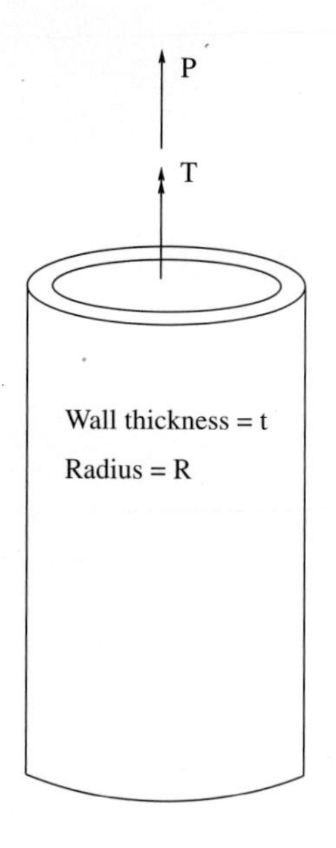


**Fig. 9.20** Thin-walled tube in torsion and axial loading.

**Example 9.11**

*Thin-walled tube in axial and torsional loading.* Consider the thin-walled tube shown in Fig. 9.20. Determine the limits on $P$ and $T$ for combined states of loading according to the Henky–von Mises condition.

*Solution*

There are two non-zero stresses in the tube. Assuming that the axis of the tube is in the $z$ direction, the axial force gives rise to

$$\sigma_{zz} = \frac{P}{A} = \frac{P}{2\pi Rt}. \tag{9.112}$$

The torque gives rise to a shear stress:

$$\sigma_{z\theta} = \frac{q}{t} = \frac{T}{2A_{\text{enclosed}}t} = \frac{T}{2\pi R^2 t}. \tag{9.113}$$

In matrix form we have:

$$\begin{bmatrix} 0 & 0 & 0 \\ 0 & 0 & T/2\pi R^2 t \\ 0 & T/2\pi R^2 t & P/2\pi Rt \end{bmatrix}_{r\theta z}. \tag{9.114}$$

Applying eqn (9.109) in cylindrical coordinates gives:

$$\frac{1}{2}\left\{\left(\frac{P}{2\pi Rt}\right)^2 + (0)^2 + \left(-\frac{P}{2\pi Rt}\right)^2\right\} + 3\left(\frac{T}{2\pi R^2 t}\right)^2 \leq \sigma_Y^2 \tag{9.115}$$

$$\frac{P^2}{(2\pi Rt\sigma_Y)^2} + \frac{T^2}{(\frac{2}{\sqrt{3}}\pi R^2 t\sigma_Y)^2} \leq 1 \tag{9.116}$$

$$\frac{P^2}{P_Y^2} + \frac{T^2}{T_Y^2} \leq 1, \tag{9.117}$$

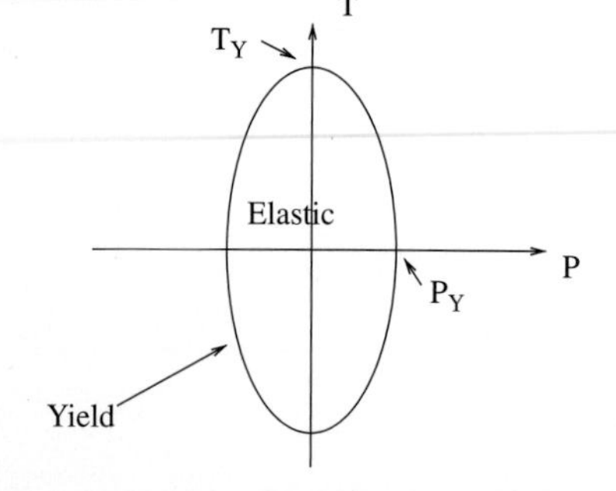


**Fig. 9.21** Elastic zone for extension and twist of a thin-walled tube.

where $P_Y = 2\pi Rt\sigma_Y$ and $T_Y = \frac{2}{\sqrt{3}}\pi R^2 t\sigma_Y$. The loci of points in the $P$-$T$ plane that correspond to yield are shown in Fig. 9.21. The interior of the ellipse corresponds to elastic states of loading, and yield occurs for any $(P, T)$ combination on the ellipse.

**Remarks:**

(1) Notice that if one biases a torsional loading with any amount of axial force, then the permissible amount of torque is decreased. Likewise if one biases an axial loading with any amount of torque, then the permissible amount of axial force decreases.

**Example 9.12**

*Thin-walled cylindrical pressure vessel.* Consider a thin-walled cylindrical pressure vessel of radius $R$ and wall thickness $t$. How much pressure can be applied before yield takes place according to the Henky–von Mises condition?

*Solution*
The state of stress in a cylindrical pressure vessel (see Section 6.2.1) is

$$\begin{bmatrix} 0 & 0 & 0 \\ 0 & \frac{pR}{t} & 0 \\ 0 & 0 & \frac{pR}{2t} \end{bmatrix}_{r\theta z} . \tag{9.118}$$

Thus by eqn (9.109) we have that

$$\left(\frac{pR}{t} - \frac{pR}{2t}\right)^2 + \left(\frac{pR}{2t} - 0\right)^2 + \left(\frac{pR}{t} - 0\right)^2 \leq 2\sigma_Y^2 \tag{9.119}$$

$$p \leq \frac{t}{R}\frac{2}{\sqrt{3}}\sigma_Y . \tag{9.120}$$

---

**Example 9.13**

*Relation of $\tau_Y$ to $\sigma_Y$ according to the Henky-von Mises condition.* Consider an experiment that produces a state of yield in pure shear. Find the relation between a measured yield stress in shear, $\tau_Y$, and the yield stress in tension, $\sigma_Y$.

*Solution*
At yield, the stress is given by

$$\begin{bmatrix} 0 & \tau_Y & 0 \\ \tau_Y & 0 & 0 \\ 0 & 0 & 0 \end{bmatrix} . \tag{9.121}$$

Evaluating the Henky–von Mises condition for this state of stress gives

$$3\tau_Y^2 = \sigma_Y^2 . \tag{9.122}$$

Thus,

$$\tau_Y = \frac{\sigma_Y}{\sqrt{3}} . \tag{9.123}$$

**Remarks:**

(1) Note that this relation is different than the one we derived in the section on Tresca's condition. The relation between the yield stress in shear and the yield stress in uniaxial tension is model-dependent – hence the difference.

---

# Chapter summary

- Vector transformation rules:

$$\begin{pmatrix} F_{x'} \\ F_{y'} \end{pmatrix} = \begin{bmatrix} \cos(\theta) & \sin(\theta) \\ -\sin(\theta) & \cos(\theta) \end{bmatrix} \begin{pmatrix} F_x \\ F_y \end{pmatrix}$$

- Stress transformation rules:

$$\begin{bmatrix} \sigma_{x'x'} & \sigma_{x'y'} \\ \sigma_{y'x'} & \sigma_{y'y'} \end{bmatrix} = \begin{bmatrix} \cos(\theta) & \sin(\theta) \\ -\sin(\theta) & \cos(\theta) \end{bmatrix} \begin{bmatrix} \sigma_{xx} & \sigma_{xy} \\ \sigma_{yx} & \sigma_{yy} \end{bmatrix} \begin{bmatrix} \cos(\theta) & -\sin(\theta) \\ \sin(\theta) & \cos(\theta) \end{bmatrix}$$

  - Double-angle form:

$$\sigma_{x'x'} = \frac{\sigma_{xx} + \sigma_{yy}}{2} + \frac{\sigma_{xx} - \sigma_{yy}}{2} \cos(2\theta) + \sigma_{xy} \sin(2\theta)$$

$$\sigma_{y'y'} = \frac{\sigma_{xx} + \sigma_{yy}}{2} - \frac{\sigma_{xx} - \sigma_{yy}}{2} \cos(2\theta) - \sigma_{xy} \sin(2\theta)$$

$$\sigma_{x'y'} = - \frac{\sigma_{xx} - \sigma_{yy}}{2} \sin(2\theta) + \sigma_{xy} \cos(2\theta)$$

- Principal angle (maximum and minimum normal stresses with no shear)

$$\tan(2\theta_p) = \frac{2\sigma_{xy}}{\sigma_{xx} - \sigma_{yy}}$$

- Maximum shear orientation (normal stress take their mean value)

$$\tan(2\theta_s) = -\frac{\sigma_{xx} - \sigma_{yy}}{2\sigma_{xy}}$$

  - Angle relation: $\theta_p - \theta_s = \pi/4$
  - Principal value form: $\tau_{\max} = (\sigma_1 - \sigma_3)/2$
- Mohr's circle: The center of the circle is the mean normal stress. The radius is the distance from the center to the point $(\sigma_{xx}, \sigma_{xy})$
- Three-dimensional case: Compute the eigenvalues and eigenvectors of $\boldsymbol{\sigma}$ from the characteristic polynomial:

$$-\lambda^3 + \mathrm{I}_\sigma \lambda^2 - \mathrm{II}_\sigma \lambda + \mathrm{III}_\sigma = 0$$

  where $\mathrm{I}_\sigma = \mathrm{trace}[\boldsymbol{\sigma}]$, $\mathrm{II}_\sigma = \frac{1}{2}[\mathrm{trace}(\boldsymbol{\sigma}^2) - (\mathrm{trace}(\boldsymbol{\sigma}))^2]$, $\mathrm{III}_\sigma = \det[\boldsymbol{\sigma}]$
- Strains transform the same way as stresses, but one needs to use the tensorial shear strain instead of the engineering shear strain.
- Tresca's yield condition: $\tau_{\max} \leq \tau_Y$
- Deviatoric stress: $\boldsymbol{s} = \boldsymbol{\sigma} - p\mathbf{1}$; pressure $p = \frac{1}{3}(\sigma_{xx} + \sigma_{yy} + \sigma_{zz})$

- Henky–von Mises yield condition:

$$\frac{1}{2}\left((\sigma_{xx}-\sigma_{yy})^2+(\sigma_{yy}-\sigma_{zz})^2+(\sigma_{zz}-\sigma_{xx})^2\right)$$
$$+3(\sigma_{xy}^2+\sigma_{yz}^2+\sigma_{zx}^2)\leq\sigma_Y^2$$

# Exercises

(9.1) The state of stress at a point in a solid is given as

$$\begin{bmatrix} 10 & 5 \\ 5 & 7 \end{bmatrix}_{xy} \text{MPa.}$$

Consider a plane passing through this point with normal vector $\boldsymbol{n} = (1/\sqrt{2})\boldsymbol{e}_x + (1/\sqrt{2})\boldsymbol{e}_y$. What are the normal and shear stresses on this plane?

(9.2) Given the following two-dimensional state of stress, find the principal values and directions.

$$\begin{bmatrix} 60 & 80 \\ 80 & -90 \end{bmatrix} \text{MPa}$$

(9.3) Given the following two-dimensional state of stress, find the maximum shear and sketch the state of stress on a properly oriented element.

$$\begin{bmatrix} 60 & 80 \\ 80 & -90 \end{bmatrix} \text{MPa}$$

(9.4) Given the following two-dimensional state of stress, find the angles of rotation which cause the normal stress in the $x'$ direction to be zero.

$$\begin{bmatrix} 60 & 80 \\ 80 & -90 \end{bmatrix} \text{MPa}$$

(9.5) Using an eigenvalue technique, find the principal values and directions for the following state of stress.

$$\begin{bmatrix} 10 & -50 \\ -50 & 5 \end{bmatrix} \text{ksi}$$

(9.6) Find the principal values of stress for the following three-dimensional state of stress. Sketch the three circles of stress on a Mohr diagram.

$$\begin{bmatrix} 10 & -50 & 0 \\ -50 & 5 & 0 \\ 0 & 0 & 60 \end{bmatrix} \text{ksi}$$

(9.7) Find the principal values and directions for the following three-dimensional state of stress. Sketch the three circles of stress on a Mohr diagram.

$$\begin{bmatrix} 10 & -50 & 2 \\ -50 & 5 & 0 \\ 2 & 0 & 60 \end{bmatrix} \text{ksi}$$

(9.8) Find the principal values of stress for the following three-dimensional state of stress. Sketch the three circles of stress on a Mohr diagram.

$$\begin{bmatrix} 10 & 0 & 10 \\ 0 & 5 & 0 \\ 10 & 0 & 60 \end{bmatrix} \text{ksi}$$

(9.9) Find the principal values of stress for the following three-dimensional state of stress. Sketch the three circles of stress on a Mohr diagram.

$$\begin{bmatrix} 10 & 2 & 1 \\ 2 & 5 & 0 \\ 1 & 0 & 6 \end{bmatrix} \text{MPa}$$

(9.10) For the state of stress that follows, find the principal stresses and principal directions. Draw the principal stresses on a properly oriented three-dimensional element.

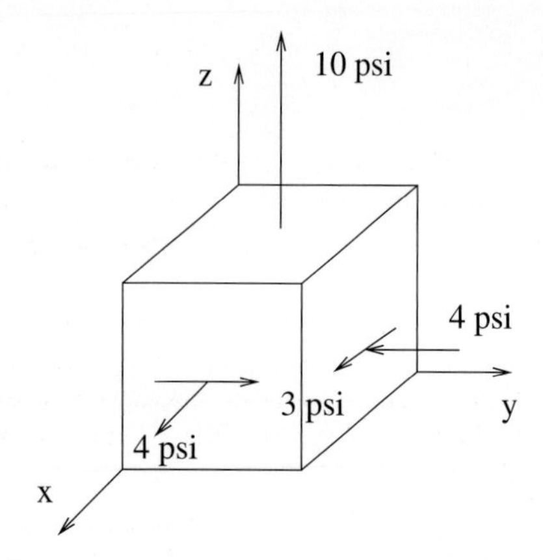


(9.11) For the beam shown below, determine the principal stresses at point A and show them on a properly oriented element. What is the maximum shear stress at point A. Assume $P = 1$ and $T = 1000$ in consistent units.

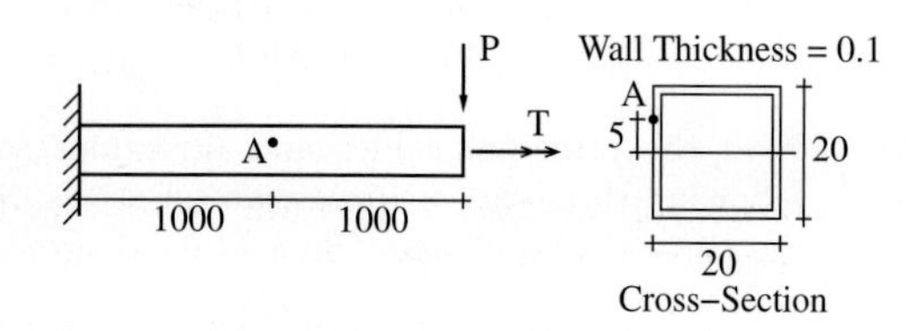


(9.12) For the state of stress shown, determine and show the magnitude and direction of the principal stresses. How large is the maximum shear stress (not necessarily in the $x - y$ plane)? In what plane does it occur? Note, $\sigma_z = \tau_{xz} = \tau_{yz} = 0$. Assume stress units of kPa.

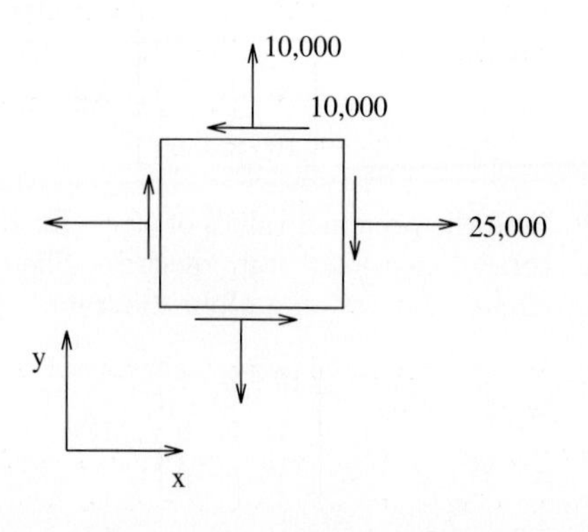


(9.13) Consider the two-dimensional state of stress shown; assume units of psi.

(a) Determine the principal stresses and show them on a properly oriented element.

(b) Determine the maximum shear stress in the plane and show this state on a properly oriented element.

(c) If the third principal stress is given as 10 psi, what is $\tau_{max}$?

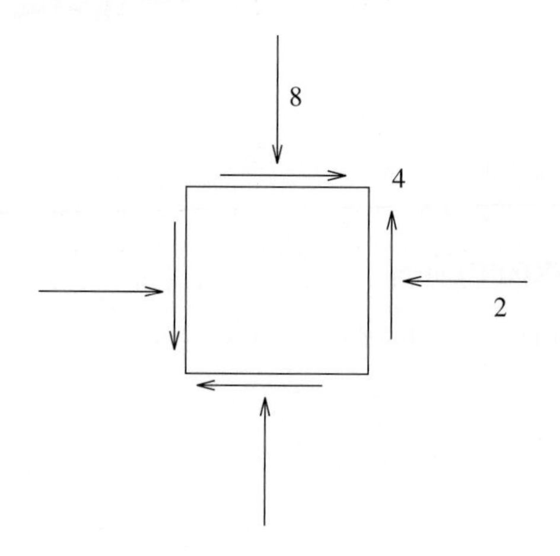


(9.14) For the two-dimensional state of plane stress

$$\boldsymbol{\sigma} = \begin{bmatrix} -a & -a \\ -a & a \end{bmatrix},$$

find the principal stresses, principal angle, maximum shear, and maximum shear angle. Show your results on properly oriented elements.

(9.15) The displacement field in a structure has been measured to be

$$u_x = Ax + Ay$$

$$u_y = Ax$$

where $A$ is a given constant. What is the maximum normal strain in the structure?

(9.16) As shown, a 0-45-90 strain-gauge rosette is applied to a thin plate of a material with Young's modulus $E = 12,500$ ksi and shear modulus $G = 5,000$ ksi, produces the following readings: $\varepsilon_{0^\circ} = 200 \times 10^{-6}$, $\varepsilon_{45^\circ} = 120 \times 10^{-6}$, $\varepsilon_{90^\circ} = -160 \times 10^{-6}$.

(a) Find the principal stresses and the maximum shear stress (assume plane stress).

(b) Check if this is a possible state of stress if the material is elastic–perfectly plastic, with a yield stress $\sigma_Y = 5$ ksi, according to the Henky–von Mises yield criterion.

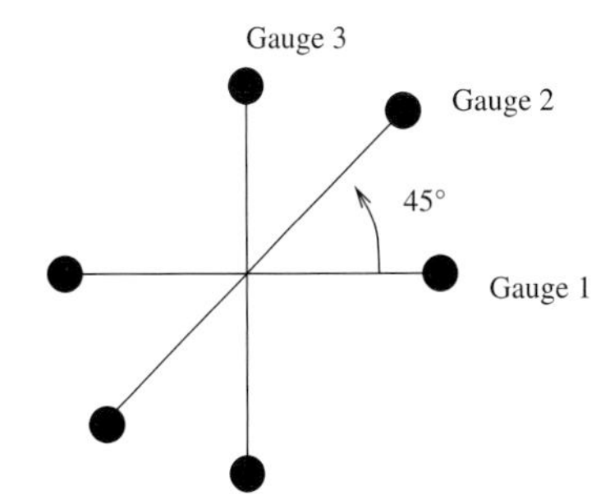


(9.17) Consider two coordinate systems that are rotated an amount $\theta$ with respect to each other. Derive an expression for $\varepsilon_{x'y'}$ in terms of $\varepsilon_{xx}$, $\varepsilon_{yy}$, $\varepsilon_{xy}$, and $\theta$.

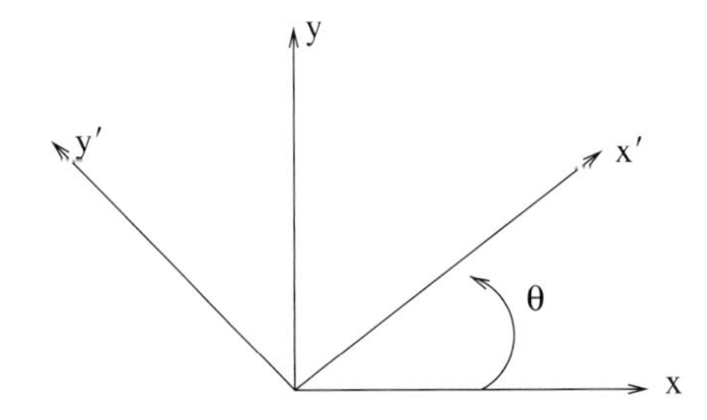


(9.18) Drive the expressions for the strain components in the $x$-$y$ coordinate system in terms of the output of a 0-60-120 strain-gauge rosette; i.e. knowing the normal strains in the three directions oriented 0, 60, and 120 degrees relative to the $x$-axis find expressions for $\varepsilon_{xx}$, $\varepsilon_{yy}$, and $\varepsilon_{xy}$.

(9.19) Derive the transformation equation for the normal strain in the $y'$ direction:

$$\varepsilon_{y'y'} = \varepsilon_{yy}\cos^2(\theta) + \varepsilon_{xx}\sin^2(\theta) - 2\varepsilon_{xy}\sin(\theta)\cos(\theta)$$

(9.20) Show that the principal axes of stress and strain coincide for isotropic linear elastic materials; see Chapter 5.

(9.21) Consider a linear elastic isotropic material with $E = 100$ MPa and $\nu = -0.1$. The state of strain is known at a particular point to be

$$\begin{bmatrix} 10 & -50 \\ -50 & 5 \end{bmatrix}_{xy} \mu\text{strain}$$

Find the principal stresses and the principal angle. Assume plane stress.

(9.22) Consider a linear elastic isotropic material with $E = 200$ MPa and $\nu = 0.3$. The state of stress is known at a particular point to be

$$\begin{bmatrix} 100 & -60 \\ -60 & 70 \end{bmatrix}_{xy} \text{MPa}$$

Find the principal strains and the principal angle. Assume plane stress.

(9.23) Shown below is a thin-walled cylindrical balloon with internal pressure $p$, radius $R$, and thickness $t$. Due to the thinness of the balloon walls they are incapable of supporting any compressive stresses. Find a relation for the maximum torque that the balloon will support. Express your answer in terms of $p$ and $R$.

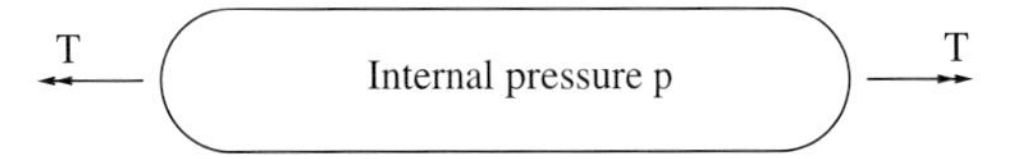


(9.24) Consider the following state of stress:

$$k\begin{bmatrix} 10 & -6 & 5 \\ -6 & 7 & 5 \\ 5 & 5 & 0 \end{bmatrix} \text{MPa.}$$

Determine the value of the parameter k for yield according to the Henky–von Mises condition. Assume $\sigma_Y = 100$ MPa.

(9.25) Is the following stress state elastic according to Tresca's condition:

$$\begin{bmatrix} 100 & -6 & 5 \\ -6 & 100 & 5 \\ 5 & 5 & 100 \end{bmatrix} \text{MPa.}$$

Assume $\tau_Y = 20$ MPa.

(9.26) Solve Exercise 9.25 using the Henky–von Mises condition with $\sigma_Y = 30$ MPa.

(9.27) What is the pressure (mean normal stress) for the following state of stresses:

$$\begin{bmatrix} 30 & -6 & 8 \\ -6 & 10 & 15 \\ 8 & 15 & 10 \end{bmatrix} \text{ksi.}$$

What is the corresponding deviatoric stress for this state of stress?

(9.28) Consider a solid round bar of radius $R$ and length $L$. Compute the total elastic energy in the bar when it is subjected to an axial end-load $P$. Compute the total deviatoric energy in the bar. Compute the total hydrostatic/volumetric energy in the bar. Do your last two expressions add up to the first?

(9.29) Consider the thin-walled tube as shown. The tube is subjected to a bending moment and a torque. The inner and outer walls are also subject to a (uniform) pressure. The tube has a yield stress in shear $\tau_Y = 1.25$ N/mm$^2$. Apply *Tresca's* condition and determine whether or not the tube will yield under the given loads.

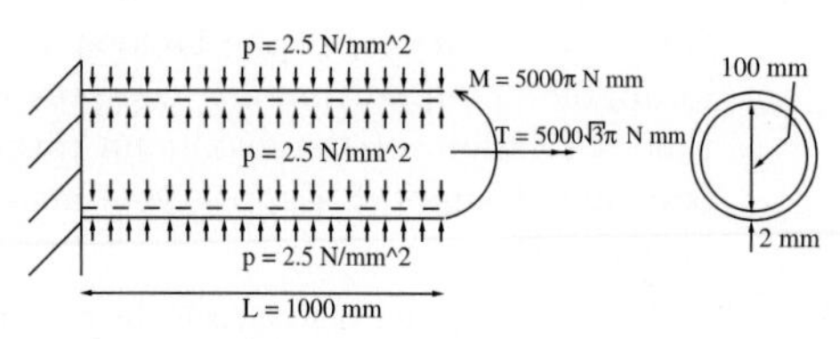


(9.30) Solve Exercise 9.29 using the Henky–von Mises conditions. Assume $\sigma_Y = 1.25\sqrt{3}$ N/mm$^2$.

(9.31) At what internal pressure does a spherical pressure vessel yield according to Tresca's condition?

(9.32) Consider a cantilevered solid round bar of length $L$ and radius $R$. Determine the allowable combinations of applied end-moment $M$ and end-torque $T$ according to the Henky–von Mises condition. Make a plot of the elastic domain in the $M$-$T$ plane, and clearly label all important points defining the elastic domain.

# Virtual Work Methods: Virtual Forces

# 10



At this point in our developments we have modestly sophisticated methods for dealing with the deformation, strain, and stress analysis of mechanical systems under the action of axial forces, axial torques, and bending loads. Through a clever combination of assumptions we know how to take a complex phenomena governed by partial differential equations and analyze them efficiently with a small set of ordinary differential equations.

In the next four chapters we will examine the development and application of energy methods that will allow us to solve a larger class of problems than we can at present. In particular, the methods we will examine will allow us to treat problems with more geometric complexity than we can easily handle with the methods we have developed to this point. The methods to be described will be presented in the context of engineering mechanics, but the reader should note that they have wide and important applications in many fields of study. In particular, these methods form the basis of the finite element method – arguably, modern engineering's most important numerical tool.

## 10.1 The virtual work theorem: Virtual force version

In this first chapter on energy methods, the technique we will look at is the method of virtual work – also known as the principle of virtual work. As the name implies we will be dealing with work-like quantities with virtual components. This is in contrast to the energy method we have been using up to this point – viz., conservation of energy. The main theorem we will use is deceptively simple and can be stated as follows:

External Virtual Work = Internal Virtual Work

Of course, to make productive use of the theorem one needs to have usable definitions of external and internal virtual work. To keep things as concrete as possible, we will introduce these definitions through an example.

Suppose we wish to analyze the two-bar truss shown in Fig. 10.1 for the horizontal deflection, $u$, at the point of application of the vertical load.

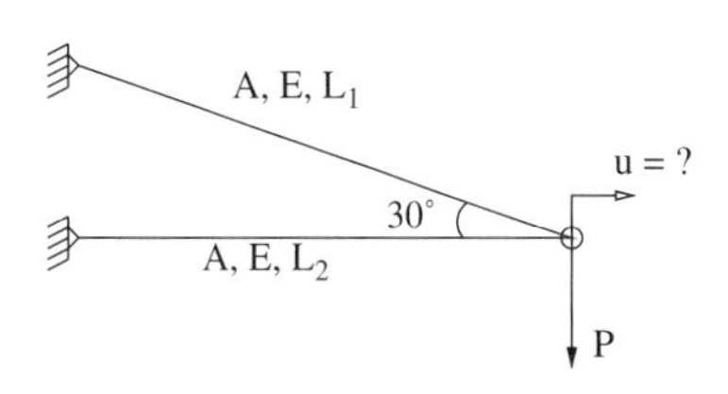


**Fig. 10.1** Two-bar truss: real system.

We will refer to the system shown in Fig. 10.1 as the real system. It is the system we wish to really analyze. In order to analyze the system using the principle of virtual work we will introduce a virtual system as shown in Fig. 10.2. The virtual system is geometrically identical to the real system; it has the same geometry and kinematic boundary conditions. It is also subject to a virtual force. This virtual force, $\bar{f}$, is applied at the point of interest and in the direction of the kinematic quantity of interest.[1]

[1] We will denote virtual force-like quantities with an over-bar. It is also very common to denote virtual quantities with a prepended $\delta$; i.e. one often writes $\delta f$ in place of $\bar{f}$.

The external virtual work (Ext. V.W.) is *defined* as the externally applied virtual force times the real displacement occurring at the point of application of the virtual force. Thus in our example,

$$\text{Ext. V.W.} = \text{virtual force} \times \text{real displacement} = \bar{f}u. \tag{10.1}$$

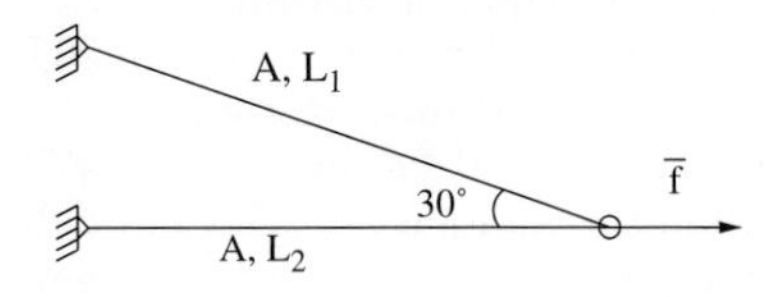


**Fig. 10.2** Two-bar truss: virtual system.

The internal virtual work (Int. V.W.) is *defined* as the integral of the stresses in the virtual system times the strains in the real system. For our two-bar truss system we have

$$\text{Int. V.W.} = \int_V \bar{\sigma}\varepsilon \, dV \tag{10.2}$$

$$= \int_{\text{Bar 1}} \bar{\sigma}\varepsilon \, dV + \int_{\text{Bar 2}} \bar{\sigma}\varepsilon \, dV \tag{10.3}$$

$$= \int_{\text{Bar 1}} \frac{\bar{R}_1}{A_1}\frac{R_1}{A_1E_1} \, dV + \int_{\text{Bar 2}} \frac{\bar{R}_2}{A_2}\frac{R_2}{A_2E_2} \, dV \tag{10.4}$$

$$= \frac{\bar{R}_1R_1L_1}{A_1E_1} + \frac{\bar{R}_2R_2L_2}{A_2E_2}. \tag{10.5}$$

In the above, $\bar{R}_i$ refers to the internal force in bar $i$ of the virtual system and $R_i$ refers to the internal force in bar $i$ of the real system.

If we put this together with the expression for the external virtual work and employ the virtual work theorem, we find the following expression for the horizontal deflection we were seeking:

$$u = \frac{1}{\bar{f}}\left[\frac{\bar{R}_1R_1L_1}{A_1E_1} + \frac{\bar{R}_2R_2L_2}{A_2E_2}\right]. \tag{10.6}$$

For our geometry, from statics we can easily show that $\bar{R}_1 = 0$ and $\bar{R}_2 = \bar{f}$. Also from statics, we have that $R_1 = 2P$ and $R_2 = -\sqrt{3}P$. Thus we have as a final answer:

$$u = -\frac{\sqrt{3}PL_2}{A_2E_2}. \tag{10.7}$$

**Remarks:**

(1) As a habit, we will always decorate quantities associated with the virtual system with an over-bar.

(2) In the scheme shown we have introduced virtual forces, and thus the method shown is often known as the method of virtual forces or the principle of virtual forces.

(3) Virtual work quantities will always be the product of a real quantity times an energetically conjugate virtual quantity – displacements times forces, stresses times strains, rotations times moments/torques. In the method of virtual forces the real quantities will be displacements, rotations, or strains and the virtual quantities will be forces, moments, or stresses.

(4) The kinematic quantity that one determines with this method will always be the displacement at the point of application of the virtual force in the direction of the virtual force; or rotation if one uses a virtual moment.

(5) The generalization of our truss-bar result to a truss system with $N$ bars is

$$u = \frac{1}{\bar{f}} \sum_{k=1}^{N} \frac{\bar{R}_k R_k L_k}{A_k E_k}. \tag{10.8}$$

(6) Note that the magnitude of the virtual force always drops out of the computations. So it is common to set it equal to 1. Sometimes, to emphasize the virtual nature of the unit force, it is written as $\bar{1}$.

(7) In systems that are simultaneously reacting to their load in multiple ways, say in torsion and bending, one can simply add up all the external virtual work quantities and set them equal to the sum of all the internal virtual work quantities; i.e., virtual work, like real work, is an additive quantity.

## 10.2 Virtual work expressions

In order to apply the principle of virtual forces to a larger variety of systems we will need additional expressions for external and internal virtual work. For each case presented, a simple example will also be given to place the concepts in concrete terms. Later, in Section 10.3, we will look at a proof of the virtual work theorem in order to better understand its meaning.

### 10.2.1 Determination of displacements

As noted above, to determine displacements at a point in a given direction we place a virtual force at that point in the direction of interest. The resultant external virtual work expression is then given by:

$$\text{Ext. V.W.} = \bar{f}u. \tag{10.9}$$

### 10.2.2 Determination of rotations

If instead one wishes to determine the rotation of a point in a given sense, we place a virtual moment at that point in the sense of interest. The resultant external virtual work expression is then given by:

$$\text{Ext. V.W.} = \bar{m}\theta, \tag{10.10}$$

where $\bar{m}$ is the applied virtual moment.

### 10.2.3 Axial rods

If an element in a mechanical system of length $L$ is loaded with axial forces, then the internal virtual work in the element is given by

$$\text{Int. V.W.} = \int_V \bar{\sigma}_{xx}\varepsilon_{xx}\,dV \tag{10.11}$$

$$= \int_V \frac{\bar{R}}{A}\varepsilon\,dV \tag{10.12}$$

$$= \int_0^L \bar{R}\varepsilon\,dx\,. \tag{10.13}$$

If and only if the system is elastic, one can also write

$$\text{Int. V.W.} = \int_0^L \bar{R}\frac{R}{AE}\,dx. \tag{10.14}$$

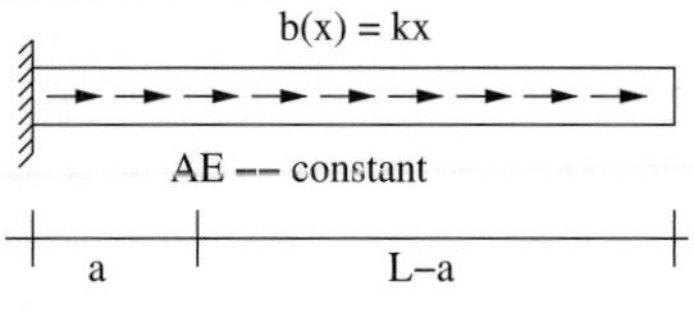

---

**Example 10.1**

*Axial rod.* As an example application, consider the rod shown in Fig. 10.3 and determine using the principle of virtual forces the displacement at $x = a$.

b(x) = kx
AE -- constant
a
L–a

**Fig. 10.3** Axial bar with linear distributed load: linear elastic (real system).

*Solution*
To begin, we construct the virtual system shown in Fig. 10.4; the virtual system is identical to the real system but with all real loads removed and a virtual force introduced at $x = a$. The external virtual work is

$$\text{Ext. V.W.} = \bar{f}u(a). \tag{10.15}$$

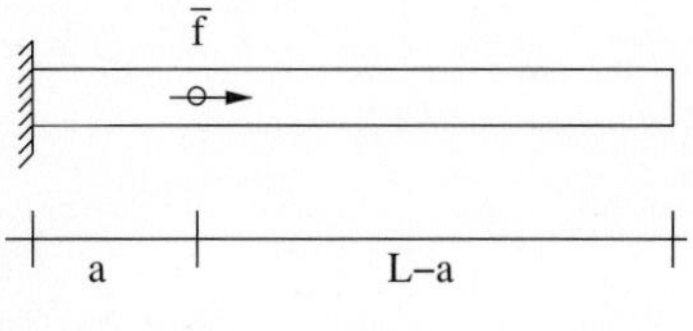

The internal virtual work is

$$\text{Int. V.W.} = \int_0^L \bar{R}\varepsilon\,dx. \tag{10.16}$$

$\bar{f}$
a
L–a

**Fig. 10.4** Virtual system for finding the displacement at $x = a$ for the bar shown in Fig. 10.3.

From equilibrium of the virtual system we have that $\bar{R} = \bar{f}H(a - x)$. From equilibrium for the real system we have that $R(x) = \frac{1}{2}k(L^2 - x^2)$. Employing the constitutive relation for the real system we find that $\varepsilon = R/AE = \frac{1}{2}k(L^2 - x^2)/AE$. Inserting these results into the virtual work theorem, we find that:

$$\bar{f}u(a) = \int_0^L \bar{R}\varepsilon\,dx \tag{10.17}$$

$$= \int_0^L \bar{f} H(a-x) \frac{1}{2AE} k(L^2 - x^2)\, dx \quad (10.18)$$

$$= \int_0^a \bar{f} \frac{1}{2AE} k(L^2 - x^2)\, dx \quad (10.19)$$

$$= \bar{f} \frac{1}{2AE} k(L^2 a - a^3/3) \quad (10.20)$$

$$u(a) = \frac{ka}{2AE}(L^2 - a^2/3). \quad (10.21)$$

**Remarks:**

(1) It is an easy matter to verify this result using the techniques developed in Chapter 2.

### 10.2.4 Torsion rods

If an element in a mechanical system of length $L$ is loaded with a torque, then the internal virtual work in the element is given by

$$\text{Int. V.W.} = \int_V \bar{\sigma}_{z\theta} \gamma_{z\theta}\, dV \quad (10.22)$$

$$= \int_V \bar{\tau} r \frac{d\phi}{dz}\, dV \quad (10.23)$$

$$= \int_0^L \frac{d\phi}{dz} \left[ \int_A \bar{\tau} r\, dA \right] dz \quad (10.24)$$

$$= \int_0^L \bar{T} \frac{d\phi}{dz}\, dz. \quad (10.25)$$

If and only if the system is elastic, one can also write

$$\text{Int. V.W.} = \int_0^L \bar{T} \frac{T}{GJ}\, dz. \quad (10.26)$$

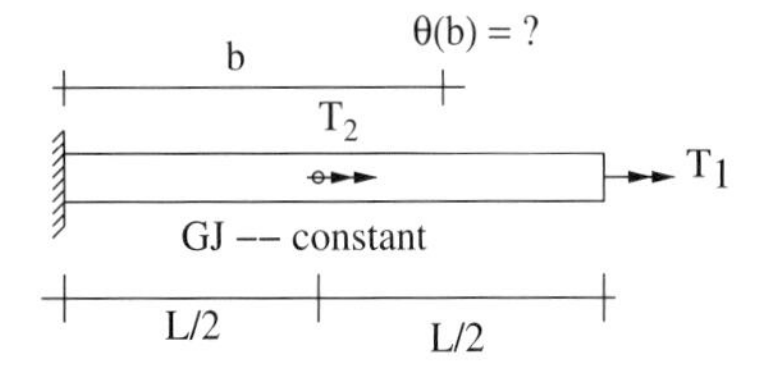


**Fig. 10.5** Torsion bar with two-point loads: linear elastic (real system).

**Example 10.2**

*Torsion rod.* As an example application, consider the torsion rod shown in Fig. 10.5, and using the principle of virtual forces (moments), determine the rotation at $z = b$.

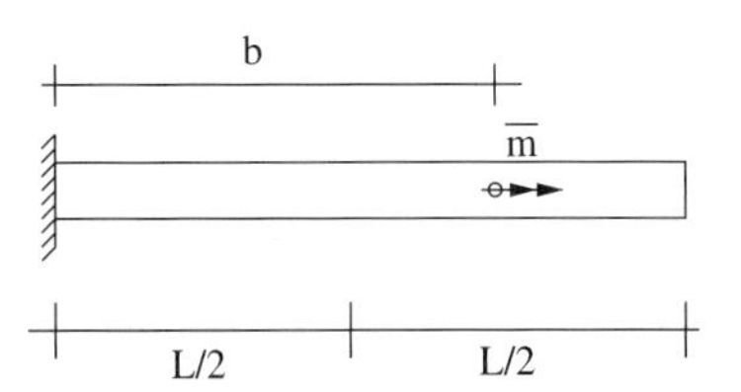


**Fig. 10.6** Virtual system for finding the rotation at $z = b$ for the bar shown in Fig. 10.5.

*Solution*

To begin, we construct the virtual system shown in Fig. 10.6; the virtual system is identical to the real system but with all real loads removed and a virtual moment introduced at $z = b$. The external virtual work is

$$\text{Ext. V.W.} = \bar{m}\theta(b). \tag{10.27}$$

The internal virtual work is

$$\text{Int. V.W.} = \int_0^L \bar{T}\frac{d\phi}{dz}\,dz. \tag{10.28}$$

From equilibrium of the virtual system we have that $\bar{T} = \bar{m}H(b-z)$. From equilibrium for the real system we have that $T(z) = (T_1 + T_2) - T_2H(z - L/2)$. Employing the constitutive relation for the real system we find that $d\phi/dz = T/GJ$. Inserting these results into the virtual work theorem, we find that

$$\bar{m}\theta(b) = \int_0^L \bar{T}\frac{d\phi}{dz}\,dz \tag{10.29}$$

$$= \int_0^L \bar{m}H(b-z)[(T_1+T_2) - T_2H(z-L/2)]/GJ\,dz \tag{10.30}$$

$$= \int_0^b \bar{m}[(T_1+T_2) - T_2H(z-L/2)]/GJ\,dz \tag{10.31}$$

$$= \bar{m}[(T_1+T_2)b - T_2(b-L/2)]/GJ \tag{10.32}$$

$$\theta(b) = \frac{T_2L}{2GJ} + \frac{T_1b}{GJ}. \tag{10.33}$$

**Remarks:**

(1) This is exactly the result we would have obtained had we followed the procedures developed in Chapter 7.

### 10.2.5 Bending of beams

If an element in a mechanical system of length $L$ is loaded with a bending moment then the internal virtual work in the element is given by

$$\text{Int. V.W.} = \int_V \bar{\sigma}_{xx}\varepsilon_{xx}\,dV \tag{10.34}$$

$$= \int_V \bar{\sigma}(-y)\kappa\,dV \tag{10.35}$$

$$= \int_0^L \kappa\left[\int_A -y\bar{\sigma}\,dA\right]dx \tag{10.36}$$

$$= \int_0^L \bar{M}\kappa\,dx. \tag{10.37}$$

If and only if the system is elastic, one can also write

$$\text{Int. V.W.} = \int_0^L \bar{M} \frac{M}{EI}\, dx. \tag{10.38}$$

### Example 10.3

*Bending of a beam.* As an example application, consider the beam shown in Fig. 10.7, and using the principle of virtual forces, determine the tip displacement, $\Delta$.

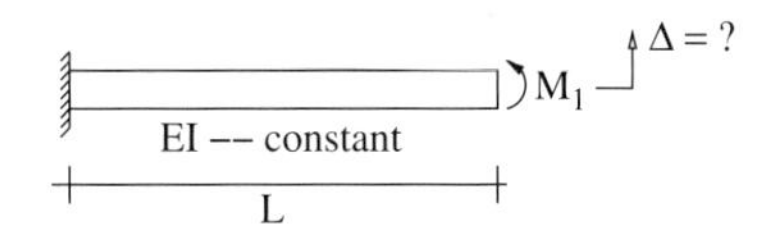


**Fig. 10.7** Beam bent by an end-moment: linear elastic (real system).

*Solution*
To begin, we construct the virtual system shown in Fig. 10.8; the virtual system is identical to the real system but with all real loads removed and a virtual force introduced at $x = L$. The external virtual work is

$$\text{Ext. V.W.} = \bar{f}\Delta. \tag{10.39}$$

The internal virtual work is

$$\text{Int. V.W.} = \int_0^L \bar{M}\kappa\, dx. \tag{10.40}$$

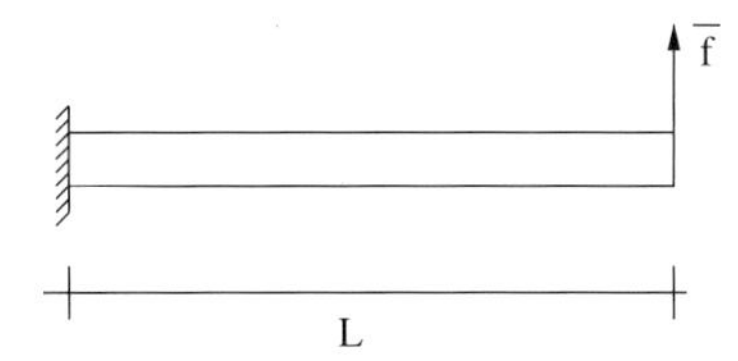


**Fig. 10.8** Virtual system for determining the tip deflection for the beam shown in Fig. 10.7.

From equilibrium of the virtual system we have that $\bar{M}(x) = \bar{f}(L - x)$. From equilibrium for the real system we have that $M(x) = M_1$. Employing the constitutive relation for the real system we find that $\kappa = M/EI$. Inserting these results into the virtual work theorem, we find that:

$$\bar{f}\Delta = \int_0^L \bar{M}\kappa\, dx \tag{10.41}$$

$$= \int_0^L \bar{f}(L - x)M_1/EI\, dx \tag{10.42}$$

$$= \frac{M_1\bar{f}}{EI}[L^2 - L^2/2] \tag{10.43}$$

$$\Delta = \frac{M_1 L^2}{2EI}. \tag{10.44}$$

**Remarks:**

(1) This result is fully consistent to what we would have obtained if we had followed our methods from Chapter 8.

### 10.2.6 Direct shear in beams (elastic only)

If an elastic element in a mechanical system of length $L$ is loaded in direct shear then the internal virtual work in the element is given by

$$\text{Int. V.W.} = \int_V \bar{\sigma}_{xy} \gamma_{xy} \, dV \tag{10.45}$$

$$= \int_V \bar{\sigma}_{xy} \frac{\sigma_{xy}}{G} \, dV \tag{10.46}$$

$$= \int_0^L \left[ \int_A \frac{\bar{V}}{A} f(y) \frac{V}{AG} f(y) \, dA \right] dx \tag{10.47}$$

$$= \alpha \int_0^L \frac{\bar{V} V}{AG} \, dx, \tag{10.48}$$

where $\alpha = \int_A f^2(y) \, dA$ and $f(y)$ is the shear stress distribution pattern for the cross-section of interest. Recall that $f(y)$ is determined from a shear flow analysis. For solid rectangular cross-sections $\alpha = 6/5$ and for solid circular cross-sections $\alpha = 10/9$.

---

## Example 10.4

*A shear beam.* As an example application, consider the beam shown in Fig. 10.9 and determine using the principle of virtual forces the tip displacement due to direct shear and bending, $\Delta$.

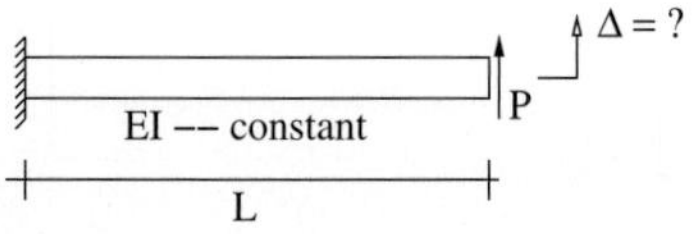

Fig. 10.9 Beam bent by an end-force: linear Elastic (real system).

*Solution*

Note that in this problem the beam is carrying its load both in shear and in bending. Thus there will be two contributions to the internal virtual work – one from the bending terms and one from the direct shear terms. To begin, we construct the virtual system shown in Fig. 10.10; the virtual system is identical to the real system but with all real loads removed and a virtual force introduced at $x = L$. The external virtual work is

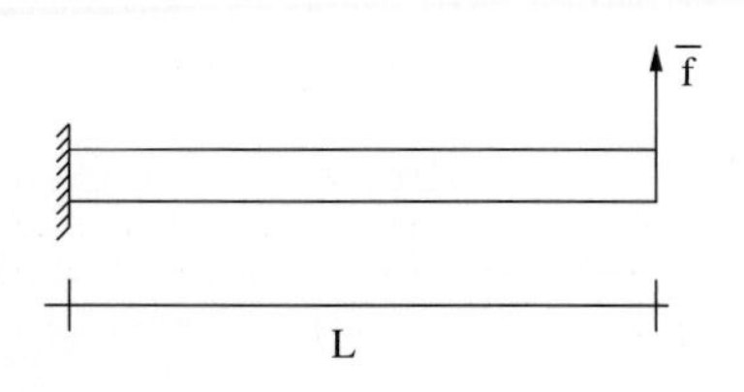


Fig. 10.10 Virtual system for determining the tip deflection for the beam shown in Fig. 10.9.

$$\text{Ext. V.W.} = \bar{f}\Delta. \tag{10.49}$$

The internal virtual work is

$$\text{Int. V.W.} = \int_0^L \bar{M}\kappa \, dx + \alpha \int_0^L \frac{\bar{V} V}{AG} \, dx. \tag{10.50}$$

From equilibrium of the virtual system we have that $\bar{M}(x) = \bar{f}(L - x)$ and $\bar{V}(x) = \bar{f}$. From equilibrium for the real system we have that $M(x) = P(L - x)$ and $V(x) = P$. Inserting these results into the virtual work theorem, we find that:

$$\bar{f}\Delta = \int_0^L \frac{\bar{M} M}{EI} + \alpha \frac{\bar{V} V}{AG} \, dx \tag{10.51}$$

$$= \int_0^L \frac{\bar{f} P}{EI}(L - x)^2 + \alpha \frac{\bar{f} P}{AG} \, dx \tag{10.52}$$

$$= \bar{f} \left[ \frac{PL^3}{3EI} + \alpha \frac{PL}{AG} \right] \tag{10.53}$$

$$\Delta = \frac{PL^3}{3EI} + \alpha \frac{PL}{AG}. \tag{10.54}$$

**Remarks:**

(1) If you compare this result to the one derived from energy conservation in Chapter 8, you will see that it is the same. The first term on the right-hand side gives the bending contribution, and the second term gives the shear "correction". As seen earlier in Chapter 8, for slender beams the direct shear contribution to the overall deflection is small in comparison to the bending contribution.

(2) This example demonstrates the remark made earlier that in systems that are simultaneously reacting to their load in multiple ways, one can simply add up all the external virtual work quantities and set them equal to the sum of all the internal virtual work quantities; i.e., virtual work, like real work, is an additive quantity.

## 10.3 Principle of virtual forces: Proof

The virtual work theorem which we stated and used in Sections 10.1 and 10.2 can be derived using the results from the prior chapters. In this section, we will give two proof-of-concept proofs to show that external virtual work is equal to internal virtual work. The proofs will also illuminate the real meaning of the principle of virtual forces.

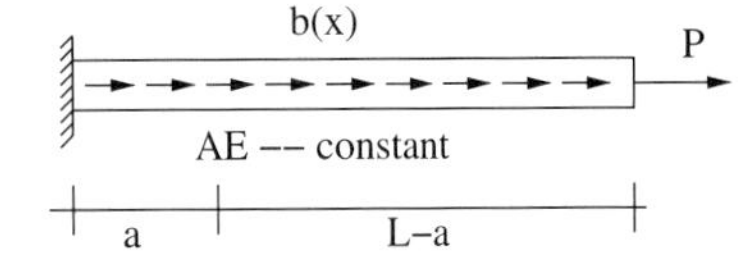


**Fig. 10.11** Axial bar with distributed load and end-load: linear Elastic (real system).

### 10.3.1 Axial bar: Proof

Let us consider the problem shown in Fig. 10.11 and determine the axial displacement at $x = a$. For the real problem, we know from Chapter 2 that we need to satisfy the equilibrium equation, $dR/dx + b(x) = 0$, the kinematic equation, $\varepsilon = du/dx$, and the constitutive equation, $\varepsilon = R/AE$. It is emphasized that these items must be satisfied with respect to the real system. Let us start with the kinematic equation for the real system:

$$\varepsilon = \frac{du}{dx}. \tag{10.55}$$

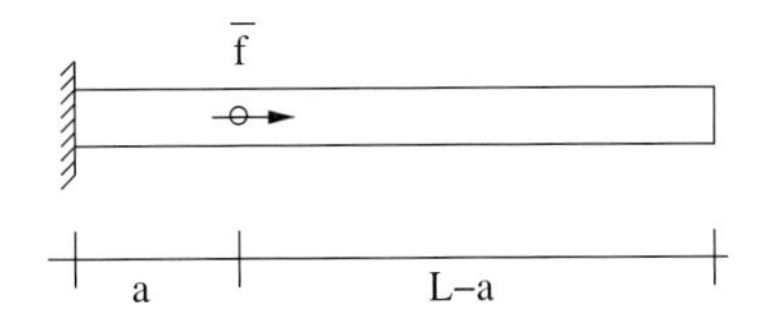


**Fig. 10.12** Virtual system for determining the displacement at $x = a$ for the bar shown in Fig. 10.11.

Using the virtual system shown in Fig. 10.12, we have from equilibrium of the virtual system that $\bar{R}(x) = \bar{f}H(a - x)$. Let us multiply both sides of eqn (10.55) by $\bar{R}(x)$ and then integrate both sides from $x = 0$ to $x = L$

$$\int_0^L \bar{R}\varepsilon\, dx = \int_0^L \bar{R}\frac{du}{dx}\, dx. \tag{10.56}$$

Equation (10.56) holds true, as it is simply an algebraic and calculus manipulation of a true statement – viz. eqn (10.55). Note that the left-hand side already gives the internal virtual work for a bar with axial forces. Let us now expand the right-hand side using the product rule of differentiation: $\frac{d(fg)}{dx} = \frac{df}{dx}g + f\frac{dg}{dx}$.

$$\int_0^L \bar{R}\varepsilon\, dx = \int_0^L \frac{d}{dx}\left(\bar{R}u\right) - \frac{d\bar{R}}{dx}u\, dx. \tag{10.57}$$

By equilibrium of the virtual system we see that $d\bar{R}/dx = -\bar{f}\delta(x-a)$. Substituting in and integrating both terms, one finds

$$\int_0^L \bar{R}\varepsilon\, dx = \bar{R}(L)u(L) - \bar{R}(0)u(0) + \bar{f}u(a). \tag{10.58}$$

Noting that $\bar{R}(L) = 0$ and $u(0) = 0$, we find that

$$\int_0^L \bar{R}\varepsilon\, dx = \bar{f}u(a). \tag{10.59}$$

Or in other words, that internal virtual work equals external virtual work.

**Remarks:**

(1) The proof of the virtual work theorem just given is, of course, specific to the problem analyzed. The result, however, is very general.

(2) More deeply, we can now understand the meaning of the principle of virtual forces. If one closely examines the proof just given, one sees that the final result, eqn (10.59), depends only on one thing about the real system: its kinematic relation. All the manipulations in the proof are either algebra, calculus, or associated with the virtual system. If fact, with respect to the virtual system, we have used only its equilibrium.

(3) The ability to apply eqn (10.59) to an actual problem requires one to know $\varepsilon(x)$ – the strains in the real system. To find these, one needs to separately apply equilibrium to the real system to find $R(x)$ and to apply the constitutive relation to the real system to find $\varepsilon(x) = R(x)/AE$. Thus one sees that the method of virtual forces is simply a way of rewriting the kinematic relation for the system. Equilibrium and the constitutive relation must be utilized separately in coming to a "final answer" in any given problem.

(4) For the virtual system we have used only the property of equilibrium. The satisfaction of the kinematic relation and the constitutive relation is not required for the virtual system. In reality, the virtual system is merely a convenient device for finding an internal force field in equilibrium.

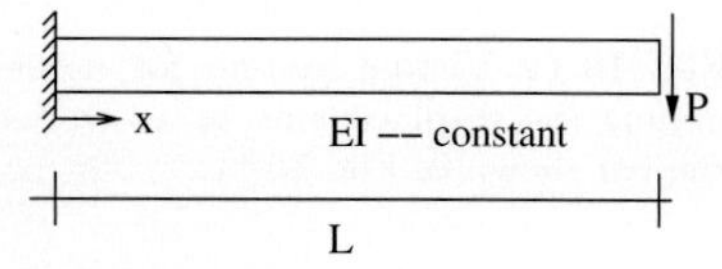


**Fig. 10.13** Cantilever beam: linear elastic (real system).

### 10.3.2 Beam bending: Proof

Let us consider the problem shown in Fig. 10.13 and determine the rotation at $x = a$. Further, let us consider only the contributions from bending; i.e. let us assume the beam is slender so that we may ignore direct shear. For the real problem, we know from Chapter 8 that we need to satisfy the equilibrium relations

$$\frac{dV}{dx} + q(x) = 0 \tag{10.60}$$

$$\frac{dM}{dx} + V(x) = 0, \tag{10.61}$$

the (moment–curvature) constitutive relation

$$M = EI\kappa, \tag{10.62}$$

and the kinematic relations

$$\kappa = \frac{d\theta}{dx} \tag{10.63}$$

$$\theta = \frac{dv}{dx}. \tag{10.64}$$

Let us start by combining the two kinematic relations into a single equation for the real system:

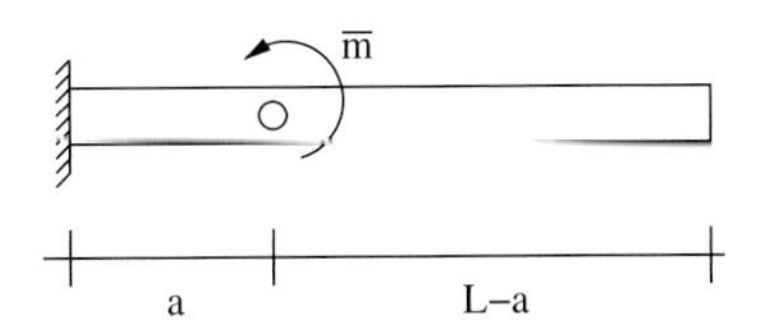


**Fig. 10.14** Virtual system for finding the rotation at $x = a$ for the beam shown in Fig. 10.13.

$$\kappa = \frac{d^2v}{dx^2}. \tag{10.65}$$

Using the virtual system shown in Fig. 10.14, we have from equilibrium of the virtual system that

$$\bar{M}(x) = \bar{m}H(a - x). \tag{10.66}$$

Let us multiply both sides of eqn (10.65) by $\bar{M}(x)$, and then integrate both sides of the result from $x = 0$ to $x = L$ to give

$$\int_0^L \bar{M}\kappa \, dx = \int_0^L \bar{M}\frac{d^2v}{dx^2} \, dx. \tag{10.67}$$

Equation (10.67) holds true, as it is simply an algebraic and calculus manipulation of a true statement – viz., $\kappa = d^2v/dx^2$. Note that the left-hand side already gives the internal virtual work for a beam in bending. Let us now expand the right-hand side using the product rule of differentiation: $\frac{d(fg)}{dx} = \frac{df}{dx}g + f\frac{dg}{dx}$.

$$\int_0^L \bar{M}\kappa \, dx = \int_0^L \frac{d}{dx}\left(\bar{M}\frac{dv}{dx}\right) - \frac{d\bar{M}}{dx}\frac{dv}{dx} \, dx. \tag{10.68}$$

By equilibrium of the virtual system we see that $d\bar{M}/dx = -\bar{m}\delta(x - a)$. Substituting in and integrating both terms, one finds

$$\int_0^L \bar{M}\kappa \, dx = \bar{M}(L)\frac{dv}{dx}(L) - \bar{M}(0)\frac{dv}{dx}(0) + \bar{m}\frac{dv}{dx}(a). \tag{10.69}$$

Noting that $\bar{M}(L) = 0$ and $\theta(0) = 0$, we find that

$$\int_0^L \bar{M}\kappa \, dx = \bar{m}\theta(a). \tag{10.70}$$

Or in other words, as before, internal virtual work equals external virtual work.

**Remarks:**

(1) As in the prior proof, the result demonstrated is particular to the problem analyzed, but it does indeed hold true more generally.

(2) Note that with respect to the real system, eqn (10.70) involves only the kinematic relations. Thus the principle of virtual forces/moments again represents only the kinematics of the real system.

(3) To profitably employ eqn (10.70) one needs to separately solve for the real moments in the beam using the equilibrium equations, and then apply the constitutive relation to find the real curvatures.

(4) Proofs that internal virtual work equals external virtual work in more general settings can be approached in a fashion similar to what we have done here. However, this rapidly becomes very tedious. In advanced courses on mechanics and structural analysis these proofs are taken up utilizing advanced concepts that make the proofs less tedious and more illuminating. Thus we omit, here, the proofs for more general cases.

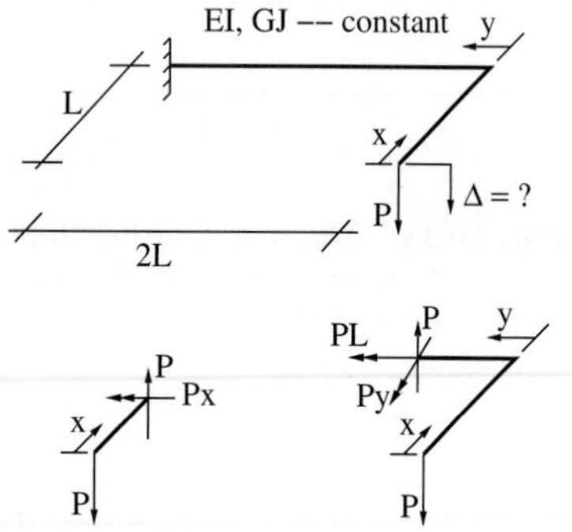

**Fig. 10.15** L-shaped hanger: linear elastic (real system).

## 10.4 Applications: Method of virtual forces

In this section, we will look at a series of examples to further illustrate the use of the method of virtual forces.

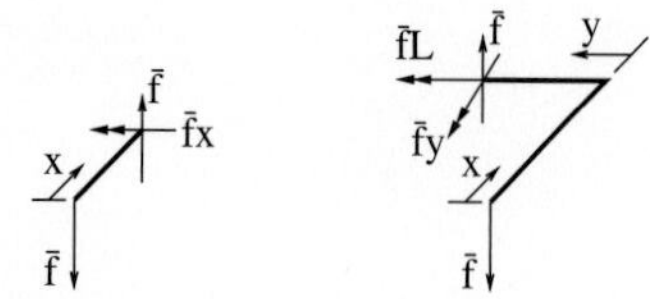

**Fig. 10.16** Virtual system for finding the tip deflection for the hanger shown in Fig. 10.15.

### Example 10.5

*L-shaped hanger.* So far our examples have been ones which we could have easily solved using techniques from earlier chapters. Let us now consider an example that would be a little difficult to deal with directly from the point of view of a differential equation formulation. For the L-shaped hanger shown in Fig. 10.15 let us determine, using the principle of virtual forces, the tip displacement, $\Delta$.

*Solution*
To begin, we construct the virtual system shown in Fig. 10.16; the virtual system is identical to the real system but with all real loads removed and a virtual force introduced at the end of the hanger arm. The external virtual work is

$$\text{Ext. V.W.} = \bar{f}\Delta. \tag{10.71}$$

The internal virtual work consists of three contributions: bending in the two segments of the L, and torsion in the section directly attached to the wall. In this example we will ignore direct shear. Assuming everything is elastic one has

$$\text{Int. V.W.} = \int_0^L \frac{\bar{M}M}{EI}\,dx + \int_0^{2L} \frac{\bar{M}M}{EI}\,dy + \int_0^{2L} \frac{\bar{T}T}{GJ}\,dy. \tag{10.72}$$

We now make section cuts in the two arms of the hanger to find that

$$M(x) = Px \qquad \bar{M}(x) = \bar{f}x \tag{10.73}$$

$$M(y) = Py \qquad \bar{M}(y) = \bar{f}y \tag{10.74}$$

$$T(y) = PL \qquad \bar{T}(y) = \bar{f}L. \tag{10.75}$$

With the aid of the principle of virtual forces, we find

$$\bar{f}\Delta = \int_0^L \frac{\bar{f}Px^2}{EI}\,dx + \int_0^{2L} \frac{\bar{f}Py^2}{EI}\,dy + \int_0^{2L} \frac{\bar{f}PL^2}{GJ}\,dy \tag{10.76}$$

$$= \bar{f}\left[\frac{PL^3}{3EI} + \frac{8PL^3}{3EI} + \frac{2PL^3}{GJ}\right] \tag{10.77}$$

$$\Delta = \frac{3PL^3}{EI} + \frac{2PL^3}{GJ}. \tag{10.78}$$

**Remarks:**

(1) To have solved this problem using the methods from the earlier chapters would have been possible but more difficult. The principle of virtual forces can be a very effective technique in problems with these sorts of geometric complexity.

---

**Example 10.6**

*Curved band.* Our last example showed the ease with which we can handle more geometric complexity if we use virtual work methods. Let us continue with another example that is not easily treated using the methods of the earlier chapters due to geometric features. Shown in Fig. 10.17(a) is a thin metal band – clamped at one end and subject to a horizontal force at the other. Let us determine, using the principle of virtual forces, the end displacement, $\Delta$.

*Solution*
To begin, we construct the virtual system shown in Fig. 10.17(b); the virtual system is identical to the real system but with all real loads removed and a virtual force introduced at the end of the band. The external virtual work is

$$\text{Ext. V.W.} = \bar{f}\Delta. \tag{10.79}$$

If we make a section cut at an arbitrary angle $\theta$ from the horizontal, we see that there are internal moments, shear forces, and axial forces. Let us assume that the band is sufficiently slender so that it is reasonable to ignore direct shear and axial contributions. Thus, the internal virtual

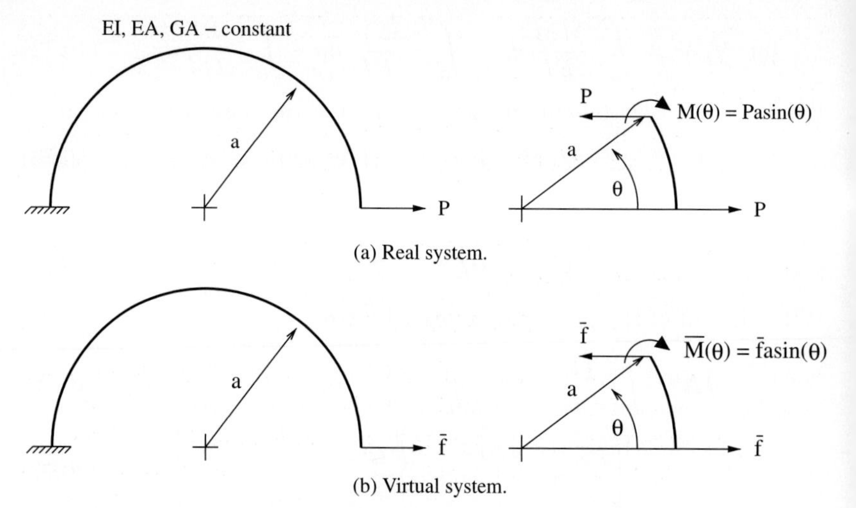


**Fig. 10.17** Curved band: linear elastic.

work consists only of a bending contribution. Assuming everything is elastic one has

$$\text{Int. V.W.} = \int_0^\pi \frac{\bar{M}M}{EI}\, a d\theta. \tag{10.80}$$

From our section cut we see that $\bar{M}(\theta) = \bar{f} a \sin(\theta)$ and $M(\theta) = Pa\sin(\theta)$. Combining now with the principle of virtual forces we find:

$$\bar{f}\Delta = \int_0^\pi P\bar{f}\sin^2(\theta)\frac{a^3}{EI}\, d\theta \tag{10.81}$$

$$= \int_0^\pi P\bar{f}\left[\frac{1}{2} - \frac{1}{2}\cos(2\theta)\right]\frac{a^3}{EI}\, d\theta \tag{10.82}$$

$$= \frac{P\bar{f}a^3}{EI}\int_0^\pi \frac{1}{2} - \frac{1}{2}\cos(2\theta)\, d\theta \tag{10.83}$$

$$= \frac{P\bar{f}a^3}{EI}\left[\frac{\theta}{2} - \frac{1}{4}\sin(2\theta)\right]_0^\pi \tag{10.84}$$

$$= \bar{f}\frac{Pa^3\pi}{2EI} \tag{10.85}$$

$$\Delta = \frac{Pa^3\pi}{2EI}. \tag{10.86}$$

**Remarks:**

(1) This example shows the application of the principle of virtual work to a situation where our methods from the earlier chapters would not have helped us directly.

## Example 10.7

*Beam supported by a truss rod.* Here we wish to analyze the system shown in Fig. 10.18 for the rotation at the point of application of the force.

*Solution*
To solve we will employ the virtual system shown in Fig. 10.19. The external virtual work will be

$$\text{Ext. V.W.} = \bar{m}\theta(3). \tag{10.87}$$

For the internal virtual work let us account for all modes of loading and determine their relative contribution to the end-rotation. Using statics, we find that the truss rod is in a state of axial load and that the beam is in a state of bending, direct shear, and axial load. Thus there will be four contributions to the internal virtual work. Using the internal force diagrams given in Figs. 10.18 and 10.19, one finds the following contributions.
Truss contribution:

$$(-\sqrt{2}\bar{m})\frac{-3\sqrt{2}P\sqrt{2}}{AE} = \frac{6\sqrt{2}P\bar{m}}{AE} \tag{10.88}$$

Axial contribution:

$$\int_0^1 \bar{m}\frac{3P}{AE}\,dy = \frac{3P\bar{m}}{AE}. \tag{10.89}$$

Direct shear contribution:

$$\alpha\int_0^1 (\bar{m})\frac{2P}{GA}\,dx = \alpha\frac{2P\bar{m}}{GA}. \tag{10.90}$$

Bending contribution:

$$\int_0^1 (-\bar{m}x)\frac{-2Px}{EI}\,dx + \int_1^3 (-\bar{m})\frac{-3P+Px}{EI}\,dx = \frac{3P\bar{m}}{EI}. \tag{10.91}$$

If we set the external virtual work equal to the sum of the internal virtual work contributions and then factor out the bending part, we have

$$\theta(3) = \frac{3P}{EI}\left[\underbrace{1}_{\text{Bending}} + \underbrace{\alpha\frac{2EI}{3GA}}_{\text{D.Shear}} + \underbrace{\frac{I}{A}}_{\text{Axial}} + \underbrace{2\sqrt{2}\frac{I}{A_T}}_{\text{Truss}}\right]. \tag{10.92}$$

If we insert the given dimensions and material properties, we find that

$$\theta(3) = \frac{3P}{EI}\left[\underbrace{1}_{\text{Bending}} + \underbrace{0.007}_{\text{D.Shear}} + \underbrace{0.003}_{\text{Axial}} + \underbrace{0.02}_{\text{Truss}}\right], \tag{10.93}$$

where it is assumed that the beam cross-section is rectangular, ($\alpha = 6/5$).

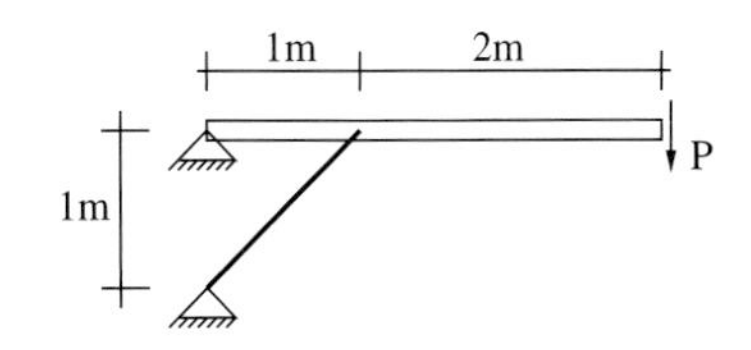


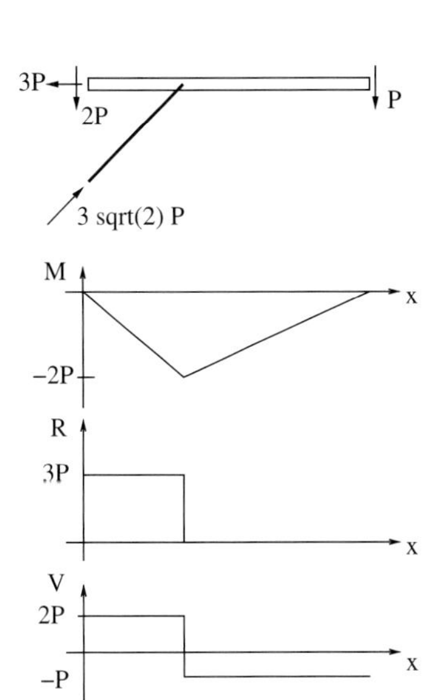


**Fig. 10.18** Beam supported by a truss rod: linear elastic (real system). $A_{\text{truss}} = 10^{-2}$ m$^2$, $A_{\text{beam}} = 2 \times 10^{-2}$ m$^2$, $I_{\text{beam}} = 0.66 \times 10^{-4}$ m$^4$, $E = 250$ GPa, $G = 100$ GPa.

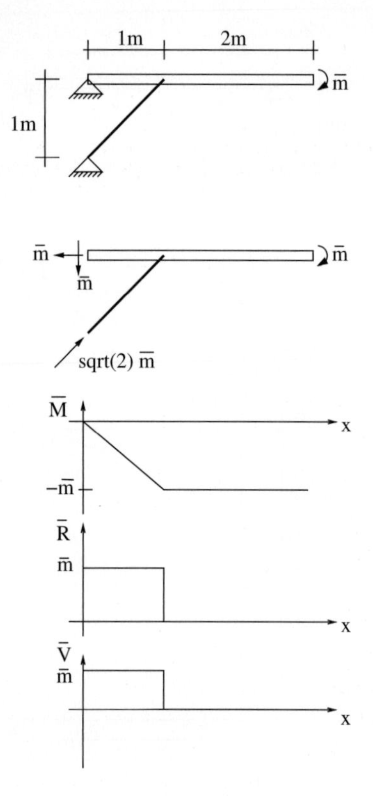


**Fig. 10.19** Virtual system for finding the end rotation for the beam in Fig. 10.18.

**Remarks:**

(1) Thus for these dimensions and material properties we would have been fully justified in ignoring direct shear and axial force contributions from the beam. The truss contributions are, however, 2% of the bending contributions, and in many situations should be retained.

## Example 10.8

*Indeterminate truss.* To apply the method of virtual forces, as illustrated by all our examples so far, one needs to be able to compute the internal forces in the real and virtual systems. If, however, one is faced with an indeterminate problem, then this will not be directly possible. To treat such problems we can look back to the method introduced in earlier chapters where redundant supports are removed and replaced by unknown reactions. The unknown reactions are then determined by imposing the kinematic constraint of zero motion at the support location. As an example application of this methodology, consider the truss system shown in Fig. 10.20(a) and determine the vertical deflection at the point of load application.

*Solution*

The system is statically indeterminate of degree one, and thus can be made determinate by removing one support, as shown in Fig. 10.20(b). In order to solve for the two unknowns in the problem, $\Delta$ and $F$, we will need two virtual systems as shown in Fig. 10.20(c). The virtual system on the left will allow us to determine an expression for $F$, and the system on the right will allow us to determine $\Delta$.

We start with an application of statics to the three systems. For the real system one has that the internal forces in the truss bars are

$$R_1 = P - F/\sqrt{2} \qquad R_2 = F \qquad R_3 = -F/\sqrt{2}. \tag{10.94}$$

For the virtual system on the left one has

$$\bar{R}_1 = -\bar{1}/\sqrt{2} \qquad \bar{R}_2 = \bar{1} \qquad \bar{R}_3 = -\bar{1}/\sqrt{2}. \tag{10.95}$$

For the virtual system on the right one has

$$\bar{R}_1 = 0 \qquad \bar{R}_2 = 0 \qquad \bar{R}_3 = \bar{1}. \tag{10.96}$$

Applying the principle of virtual forces using the virtual system on the left gives

$$\Delta_F \bar{1} = \frac{(-\bar{1}/\sqrt{2})(P - F/\sqrt{2})L}{AE} + \frac{\bar{1}F\sqrt{2}L}{AE} + \frac{(-\bar{1}/\sqrt{2})(-F/\sqrt{2})L}{AE}. \tag{10.97}$$

Since we know that $\Delta_F$ is zero, we can use this equation to solve for $F$:

$$F = \frac{P}{2 + \sqrt{2}}. \tag{10.98}$$

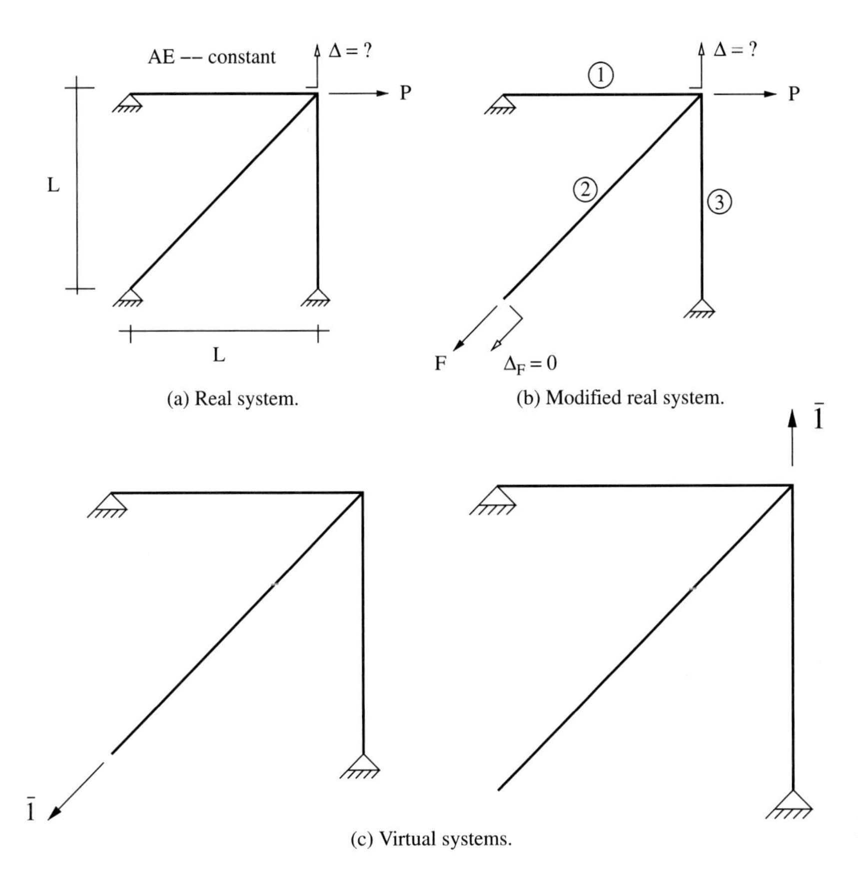


**Fig. 10.20** Indeterminate truss system: linear elastic.

Applying the principle of virtual forces using the virtual system on the right gives:

$$\Delta\bar{1} = \frac{0(P - F/\sqrt{2})L}{AE} + \frac{0F\sqrt{2}L}{AE} + \frac{\bar{1}(-F/\sqrt{2})L}{AE} \tag{10.99}$$

and upon use of eqn (10.98) we arrive at

$$\Delta = -\frac{PL}{AE}\frac{1}{2+\sqrt{2}}. \tag{10.100}$$

## Chapter summary

- Virtual work theorem: Ext. V.W. = Int. V.W.
- The principle of virtual forces is an alternative expression of the kinematic relations for a mechanical system.
- Ext. V.W. (virtual forces): $f\Delta$ or $\bar{m}\theta$

- Int. V.W. (virtual forces):
  - Axial forces: $\int_L \bar{R}\varepsilon \, dx$, (elastic) $\int_L \bar{R}\frac{R}{AE} \, dx$
  - Torsion: $\int_L \bar{T}\phi' \, dz$, (elastic) $\int_L \bar{T}\frac{T}{GJ} \, dz$
  - Bending: $\int_L \bar{M}\kappa \, dx$, (elastic) $\int_L \bar{M}\frac{M}{EI} \, dx$
  - Direct shear (elastic only): $\alpha \int_L \bar{V}\frac{V}{AG} \, dx$ ($\alpha = 6/5$, rectangular; $\alpha = 10/9$, solid round)

# Exercises

(10.1) A solid circular bar is bent 90° at two locations and is built-in at one end. (a) Determine a formula for the vertical deflection at the point of load application. Assume A, I, J, E, and G are constants. (b) Let $L = 200$ mm and the diameter of the bar be $d = 30$ mm. What is the percent contribution to the total deflection from axial loading, bending, torsion, and direct shear? Assume $E/G = 2$. (c) Repeat part (b) with $L = 500$ mm and $d = 10$ mm.

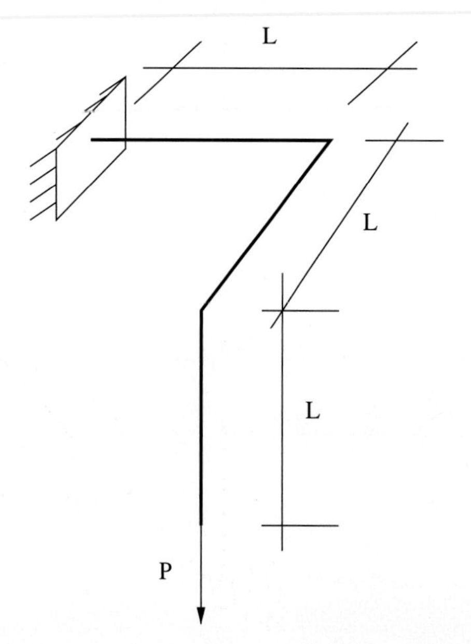


(10.2) Derive the formula for the tip deflection of a cantilever beam with a transverse load at the tip using virtual work. Assume linear elasticity and ignore direct shear.

(10.3) Find the vertical deflection at the tip of the structure shown below. Assume all sections are slender.

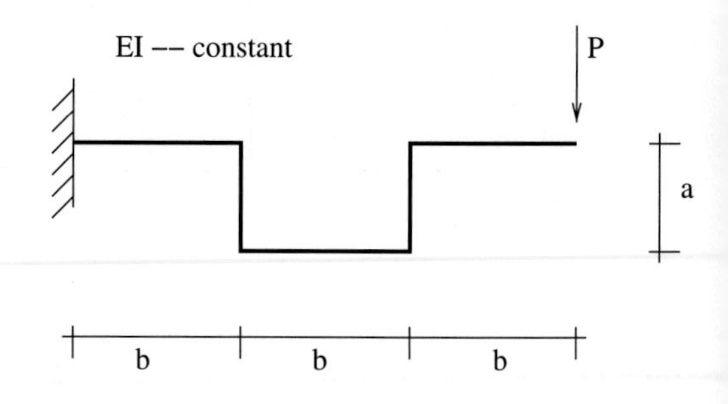


(10.4) Determine the vertical deflection at the upper point of loading for the structure shown. Ignore direct shear effects. (Recall that $\sin^2(\theta) = \frac{1}{2} - \frac{1}{2}\cos(2\theta)$, $\cos^2(\theta) = \frac{1}{2} + \frac{1}{2}\cos(2\theta)$.)

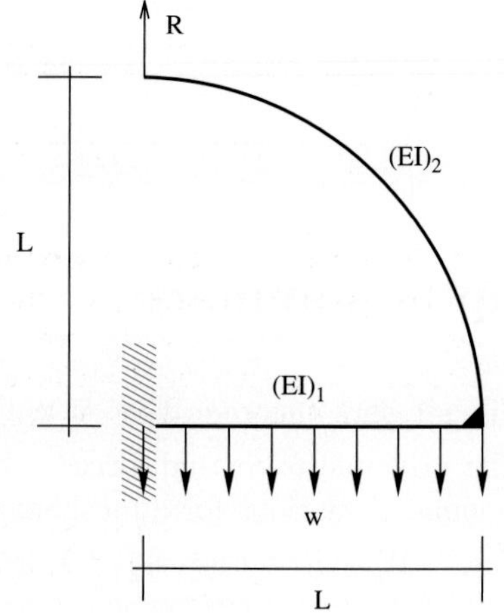

(10.5) Shown is a serpentine spring, composed of $N$ "hairpin" segments of a round wire. Determine an expression for the torsional stiffness of the spring. Express your answer in terms of $E$, $I$, $J$, $G$, $a$, $L$, $N$.

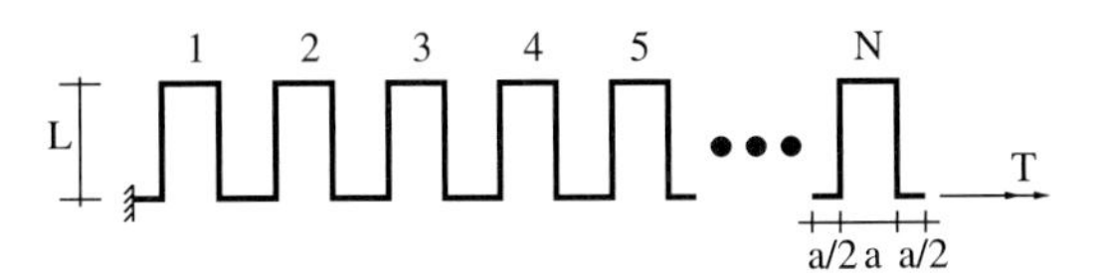


(10.6) Using virtual work, find the horizontal deflection at point $A$ for the elastic system shown below. Assume all members are slender and that the section properties are the same constant throughout.

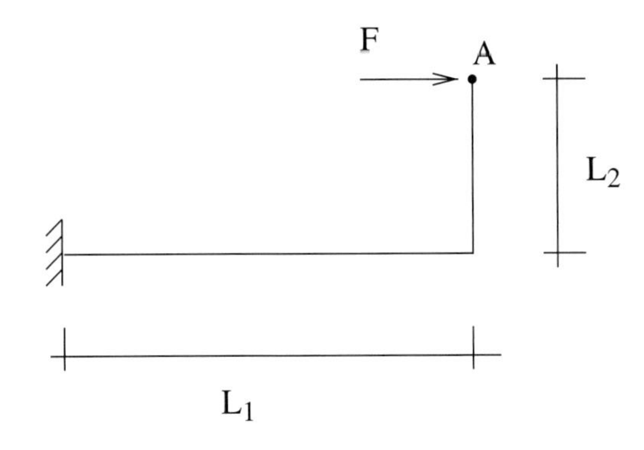


(10.7) By exploiting symmetry, find the beam deflections at the load points. Assume a flexural rigidity EI and ignore direct shear effects.

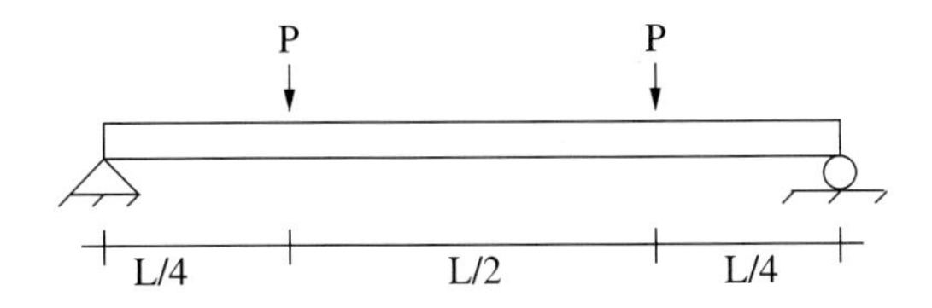


(10.8) For the structure shown, determine the deflection at the center of the beam. Assume the truss members have a cross-sectional area $A$ and a Young's modulus $E$; assume the beam has a flexural rigidity $EI$. Ignore direct shear in the beam.

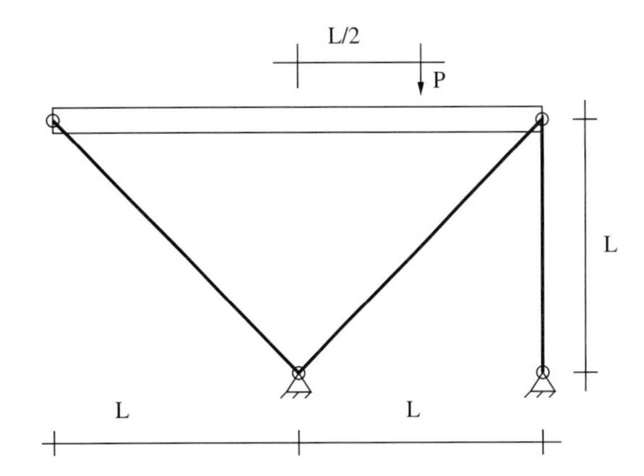


(10.9) Using the geometry and load in Exercise 10.1(c), determine the deflections at the load point perpendicular to the direction of $P$.

(10.10) Using the geometry and load in Exercise 10.4, determine the rotation at the point where the curved arc meets the horizontal beam.

(10.11) For the truss shown, find the horizontal deflection at the load point.

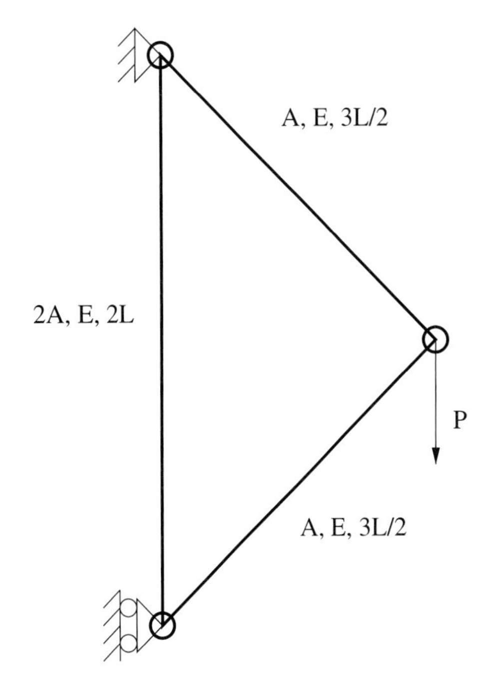

(10.12) Using the geometry and load in Exercise 10.11, determine the vertical deflection at the load point.

(10.13) Using the geometry and load in Exercise 10.11, determine the vertical motion of the roller.

(10.14) Find the the deflection at the hinge in the beam shown. Assume EI constant, and that the beam is slender.

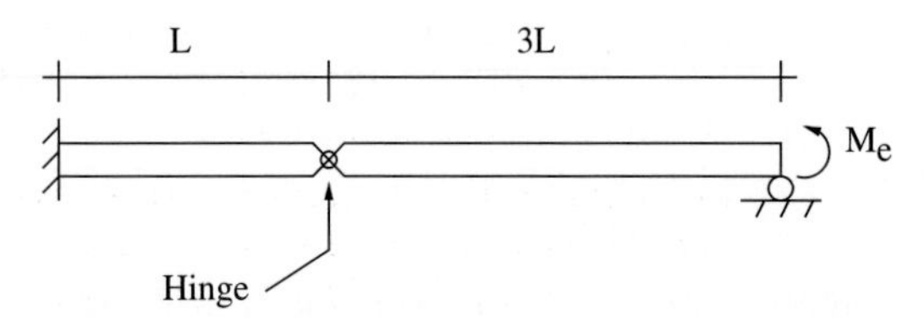


(10.15) Using the geometry and load in Exercise 10.14, determine the rotation at the point of application of the end moment, $M_e$.

(10.16) Find the rotation at point B in the slender frame below. Assume EI constant.

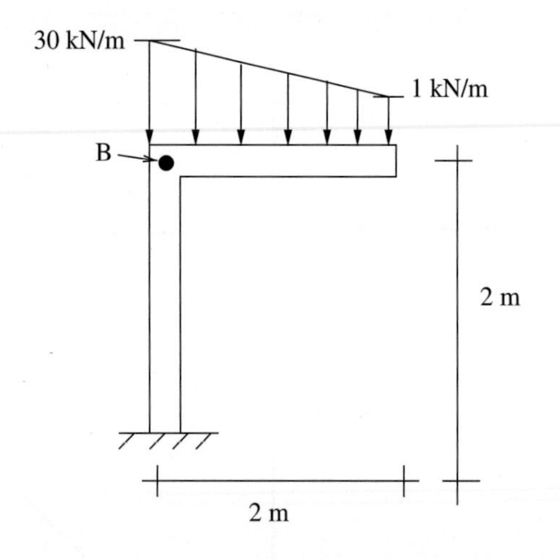


(10.17) Using the geometry and loading in Exercise 10.16, determine the horizontal deflection at point B.

(10.18) Consider the following U-shaped structure. (a) Find the formula for the change in separation of the load points. Assume EI constant and ignore direct shear effects. (b) Assume the structure is formed from a solid round wire with $L = 100$ mm, $R = 15$ mm, and a diameter of $d = 2$ mm. Find the stiffness of the U (for opening).

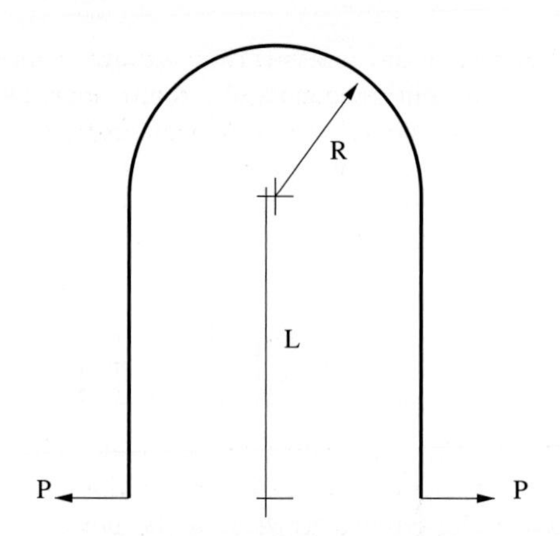


(10.19) For the frame shown below, determine the deflection in the direction of the applied force $P$. Do not ignore any terms. Assume all elastic and geometric properties are given constants.

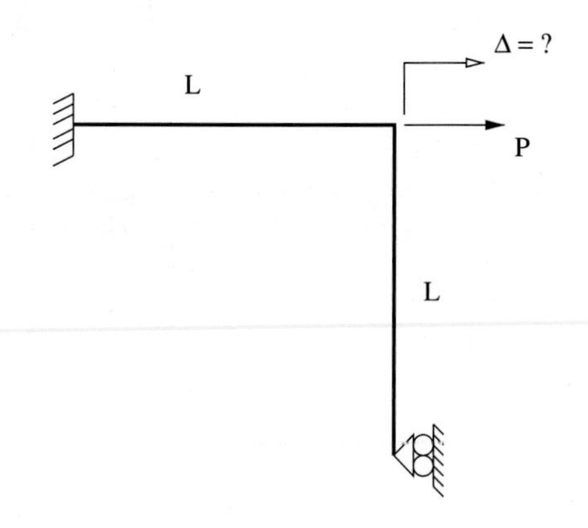


(10.20) Determine the vertical displacement at the point of load for the structure shown. Assume all section properties are constants; the length of each arm is $L$, and they meet at 90°. Neglect the effects of direct shear.

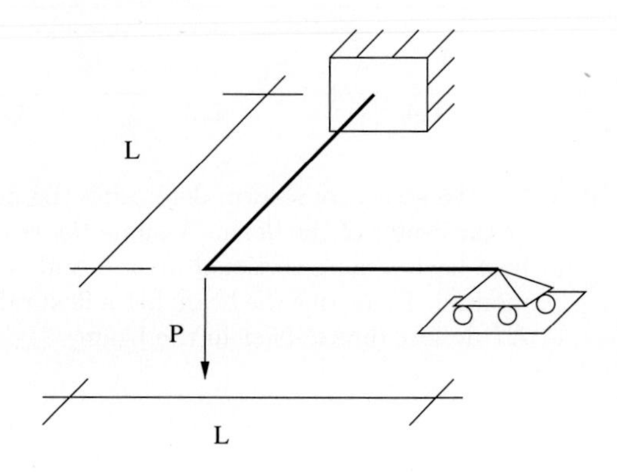

(10.21) Find the rotation at the point of application of the end-moment, $M_e$, in the beam shown below. Assume EI and GA are constants. Include the effects of direct shear.

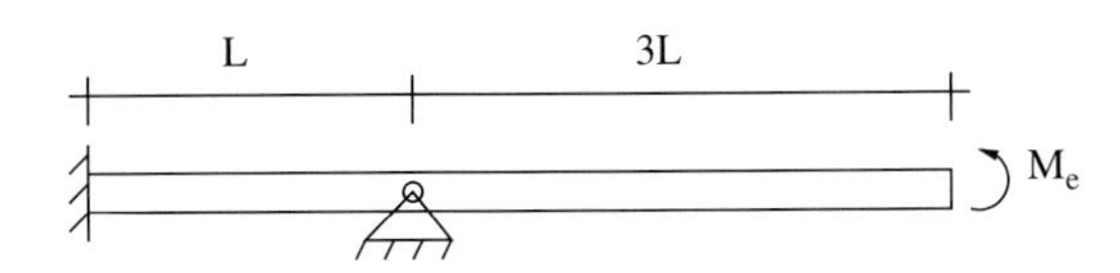


(10.22) Using the geometry and load in Exercise 10.21, determine the deflection at the load point.

(10.23) Using the geometry and load in Exercise 10.21, determine the rotation at the location of the pin support.

(10.24) Using the geometry and load in Exercise 10.21, determine the reaction at the location of the pin support.

(10.25) Derive the virtual work equation that is used to solve Exercise 10.14. (Hint: Start with the relation $\kappa = d^2v/dx^2$.)

(10.26) Derive the virtual work equation that would be used to determine $\theta$ in the exercise shown. Assume the bar is linear elastic and circular. (Hint: start with the basic kinematic relation for torsion.)

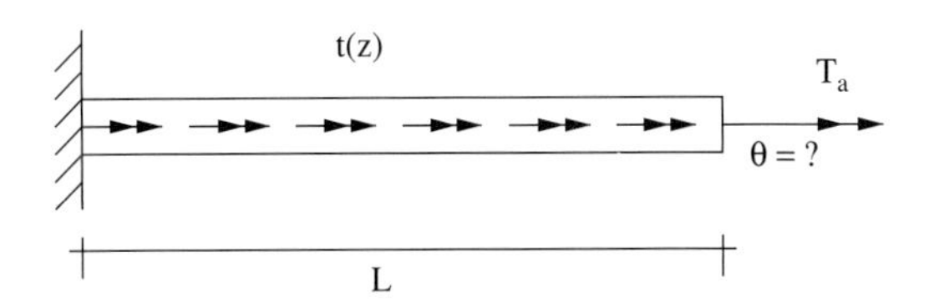


(10.27) Consider Example 10.7. Assume that in addition to the indicated load there is a temperature increase of 50°C and that the coefficient of thermal expansion for the beam and truss bar material is $25 \times 10^{-6}$ /C. Find the vertical deflection at the point of load application.

(10.28) In the statically indeterminate system shown: (a) find the force in the truss bar; (b) find the deflection $\Delta$ at the point of load application.

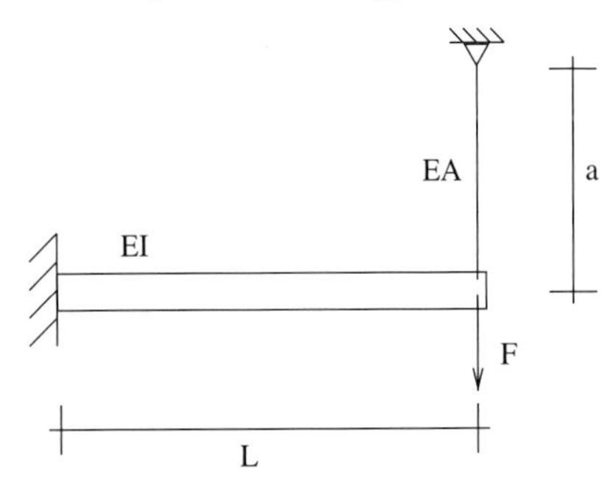

# 11 Potential-Energy Methods



In this chapter we will examine the use of potential-energy methods to solve problems similar to those we looked at in Chapter 10. In fact, we will see in the second part of this chapter that the methods we will develop will lead to equations that are strikingly similar to those used in the method of virtual forces. This similarity is not accidental as potential energy methods are intimately related virtual work methods. Thus the techniques we will use in this chapter can be used as an alternate way of understanding virtual work. It is, however, noted that the concepts of virtual work are more general than potential energy methods – potential-energy methods apply only to conservative systems, while virtual work methods can be applied to conservative and non-conservative systems.

## 11.1 Potential energy: Spring-mass system

As an introduction to potential-energy methods let us recall the familiar example of a mass with weight $P$ resting on top of a linear spring with spring constant $k$, as shown in Fig. 11.1. Before we place the mass on the spring, let us assume that it has a length $z_o$. After we place the mass on the spring, the spring will compress an amount $\Delta$ and the mass will come to a static equilibrium position $z = z_o - \Delta$. We know for static equilibrium of the mass that the sum of the gravitational force and the spring force (sum of the forces in the $z$-direction) must be zero:

$$k\Delta - P = 0. \tag{11.1}$$

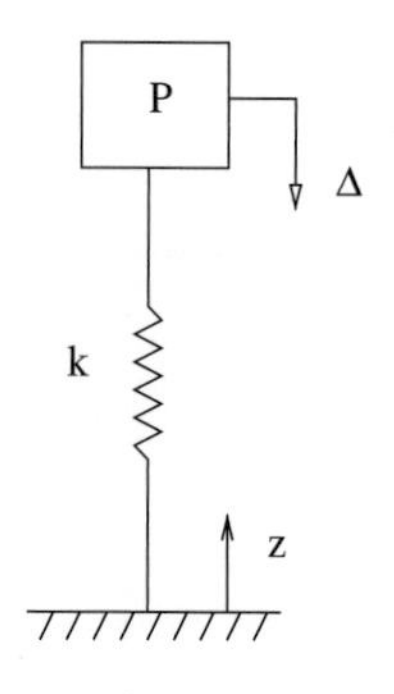


**Fig. 11.1** Spring-mass system.

Both of the forces acting on the mass happen to be conservative. From Chapter 1, we know that the gravitational force can be expressed as the potential

$$\Pi_{\text{gravity}} = Pz = P(z_o - \Delta), \tag{11.2}$$

where gravity is acting downwards and thus the gravity force acting on the mass is $F_{\text{gravity}} = -d\Pi/dz = -P$. The spring force itself is $F_{\text{spring}} = k\Delta = k(z_o - z)$; i.e. for positive motion $\Delta$ the spring pushes up on the mass. If we want, we can also express this force in terms of the potential

$$\Pi_{\text{spring}} = \frac{1}{2}k(z_o - z)^2 = \frac{1}{2}k\Delta^2 \tag{11.3}$$

Note that $-d\Pi_{\text{spring}}/dz = k(z_o - z)$ as desired. Also observe that the potential energy for the spring is simply the stored elastic energy in the spring.

The total potential energy of the system is given by

$$\Pi(z) = \Pi_{\text{spring}}(z) + \Pi_{\text{gravity}}(z) = \frac{1}{2}k(z_o - z)^2 + Pz \tag{11.4}$$

and the force balance will be given by

$$F_{\text{spring}} + F_{\text{gravity}} = -\frac{d\Pi_{\text{spring}}}{dz} - \frac{d\Pi_{\text{gravity}}}{dz} = -\frac{d\Pi}{dz} = 0. \tag{11.5}$$

This last expression is our main result. It says that for a static equilibrium, the total potential energy of a system must be stationary. In other words, we can express static equilibrium directly by saying the sum of the forces must be zero or indirectly by asserting that the total potential energy of a conservative system must be stationary. This is known as the *principle of stationary potential energy.*

**Remarks:**

(1) It is usually more convenient to express the potentials in terms of the displacement of the system as opposed to in terms of the absolute position of the system. In this case we will have

$$\Pi(\Delta) = \Pi_{\text{spring}}(\Delta) + \Pi_{\text{gravity}}(\Delta) = \frac{1}{2}k\Delta^2 - P\Delta. \tag{11.6}$$

In eqn (11.6) we have omitted the constant term $Pz_o$, since upon differentiation it will disappear. Requiring stationarity with respect to $\Delta$ will give $d\Pi/d\Delta = k\Delta - P = 0$ as expected.

(2) The potential energy of an elastic system will always be its stored elastic energy.

(3) All loads which are constant can always be modeled as being provided by gravitational weights. For an arbitrary constant load $F$, the potential of the load can always be expressed as $\Pi_{\text{load}} = -F\Delta$, where $\Delta$ is the motion at the point of application of the load in the direction in which the load is applied.

(4) It is important to pay attention to what it means to require the total potential energy of a system to be stationary. This principle has been derived from the notion of static equilibrium. Thus to come to a complete solution to a given mechanical problem will also require the separate application of kinematic and constitutive relations. In the example we have used to develop this principle, we have used a constitutive relation for the spring with parameter $k$. The kinematics of the problem have also been employed in the hidden assumption that the displacement of the mass and the spring are the same. So in summary, the principle of stationary potential energy is equivalent to equilibrium, and to solve any problem using the principle of stationary potential energy will, in general, require the separate application of kinematic relations.

The constitutive relation will *a priori* be embedded in the definition of the stored energy, and need not be introduced separately.

(5) In many texts one will see the phrase *minimum potential energy.* The notion of stationary potential energy is slightly more general. Stationary potential energy is equivalent to the notion of static equilibrium and this equilibrium can be stable, unstable, or neutral. Minimum potential energy is associated with stable static equilibria. We will investigate this issue in more detail in Chapter 12. All the problems in the present chapter, however, are associated with stable equilibria and thus are in fact minimum problems.

## 11.2 Stored elastic energy: Continuous systems

To be able to exploit the concept of stationary potential energy for structural systems we will need expressions for the stored elastic energy, $W$, for various load-bearing cases. The needed expressions were developed in the prior chapters based upon the notion that the total stored energy in a body is given by integrating the strain energy density over the volume. In the general linear elastic setting the strain energy density is recalled to be

$$w = \frac{1}{2}(\sigma_{xx}\varepsilon_{xx} + \sigma_{yy}\varepsilon_{yy} + \sigma_{zz}\varepsilon_{zz} + \sigma_{xy}\gamma_{xy} + \sigma_{yz}\gamma_{yz} + \sigma_{zx}\gamma_{zx}). \quad (11.7)$$

For the different types of loadings we study, this expression, when integrated over the volume of a load bearing member, reduces to:
Axial forces:

$$W = \int_V \frac{1}{2}\sigma_{xx}\varepsilon_{xx}\,dV = \int_L \int_A \frac{1}{2}E\varepsilon^2\,dA\,dx \quad (11.8)$$

$$= \int_0^L \frac{1}{2}AE\left(\frac{du}{dx}\right)^2 dx. \quad (11.9)$$

Torsional loads:

$$W = \int_V \frac{1}{2}\sigma_{z\theta}\gamma_{z\theta}\,dV = \int_L \int_A \frac{1}{2}G\gamma^2\,dA\,dz \quad (11.10)$$

$$= \int_0^L \int_A \frac{1}{2}Gr^2\left(\frac{d\phi}{dz}\right)^2 dA\,dz \quad (11.11)$$

$$= \int_0^L \frac{1}{2}GJ\left(\frac{d\phi}{dz}\right)^2 dz. \quad (11.12)$$

Bending loads:

$$W = \int_V \frac{1}{2}\sigma_{xx}\varepsilon_{xx}\,dV = \int_L \int_A \frac{1}{2}E\varepsilon^2\,dA\,dx \quad (11.13)$$

$$= \int_0^L \int_A \frac{1}{2} E y^2 \left(\frac{d^2 v}{dx^2}\right)^2 dA\, dx \quad (11.14)$$

$$= \int_0^L \frac{1}{2} EI \left(\frac{d^2 v}{dx^2}\right)^2 dx. \quad (11.15)$$

**Remarks:**

(1) We have not provided an expression for the stored energy due to direct shear. You will notice that all of the expressions are given in terms of the relevant measure of strain: normal strain $du/dx$, twist rate $d\phi/dz$, and curvature $d^2v/dx^2$. With our assumed beam kinematics we do not have such a measure of strain for direct shear. Thus we are not able to provide, within the assumptions of our beam theory, an expression for the stored energy in direct shear in terms of a kinematic measure.

(2) In a linear system that is carrying its load in multiple ways, one can simply add up all the different stored energy terms to compute the total stored energy. This functions just as it did with virtual work.

To illustrate the use of these expressions let us look at two simple examples, in which we will assume that we are dealing with dead loads. We will also examine only problems associated with point loads, which will greatly simplify the computation of the stored elastic energies.

### Example 11.1

*Bar with two axial forces* For the bar shown in Fig. 11.2, determine the relation between the applied forces ($P_1$ and $P_2$) and the resulting displacements ($\Delta_1$ and $\Delta_2$).

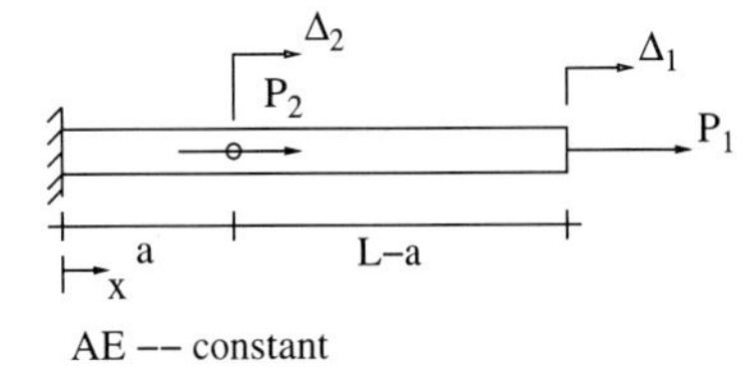


**Fig. 11.2** Bar with two axial forces.

*Solution*

The potential energy for the total system is given by:

$$\Pi = \int_0^L \frac{1}{2} AE\varepsilon^2\, dx - P_1\Delta_1 - P_2\Delta_2. \quad (11.16)$$

In order to further the analysis we note that between the loads the displacement field is linear and thus the strains are constant in each segment. Additionally, the strains are expressible in terms of the displacements $\Delta_1$ and $\Delta_2$. Thus,

$$\Pi(\Delta_1, \Delta_2) = \int_0^a \frac{1}{2} AE \left(\frac{\Delta_2}{a}\right)^2 dx + \int_a^L \frac{1}{2} AE \left(\frac{\Delta_1 - \Delta_2}{L - a}\right)^2 dx$$
$$- P_1\Delta_1 - P_2\Delta_2. \quad (11.17)$$

For equilibrium, the potential energy will be stationary (according to the principle of stationary potential energy). This results in two equations:

$$\frac{\partial \Pi}{\partial \Delta_1} = 0 \tag{11.18}$$

$$\frac{\partial \Pi}{\partial \Delta_2} = 0. \tag{11.19}$$

Computing the indicated derivatives gives two linear relations between the applied forces and the resulting displacements:

$$\frac{AE(\Delta_1 - \Delta_2)}{L-a} = P_1 \tag{11.20}$$

$$\frac{AE\Delta_2}{a} - \frac{AE(\Delta_1 - \Delta_2)}{L-a} = P_2\,. \tag{11.21}$$

**Remarks:**

(1) The solution as given provides the required loads for known displacements. One can of course invert the relations (two equations in two unknowns) to find the displacements for given forces.

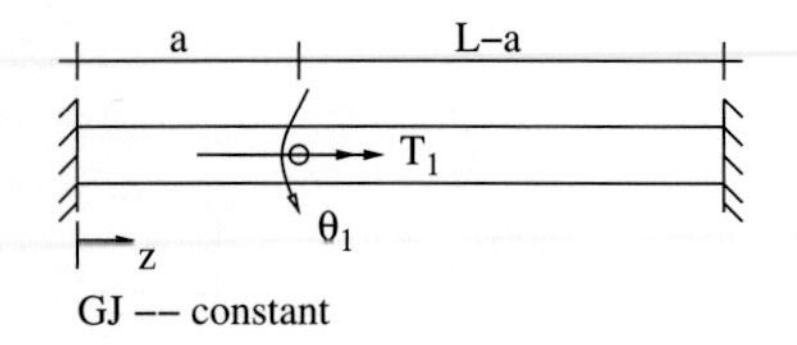


**Fig. 11.3** Rod with a single-point torque.

### Example 11.2

*Statically indeterminate rod with a point torque.* For the rod shown in Fig. 11.3 determine the relation between the applied torque, $T_1$, and the resulting rotation at the point of application, $\theta_1$.

*Solution*

The potential energy for the total system is given by:

$$\Pi = \int_0^L \frac{1}{2} GJ \left(\frac{d\phi}{dz}\right)^2 dz - T_1\theta_1. \tag{11.22}$$

We note that between the load and the supports the rotation field is linear and thus the twist rates are constant in each segment. This holds, since there are no distributed torques on the rod. The twist state is thus expressible in terms of the rotation $\theta_1$ and the location of the applied torque. This leads to

$$\Pi(\theta_1) = \int_0^a \frac{1}{2} GJ \left(\frac{\theta_1}{a}\right)^2 dz + \int_a^L \frac{1}{2} GJ \left(\frac{-\theta_1}{L-a}\right)^2 dz - T_1\theta_1. \tag{11.23}$$

For equilibrium, the potential energy will be stationary (according to the principle of stationary potential energy). This results in the equation:

$$\frac{d\Pi}{d\theta_1} = 0. \tag{11.24}$$

Computing the indicated derivative gives

$$T_1 = GJ\frac{\theta_1}{a} + GJ\frac{\theta_1}{L-a}. \tag{11.25}$$

**Remarks:**

(1) Note that the static indeterminacy did not create any difficulties for the application of the principle of stationary potential energy. Static indeterminacy is naturally handled by the kinematic nature of the principle.

## 11.3 Castigliano's first theorem

The results of the last two examples illustrate a theorem known as Castigliano's first theorem:

If the total stored energy, $W$, in an elastic body is expressed in terms of the displacements/rotations at the points of application of isolated forces/torques, then those forces/torques are given by the derivatives of the stored energy with respect to the corresponding displacements/rotations.

To express this statement in equation form, suppose one has an elastic body subject to point forces $P_k$ and point torques $M_k$. Let the motion at these locations be given by $\Delta_k$ and $\theta_k$. If the total stored elastic energy in the body is given by $W(\Delta_1, \Delta_2, \cdots, \theta_1, \theta_2, \cdots)$, then

$$P_k = \frac{\partial W}{\partial \Delta_k} \tag{11.26}$$

$$M_k = \frac{\partial W}{\partial \theta_k}. \tag{11.27}$$

**Remarks:**

(1) To solve problems in mechanics we generally need to use equilibrium, kinematic, and constitutive relations. The principle of stationary potential energy and Castigliano's first theorem are really re-statements of the equilibrium relations for a system. Thus to come to a "solution" of a problem we must independently determine the kinematic response of the system in order to compute the integrand for the stored energy. Likewise the constitutive relation is already buried in the stored energy relation via the explicit presence of the material constants in the integrals.

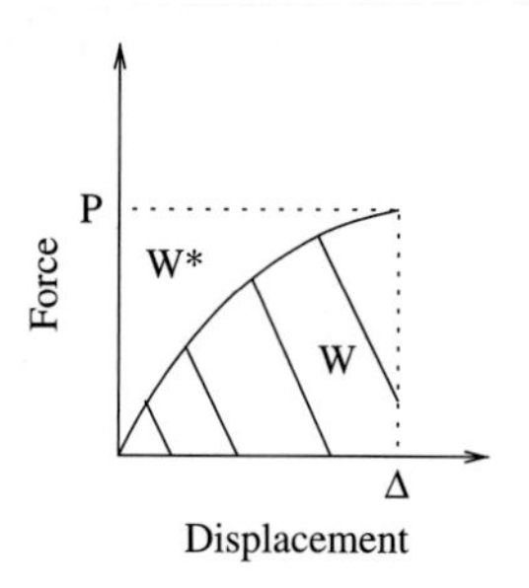


**Fig. 11.4** Response of a non-linear spring.

## 11.4 Stationary complementary potential energy

Every elastic system can have its energetics described in two ways: energy and complementary energy. Consider a *non-linear* elastic spring whose force-displacement response is shown in Fig. 11.4. If the spring is extended with a force $P$ to a displacement $\Delta$, the energy stored in the spring is given by the area under the curve, $W(\Delta)$. The complementary energy in the spring is defined to be the area above the curve $W^*(P)$. The two are related as

$$W(\Delta) + W^*(P) = P\Delta\,. \tag{11.28}$$

If the spring is *linear* elastic with spring constant $k$, then $W(\Delta) = \frac{1}{2}k\Delta^2$ (as before) and $W^*(P) = \frac{1}{2}\frac{1}{k}P^2$.

**Remarks:**

(1) In linear elastic systems, $W$ and $W^*$ are numerically equivalent. In non-linear elastic systems they are not.

(2) The relation between energy and complementary energy defined here is analogous to the relations between different types of state functions in thermodynamics. In fact, eqn (11.28) is nothing more that the Legendre transformation that one studies in thermodynamics.

With a definition of complementary stored energy, we can now define complementary potential energy. Consider again the linear spring shown in Fig. 11.1, but this time subjected to a fixed displacement $\Delta$ instead of a fixed load $P$. The complementary potential energy of the total system is composed of the stored complementary energy of the spring, $\frac{1}{2k}P^2$, and the complementary potential energy of the load which is again given as $-P\Delta$. Thus we have for the total complementary potential energy:

$$\Pi^*(P) = \frac{1}{2k}P^2 - P\Delta, \tag{11.29}$$

where the independent variable is the force $P$ in the spring. The principle which we wish to exploit is that conservative systems have stationary complementary potential energy when they satisfy their fundamental kinematic relations. Thus, when the system is in a kinematically compatible configuration we have

$$\frac{d\Pi^*}{dP} = 0. \tag{11.30}$$

For the linear spring system, this leads to

$$0 = \frac{d\Pi^*}{dP} = \frac{P}{k} - \Delta. \tag{11.31}$$

**Remarks:**

(1) As with the concept of stationary potential energy, it is important to notice what type of relation we have arrived at via the *principle of stationary complementary potential energy*: kinematic, constitutive, or equilibrium. Equation (11.31) says that the spring displacement, $P/k$, is equal to the applied displacement (where use has been made of the constitutive parameter for the spring, $k$). Thus, by finding the stationary complementary potential energy states of the system we have arrived at a statement of kinematic compatibility with aid of the constitutive relation. The notion of stationary complementary potential energy is equivalent to asserting that a system satisfies its relevant kinematic relations and to solve any problem using the principle of stationary complementary potential energy will require the separate application of the equilibrium equations. The constitutive relation is already included in the analysis by way of the definition of the complementary stored energy.

(2) Note that the (negative) derivative of each contribution to the total complementary potential energy gives the motion of the individual elements of the mechanical system. For the system to remain connected, kinematically compatible, the motions must sum to zero.

(3) This kinematic interpretation of the principle of stationary complementary potential energy shows, also, that this concept is equivalent (for elastic systems) to the principle of virtual forces.

(4) Later, in Chapter 13 we will see that the principle of stationary potential energy is related to a concept known as the principle of virtual displacements.

## 11.5 Stored complementary energy: Continuous systems

To exploit the principle of stationary complementary potential energy we will need expressions for the stored complementary energy in various types of structural systems. In Section 11.2 we derived the expressions for the stored elastic energy, and these can be easily converted to complementary energy expressions using the notion that complementary energy is represented by the area above the response curve of a system, while the area under the response curve is the energy in the system. If we consider only linear elastic systems, then for our various types of load-bearing members we find:

Axial forces:

$$W^* = \int_V \frac{1}{2}\sigma_{xx}\varepsilon_{xx}\,dV = \int_L \int_A \frac{1}{2E}\sigma^2\,dA\,dx \tag{11.32}$$

$$= \int_0^L \frac{1}{2}\frac{R^2}{AE}\,dx. \tag{11.33}$$

Torsional loads:

$$W^* = \int_V \frac{1}{2}\sigma_{z\theta}\gamma_{z\theta}\, dV = \int_L \int_A \frac{1}{2G}\tau^2\, dA\, dz \tag{11.34}$$

$$= \int_0^L \frac{1}{2}\frac{T^2}{GJ} dz. \tag{11.35}$$

Bending loads:

$$W^* = \int_V \frac{1}{2}\sigma_{xx}\varepsilon_{xx}\, dV = \int_L \int_A \frac{1}{2E}\sigma^2\, dA\, dx \tag{11.36}$$

$$= \int_0^L \frac{1}{2}\frac{M^2}{EI}\, dx\,. \tag{11.37}$$

Direct shear:

$$W^* = \int_V \frac{1}{2}\sigma_{xy}\gamma_{xy}\, dV = \int_L \int_A \frac{1}{2G}\tau^2\, dA\, dx \tag{11.38}$$

$$= \alpha \int_0^L \frac{1}{2}\frac{V^2}{GA}\, dx\,. \tag{11.39}$$

**Remarks:**

(1) As with "regular" energy, if an element in a linear system is carrying its load in multiple ways, then one can simply add up all the terms to find the total stored complementary energy.

Let us now consider two examples using the concept of stationary complementary potential energy.

---

**Example 11.3**

*Bar with two axial forces revisited.* In this example we will re-solve Example 11.1 using stationary complementary potential energy.

*Solution*

The complementary potential energy for the system is given as

$$\Pi^* = \int_0^L \frac{1}{2}\frac{R^2}{AE}\, dx - P_1\Delta_1 - P_2\Delta_2. \tag{11.40}$$

By separately using the equilibrium equations for the system, we find the internal forces to be $P_1 + P_2$ and $P_1$ in the two segments of the bar. Thus,

$$\Pi^*(P_1, P_2) = \int_0^a \frac{1}{2}\frac{(P_1 + P_2)^2}{AE}\, dx + \int_a^L \frac{1}{2}\frac{P_1^2}{AE}\, dx - P_1\Delta_1 - P_2\Delta_2. \tag{11.41}$$

The complementary potential energy should be stationary according to the principle of stationary complementary potential energy. This results in two equations:

$$\frac{\partial \Pi^*}{\partial P_1} = 0 \tag{11.42}$$

$$\frac{\partial \Pi^*}{\partial P_2} = 0. \tag{11.43}$$

Computing the indicated derivatives gives two linear relations between the applied forces and the resulting displacements:

$$\frac{(P_1 + P_2)a}{AE} + \frac{P_1(L-a)}{AE} = \Delta_1 \tag{11.44}$$

$$\frac{(P_1 + P_2)a}{AE} = \Delta_2. \tag{11.45}$$

**Remarks:**

(1) This result is entirely equivalent to eqns (11.20) and (11.21), which we obtained using stationary potential energy.

(2) Note that to solve the problem using stationary complementary potential energy, we needed to separately invoke the equilibrium equations to determine $R(x)$ in terms of the applied loads. The constitutive relation is already embedded in the complementary energy expression. Thus the principle of stationary complementary potential energy represents the governing kinematic relations of the problem. It is, in the elastic case, equivalent to the method of virtual forces introduced in Chapter 10.

### Example 11.4

*Cantilever beam.* Figure 11.5 shows a cantilever beam with an end-load. Find the deflection of the tip of the beam.

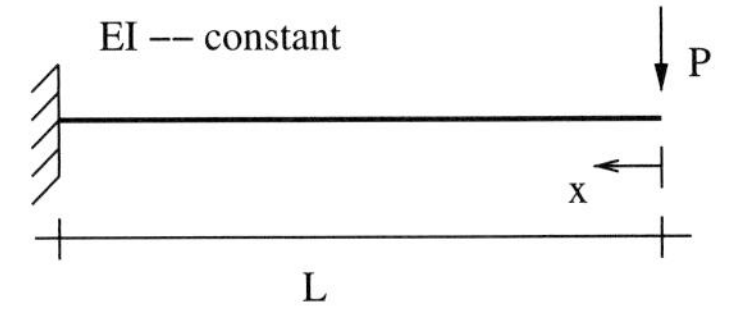


**Fig. 11.5** Cantilever beam.

*Solution*

First let us construct the complementary potential for the system. If we ignore direct shear, then we have

$$\Pi^* = \int_0^L \frac{1}{2}\frac{M^2}{EI}\,dx - P\Delta. \tag{11.46}$$

From the equilibrium equations for the beam one finds that $M(x) = Px$; thus

$$\Pi^*(P) = \int_0^L \frac{1}{2}\frac{(Px)^2}{EI}\,dx - P\Delta. \tag{11.47}$$

From the requirement of stationarity with respect to $P$, one can compute that

$$\Delta = \int_0^L \frac{1}{2} 2\frac{Px^2}{EI}\,dx = \frac{PL^3}{3EI}. \tag{11.48}$$

**Remarks:**

(1) Notice that we recover the same answer that we found when we solved this type of problem via direct integration of the ordinary differential equation $EIv'''' = q$.

## 11.6 Castigliano's second theorem

The results of the last two examples illustrate a theorem known as Castigliano's second theorem:

If the total stored complementary energy, $W^*$, in an elastic body is expressed in terms of isolated applied forces/torques, then the displacements/rotations at the points of application of the loads are given by the derivatives of the stored complementary energy with respect to the corresponding forces/torques.

To express in equation form, suppose one has an elastic body subject to point forces $P_k$ and point torques $M_k$. Let the motion at these locations be given by $\Delta_k$ and $\theta_k$. If the total stored complementary elastic energy in the body is given by $W^*(P_1, P_2, \cdots, M_1, M_2, \cdots)$, then

$$\Delta_k = \frac{\partial W^*}{\partial P_k} \tag{11.49}$$

$$\theta_k = \frac{\partial W^*}{\partial M_k}. \tag{11.50}$$

**Remarks:**

(1) As seen throughout, to solve problems in mechanics we need to use equilibrium, kinematic, and constitutive relations. The principle of stationary complementary potential energy and Castigliano's second theorem are really restatements of the kinematic relations for a system. Notice that to come to a "solution" of a problem we must independently apply equilibrium of the system in order to compute the integrand for the stored complementary energy. Likewise the constitutive relation is already embedded in the relation via the explicit presence of the material constants in the integrals.

Let us now consider a series of examples using Castigliano's second theorem.

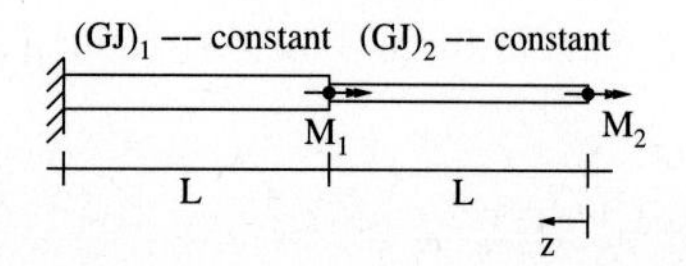


**Fig. 11.6** Torsion bar with two applied loads.

**Example 11.5**

*Torsion bar with two loads.* Shown in Fig. 11.6 is a stepped torsion bar with two applied loads. Find the rotation of the bar at $z = L$.

*Solution*
First write down the complementary stored energy as

$$
\begin{aligned}
W^*(M_1, M_2) &= \int_0^{2L} \frac{T^2}{2GJ}\, dz \\
&= \int_0^{L} \frac{M_2^2}{2G_2J_2}\, dz + \int_L^{2L} \frac{(M_1+M_2)^2}{2G_1J_1}\, dz.
\end{aligned}
\tag{11.51}
$$

The desired rotation is given via Castigliano's second theorem as

$$
\theta_1 = \frac{\partial W^*}{\partial M_1} = \int_L^{2L} \frac{2(M_1+M_2)}{2G_1J_1}\, dz = \frac{(M_1+M_2)L}{G_1J_1}. \tag{11.52}
$$

**Remarks:**

(1) The sense of the rotation computed is the same as the sense of the applied torque. This is just as we had when using the method of virtual forces.

(2) To be able to complete the computation, we had to solve separately for $T(z)$ using the governing equilibrium equations.

---

**Example 11.6**

*Mid-span deflection in a cantilever beam.* Consider the cantilever beam shown in Fig. 11.5, and find the upward deflection of the beam at mid-span using Castigliano's second theorem.

*Solution*
First note that there is no load at mid-span. Thus we can not just form the complementary stored energy and take its derivative with respect to the corresponding load – as it does not exist. To be able to apply Castigliano's second theorem to this problem we will need to modify the problem by adding a force, $f$, at $x = L/2$. At the end of the solution we will make use of the fact that we know that $f = 0$. The modified system is shown in Fig. 11.7. The complementary stored energy is given as

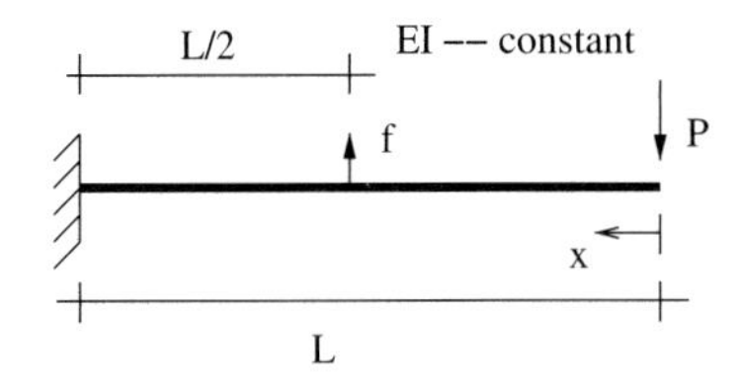

**Fig. 11.7** Cantilever with added force at mid-span.

$$
W^*(f, P) = \int_0^L \frac{1}{2}\frac{M^2}{EI}\, dx, \tag{11.53}
$$

where by equilibrium $M(x) = -Px + f\langle x - \frac{L}{2}\rangle$. The desired deflection is then given by

$$
v\left(\frac{L}{2}\right) = \frac{\partial W^*}{\partial f}(0, P) = \int_0^L \left[\frac{M}{EI}\frac{\partial M}{\partial f}\right]_{f=0} dx \tag{11.54}
$$

$$= \int_0^L \left[ \frac{-Px + f\langle x - \frac{L}{2}\rangle}{EI} \langle x - \frac{L}{2}\rangle \right]_{f=0} dx \quad (11.55)$$

$$= \int_{\frac{L}{2}}^L \frac{-Px}{EI}(x - \frac{L}{2})\, dx \quad (11.56)$$

$$= -\frac{5PL^3}{48EI}. \quad (11.57)$$

**Remarks:**

(1) When applying Castigliano's second theorem in problems where there is no load at the location of interest, one must first introduce an extra load at that location. After the taking of the required derivative, one can then set this extra load to zero. Remember to set it to zero after taking the derivative not before; otherwise you will always get zero for an answer.

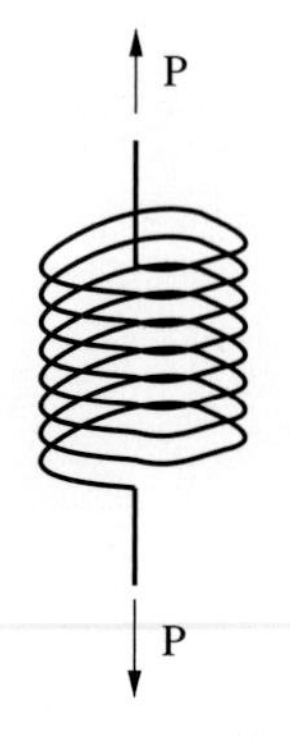


Fig. 11.8 Helical spring with eight coils.

### Example 11.7

*Closely coiled helical spring.* Figure 11.8 shows a closely coiled helical spring. It is made of a thin wire of radius $r$ that has been wound into a coil of radius $R$. Using Castigliano's second theorem, find the spring constant for the spring.

*Solution*

The spring constant for the spring gives the relationship between the applied force and the extension of the spring $\Delta = P/k$. If we make a free-body diagram of the spring as shown in Fig. 11.9, then we see that the wire is in a state of torsion and direct shear. As long as the length of the wire is large in comparison to the radius of the wire we are safe in ignoring the direct shear contribution to the overall deflection. Thus we can write

$$W^*(P) = \int_{\text{arc-length}} \frac{1}{2}\frac{T^2}{GJ}\, ds\,, \quad (11.58)$$

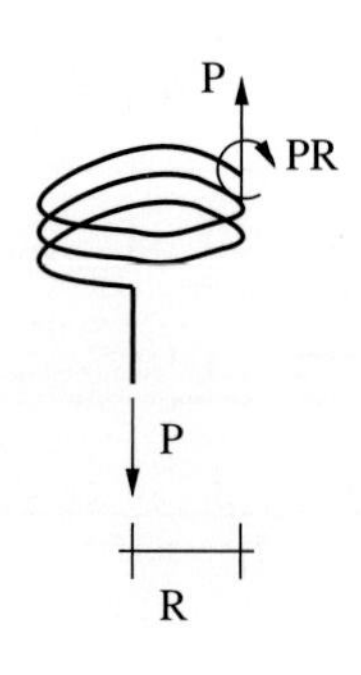


Fig. 11.9 Free-body diagram of helical spring.

where the integral is taken over the helical contour of the wire, $T = PR$ from our free-body diagram, $J = \pi r^4/2$, and $ds$ is the arc-length integration parameter. The total arc-length of the wire $L = 2\pi R N_{\text{coils}}$, where $N_{\text{coils}}$ is the number of complete coils in the spring. Substituting this into the expression for the stored complementary energy gives

$$W^*(P) = \frac{2P^2R^3N_{\text{coils}}}{Gr^4}\,. \quad (11.59)$$

We can now apply Castigliano's second theorem to find

$$\Delta = \frac{\partial W^*}{\partial P} = \frac{4R^3N_{\text{coils}}}{Gr^4}P \quad (11.60)$$

and we can identify the expression for the spring constant of a closely coiled helical spring as

$$k = \frac{Gr^4}{4R^3 N_{\text{coils}}}. \tag{11.61}$$

**Remarks:**

(1) As with the method of virtual forces, the concept of stationary complementary potential can be used profitably on geometrically complex systems.

---

### Example 11.8

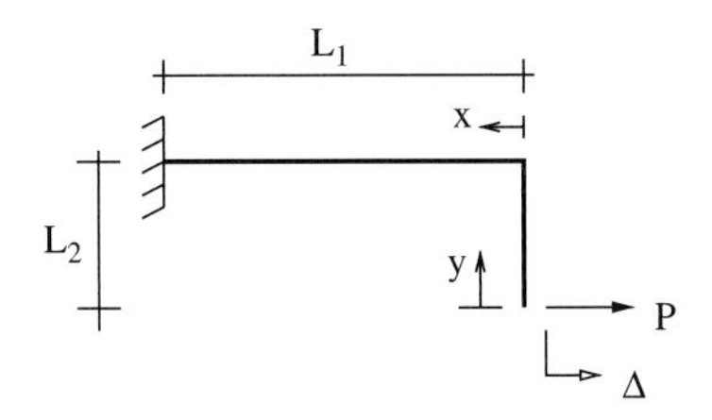


**Fig. 11.10** Angle frame with $EI$, $AE$, $GA$ constant.

*Angle frame.* Consider the frame shown in Fig. 11.10 and find the tip-deflection in the direction of the applied load. Do not assume that the members are slender.

*Solution*

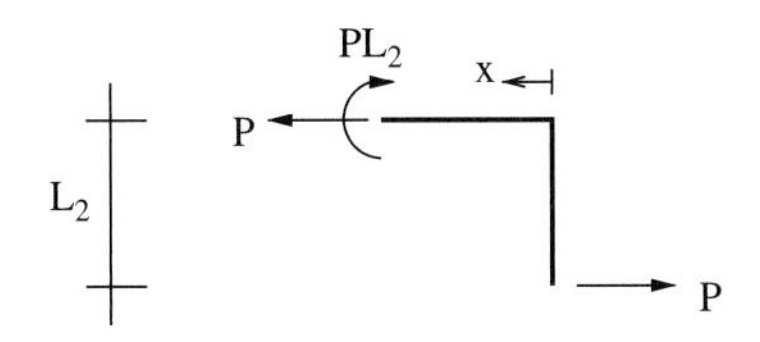


**Fig. 11.11** Free-body diagram of horizontal segment of angle frame.

The frame is composed of two segments. From a free-body diagram as shown in Fig. 11.11 we see that the horizontal segment is in a state of axial load and bending. From a free-body diagram of the vertical segment, Fig. 11.12, we see that it is in a state of bending and direct shear. Thus for the total complementary stored energy we find

$$\begin{aligned} W^* = &\int_0^{L_2} \frac{M^2}{2EI}\, dx + \int_0^{L_2} \frac{R^2}{2AE}\, dx \\ &+ \int_0^{L_1} \frac{M^2}{2EI}\, dy + \alpha \int_0^{L_1} \frac{V^2}{2GA}\, dy. \end{aligned} \tag{11.62}$$

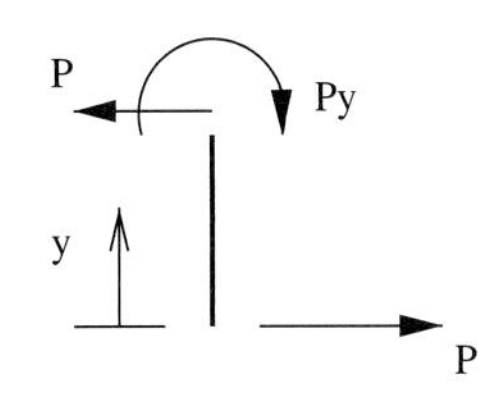


**Fig. 11.12** Free-body diagram of vertical segment of angle frame.

The desired deflection is given by

$$\begin{aligned} \Delta = \frac{\partial W^*}{\partial P} = &\int_0^{L_1} \frac{M}{EI}\frac{\partial M}{\partial P}\, dx + \int_0^{L_1} \frac{R}{AE}\frac{\partial R}{\partial P}\, dx \\ &+ \int_0^{L_2} \frac{M}{EI}\frac{\partial M}{\partial P}\, dy + \alpha \int_0^{L_2} \frac{V}{GA}\frac{\partial V}{\partial P}\, dy. \end{aligned} \tag{11.63}$$

From equilibrium applied to the free body diagrams one has that $M(x) = PL_2$, $R(x) = P$, $M(y) = Py$, and $V(y) = P$. Inserting these expressions and their derivatives, one finds:

$$\Delta = \frac{PL_1}{AE} + \frac{PL_2^2 L_1}{EI} + \frac{PL_2^3}{3EI} + \alpha \frac{PL_2}{GA}. \tag{11.64}$$

**Remarks:**

(1) The final result is observed to be a superposition of the extension of the horizontal arm, the rotation of the horizontal arm times the

length of the vertical arm, the bending of the vertical arm, and the shear deformation of the vertical arm.

### Example 11.9

*Beam and truss system.* Consider the beam and truss system shown in Fig. 11.13(a). Find the downward deflection at the point where the load is applied. Assume that all members are slender.

*Solution*

Free-body diagrams of the system, Fig. 11.13(b), show that the beam is in a state of bending with $M(x) = F\sqrt{2}x$ and that the truss bars have internal forces of $R_1 = F\sqrt{2}$ and $R_2 = F$. Since all members are slender, we will ignore the direct shear and axial effects in the beam. The total stored complementary energy in the system is

$$W^* = \int_0^{\sqrt{2}L} \frac{M^2}{2EI}\,dx + \int_0^{\sqrt{2}L} \frac{R_1^2}{2AE}\,dy + \int_0^{L} \frac{R_2^2}{2AE}\,dz. \tag{11.65}$$

The desired deflection is then given as

$$\Delta = \frac{\partial W^*}{\partial P} = F\left[\frac{4\sqrt{2}L^3}{3EI} + \frac{(1+2\sqrt{2})L}{AE}\right]. \tag{11.66}$$

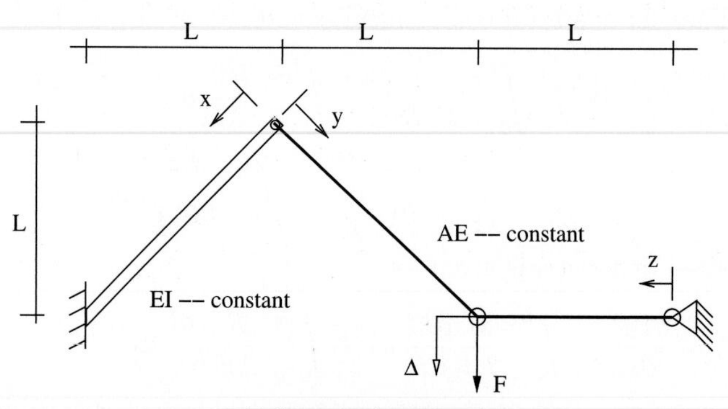


(a) System to be analyzed.

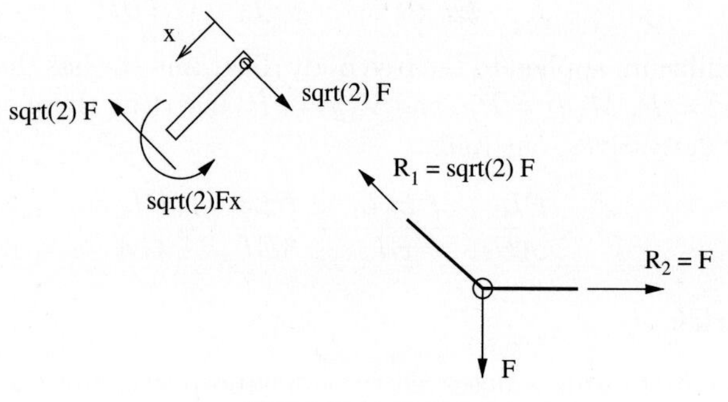


(b) Free-body diagrams.

**Fig. 11.13** Beam and truss system.

**Remarks:**

(1) In this example the system was composed of different types of load bearing elements. Notwithstanding, the principle of stationary complimentary potential energy still holds as long as one accounts for all contributions to the complementary potential energy.

---

### Example 11.10

*Statically indeterminate truss.* Consider now the statically indeterminate truss shown in Fig. 11.14(a). Find the horizontal deflection at the point of application of the load.

*Solution*

The system is statically indeterminate; thus, we cannot *a priori* solve for the equilibrium state of the system in terms of the applied load alone. We will instead follow the assumed reaction force methodology which we have used previously; viz., we will remove one of the supports and make the system determinate. At the end of the solution procedure we will enforce the kinematic constraint that the motion is zero at the support which we removed.

Figure 11.14(b) shows our truss structure with one support removed. From statics we easily see that the bar forces are $R_1 = P - F/\sqrt{2}$, $R_2 = F$, and $R_3 = -F/\sqrt{2}$. The stored complementary energy is

$$W^* = \int_0^L \frac{R_1^2}{2AE}\,dx + \int_0^{\sqrt{2}L} \frac{R_2^2}{2AE}\,dy + \int_0^L \frac{R_3^2}{2AE}\,dz. \tag{11.67}$$

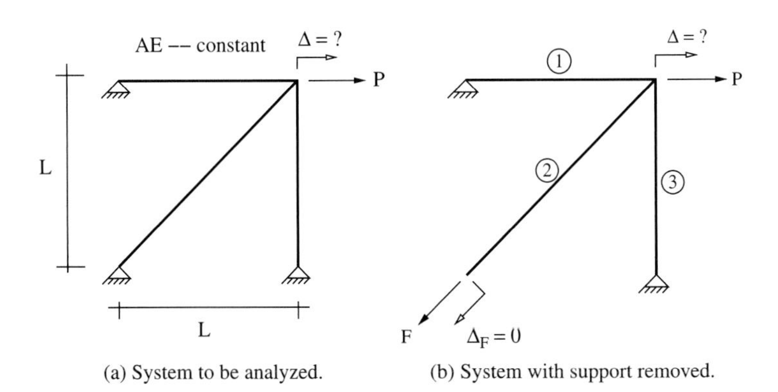


**Fig. 11.14** Indeterminate truss system.

Taking the required derivative we obtain

$$\Delta = \frac{\partial W^*}{\partial P} = \int_0^L \frac{R_1}{AE} \underbrace{\frac{\partial R_1}{\partial P}}_{=1} dx + \int_0^{\sqrt{2}L} \frac{R_2}{AE} \underbrace{\frac{\partial R_2}{\partial P}}_{=0} dy$$

$$+ \int_0^L \frac{R_3}{AE} \underbrace{\frac{\partial R_3}{\partial P}}_{=0} dz \tag{11.68}$$

$$= \frac{(P - F/\sqrt{2})L}{AE}. \tag{11.69}$$

This answer involves the unknown support reaction $F$. To eliminate $F$, we can solve for it using the fact that the displacement at the support is to be zero. Thus we can use the fact that $\partial W^*/\partial F = 0$. This gives:

$$0 = \int_0^L \frac{R_1}{AE} \underbrace{\frac{\partial R_1}{\partial F}}_{=-1/\sqrt{2}} dx + \int_0^{\sqrt{2}L} \frac{R_2}{AE} \underbrace{\frac{\partial R_2}{\partial F}}_{=1} dy$$

$$+ \int_0^L \frac{R_3}{AE} \underbrace{\frac{\partial R_3}{\partial F}}_{=-1/\sqrt{2}} dz \tag{11.70}$$

$$= \frac{FL}{AE}\left[\frac{1}{2} + \sqrt{2} + \frac{1}{2}\right] - \frac{PL}{\sqrt{2}AE} \tag{11.71}$$

$$F = \frac{P}{2 + \sqrt{2}}. \tag{11.72}$$

Substituting this result back into our expression for the displacement gives the final result of

$$\Delta = \frac{PL}{AE}\frac{2\sqrt{2}+1}{2\sqrt{2}+2}. \tag{11.73}$$

**Remarks:**

(1) The same mechanical system was examined in Example 10.8 for the vertical motion at the load point using the method of virtual forces. Note the similarity in analysis methodology.

## 11.7 Stationary potential energy: Approximate methods

For the problems we have treated so far, we have taken a set of governing equations and solved them exactly. This has been done either by way of direct integration, the method of virtual forces, or through potential energy methods. In a large class of engineering problems, it is rather difficult to find exact solutions and, further, exact solutions are often not needed; approximate solutions will suffice. After all, the governing equations really only represent a model of reality. The notion that the solution to a problem is related to the stationarity of a certain quantity provides a natural setting for finding approximate solutions. Instead of exactly finding the stationarity conditions we can consider finding approximate stationarity conditions. In this section we will examine using stationary potential energy to find approximate solutions.

We will introduce the basic concepts via the problem illustrated in Fig. 11.15. Let us determine the deflection of the beam. The potential energy for this system is given by

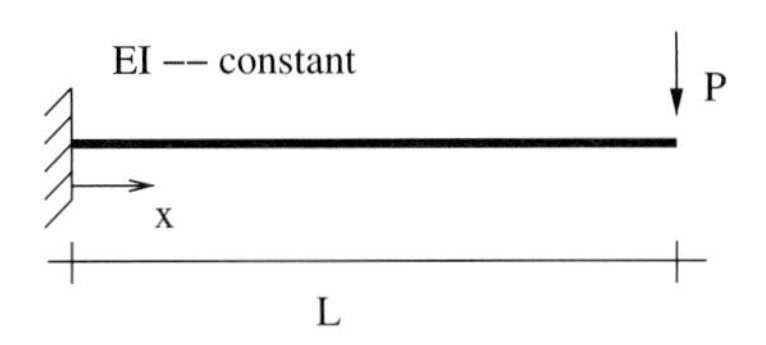


**Fig. 11.15** Cantilever beam with a dead-load.

$$\Pi(v(x)) = \int_0^L \frac{1}{2}EI\left(v''\right)^2\, dx + Pv(L) \tag{11.74}$$

and we know from the principle of stationary potential energy that $\Pi$ takes on a stationary value at equilibrium. Notice that $\Pi$ is a function of a function; it is what is sometimes called a functional. The true equilibrium displacement field, $v(x)$, for the beam will be the one that makes $\Pi$ stationary. $v(x)$, however, cannot be arbitrarily chosen, since $v(x)$ must *a priori* satisfy the kinematic boundary conditions. Thus we should only consider functions in the set:

$$\mathcal{S} = \{v(x) \mid v(0) = 0 \text{ and } v'(0) = 0\}\,. \tag{11.75}$$

The set of functions $\mathcal{S}$ is known as the space of trial solutions – the true equilibrium displacement field lies someplace within this set of functions.

We could try to solve the problem by simply checking all the functions $v(x) \in \mathcal{S}$ until we find one that makes $\Pi$ stationary. This, however, is a rather daunting task, since $\mathcal{S}$ is infinite dimensional. If we were to follow the standard methods learned in calculus, we would be tempted to take the derivative of $\Pi$ with respect to the function $v(x)$, set the result equal to zero, and then solve for the appropriate function within $\mathcal{S}$. The taking of the derivative of a functional with respect to a function, however, is a somewhat sophisticated mathematical concept outside the scope of this book. What we can do, however, is solve the stationarity problem approximately by searching only for a stationary state over a well-chosen subset of $\mathcal{S}$. This will result in an approximate solution for the equilibrium displacement field.

To be specific, let us consider the subset of functions

$$\widetilde{\mathcal{S}}_1 = \left\{v(x) \mid v(x) = Cx^2\,, \quad C \in \mathbb{R}\right\} \subset \mathcal{S}. \tag{11.76}$$

The set $\widetilde{\mathcal{S}}_1$ is composed of a set of parabolas parameterized by the parameter $C$ which is an arbitrary real number.[1] Finding a stationary state of $\Pi$ over $\widetilde{\mathcal{S}}_1$ is substantially easier, since all we need to determine is the value of $C$ which makes $\Pi$ stationary when evaluated on functions in $\widetilde{\mathcal{S}}_1$. Let us plug our approximation into the expression for the potential energy:

[1] $\widetilde{\mathcal{S}}_1$ is a one-dimensional set; one can specify any element of the set by giving a value to a single scalar parameter.

$$\Pi(C) = \int_0^L \frac{1}{2}EI4C^2\, dx + PCL^2 \tag{11.77}$$

The functional $\Pi$ now becomes just an ordinary function of $C$. We can now use standard calculus to find the stationary condition:

$$0 = \frac{d\Pi}{dC} = 4EICL + PL^2 . \tag{11.78}$$

Solving for $C$ gives $C = -PL/4EI$. Thus we find an approximation for the deflection

$$v_1(x) \approx -\frac{PLx^2}{4EI} . \tag{11.79}$$

**Remarks:**

(1) The given solution was obtained by making the potential energy stationary, not over all trial solutions, but rather over a small subset of trial solutions. This produces only an approximate answer. The better the guess for the functional form of the solution the better the quality of the approximation. If the guess for the form of the approximate solution contains the exact solution, then one will obtain the exact answer.

(2) Notice that this methods yields a solution for the deformation of the beam at all points – not just at a single point. This permits one to also compute approximate values for bending moments, etc.

(3) To have a sense of the quality of the approximate solution, let us evaluate the solution at $x = L$. From the approximation we obtain a tip deflection $\Delta = PL^3/4EI$. The exact answer is $\Delta = PL^3/3EI$. Thus the deflection error at the end of the cantilever is about 25%.

(4) Notice that each function in $\widetilde{\mathcal{S}}_1$ satisfies the conditions for functions in $\mathcal{S}$. Thus $\widetilde{\mathcal{S}}_1$ is a subset of $\mathcal{S}$. Failure to adhere to this requirement will produce erroneous results.

(5) Good use of this methodology requires an intuitive sense of the deflected shape of a structure. The better the guess for the deflected shape, the better the result. If we had instead chosen as our space of approximate trial solutions[2]

$$\widetilde{\mathcal{S}}_2 = \{v(x) \mid v(x) = C_1x^2 + C_2x^3\} \subset \mathcal{S} , \tag{11.80}$$

then $\Pi$ would have become a function of $C_1$ and $C_2$; i.e. $\Pi(C_1, C_2)$. Finding the stationary state would then require solving

$$0 = \frac{\partial \Pi}{\partial C_1} \tag{11.81}$$

$$0 = \frac{\partial \Pi}{\partial C_2} \tag{11.82}$$

for $C_1$ and $C_2$. This would have resulted in $C_1 = -PL/2EI$ and $C_2 = P/6EI$ and a $v_2(x) \approx -PLx^2/2EI + Px^3/6EI$, which happens to be the exact answer.

[2] $\widetilde{\mathcal{S}}_2$ is a two-dimensional set; one can specify any element in the set by providing numerical values for two scalar parameters.

(6) The relation between $\mathcal{S}$, $\widetilde{\mathcal{S}}_1$, and $\widetilde{\mathcal{S}}_2$ is depicted in Fig. 11.16. Our first selection was a one-dimensional subset of the space of trial solutions and did not contain the exact solution. Our second selection was a two-dimensional subset of $\mathcal{S}$ and happened to contain the exact solution. The basic property of the method is that it gives you the best answer within the set of functions you have picked. Thus one recovers the exact solution if it is contained within one's guess. If not, one ends up computing the best possible solution of the assumed form – the one with minimum error.

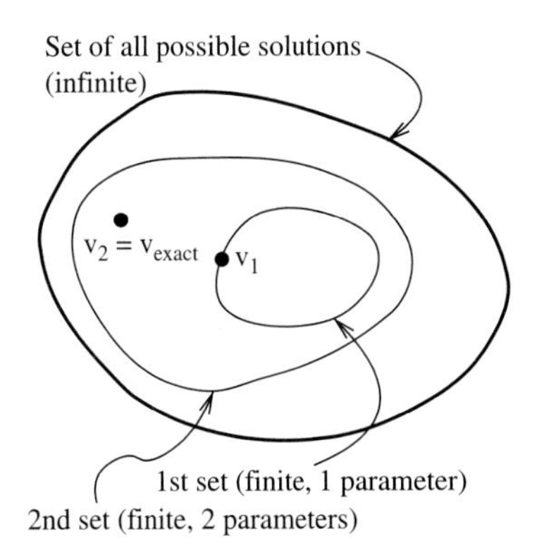


**Fig. 11.16** Geometric relation between $\mathcal{S}$, $\widetilde{\mathcal{S}}_1$ and $\widetilde{\mathcal{S}}_2$, and the exact solution for the cantilever beam.

## Example 11.11

*Built-in beam with a distributed dead-load.* Consider the beam shown in Fig. 11.17. Find an approximation for the deflection of the beam.

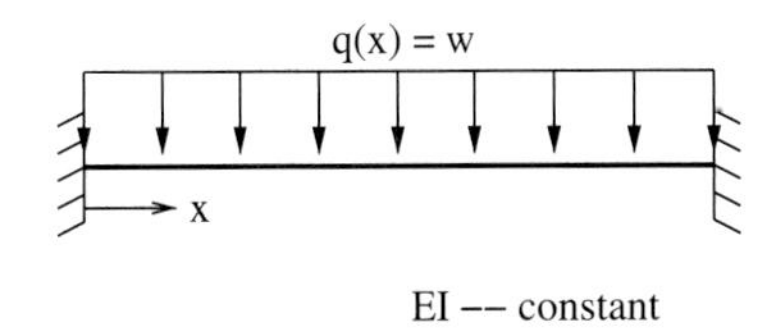


**Fig. 11.17** Built-in beam with a distributed dead-load.

*Solution*
The space of trial solutions contains functions that have zero displacement and rotation at both ends:

$$\mathcal{S} = \{v(x) \mid v(0) = v'(0) = v(L) = v'(L) = 0\}. \tag{11.83}$$

Since the beam is uniformly loaded we can guess that the displacement will be symmetric with respect to $x = L/2$, having maximum displacement in the center, as has been sketched in Fig. 11.18. A reasonable guess for the form of this function would be

$$\widetilde{\mathcal{S}} = \{v(x) \mid v(x) = C\left(\cos(2\pi x/L) - 1\right)\}. \tag{11.84}$$

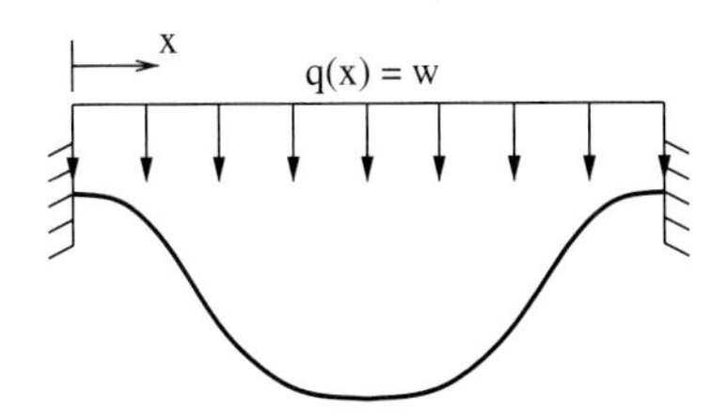


**Fig. 11.18** Guess for deflected shape of beam in Fig. 11.17.

Note that $\widetilde{\mathcal{S}} \subset \mathcal{S}$ and contains only a single parameter. To proceed further, we need an expression for the potential energy for the system and in particular for the distributed dead-load. If we consider a small segment of length $dx$ along the beam, then the load on this segment will be $w\,dx$. Thus the potential for the load over this small segment will be $wv\,dx$, and the total potential for the load will be $\int_0^L wv\,dx$. With this, our condition for stationarity becomes

$$\begin{aligned} 0 &= \frac{d}{dC}\left[\int_0^L \frac{1}{2}EIC^2\frac{16\pi^4}{L^4}\cos^2(2\pi x/L)\,dx\right. \\ &\quad \left. + \int_0^L wC\left(\cos(2\pi x/L) - 1\right)\,dx\right] \end{aligned} \tag{11.85}$$

$$
\begin{aligned}
&= \int_0^L EIC\frac{16\pi^4}{L^4}\cos^2(2\pi x/L)\,dx \\
&\quad + \int_0^L w\,(\cos(2\pi x/L)-1)\,dx && (11.86)\\
&= EIC\frac{16\pi^4}{L^4}\frac{1}{2}L - wL && (11.87)\\
C &= \frac{wL^4}{8\pi^4 EI}. && (11.88)
\end{aligned}
$$

The final approximate solution is:

$$
v(x) \approx \frac{wL^4}{8\pi^4 EI}\left(\cos(2\pi x/L)-1\right). \tag{11.89}
$$

**Remarks:**

(1) As a check on the accuracy of the solution we can evaluate the displacement at the mid-span. From the approximate solution we find $\Delta = -wL^4/4\pi^4 EI$. The exact solution gives $\Delta = -wL^4/384EI$. Thus the deflection error at the mid-span is only 1.4%. A good guess can produce excellent answers.

## 11.8 Ritz's method

The method illustrated for finding approximate solutions is often known as Ritz's method, by which one constructs an $N$-dimensional space of approximate solutions of the form

$$
\widetilde{\mathcal{S}} = \left\{ v(x) \;\middle|\; v(x) = \sum_{i=1}^{N} C_i f_i(x) \right\} \subset \mathcal{S}. \tag{11.90}
$$

The functions $f_i(x)$ must be known and selected to give flexibility to the solution and at the same time satisfy the kinematic boundary conditions. The stationary conditions for the total potential energy result in $N$ simultaneous equations in the $N$ unknowns $C_1, C_2, \cdots, C_N$:

$$
\frac{\partial \Pi}{\partial C_i} = 0 \quad i = 1, 2, \cdots, N. \tag{11.91}
$$

**Remarks:**

(1) For torsion problems, the space of solutions will be for the rotation field $\phi(z)$. Thus one will be guessing a functional form for rotations rather than transverse displacements.

(2) For problems with axial forces, the space of solutions will be for the axial displacement field $u(x)$ and one will be guessing a functional form for the axial displacements.

(3) In problems where the system is carrying its load in multiple ways one can still employ Ritz's methods by guessing forms for all the relevant motions and including all the types of potential energy appearing in the system.

(4) In general, for stable mechanical systems, adding more terms (increasing $N$) will always improve the approximation.

**Example 11.12**

*Tension–compression bar with three point loads.* Consider the tension–compression bar shown in Fig. 11.19. The bar is loaded with three point loads $P_1$, $P_2$, and $P_3$ at $x_1 = L/4$, $x_2 = L/2$, and $x_3 = 3L/4$, respectively. Find an approximation for the axial motion of the bar.

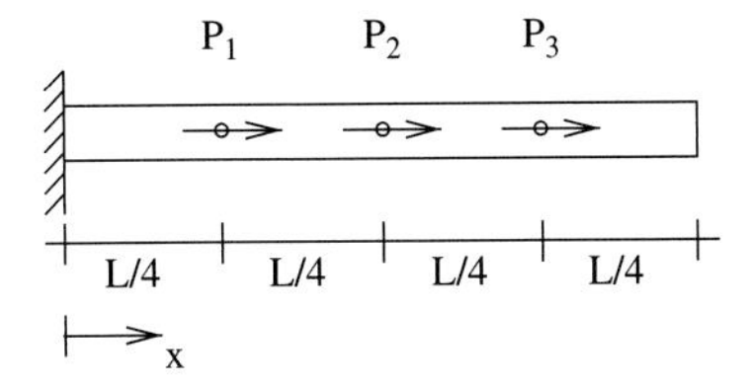


**Fig. 11.19** Tension–compression bar with three point loads, constant $AE$.

*Solution*

The potential energy for this system is given by

$$\Pi(u(x)) = \int_0^L \frac{1}{2} AE(u')^2 \, dx - \sum_{j=1}^{3} P_j v(x_j). \tag{11.92}$$

Let us look for an approximate solution of the form

$$u(x) = \sum_{i=1}^{N} C_i f_i(x), \tag{11.93}$$

where for our approximation functions $f_i(x)$ we will choose the polynomials

$$f_i(x) = \left(\frac{x}{L}\right)^i. \tag{11.94}$$

Note that each function satisfies the kinematic boundary condition for the problem; i.e. $f_i(0) = 0$ for all $i$.

In this case, we have

$$\begin{aligned} \Pi(C_1, C_2, \ldots, C_N) = \int_0^L \frac{1}{2} AE \left( \sum_{i=1}^{N} C_i f_i'(x) \right)^2 dx \\ - \sum_{j=1}^{3} P_j \left( \sum_{i=1}^{N} C_i f_i(x_j) \right). \end{aligned} \tag{11.95}$$

The potential energy will be stationary if

$$\frac{\partial \Pi}{\partial C_k} = 0 \tag{11.96}$$

for each $k = 1, \ldots, N$. In order to execute the needed derivatives, it is useful to note that for a fixed $k$:

$$\frac{\partial}{\partial C_k} \sum_{i=1}^{N} C_i f_i(x) = \frac{\partial}{\partial C_k} \sum_{i=1}^{k-1} C_i f_i(x) + \frac{\partial}{\partial C_k}(C_k f_k(x)) + \frac{\partial}{\partial C_k} \sum_{i=k+1}^{N} C_i f_i(x). \tag{11.97}$$

On the right-hand side the first sum does not involve $C_k$ and thus the derivative is equal to zero; the same holds for the last term on the right-hand side. Only the middle term on the right-hand side involves $C_k$, and the indicated derivative yields the simple result:

$$\frac{\partial}{\partial C_k} \sum_{i=1}^{N} C_i f_i(x) = f_k(x). \tag{11.98}$$

If one exploits this result, one finds (after a little manipulation) that eqn (11.96) gives:

$$\sum_{i=1}^{N} \left( \int_0^L f_k'(x) AE f_i'(x)\, dx \right) C_i = \sum_{j=1}^{3} P_j f_k(x_j). \tag{11.99}$$

This is a system of $N$ linear equations for the $N$ unknown parameters $C_i$. In matrix form, this gives

$$\boldsymbol{KC} = \boldsymbol{F}, \tag{11.100}$$

where the entries of the matrix $\boldsymbol{K}$ are given by

$$K_{ki} = \int_0^L f_k'(x) AE f_i'(x)\, dx = \frac{ki}{k+i-1} \frac{AE}{L} \tag{11.101}$$

and the entries of the vector $\boldsymbol{F}$ are given by

$$F_k = \sum_{j=1}^{3} P_j f_k(x_j) = P_1 \left(\frac{1}{4}\right)^k + P_2 \left(\frac{1}{2}\right)^k + P_3 \left(\frac{3}{4}\right)^k. \tag{11.102}$$

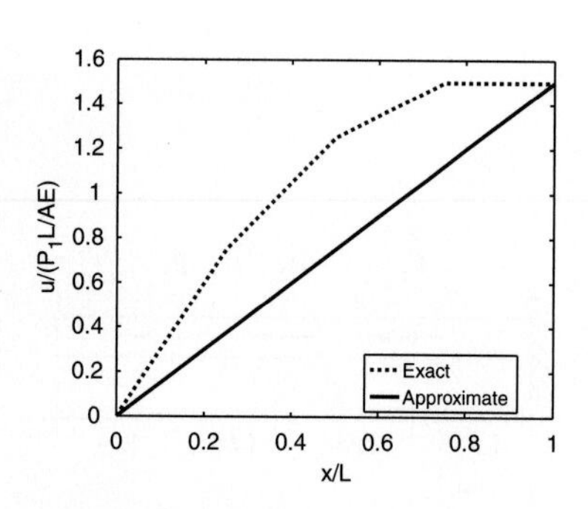


**Fig. 11.20** Approximate and exact solutions for $N = 1$.

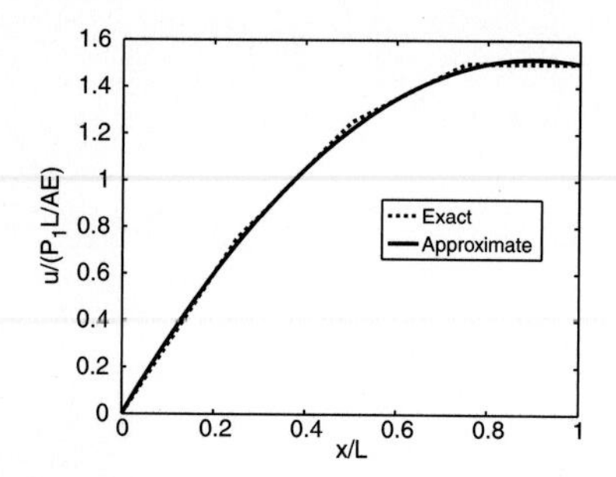


**Fig. 11.21** Approximate and exact solutions for $N = 2$.

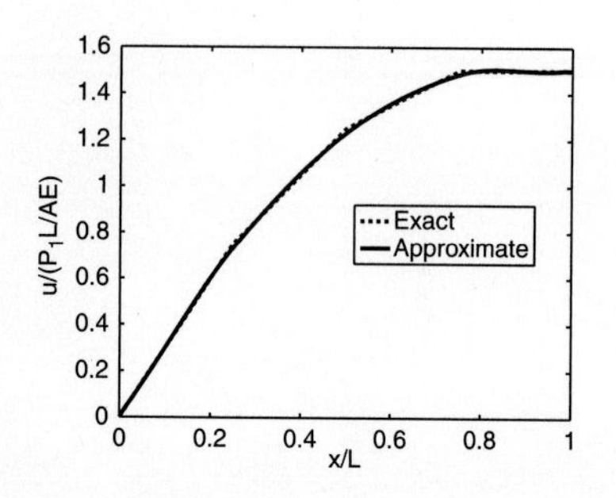


**Fig. 11.22** Approximate and exact solutions for $N = 10$.

**Remarks:**

(1) The final set of matrix equations can be used to compute numerical solutions. For small values of $N$ this is tractable by hand. For larger values a small computer program can be used to solve the equations.

(2) Increasing the number $N$ increases the accuracy of the approximation. Figures 11.20–11.22 show the solution for $N = 1$, 2, and 10, respectively, for the case where $P_1 = P_2 = P_3$. Also shown is the exact solution.

**Example 11.13**

*Cantilever with general loading.* Let us consider a beam with known kinematic boundary conditions. The beam is subjected to $N_p$ point loads $P_l$ $(l = 1, \ldots, N_p)$ at locations $x_l^p$ $(l = 1, \ldots, N_p)$ and $N_m$ point moments $M_k$ $(k = 1, \ldots, N_m)$ at locations $x_k^m$ $(k = 1, \ldots, N_m)$. Find an approximate expression for the deflection of the beam.

*Solution*
The total potential energy is

$$\Pi(v(x)) = \int_0^L \frac{1}{2} EI (v'')^2 \, dx - \sum_{l=1}^{N_p} P_l v(x_l^p) - \sum_{k=1}^{N_m} M_k v'(x_k^m). \quad (11.103)$$

Let us find the stationary solution over all functions $v(x)$ in the set:

$$\widetilde{\mathcal{S}} = \left\{ v(x) \;\Big|\; v(x) = \sum_{j=1}^{N} C_j f_j(x), \quad \{C_j\}_{j=1}^N \in \mathbb{R} \right\}, \quad (11.104)$$

where each function $f_j(x)$ is taken as known and assumed to satisfy the kinematic boundary conditions for the beam. Then,

$$\Pi(v(x)) \Rightarrow \Pi(C_1, \ldots, C_N) = \int_0^L \frac{1}{2} EI \left( \sum_{j=1}^{N} C_j f_j''(x) \right)^2 dx$$
$$- \sum_{l=1}^{N_p} P_l \left( \sum_{j=1}^{N} C_j f_j(x_l^p) \right) - \sum_{k=1}^{N_m} M_k \left( \sum_{j=1}^{N} C_j f_j'(x_k^m) \right). \quad (11.105)$$

The stationary conditions with respect to the parameters $\{C_i\}_{i=1}^N$ require

$$\frac{\partial \Pi}{\partial C_i} = 0 \qquad (i = 1, \ldots, N). \quad (11.106)$$

This yields

$$0 = \int_0^L 2 \cdot \frac{1}{2} EI \left( \sum_{j=1}^{N} C_j f_j''(x) \right) \cdot f_i''(x) \, dx$$
$$- \sum_{l=1}^{N_p} P_l f_i(x_l^p) - \sum_{k=1}^{N_m} M_k f_i'(x_k^m) \quad (11.107)$$

$$\Rightarrow \sum_{j=1}^{N} \left( \int_0^L f_i''(x) EI f_j''(x) \, dx \right) C_j = \sum_{l=1}^{N_p} P_l f_i(x_l^p)$$
$$+ \sum_{k=1}^{N_m} M_k f_i'(x_k^m) \quad (11.108)$$

$$\Rightarrow \sum_{j=1}^{N} K_{ij} C_j = F_i \tag{11.109}$$

$$\Rightarrow \boldsymbol{K}\boldsymbol{C} = \boldsymbol{F}. \tag{11.110}$$

The entries of the matrix $\boldsymbol{K}$ and the vector $\boldsymbol{F}$ are

$$K_{ij} = \int_0^L f_i''(x) EI f_j''(x)\, dx \;, \tag{11.111}$$

$$F_i = \sum_{l=1}^{N_p} P_l f_i(x_l^p) + \sum_{k=1}^{N_m} M_k f_i'(x_k^m) \;, \tag{11.112}$$

respectively. By solving the linear system of equations $\boldsymbol{K}\boldsymbol{C} = \boldsymbol{F}$ for the vector of *generalized displacements* $\boldsymbol{C}$, one obtains the approximate solution

$$v(x) \approx \sum_{i=1}^{N} C_i f_i(x) \;. \tag{11.113}$$

**Remarks:**

(1) The final result is an $N$-dimensional approximation to the true solution.

(2) A judicious choice for the Ritz functions $f_i(x)$ can lead to a very good answer with only a few functions.

(3) The accuracy of the method in general increases with increasing $N$.

## 11.9 Approximation errors

As we have seen, a reasonable guess to the deflected shape of a system can provide an excellent approximation to the exact solution when using the principle of stationary potential energy. Also, using a higher dimensional approximation space leads to better answers. However, without knowing the exact solution ahead of time it is not easy to assess how much error one has made in any given problem. A complete discussion of approximation errors is beyond the scope of this book. Notwithstanding, it is useful to understand a few basic concepts.

### 11.9.1 Types of error

In the problems we have been treating we have been solving for a function which is an approximation to another function. In order to be concrete, let us call $v_a(x)$ the approximate solution and $v_e(x)$ the exact solution. The *pointwise* error in the approximation is

$$e(x) = v_a(x) - v_e(x). \tag{11.114}$$

It is in fact a function of position. At some points it may be zero, while at others it may be large. It can vary from point to point. The *relative* pointwise error is

$$e_r(x) = \frac{v_a(x) - v_e(x)}{v_e(x)} \tag{11.115}$$

and it simply expresses the error in the approximation in relative terms – i.e. as a percentage of the exact solution. It too is a function of position. These expressions for the error give a lot of detail about the approximate solution, but they are also somewhat inconvenient in that one cannot simply say, for example, "my approximate solution has 10% error". There is no single number to speak of since the errors are functions. Also, the relative error expression is undefined when the exact solution is zero. For this reason, we often introduce other measures for the error.

The most common measure of error is known as the $L^2$ error, and for a given function it is a scalar number. In the present context it is quite similar to the error one computes when performing a least-squares fit – i.e. it represents a squared difference. The $L^2$ measure or norm of a function $f(x)$ is defined as:

$$\|f(x)\| = \sqrt{\int_0^L f^2(x)\,dx} \tag{11.116}$$

and the $L^2$ error is defined as

$$E = \|e(x)\| = \|v_a(x) - v_e(x)\| = \sqrt{\int_0^L (v_a(x) - v_e(x))^2\,dx}\,; \tag{11.117}$$

i.e. to compute the $L^2$ error one squares the pointwise error, integrates it over the system, and then takes the square-root. The result is a single number which we can refer to as the error in the approximation. In order to compute a relative $L^2$ error, one can compute

$$E_r = \frac{\|e(x)\|}{\|v_e(x)\|}\,. \tag{11.118}$$

This expression also has the advantage that it is well-defined except in the case where the exact solution is identically zero everywhere – a case rarely of interest.

### 11.9.2 Estimating error in Ritz's method

The objective of Ritz's method is to reduce the error in the approximation. With respect to our new definitions, this means that what we would like to happen is that $E \to 0$ as $N \to \infty$. Knowing how $E$ depends upon $N$ (once the approximation functions have been chosen) is crucial to being able to estimate the error that one has made. This is an advanced topic which we will not treat in extensive detail. But it should be observed that the general characteristic of Ritz's method applied to elastic systems is that convergence to zero error occurs starting with the most significant digit in $v_a(x)$ and continuing to the least significant digit. Thus one can perform computations with increasing values of $N$ and compare solutions from one value to another. As the digits of the approximate solution take on the values of the exact solution they will stop changing as $N$ increases. In this way one can estimate *how may digits of accuracy* one has in the approximation, and thus one can also estimate the percent error one has made. This can be done on a pointwise basis using $v_a(x)$ or on an average basis using $\|v_a(x)\|$.

### Example 11.14

*Double built-in beam with a point load.* Consider a beam that is built in at both ends and subjected to a point force $P$ at mid-span. Approximately solve for the beam's deflection and estimate the number of correct digits.

*Solution*

The solution space for this problem was given in eqn (11.83). As an approximation space we will use functions in the set

$$\widetilde{\mathcal{S}} = \left\{ v(x) \,\middle|\, v(x) = \sum_{i \in \text{odd}} C_i[\cos(2\pi i x/L) - 1] \right\}. \tag{11.119}$$

Note that we restrict our sum to functions which are symmetric about the center of the span, and each function individually satisfies the kinematic boundary conditions. Employing the principle of stationary potential energy one can solve for the $C_i$ by hand to give:

$$C_i = -\frac{4PL^3}{EI(2\pi i)^4}. \tag{11.120}$$

In order to estimate the error in the approximate solution, we compute the $L^2$ norm of the approximate deflection as a function of the number of terms in the approximation. Doing so gives:

$$\|v(x)\| = \frac{|P|L^{7/2}}{4EI\pi^4}\sqrt{\frac{1}{2}\sum_{i \in \text{odd}} \frac{1}{i^8} + \left(\sum_{i \in \text{odd}} \frac{1}{i^4}\right)^2}. \tag{11.121}$$

Table 11.1 shows the result of this computation, where the first column indicates how many terms we have used in our approximation and the second column gives values of the $L^2$ norm divided by $|P|L^{7/2}/4EI\pi^4$. The underline indicates the digits which remain unchanged as we add new terms to our approximation. We can assume these digits are converged.

**Table 11.1** Convergence behavior of approximate solution.

| No. Terms | Norm |
|---|---|
| 1 | 1.22474487139 |
| 2 | 1.23487650463 |
| 3 | 1.23618903404 |
| 4 | 1.23653070762 |
| 5 | 1.23665574650 |
| 6 | 1.23671178040 |
| 7 | 1.23674050485 |
| 8 | 1.23675671035 |
| 9 | 1.23676653310 |
| 10 | 1.23677282839 |
| Exact | 1.23678983555 |

**Remarks:**

(1) From the result we can conclude with some confidence, for example, that with only three terms our result is accurate to three digits or has error less than 0.1%. Also shown in Table 11.1 is the exact solution computed by solving the governing differential equation. This verifies our conclusion.

(2) This type of analysis is not an exact error analysis but suffices in most situations.

(3) If we had not restricted $i$ to be odd, we would have had essentially the same result. The main difference would have been that half of the generalized displacements would have been zero.

### 11.9.3 Selecting functions for Ritz's method

An important aspect of selecting the functions for Ritz's method, beyond the requirement that they satisfy the kinematic boundary conditions, is that with increasing $N$ the new functions add to the approximation space without overlapping too much with the other functions. Mathematically this is expressed by trying to use functions that are as *orthogonal* to each other as possible. With ordinary vectors, say $\boldsymbol{a} = a_1\boldsymbol{e}_1 + a_2\boldsymbol{e}_2$ and $\boldsymbol{b} = b_1\boldsymbol{e}_1 + b_2\boldsymbol{e}_2$, we define orthogonality as the requirement that their inner (or dot) product be zero:

$$\langle \boldsymbol{a}, \boldsymbol{b} \rangle = a_1 b_1 + a_2 b_2 = \|\boldsymbol{a}\| \, \|\boldsymbol{b}\| \cos(\theta_{ab}) = 0, \tag{11.122}$$

where we have utilized the corresponding norm for such vectors $\|\boldsymbol{a}\| = \sqrt{\langle \boldsymbol{a}, \boldsymbol{a} \rangle}$, and $\theta_{ab}$ is the angle between the two vectors. When we are measuring error with the $L^2$ norm we also have a corresponding $L^2$ inner product between functions. If $f(x)$ and $g(x)$ are two functions, then their $L^2$ inner product is given as

$$\langle f(x), g(x) \rangle = \sqrt{\int_0^L f(x) g(x) \, dx}. \tag{11.123}$$

The abstract angle between two functions is defined via

$$\cos(\theta_{fg}) = \frac{\langle f(x), g(x) \rangle}{\|f(x)\| \, \|g(x)\|}. \tag{11.124}$$

So orthogonality between functions occurs when their inner product is zero – just as with ordinary vectors.

**Remarks:**

(1) In certain problems it is convenient to choose a set of orthogonal functions but in others it is not. Even if full orthogonality can

not be achieved, it is helpful if the chosen functions are mostly orthogonal; i.e. their inner products are close to zero.

(2) Common functions which are orthogonal over the interval $[0, L]$ are trigonometric functions such as $f_n(x) = \cos(n\pi x/L)$ and $f_m = \sin(m\pi x/L)$.

(3) Polynomials such as $f_i = (x/L)^i$ are not orthogonal over $[0, L]$ and if one uses many of them, one will encounter numerical difficulties due to their non-orthogonal nature.

(4) Polynomials which are orthogonal over the interval $[0, L]$ would be Legendre polynomials. The first three of these are $f_1(x) = 1$, $f_2(x) = 2x/L - 1$, and $f_3 = (3/2)(2x/L - 1)^2 - 1/2$.

(5) For hand solutions, one usually picks only one or two functions based on intuition and does not worry about orthogonality. Orthogonality is more important when using many functions.

# Chapter summary

- Potential energy: $\Pi_{\text{total}} = \Pi_{\text{elastic}} + \Pi_{\text{load}}$
- Stationary potential energy is an alternative statement of equilibrium for a system.
- Potential energy of an elastic system is equal to its stored elastic energy:
  - Axial forces: $W = \int_L \frac{1}{2} AE(du/dx)^2\, dx$
  - Torsion loads: $W = \int_L \frac{1}{2} GJ(d\phi/dz)^2\, dz$
  - Bending loads: $W = \int_L \frac{1}{2} EI(d^2v/dx^2)^2\, dx$
- Potential energy of loads:
  - Dead loads (point): $-P\Delta$ or $-M\theta$
  - Dead loads (distributed): $-\int_L bu\, dx$, $-\int_L t\phi\, dz$, $-\int_L qv\, dx$
- Castigliano's first theorem: $P_k = \partial W/\partial \Delta_k$ and $M_k = \partial W/\partial \theta_k$
- Method of Ritz: Select a subset $\widetilde{\mathcal{S}}$ of the space of trial solutions $\mathcal{S}$, that is parameterized by a finite number of parameters, and find the stationary point of the system's total potential energy with respect to these parameters.
- $L^2$ norm $\|f(x)\| = \sqrt{\int f^2\, dx}$
- $L^2$ error $E = \|v_a - v_e\|$
- Complementary potential energy: $\Pi^*_{\text{total}} = \Pi^*_{\text{elastic}} + \Pi^*_{\text{load}}$
- Stationary complementary potential energy is an alternative statement of kinematic compatibility of a system.
- Complementary potential energy of an elastic system is equal to its complementary stored energy: $W^* = -W + P\Delta$:

- Axial forces: $W^* = \int_L \frac{1}{2} R^2/AE\,dx$
- Torsion loads: $W^* = \int_L \frac{1}{2} T^2/GJ\,dz$
- Bending loads: $W^* = \int_L \frac{1}{2} M^2/EI\,dx$
- Direct shear loads: $\alpha \int_L \frac{1}{2} V^2/GA\,dx$

• Complementary potential energy of loads:
  - Dead loads (point): $-P\Delta$ or $-M\theta$

• Castigliano's second theorem: $\Delta_k = \partial W/\partial P_k$ and $\theta_k = \partial W/\partial M_k$

# Exercises

(11.1) For the truss shown below, use Castigliano's first theorem to find the deflection at the load point in the direction of the load. Let $L = 24$ in, $AE = 15 \times 10^6$ lb, and $\boldsymbol{P} = 100\boldsymbol{e}_x + 20\boldsymbol{e}_y$ lb.

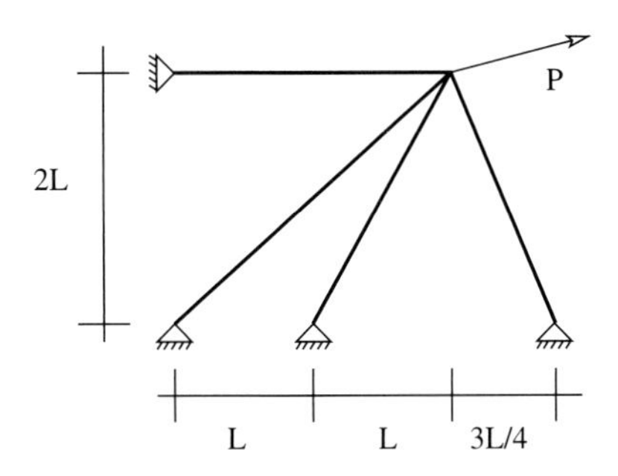


(11.2) Consider the truss system shown. Using the principle of stationary potential energy, determine the vertical deflection at the point of application of the load.

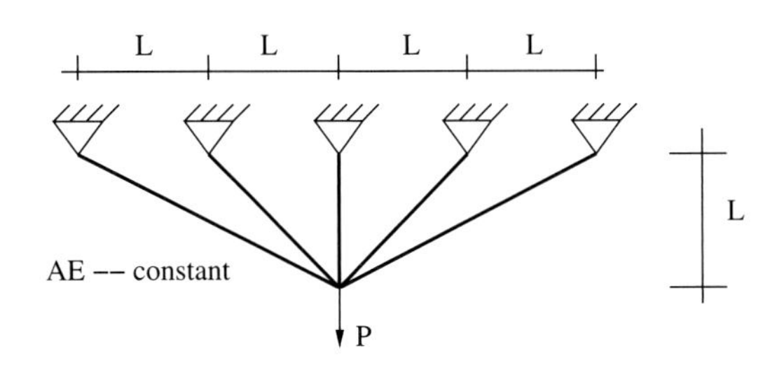


(11.3) For the stepped torsion bar, determine the rotation at the point of application of the load.

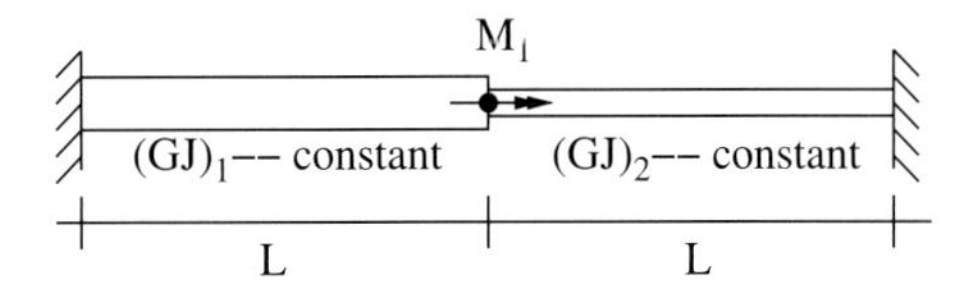


(11.4) For the truss given in Exercise 10.11 of Chapter 10, find the vertical motion of the roller and the horizontal and vertical motions of the load point using the principle of stationary potential energy. (Hint: you will end up with a system of three equations in three unknowns that you will need to solve.)

(11.5) An elastic circular bar is fixed at one end and attached to a rubber grommet at the other end. The grommet functions as a torsional spring with spring constant $k$. What magnitude torque $T_a$ is required in the center of the bar to rotate the center an amount $\theta_a$? Assume a constant shear modulus $G$ and polar moment of inertia $J$. Use Castigliano's first theorem to solve this exercise.

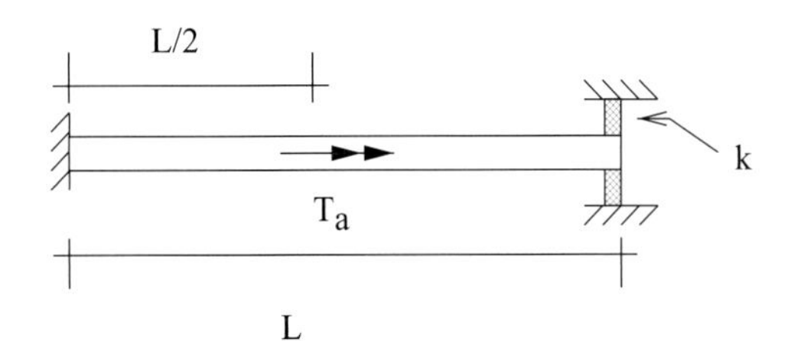


(11.6) Consider an inhomogeneous bar with $AE(x) = C + Dx$, where $C$ and $D$ are given constants. How much force is required to displace the end of the bar an amount $\Delta$? Use Castigliano's second theorem to solve this exercise.

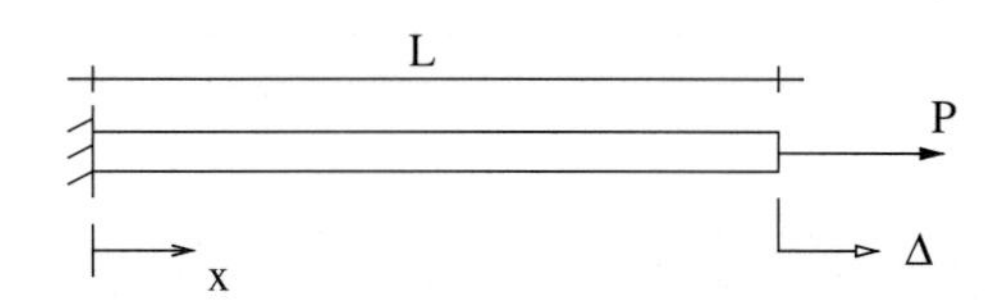


(11.7) Solve Exercise 10.1 from Chapter 10 using the principle of stationary complementary potential energy.

(11.8) Solve Exercise 10.5 from Chapter 10 using the principle of stationary complementary potential energy.

(11.9) For the round elastic torsion bar shown, determine the rotation at the point of application of the torque using Castigliano's second theorem. Assume $GJ(z) = A\exp[z/L]$, where $A$ is a given constant.

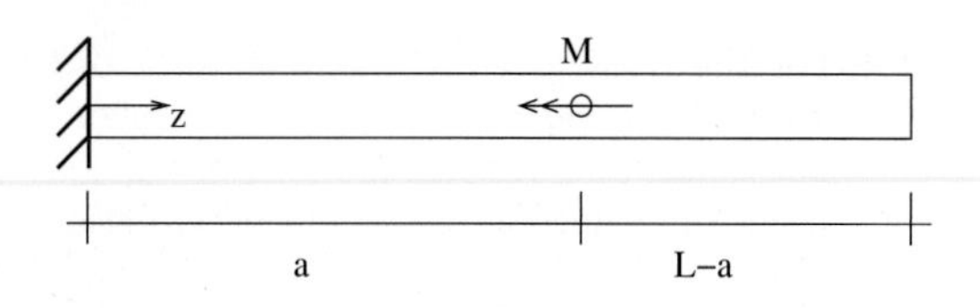


(11.10) For the system shown, use Castigliano's second theorem to determine the deflection of the disk in the direction of the load at the point where the load is applied. You may assume that $GJ$, $EI$, $AE$, and $\alpha AG$ are all given and constant.

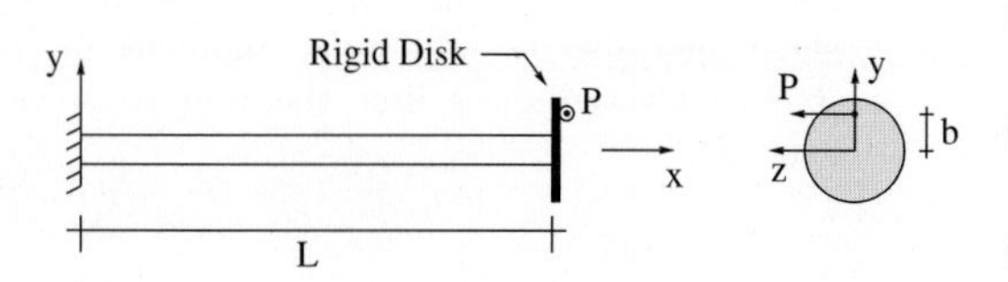


(11.11) Solve Exercise 10.3 from Chapter 10 using Castigliano's second theorem.

(11.12) Solve Exercise 10.4 from Chapter 10 using Castigliano's second theorem.

(11.13) Shown is an S-shaped hanger made of a slender curved beam. Using Castigliano's second theorem, find the stiffness of the hanger.

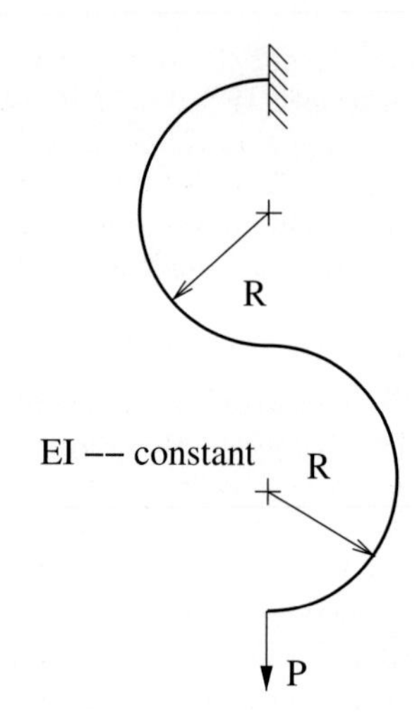


(11.14) Solve Exercise 10.6 from Chapter 10 using Castigliano's second theorem.

(11.15) Solve Exercise 10.8 from Chapter 10 using Castigliano's second theorem.

(11.16) Shown below is a circular ring structure subjected to diametrically opposed forces. Using Castigliano's second theorem, determine the stiffness of the ring. Include only bending energy contributions; assume moduli and section properties are constants.

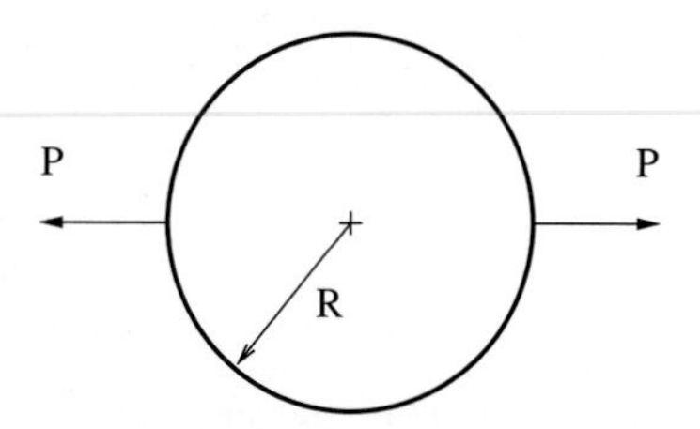


(11.17) Solve Exercise 10.16 from Chapter 10 using Castigliano's second theorem.

(11.18) Solve Exercise 10.18 from Chapter 10 using Castigliano's second theorem.

(11.19) Solve Exercise 10.19 from Chapter 10 using Castigliano's second theorem; ignore direct shear.

(11.20) Solve Exercise 10.20 from Chapter 10 using Castigliano's second theorem.

(11.21) Consider a cantilever beam with a uniform distributed load $q(x) = q_o$. Assume a deflection solution of the form $v(x) = Cx^2$ and determine an approximate solution using the principle of stationary potential energy. Compare the tip deflection to the exact solution.

(11.22) Consider a simply supported beam, pinned at both ends, of length $L$ with equal transverse loads of magnitude $P$ in the positive and negative directions at $x = L/4$ and $x = 3L/4$, respectively. By approximately minimizing the potential energy of the system with a single function find the displacement field for the beam. Compare your approximation to the exact answer.

(11.23) Solve Exercise 10.7 from Chapter 10 by approximately solving the principle of stationary potential energy with a two term approximation. Compare your answer to the exact deflection of the beam computed by solving the governing ordinary differential equation.

(11.24) Consider a round elastic bar of length $L$ with constant shear modulus, $G$, and polar moment of inertia, $J$. The bar is built-in at both ends and subject to a spatially varying distributed torsional load

$$t(z) = p\sin(\frac{2\pi}{L}z)\,,$$

where $p$ is a constant with units of torque per unit length. Find an approximate expression for the rotation field using the principle of stationary potential energy. Compare your result to the exact solution.

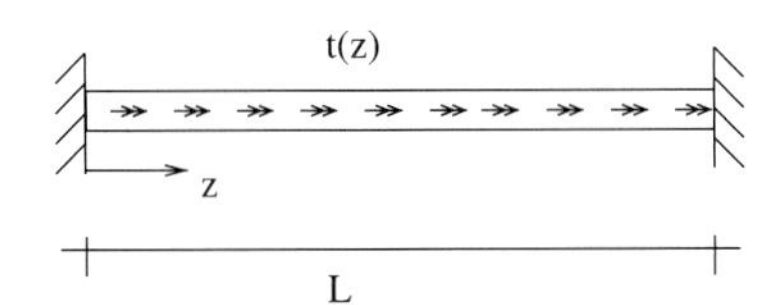


(11.25) Consider the torsion bar shown below. Determine the end-rotation of the bar by approximately solving the principle of stationary potential energy. Compare your result to the exact solution. Assume $GJ$ is a constant.

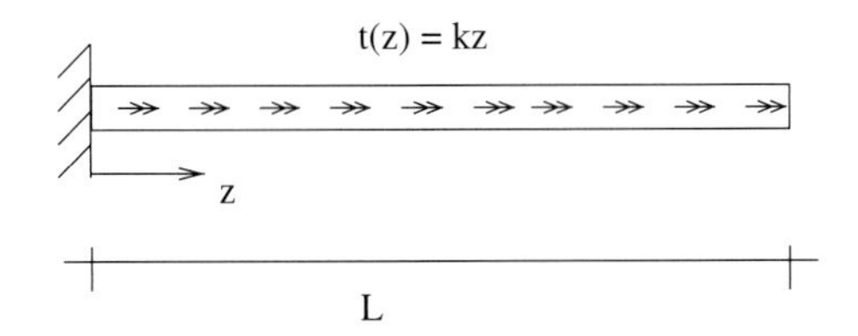


(11.26) The stepped beam shown is to be designed such that the tip rotation does not exceed $\theta_{\max}$. $L_1$ and $L_2$ are given fixed values. Find a formula for the minimum acceptable values for $(EI)_1$ and $(EI)_2$. Develop your result by approximately solving the principle of stationary potential energy.

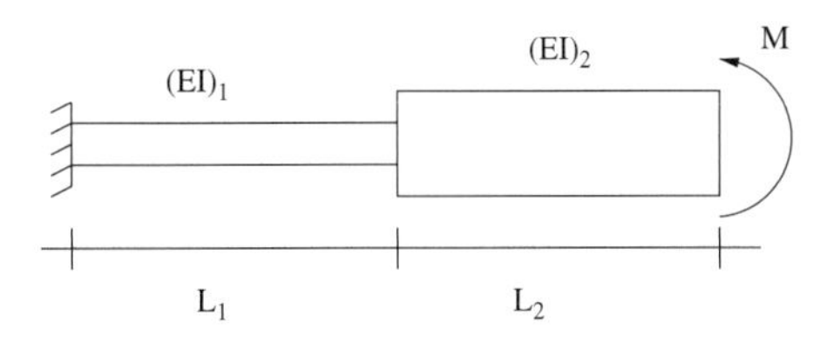


(11.27) Approximately solve Exercise 11.16 using the principle of stationary potential energy.

(11.28) Consider a doubly built-in beam of length $L$ with a transverse load of magnitude $P$ in the positive direction at $x = L/2$. (a) By approximately minimizing the potential energy of the system find the displacement field for the beam: use an approximation space with one parameter. (b) Compare your approximation to the exact answer with an accurate plot. (c) Assess your accuracy by computing the relative displacement error at the middle of the beam (as a percentage).

(11.29) Carefully derive the matrix equations that would result from using the method of Ritz on an elastic tension-compression bar problem fixed at its right end and subject to a distributed force $b(x)$.

(11.30) Consider a linear elastic bar with cross-sectional properties $AE = 330 \times 10^6$ lbf and length 4 ft which is built-in at both ends. The bar is loaded with a distributed force $b(x) = 100$ kips/ft. Solve for the displacement and strain fields in the bar using the method of Ritz and the approximation functions $f_n(x) = \sin(n\pi x/L)$ for $n = 1, 2, 3, \ldots$. How many terms in the expansion are required to reduce the relative $L^2$ error in the displacements to 1%? To compute the errors you can use an approximate quadrature – something simple, like Riemann sums. How many terms are needed to achieve the same with respect to the strains? (Hint: A small computer program is helpful for this exercise.)

(11.31) Consider the elastic rod shown with $AE$ a constant. Using an approximate potential energy method find the displacement field of the rod; use the space of approximate solutions $\widetilde{\mathcal{S}} = \{u(x) \mid u(x) = C\sin(2\pi x/L)\}$. Also determine the strain at the center of the rod.

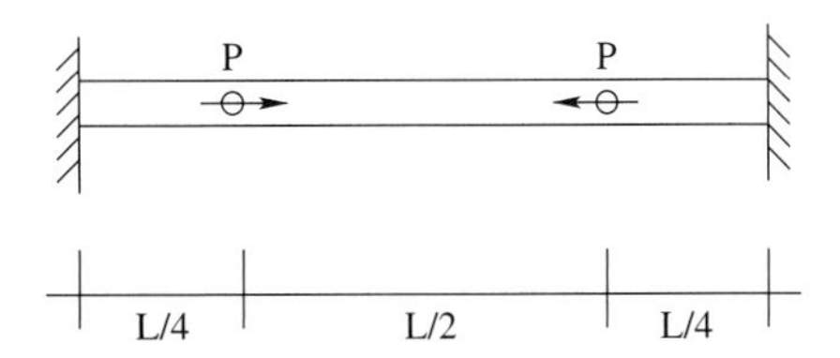

(11.32) Consider the beam shown and find its tip deflection using an approximate potential energy method; assume deflections of the form $v(x) = \sum_{j=1}^{N} C_j (x/L)^{j+1}$. Assume: $EI = 10^6$ lbf $\cdot$ in$^2$, $P = 23$ lbf, $L = 55$ in, $a = 5$ in, and $q_o = 0.70$ lbf/in. (Hint: A small computer program is helpful for this exercise.)

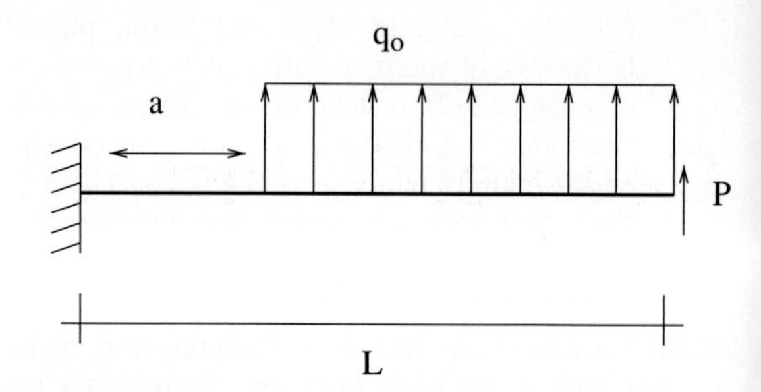

# Geometric Instability

# 12

In Chapter 9 we examined the multi-axial failure of systems under combined states of stress. The failure in these cases was associated with the load-bearing capacity of the material – i.e. with the yield limit of the material. This, however, is not the only way in which a mechanical system can fail. A second major type of failure mechanism is geometric instability – also known as buckling. A geometric instability (or buckling instability) in a system allows the system to undergo very large and sudden deformations when a critical amount of load is applied to the system. Figure 12.1, for example, shows a beam that is pinned at the top and clamped at the bottom with a small amount of load applied axially. With the application of a specific amount of load, which is well below the axial yield load for the beam, it buckles suddenly to the right, as shown in Fig. 12.2. When designing a load carrying structure, one often has to check not only for yielding, but also for buckling. In this chapter we will first examine the general mechanical principles associated with buckling failure. Then we will apply this understanding to the buckling failure of beams with compressive axial loads.

**Fig. 12.1** Pin-clamped beam subject to a load below the buckling load.

## 12.1 Point-mass pendulum: Stability

To understand the concept of stability, let us start by examining a simple mechanical system – a point-mass pendulum. The system as shown in Fig. 12.3 consists of a point mass supported by a rigid massless bar fixed to a frictionless pivot. The total potential energy of the system is given solely by the gravitational potential energy of the mass:

$$\Pi(\theta) = MgL(1 - \cos(\theta)). \tag{12.1}$$

Note that there are no elastic storage elements in the system. One should also recall that the negative derivative of the potential will give the total torque on the pendulum; i.e.

$$T = -\frac{d\Pi}{d\theta} = -MgL\sin(\theta). \tag{12.2}$$

Because the system is conservative we know the potential energy must be stationary for equilibrium – principle of stationary potential energy. This implies $T = 0$, and consequently that $\theta = 0$ or $\theta = \pi$. The first of these solutions corresponds to the pendulum pointing straight down and the other straight up; see Fig. 12.4. Both solutions are equilibria.

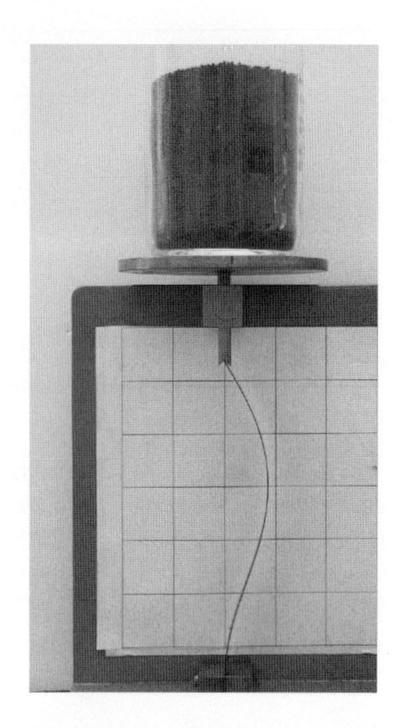

**Fig. 12.2** Pin-clamped beam subject to a load above the buckling load.

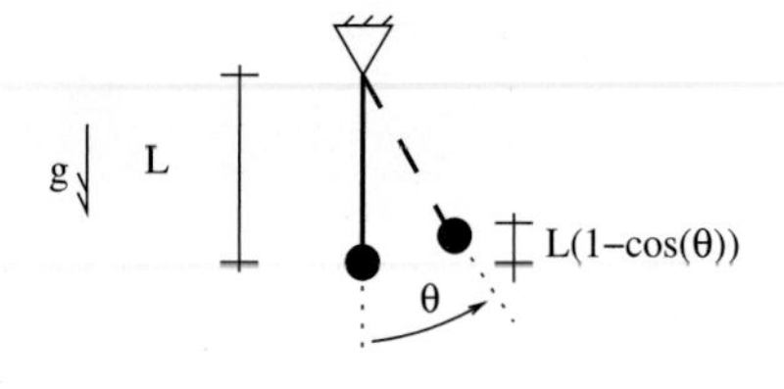


**Fig. 12.3** Point-mass pendulum geometry.

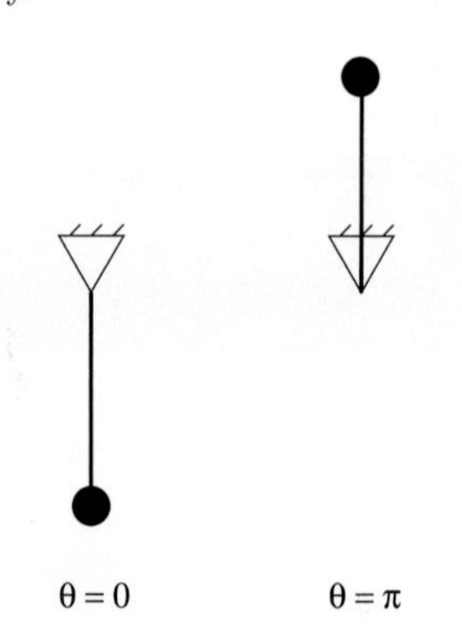


**Fig. 12.4** Equilibrium configurations of a point-mass pendulum.

Intuitively we know that the $\theta = 0$ solution is stable and that the $\theta = \pi$ solution is unstable. Our intuition is based on the basic observation that any small perturbation of the position of the pendulum, when pointing up, will cause it to fall away from the computed equilibrium. Whereas any small perturbation of the position of the pendulum, when pointing down, will cause it to return to the equilibrium. We can use our intuition as the basis for a sound mathematical definition of stability.

Consider first the case of $\theta = 0$ and a small perturbation of the position of the pendulum by an amount $|\delta\theta| \ll 1$. If $\delta\theta > 0$, then $T(0 + \delta\theta) < 0$. This implies that a positive perturbation of the position of the system will result in a negative torque being applied to the pendulum. Thus the torque (from the gravitational forces) opposes the perturbation and brings the system back towards the equilibrium position $\theta = 0$. Likewise, if $\delta\theta < 0$, then $T(0 + \delta\theta) > 0$. Thus this perturbation is also opposed by the loading system. The result is the conclusion that the equilibrium solution $\theta = 0$ is stable; i.e. all perturbations of the system away from the equilibrium position are opposed by the loading system. For the second equilibrium solution $\theta = \pi$ we have a completely different situation. Consider again the perturbations $|\delta\theta| \ll 1$. If $\delta\theta > 0$, then $T(\pi + \delta\theta) > 0$. This implies that a positive perturbation of the position of the system will result in a positive torque being applied to the pendulum. Thus the torque (from the gravitational forces) works in concert with the perturbation and brings the system further away from the equilibrium position $\theta = \pi$. Likewise, if $\delta\theta < 0$, then $T(\pi + \delta\theta) < 0$. Thus this perturbation is also reinforced by the gravitational load.

These results can be succinctly summarized by the condition that the sign of the second derivative of the potential energy can be used as a test of stability. If we evaluate the second derivative of the potential energy *at an equilibrium position*, then we may conclude:

$$\frac{d^2\Pi}{d\theta^2} > 0 \quad \Rightarrow \quad \text{stable equilibrium,} \tag{12.3}$$

$$\frac{d^2\Pi}{d\theta^2} = 0 \quad \Rightarrow \quad \text{neutral equilibrium,} \tag{12.4}$$

$$\frac{d^2\Pi}{d\theta^2} < 0 \quad \Rightarrow \quad \text{unstable equilibrium.} \tag{12.5}$$

Note further that a stable equilibrium also implies that the potential energy is minimal at equilibrium.

## 12.2 Instability: Rigid links

As a next step towards understanding buckling let us look at a slightly more complex system. The system will be composed of a single rigid link and a linear torsional spring as shown in Fig. 12.5. For small values of the load $P$ the *stable equilibrium* position of the link is vertical, or, in other words, the rotation $\theta$ of the link about the pivot is zero. If the load is progressively increased, we will reach a critical load at which this equilibrium position becomes unstable and the link (under any small perturbation) will *buckle* sideways. To see this more quantitatively, consider the free-body diagram shown in Fig. 12.6. If we consider moment balance about the pin, we find

$$0 = \sum M_{\text{pin}} = PL\sin(\theta) - k\theta. \tag{12.6}$$

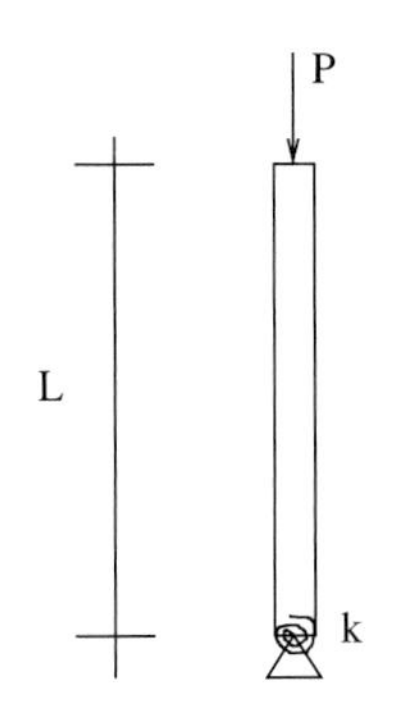


**Fig. 12.5** Rigid link supported by a torsional spring.

Equation (12.6) admits two solutions:

(1) For any $P$, $\theta = 0$ satisfies the equilibrium equation. Thus the vertical position of the link is always an equilibrium solution.

(2) The second solution to the equilibrium equation is

$$P = \frac{k}{L}\frac{\theta}{\sin(\theta)}. \tag{12.7}$$

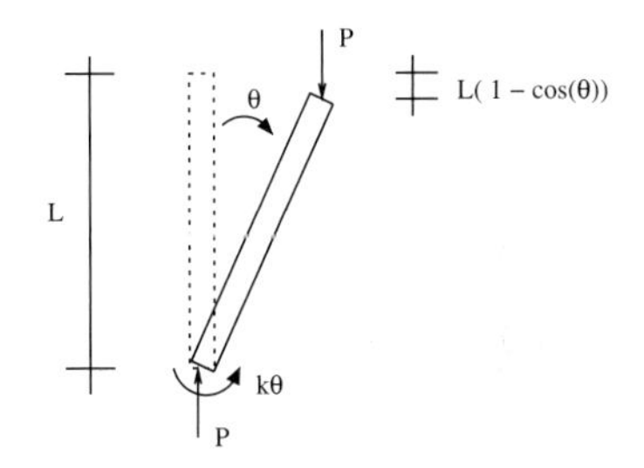


**Fig. 12.6** Free-body diagram of rigid link in a buckled state.

A plot of the two solutions (see Fig. 12.7) displays the characteristic features of a buckling solution.

(1) For values of $P < \frac{k}{L}$, there is only one possible solution

$$\theta = 0, \tag{12.8}$$

which is represented by the vertical curve in Fig. 12.7.

(2) For values of $P > \frac{k}{L}$ there are multiple solutions possible. One possibility is $\theta = 0$; the other two are are represented by the rotation values given by the nearly horizontal curve (plus and minus rotations for a give value of $P$).

(3) The load value at which multiple solutions first appear is known as the critical load (or critical buckling load). In our example, we have $P_{\text{crit}} = \frac{k}{L}$.

(4) Though we have not yet proved it, the solution $\theta = 0$ is unstable for $P > P_{\text{crit}}$. The solution given by the nearly horizontal curve is stable for $P > P_{\text{crit}}$. Thus as the load is increased from zero to above $P_{\text{crit}}$ the link will spontaneously rotate about the pin. Due to the very shallow curvature of the second solution, the rotation observed in practice is visually very dramatic. Because the motions associated with buckled states are typically large, buckling is usually considered to be synonymous with failure – buckling failure.

### 12.2.1 Potential energy: Stability

For our system there are two contributions to the potential energy: (1) the potential of the load and (2) the potential energy of the spring. Assuming that the load is a dead-load (i.e. is just a static weight applied to the top of thc link), we have that

$$\Pi_{\text{load}} = -PL(1 - \cos(\theta)). \tag{12.9}$$

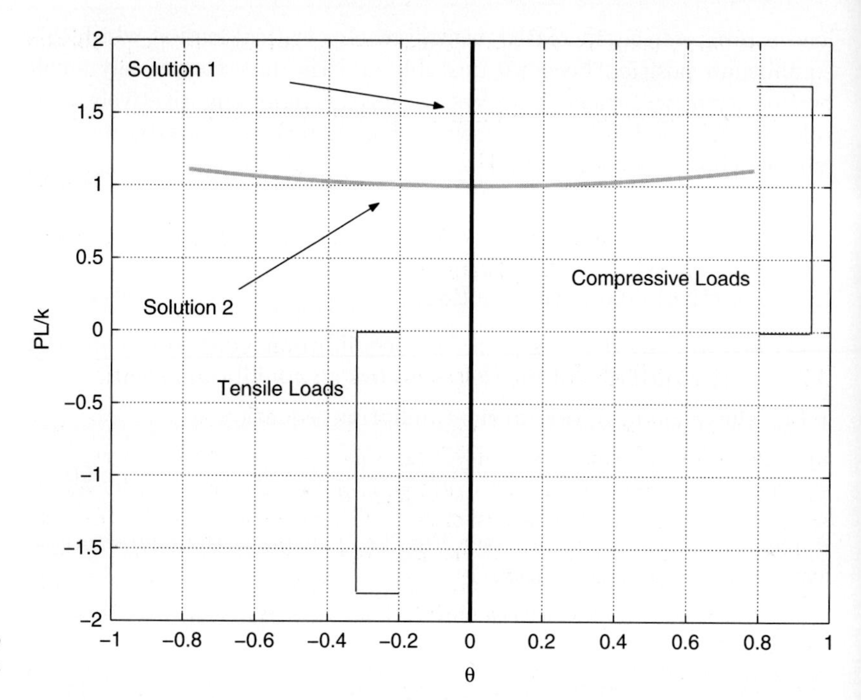


**Fig. 12.7** Equilibrium solutions to eqn (12.6). The rotation $\theta$ is shown in radians.

Thus the load's potential energy decreases with buckling rotation. The potential energy of the spring (which we assume to be linear) is given by the expression

$$\Pi_{\text{spring}} = \frac{1}{2}k\theta^2. \tag{12.10}$$

The total potential energy is given by

$$\Pi(\theta) = \Pi_{\text{spring}} + \Pi_{\text{load}}. \tag{12.11}$$

For the system to be in equilibrium, the first derivative of the potential needs to be zero. This yields

$$0 = \frac{d\Pi}{d\theta} = k\theta - PL\sin(\theta), \tag{12.12}$$

which is nothing more that the moment equilibrium equation for the system.

To assess the stability we need to examine the value of the second derivative of the potential energy along the various equilibrium solutions. The second derivative is given by:

$$\frac{d^2\Pi}{d\theta^2} = k - PL\cos(\theta). \tag{12.13}$$

We are now in a position to test the stability of the two equilibrium solutions we have computed.

(1) Along the equilibrium solution $\theta = 0$, we find $d^2\Pi/d\theta^2 = k - PL$. Thus, as long as $P < k/L$, we have a stable solution. For $P > k/L$,

we see that $d^2\Pi/d\theta^2 < 0$. Thus $\theta = 0$ is an unstable equilibrium for $P > P_{\text{crit}}$.

(2) Along the second equilibrium solution, we find that

$$\frac{d^2\Pi}{d\theta^2} = k\left(1 - \theta\cot(\theta)\right), \tag{12.14}$$

which is positive for $\theta \neq 0$. Thus the second solution, which appears for $P > P_{\text{crit}}$, is stable.

### 12.2.2 Small deformation assumption

If one is only interested in finding the critical load of a system and an indication of the post buckling deformation pattern, then it is appropriate to employ a small deformation assumption to the analysis and linearize all terms associated with large motion. In the case of the rigid link example of the previous sections, this amounts to replacing the trigonometric functions with appropriate linearizations. In our example we can substitute $\sin(\theta) \approx \theta$, and the equilibrium equation becomes:

$$k\theta - PL\theta = 0. \tag{12.15}$$

As before, we find two solutions:

(1) $\theta = 0$ for any values of $P$; this is the trivial solution.

(2) $P = \frac{k}{L}$ and $\theta$ arbitrary. This solution is the linearization of the second equilibrium solution from above. It provides the critical load, as it tells us that this value of the load is associated with some non-zero rotation about the pin. It unfortunately does not tell us the actual magnitude of the rotation. This information is lost by the linearization process. Notwithstanding, the buckling load is determined properly.

The utility of linearizing the equations becomes all the more apparent when one deals with systems more complex than a single rigid link.

**Example 12.1**

*Multiple rigid links and springs.* Consider the system shown in Fig. 12.8. It is composed of two rigid links that are connected by a torsional spring. The bottom support is also composed of a torsional spring. What is the buckling load for the system?

*Solution*

The system is shown in Fig. 12.8 with its degrees of freedom defined. In contrast to our first system, this system has multiple degrees of freedom. Nonetheless, the techniques developed in the previous sections can be applied. Let us first find the equilibrium equations for the system (in

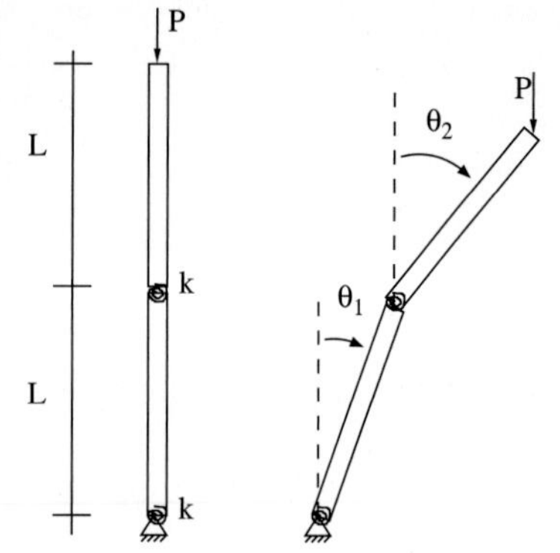

**Fig. 12.8** Two rigid links with two torsional springs.

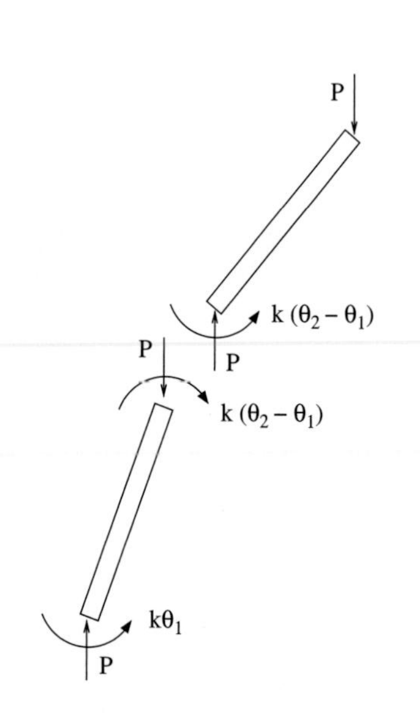


**Fig. 12.9** Free-body diagrams for two rigid links with two torsional springs.

linearized form) and then search for load values at which we can have non-zero motion of the system. Figure 12.9 shows free-body diagrams of the two links. If we write down moment equilibrium for each link we find for the lower link that

$$0 = k\theta_1 - k(\theta_2 - \theta_1) - PL\theta_1 \tag{12.16}$$

and for the upper link that

$$0 = k(\theta_2 - \theta_1) - PL\theta_2. \tag{12.17}$$

If we divide both equations through by $L$, we can arrange the equations in the following form:

$$\begin{bmatrix} \frac{2k}{L} - P & -\frac{k}{L} \\ -\frac{k}{L} & \frac{k}{L} - P \end{bmatrix} \begin{pmatrix} \theta_1 \\ \theta_2 \end{pmatrix} = \begin{pmatrix} 0 \\ 0 \end{pmatrix}. \tag{12.18}$$

This is a system of homogeneous equations and it has the trivial solution $(\theta_1, \theta_2)^T = (0, 0)^T$. This, of course, is the "un"-buckled solution. Our objective is to find load values, $P$, at which this system of equations has non-trivial solutions. For a linear system of homogeneous equations to have a non-trivial solution, we know from linear algebra that the determinant of the matrix must equal zero. Applying this condition, we find a quadratic equation for $P$:

$$0 = \left(\frac{2k}{L} - P\right)\left(\frac{k}{L} - P\right) - \left(-\frac{k}{L}\right)\left(-\frac{k}{L}\right) \tag{12.19}$$

$$0 = P^2 - \frac{3k}{L}P + \frac{k^2}{L^2}. \tag{12.20}$$

This equation has two solutions $P = \frac{k}{L}\frac{3 \pm \sqrt{5}}{2}$. For each of these values the system admits non-trivial motion of the system. The lesser of the two values gives the critical load for the system. Thus,

$$P_{\text{crit}} = \frac{k}{L}\frac{3 - \sqrt{5}}{2}. \tag{12.21}$$

**Remarks:**

(1) The problem above should be recognized as the classical eigenvalue problem from linear algebra. The load values leading to non-trivial motions of the system are the system eigenvalues.

(2) The eigenvectors associated with each eigenvalue provide information on the expected buckling shapes. Since the eigenvectors have arbitrary magnitude, we obtain information only on the shape, not the magnitude of deformation. For our solution, we find the following two eigenvectors:

$$\boldsymbol{v}_1 = \begin{pmatrix} 1 \\ \frac{1+\sqrt{5}}{2} \end{pmatrix} \qquad \boldsymbol{v}_2 = \begin{pmatrix} 1 \\ \frac{1-\sqrt{5}}{2} \end{pmatrix}, \tag{12.22}$$

which are sketched in Fig. 12.10. From Fig. 12.10 (left) we see that for our system the expected buckling mode will be a leaning-over motion as opposed to the scissor-mode associated with the second eigenvector.

(3) The larger eigenvalue is associated with a non-trivial motion of the system. However, the buckling mode associated with this eigenvalue will rarely be seen in practice. This is because once the load reaches the first eigenvalue, the system will start to buckle into the mode associated with the smaller of the two eigenvalues. In some special cases, such as with dynamically applied loads it may be possible to increase the load up to the higher eigenvalue before the system begins to buckle. In such a special situation it may be possible to drive the system into the second buckling mode.

(4) Figure 12.11 sketches the (full non-linear) equilibrium solutions to this problem in a manner similar to that shown in Fig. 12.7. Plotted on the horizontal axis is the norm of the motion. The vertical line represents the trivial solution which becomes unstable for all $P > P_{\mathrm{crit}}$. The lower nearly horizontal solution is associated with the first buckling mode and its zero crossing defines the critical load for the system. The upper nearly horizontal solution is associated with the second buckling mode.

(5) To verify that the various solutions are stable or unstable is more involved when a system has multiple degrees of freedom. The conceptual idea is similar to the case of a single degree of freedom

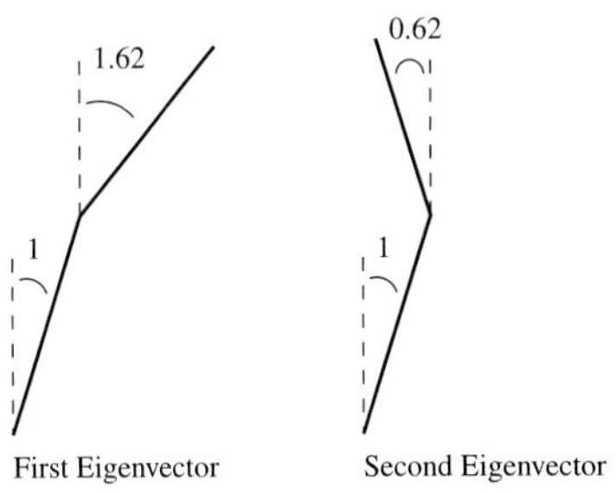

1.62 1 First Eigenvector

0.62 1 Second Eigenvector

**Fig. 12.10** Sketch of eigenvectors.

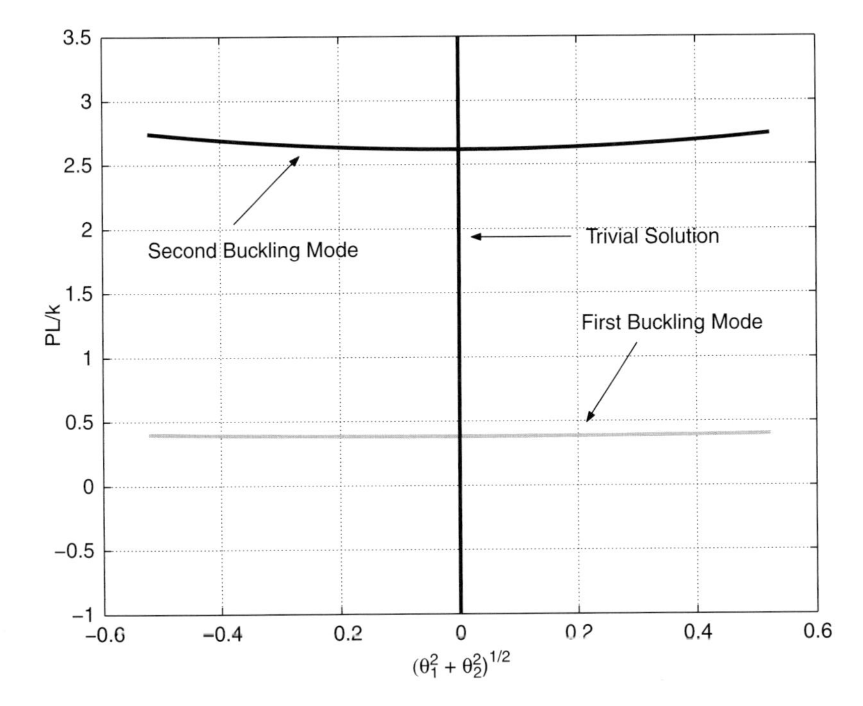


**Fig. 12.11** Equilibrium solutions for Example 12.1. Rotations are given in radians.

system but the reasoning and mathematics are slightly more complex. When a system has $N$ degrees of freedom, then there are $N^2$ possible second derivatives of the potential energy $\partial^2\Pi/\partial x_i \partial x_j$, where we have used the labels $x_i$ and $x_j$ to denote a degree of freedom and $i, j = 1, \ldots, N$. The extension of the idea that the second derivative should be positive for a stable equilibrium solution is that the $N$ by $N$ matrix of partial derivatives of the potential energy, $H_{ij} = \partial^2\Pi/\partial x_i \partial x_j$, should be positive definite. The matrix $H_{ij}$ is known as the Hessian matrix of the system; note also that it is symmetric for reasonable[1] potential energy functions.

[1] This amounts to saying that the mixed second partial derivatives of $\Pi$ are equal. This is guaranteed, for example, if $\Pi$ and its first and second derivatives are all continuous.

## 12.3 Euler buckling of beam-columns

In this section we will extend the ideas developed in Section 12.2 to the buckling of beams that are subject to compressive axial loads – i.e. columns. A column is similar to the multiple rigid link system which we have already examined. The primary difference is that it can bend at an infinite number of locations as opposed to a finite number of pre-specified joints. In this sense we should be able to write down the equilibrium equations for a beam with an axial load, and then search for axial load values that lead to non-trivial transverse motions. Because the beam can bend at an infinite number of points we should expect to find an infinite number of such axial loads. We will take the smallest of such loads to be the critical load for the system.

### 12.3.1 Equilibrium

Let us begin by deriving the equilibrium equations for a beam where we account for the possibility of buckling motions. Consider a beam with an axial load and consider the equilibrium of a differential element cut from the beam as shown in Fig. 12.12. The expression for vertical force equilibrium leads to[2]

$$0 = \sum F_y = -V(x) + q(x)\Delta x + V(x + \Delta x). \tag{12.23}$$

[2] In eqn (12.23) we have ignored the fact that the distributed load is a function of $x$. One can always appeal to the mean-value theorem to make this part of the analysis more precise; however, the final result does not change.

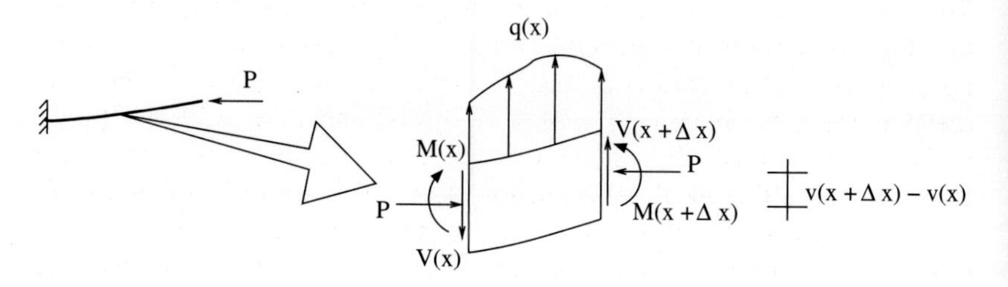


**Fig. 12.12** Equilibrium of a buckled beam.

Dividing through by $\Delta x$ and taking the limit as $\Delta x \to 0$ gives

$$\frac{dV}{dx} + q(x) = 0. \tag{12.24}$$

This is identical to the expression we found earlier in Chapter 8. If we now consider moment equilibrium about the $z$-axis through the left-side of the differential element we find:

$$\begin{aligned} 0 = \sum M_z = &- M(x) + V(x)\Delta x + M(x + \Delta x) \\ &+ P[v(x + \Delta x) - v(x)]. \end{aligned} \tag{12.25}$$

Note that in this expression we have omitted the term associated with the distributed load, $q(x)$, as it drops out just as it did in Chapter 8. Dividing through by $\Delta x$ and taking the limit at $\Delta x \to 0$ gives

$$\frac{dM}{dx} + V + P\frac{dv}{dx} = 0. \tag{12.26}$$

Here, in contrast to Chapter 8, we have an additional term associated with the axial load, $P$.

Let us now combine our two new equilibrium equations with the moment–curvature relation, $M = EI\kappa$, and the kinematic equation, $\kappa = v''$. Substituting one into the other leads to the following result:

$$EIv'''' + Pv'' = q, \tag{12.27}$$

where we have assumed that $EI$ is not a function of $x$.

### 12.3.2 Applications

The application of eqn (12.27) will closely follow the steps used in the analysis of systems composed of rigid links. For each particular system we will seek the minimum value of the axial load, $P$, at which it is possible to have a non-zero solution $v(x)$ to the equilibrium equations for the beam as represented by eqn (12.27).

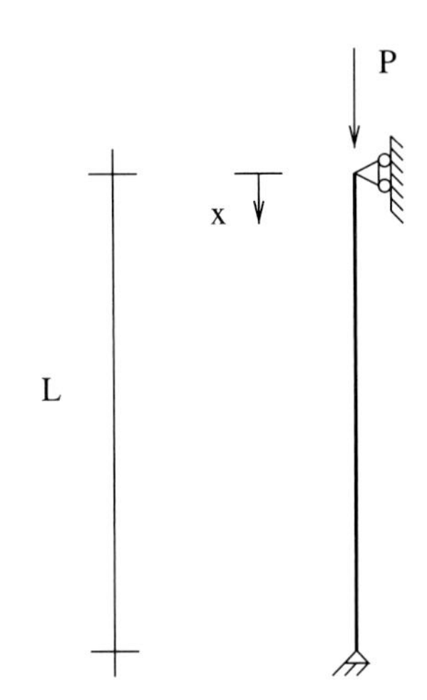


**Fig. 12.13** Pin-pin column with an axial load.

**Example 12.2**

*Pin–pin column.* Consider the column shown in Fig. 12.13; see also Fig. 12.14. Determine its critical buckling load.

*Solution*

To look for equilibrium solutions of this problem we need to identify the appropriate boundary conditions and the distributed load. Since the only applied load is the axial compression, we have that $q(x) = 0$. The ends are pinned, so we have that $v(0) = v''(0) = 0$ and that $v(L) = v''(L) = 0$. Due to the term $Pv''$ in eqn (12.27), we cannot simply integrate the ordinary differential equation four times like we did in Chapter 8. Here we will need to employ the classical technique for solving ordinary differential equations with constant coefficients. To this end, we will

**Fig. 12.14** Pin-pin beam subject to a load below the buckling load.

assume a solution of the form $v(x) = Ae^{sx}$. Substituting into eqn (12.27) we find the characteristic equation for the parameter $s$:

$$s^2(EIs^2 + P) = 0. \tag{12.28}$$

The solutions to this equation are $s = 0,\ 0,\ i\lambda$, and $-i\lambda$, where $\lambda = \sqrt{P/EI}$ and $i = \sqrt{-1}$. For each root we obtain one possible solution to our ordinary differential equation. Because of the repeated root we need to employ variation of parameters. After doing so we find that our (homogeneous) solution[3] is of the general form:

$$v(x) = A_1 \cos(\lambda x) + A_2 \sin(\lambda x) + A_3 x + A_4. \tag{12.29}$$

We can determine the unknown coefficients through the application of the boundary conditions.

$$v(0) = 0 \Rightarrow A_1 + A_4 = 0 \tag{12.30}$$

$$v''(0) = 0 \Rightarrow -A_1\lambda^2 = 0 \tag{12.31}$$

$$v(L) = 0 \Rightarrow A_1 \cos(\lambda L) + A_2 \sin(\lambda L) + A_3 L + A_4 = 0 \tag{12.32}$$

$$v''(L) = 0 \Rightarrow -A_1\lambda^2 \cos(\lambda L) - A_2\lambda^2 \sin(\lambda L) = 0. \tag{12.33}$$

[3] There is no particular solution to find in this problem, as the differential equation is homogeneous.

Equation (12.31) implies that $A_1 = 0$. Combined with eqn (12.30), we see that $A_4 = 0$. These two results with eqns (12.32) and (12.33) show that $A_3 = 0$, also. We are thus left with the result that

$$v(x) = A_2 \sin(\lambda x), \tag{12.34}$$

where

$$A_2 \sin(\lambda L) = 0. \tag{12.35}$$

Looking at this we see two basic solution classes:

(1) $A_2 = 0$ and $P$ is arbitrary. This is the trivial solution associated with no buckling deformation.

(2) $\sin(\lambda L) = 0$ and $A_2$ is arbitrary. This is the buckling solution. $\sin(\lambda L)$ will be zero whenever $\lambda L = n\pi$, where $n$ is an integer. Since $\lambda = \sqrt{P/EI}$, we see that there is an infinite number of buckling loads,

$$P_n = \frac{n^2\pi^2 EI}{L^2} \qquad n = 1, 2, 3, \cdots. \tag{12.36}$$

The critical buckling value is taken as the minimum element of this sequence; viz.

$$P_{\text{crit}} = P_1 = \frac{\pi^2 EI}{L^2}. \tag{12.37}$$

**Remarks:**

(1) Reiterating the basic procedure: (1) We examine the equilibrium of the system under consideration. (2) We look for values of the load

parameter at which non-trivial solutions exist. (3) The minimum of these load parameters is taken as the critical load.

(2) The basic statements about stable and unstable solutions which we proved for the rigid link example carry over to the column case without modification. The trivial solution is stable up to the first buckling load. For $P > P_{\mathrm{crit}}$ the trivial solution becomes unstable.

(3) The buckling mode associated with the critical load according to eqn (12.34) is a half-period sine wave as shown in Fig. 12.15. The higher buckling loads are associated with higher-frequency sine-waves.

(4) The critical load associated with pin-pin buckling of a column is often known as the *Euler buckling load.*

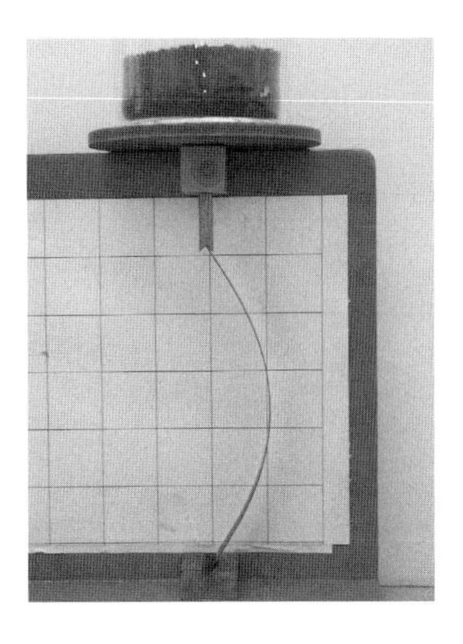

**Fig. 12.15** Pin-pin beam subject to a load above the buckling load.

### Example 12.3

*Pin-clamped column.* Consider a pin-clamped column of length $L$ as shown in Fig. 12.1. Determine its critical buckling load.

*Solution*

The difference between this problem and the prior example is only in the boundary conditions. There is no distributed load; i.e. $q(x) = 0$. The top end is pinned so we have that $v(0) = v''(0) = 0$, and the bottom end is clamped so we that $v(L) = v'(L) = 0$. We can now reuse the homogeneous solution, eqn (12.29), which we found in the prior example, and search for values of the load parameter which allow for non-trivial solutions. Application of our new boundary conditions gives

$$v(0) = 0 \Rightarrow A_1 + A_4 = 0 \tag{12.38}$$

$$v''(0) = 0 \Rightarrow -A_1\lambda^2 = 0 \tag{12.39}$$

$$v(L) = 0 \Rightarrow A_1\cos(\lambda L) + A_2\sin(\lambda L) + A_3 L + A_4 = 0 \tag{12.40}$$

$$v'(L) = 0 \Rightarrow -A_1\lambda\sin(\lambda L) + A_2\lambda\cos(\lambda L) + A_3 = 0. \tag{12.41}$$

Equation (12.39) implies that $A_1 = 0$. Combined with eqn (12.38), we see that $A_4 = 0$. The solution is seen to be

$$v(x) = A_2\sin(\lambda x) + A_3 x, \tag{12.42}$$

where eqns (12.40) and (12.41) require

$$\begin{bmatrix} \sin(\lambda L) & L \\ \lambda\cos(\lambda L) & 1 \end{bmatrix} \begin{pmatrix} A_2 \\ A_3 \end{pmatrix} = \begin{pmatrix} 0 \\ 0 \end{pmatrix}. \tag{12.43}$$

Looking at this, we again see two basic solution classes:

(1) The trivial solution where $A_2 = A_3 = 0$ and $P$ is arbitrary. This is the solution associated with no buckling deformation.

(2) The other solution class will occur when the homogeneous system of equations (12.43) has a non-trivial solution. This will happen when the coefficient matrix has zero determinant. Taking the determinant one finds

$$\tan(\lambda L) - \lambda L = 0. \tag{12.44}$$

There is an infinite sequence of values of $\lambda L$ which satisfies this equation. The smallest (non-zero) value for which this occurs is $\lambda L = 4.493$. Thus we have

$$P_{\text{crit}} = \frac{(4.493)^2 EI}{L^2}. \tag{12.45}$$

**Remarks:**

(1) Notice that the additional boundary restraint associated with clamping one end of the column increases the critical buckling load by a factor of about 2 over the Euler buckling load.

(2) It is typical to phrase buckling loads in terms of the Euler buckling load (the value for the pin-pin case). For the clamped-clamped case the buckling load is four times the Euler load. For the free-clamped case the buckling load is 1/4 of the Euler load.

(3) It is important to ascertain the exact boundary conditions when computing buckling loads. In particular, if the column in question can bend about two axes then the moment of inertia used in the buckling formulae should be the minimum of the two principal moments of inertia.

### 12.3.3 Limitations to the buckling formulae

The analysis carried out for the buckling of columns has three principal limitations that should be kept in mind:

(1) The analysis carried out utilizes linearized equations. Thus the analysis can only provide the buckling loads. It cannot provide quantitative information about the post-buckled condition of a column. It can only provide qualitative information about the general shape of the column after buckling. This is given, as in the rigid link case, by the associated eigenvectors (or more properly in this setting, eigenfunctions).

(2) The analysis carried out assumed an elastic response of the system. Thus one must check the validity of $P_{\text{crit}}$ by ensuring that $\sigma_{\text{crit}} = P_{\text{crit}}/A < \sigma_Y$. In the pin-pin case, for example, we have

$$\sigma_{\text{crit}} = \frac{\pi^2 EI}{AL^2} = \frac{\pi^2 E}{(L/r)^2}, \tag{12.46}$$

where $r = \sqrt{I/A}$ is known as the radius of gyration. Columns for which the critical stress is less than the yield stress are known as *slender* columns. The governing parameter for slenderness is seen to be $L/r$. A plot of $\sigma_{\text{crit}}$ in eqn (12.46) as a function of $L/r$ is often known as Euler's hyperbola. In a column that is not slender, yielding will precede buckling. The analysis of post-yield buckling is left to more advanced texts.

(3) The last important limitation that one should be aware of regarding the analysis presented is that throughout we have assumed that the axial loads are always perfectly applied. In reality, loads are never perfectly placed. This limitation is taken up in more detail in the next section.

## 12.4 Eccentric loads

Eccentric loads on a column are loads that are not perfectly aligned with the centroid of the cross-section. In such situations we encounter a buckling-like behavior. As load is applied, the column appears to remain in a stable trivial equilibrium state. Then when the load reaches some critical value the column undergoes sudden transverse motion. This appears on the surface to behave just like buckling. However, it is associated with only a single equilibrium solution. We will begin our quantitative discussion of eccentric loads by first looking at a rigid link system; this will be followed by studying an eccentric load on a pin-pin column.

### 12.4.1 Rigid links

Consider the rigid link shown in Fig. 12.16(a). The link is of length $L$, is supported by a linear torsional spring with spring constant $k$, and the

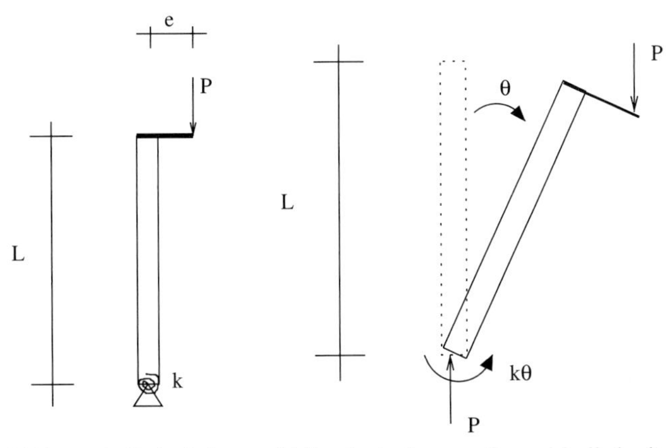


(a) Eccentrically loaded rigid link supported by a torsional spring.

(b) Free body diagram of eccentrically loaded rigid link.

**Fig. 12.16** Single rigid link with an eccentric load.

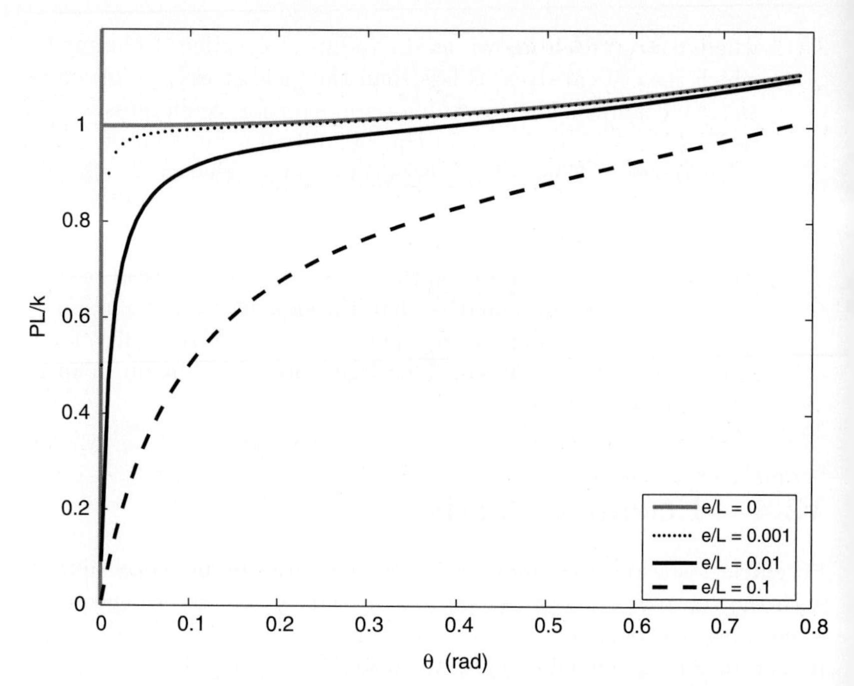


**Fig. 12.17** Equilibrium solutions for an eccentrically loaded rigid link. Eccentricity ratios shown from the lower curve to the upper curve are: 0.1, 0.01, 0.001, 0. The solution for zero eccentricity also includes the trivial vertical solution $\theta = 0$.

load has an eccentricity $e$. Using the free-body diagram in Fig. 12.16(b), we have by moment equilibrium that:

$$0 = \sum M_{\text{pin}} = k\theta - P\left[L\sin(\theta) + e\cos(\theta)\right]. \tag{12.47}$$

For non-zero eccentricity, i.e. $e \neq 0$, this equation has only one solution:

$$P = \frac{k}{L}\frac{\theta}{\sin(\theta) + \frac{e}{L}\cos(\theta)}. \tag{12.48}$$

To gain an understanding of the solution and its relation to buckling, the solution has been has been plotted for various values of $e/L$ in Fig. 12.17. Also, plotted in this figure are the two solutions for zero eccentricity. As can been seen, for small values of the eccentricity the solution closely follows the stable equilibrium solution of the zero eccentricity case. The system stays nearly vertical and then suddenly, as $P_{\text{crit}}$ (of the $e = 0$ case) is approached, the system rotates substantially. For larger eccentricities, the deviation is larger but the trend is similar.

**Remarks:**

(1) Systems with small eccentricities can be effectively analyzed by assuming that the load placement is perfect. The buckling load is seen to operate as an upper bound to the load at which the system will begin to undergo large motions.

### 12.4.2 Euler columns

The analysis of the prior section can be carried over to the beam-column case. Figure 12.18 shows a pin-pin column with an eccentrically applied axial load. The eccentricity provides a moment at the pin location; thus the boundary conditions for this problem are given by

$$v(0) = 0 \tag{12.49}$$

$$EIv''(0) = -Pe \tag{12.50}$$

$$v(L) = 0 \tag{12.51}$$

$$EIv''(L) = -Pe. \tag{12.52}$$

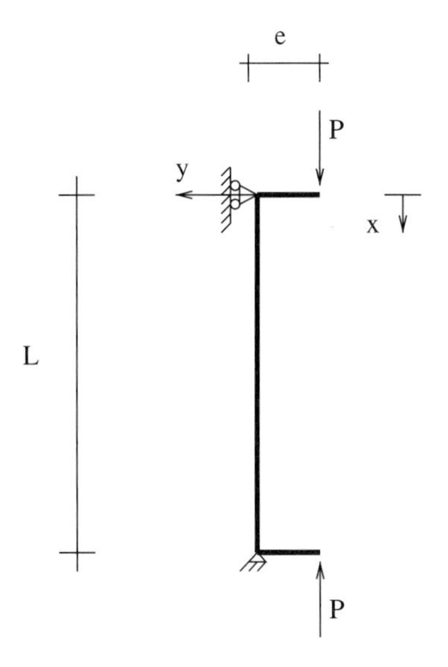

**Fig. 12.18** Pin-pin column with an eccentric axial load.

The distributed load is again $q(x) = 0$. Because of this we can reuse our homogeneous solution, eqn (12.29). Applying our boundary conditions we find:

$$v(0) = 0 \quad \Rightarrow A_1 + A_4 = 0 \tag{12.53}$$

$$EIv''(0) = -Pe \quad \Rightarrow EI(-A_1\lambda^2) = -Pe \tag{12.54}$$

$$v(L) = 0 \quad \Rightarrow A_1 \cos(\lambda L) + A_2 \sin(\lambda L) + A_3 L + A_4 = 0 \tag{12.55}$$

$$EIv''(L) = -Pe \quad \Rightarrow - EI\lambda^2(A_1 \cos(\lambda L) + A_2 \sin(\lambda L)) = -Pe. \tag{12.56}$$

Equation (12.54) tells us that $A_1 = e$. Combined with eqn (12.53) we see that $A_4 = -e$. Equation (12.56) can now be solved to show that $A_2 = e \tan(\lambda L/2)$. Equation (12.55) then reveals that $A_3 = 0$. Combining these results together we find a single equilibrium solution for $e \neq 0$:

$$v(x) = e \left[ (\cos(\lambda x) - 1) + \tan\left(\frac{\lambda L}{2}\right) \sin(\lambda x) \right]. \tag{12.57}$$

**Remarks:**

(1) Notice that when $\lambda L/2$ is small the (transverse) deformation of the column is $O(e)$. However, as $(\lambda L/2) \to \pi/2$, which is the same as $P \to P_{\text{crit,e=0}}$, the magnitude of the deflection increases rapidly.

(2) Just as in the eccentrically loaded rigid link case, the solution closely tracks the stable equilibrium solution to the $e = 0$ problem. Hence, the critical loads from the ideal case can be used as effective upper bounds for the eccentrically loaded cases.

(3) As before, one should check the validity of the analysis. In particular, one should check the validity of the elasticity assumption. The maximum stresses will occur at mid-span where the bending moment is largest.

$$\sigma_{\max} = \frac{P}{A} + \frac{M_{\max} y_{\max}}{I}. \tag{12.58}$$

Rearranging terms and using our deflection solution one can show:

$$\sigma_{\max} = \sigma_N \left[1 + \frac{e y_{\max}}{r^2} \sec\left(\sqrt{\frac{\sigma_N}{E}} \frac{(L/r)}{2}\right)\right], \tag{12.59}$$

where $\sigma_N = P/A$ and $y_{\max}$ is the distance from the neutral axis to the compression chord. For our solution to make sense we need to always check that $\sigma_{\max} < \sigma_Y$.

## 12.5 Approximate solutions

It is also possible to solve buckling problems using approximation methods. In fact, for many practical problems this is the only possibility, as it is often not feasible to exactly solve the governing equilibrium equations. As we know, equilibrium equations can be obtained from the stationary conditions of a (conservative) system's total potential energy. If we approximately solve the stationary potential energy problem, then we will obtain an approximate equilibrium solution. In the setting of this chapter, this will lead to approximate buckling loads. Just as seen above, the resulting mathematical structure will be that of an eigenvalue problem.

Our main ingredient for the solution of approximate buckling problems will be the system's total potential energy. If we restrict ourselves to the buckling of slender beam-columns with an end-load $P$, then the required potential energy is given by

$$\Pi = \Pi_{\text{beam}} + \Pi_{\text{load}}, \tag{12.60}$$

where

$$\Pi_{\text{beam}}(v(x)) = \int_0^L \frac{1}{2} EI(v'')^2 \, dx \tag{12.61}$$

and

$$\Pi_{\text{load}} = -P \cdot \text{ displacement at the load point} \tag{12.62}$$

$$= -P \int_0^L \frac{1}{2} \theta^2 \, dx, \tag{12.63}$$

$$= -P \int_0^L \frac{1}{2} (v')^2 \, dx. \tag{12.64}$$

The expression for the potential energy associated with the elastic energy of the beam is quite familiar now; however, the expression for the potential for the load is a bit more tricky. The fundamentals, however, are the same as before: *the potential of the load is minus the magnitude of the load times the displacement at the load point in the direction of the load.* Consider, for example, a beam-column as in Fig. 12.19. If we look at an infinitesimal segment of the beam-column of length $dx$, then this segment will rotate (during buckling) by an angle $\theta = v'$. The resultant vertical drop associated with the segment is equal to $-\frac{1}{2}\theta^2\,dx$; here we have assumed that $1-\cos(\theta)\approx\frac{1}{2}\theta^2$. To find the vertical drop at the end of the beam-column, we need to add up the contributions from each segment of the structure; i.e., one must integrate the expression over the interval $[0, L]$. This results in the final expression in eqn (12.64) and the total potential energy will be given by

$$\Pi(v(x)) = \Pi_{\text{beam}} + \Pi_{\text{load}} \tag{12.65}$$

$$= \int_0^L \frac{1}{2}EI(v'')^2\,dx - P\int_0^L \frac{1}{2}(v')^2\,dx\,. \tag{12.66}$$

To estimate buckling loads and buckling modes/shapes, one can apply Ritz's method to this potential energy. The procedure is identical to what we saw in Section 11.7, and we will illustrate it by several examples.

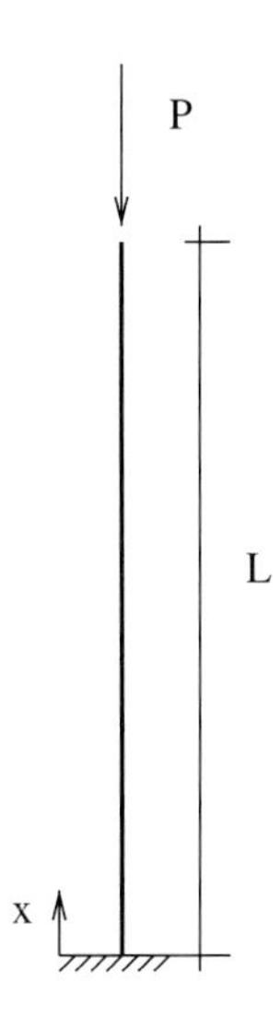


**Fig. 12.19** Clamped-free beam-column subjected to an axial compression.

**Example 12.4**

*Clamped-free beam-column buckling.* Consider a beam-column which is built-in at one end and free from kinematic boundary conditions at the other. The system is subjected to an axial (compressive) load $P$; see Fig. 12.19. Find the system's critical load via approximation.

*Solution*

Let us consider a simple one-parameter approximation for $v(x)$ of the form: $v(x)\approx Cx^2$. Inserting this expression into eqn (12.66) yields:

$$\Pi(C) = \int_0^L \frac{1}{2}EI4C^2\,dx - P\int_0^L \frac{1}{2}4C^2x^2\,dx. \tag{12.67}$$

The stationary conditions for this potential energy are easily found via conventional calculus as:

$$\frac{d\Pi}{dC} = 4EICL - P4C\cdot\frac{1}{3}L^3 \tag{12.68}$$

$$= C\left[4EIL - P\frac{4}{3}L^3\right] = 0\,. \tag{12.69}$$

There are two possible solutions to this (equilibrium) equation. The first is $C=0$; this is the trivial solution with $v(x)=0$ corresponding to any load $P$. The second occurs when $P=3EI/L^2$. At this value of the load, $C$ can take on any arbitrary value. This is the critical load at which instability occurs.

**Remarks:**

(1) The exact critical buckling load for this configuration is, $P_{\text{cr}}^{\text{exact}} = \pi^2EI/4L^2$. The relative error in the approximation is:

$$e_r = \frac{\pi^2/4-3}{\pi^2/4} = 1-\frac{12}{\pi^2} = 21\%\,. \tag{12.70}$$

(2) The structure of the equilibrium equations does not change from what we saw when we computed exact buckling loads. The equilibrium equations possess trivial solutions for arbitrary loads, and for certain loads they possess non-trivial solutions.

(3) In this example only one critical load appears, since our approximation space was one-dimensional. For each dimension of the approximation space one will normally compute one buckling load. The minimum of these loads is an approximation to the critical load for the system.

### Example 12.5

*Clamped-free buckling: Two parameters.* Improve the solution in Example 12.4 by finding a two-parameter approximate solution.

*Solution*

Consider the two-parameter polynomial approximation $v(x) \approx C\,(x/L)^2 + D\,(x/L)^3$, which satisfies the kinematic boundary conditions. If we insert this into the expression for the potential energy, then

$$\begin{aligned} \Pi(C,D) = &\int_0^L \frac{1}{2} EI \left[ \frac{2C}{L^2} + \frac{6D}{L^2}\left(\frac{x}{L}\right)\right]^2 dx \\ &-P\int_0^L \frac{1}{2}\left[\frac{2C}{L}\left(\frac{x}{L}\right) + \frac{3D}{L}\left(\frac{x}{L}\right)^2\right]^2 dx \,. \end{aligned} \tag{12.71}$$

The stationary conditions for this potential are

$$\left.\begin{aligned} \frac{\partial \Pi}{\partial C} = 0 \\ \frac{\partial \Pi}{\partial D} = 0 \end{aligned}\right\} \Rightarrow \begin{bmatrix} \frac{4EI}{L^3} - \frac{P}{L}\frac{4}{3} & \frac{6EI}{L^3} - \frac{P}{L}\frac{3}{2} \\ \frac{6EI}{L^3} - \frac{P}{L}\frac{3}{2} & \frac{12EI}{L^3} - \frac{P}{L}\frac{9}{5} \end{bmatrix} \begin{bmatrix} C \\ D \end{bmatrix} = \begin{bmatrix} 0 \\ 0 \end{bmatrix}. \tag{12.72}$$

Equation (12.72) is a system of homogeneous linear equations in two unknowns. One solution is the trivial solution $C = D = 0$ for arbitrary values of the load. The case in which we are interested is the one where $C$ and/or $D$ are possibly non-zero, as this corresponds to a bending of the beam-column – a state of buckling. For this to occur the coefficient matrix in eqn (12.72) must have zero determinant. Computing the determinant of the matrix and setting it equal to zero results in a second-order polynomial (the characteristic polynomial) in $P$ which we can solve to determine two buckling loads: $P_1$, $P_2$. The minimum of these will be the critical load for the system.

In order to emphasize better that this is an eigenvalue problem we first rewrite eqn (12.72) in the form

$$\left( \begin{bmatrix} 4 & 6 \\ 6 & 12 \end{bmatrix} - \frac{PL^2}{EI} \begin{bmatrix} 4/3 & 3/2 \\ 3/2 & 9/5 \end{bmatrix} \right) \begin{bmatrix} C \\ D \end{bmatrix} = \begin{bmatrix} 0 \\ 0 \end{bmatrix}. \tag{12.73}$$

If we now introduce the following definitions:

$$\boldsymbol{A} = \begin{bmatrix} 4 & 6 \\ 6 & 12 \end{bmatrix}, \tag{12.74}$$

$$\boldsymbol{B} = \begin{bmatrix} 4/3 & 3/2 \\ 3/2 & 9/5 \end{bmatrix}, \tag{12.75}$$

$$\lambda = \frac{PL^2}{EI}, \tag{12.76}$$

$$\boldsymbol{x} = \begin{bmatrix} C \\ D \end{bmatrix}, \tag{12.77}$$

then our equilibrium equations have the compact form

$$(\boldsymbol{A} - \lambda \boldsymbol{B})\, \boldsymbol{x} = \boldsymbol{0}, \tag{12.78}$$

which is known as a *generalized eigenvalue problem.* In the case where $\boldsymbol{B}$ is the identity matrix, one has $\boldsymbol{A}\boldsymbol{x} = \lambda \boldsymbol{x}$, which is the standard eigenvalue problem. The two eigenvalues are computed by setting the determinant of $\boldsymbol{A} - \lambda \boldsymbol{B}$ zero which results in a quadratic polynomial for $\lambda$. The solutions are

$$\lambda = \frac{PL^2}{EI} = 2.49,\ 32.2. \tag{12.79}$$

Since the smallest buckling load is the critical load,

$$P_{\text{cr}} = 2.49 \frac{EI}{L^2}. \tag{12.80}$$

**Remarks:**

(1) In this problem we have chosen non-dimensional forms for the functions in Ritz's method. This turns out to be advantageous numerically and in terms of understanding. It also results in a situation where all the Ritz parameters have the same dimensions – that of displacement. For this reason, the coefficients $C$ and $D$ are often referred to as generalized displacements.

(2) With respect to the exact solution, our approximate result for $P_{\text{cr}}$ has a relative error of only 0.75%.

(3) If one computes the associated eigenvector $[C, D]^T$, then one can plot the basic shape of the buckling mode as $v(x) = C(x/L)^2 + D(x/L)^3$. Doing so results in $[C, D]^T = [-1.30, 0.392]^T$, and the corresponding buckling mode is shown in Fig. 12.20.

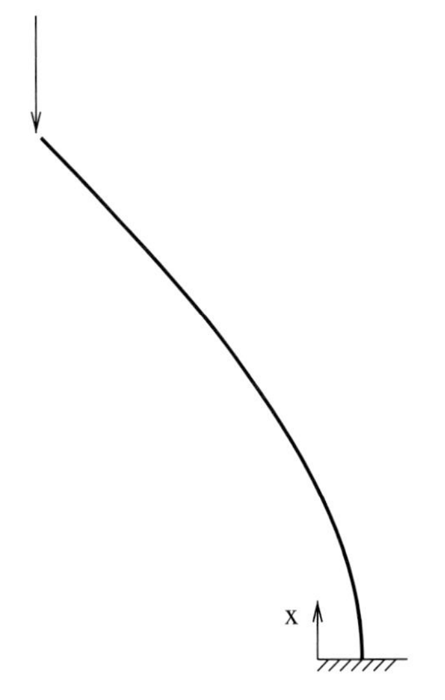


**Fig. 12.20** Approximate buckling mode from two-parameter solution.

**Example 12.6**

*Clamped-free beam-column with mid-span spring support.* Consider a clamped-free beam-column, but now with a lateral spring support at mid-span as shown in Fig. 12.21. Determine the buckling load for this system using an approximation method.

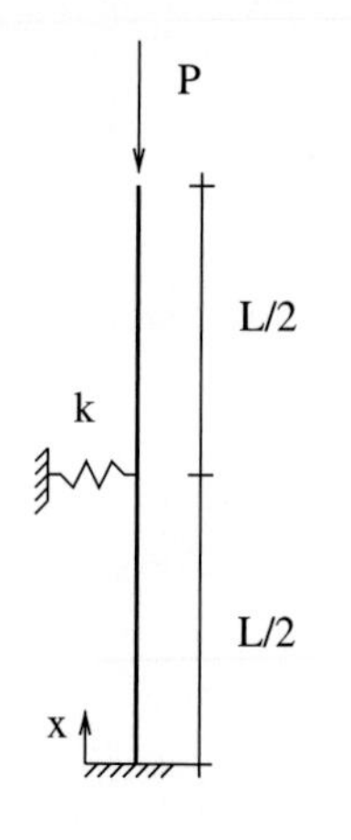


**Fig. 12.21** Clamped-free beam-column with spring support.

*Solution*

The calculation involves only a slight modification of our previous example. One only has to account for the additional contribution of the elastic potential energy associated with the spring. This is simply added to the total potential energy:

$$\Pi = \Pi_{\text{beam}} + \Pi_{\text{spring}} + \Pi_{\text{load}}, \tag{12.81}$$

where

$$\Pi_{\text{spring}} = \frac{1}{2}k\left[v\left(\frac{L}{2}\right)\right]^2. \tag{12.82}$$

If we use the same approximation space as in Example 12.5, then

$$\Pi_{\text{spring}} = \frac{1}{2}k\left[C\left(\frac{1}{2}\right)^2 + D\left(\frac{1}{2}\right)^3\right]^2. \tag{12.83}$$

Taking partial derivatives with respect to $C$ and $D$ of this added term yields a new contribution to the equilibrium equations in terms of the spring constant $k$:

$$\left(\frac{kL^3}{64EI}\begin{bmatrix} 4 & 2 \\ 2 & 1 \end{bmatrix} + \begin{bmatrix} 4 & 6 \\ 6 & 12 \end{bmatrix} - \frac{PL^2}{EI}\begin{bmatrix} 4/3 & 3/2 \\ 3/2 & 9/5 \end{bmatrix}\right)\begin{bmatrix} C \\ D \end{bmatrix} = \begin{bmatrix} 0 \\ 0 \end{bmatrix}. \tag{12.84}$$

Taking the determinant of the (entire) coefficient matrix and setting it equal to zero yields a second-order polynomial for the eigenvalues – the smallest of which is the critical load. The corresponding eigenvector then defines the general buckled shape of the system.

**Remarks:**

(1) Unlike the prior two examples, an exact solution of the governing differential equation is rather involved. The computation of an approximate solution is much more feasible.

### 12.5.1 Buckling with distributed loads

End-loads are not the only types of load that can appear in buckling problems. For example, one can also have distributed axial loads, $b(x)$, along the length of the beam-column. In this case the total potential energy for the beam-column will be given as

$$\Pi = \int_0^L \frac{1}{2}EI(v'')^2\,dx - \int_0^L \left[b(x)\left\{\int_0^x \frac{1}{2}(v'(\bar{x}))^2\,d\bar{x}\right\}dx\right], \tag{12.85}$$

where the first integral is the elastic potential energy for the beam-column and the second term represents the potential of the load. The expression for the potential of the load appears rather complex, but can be justified if examined closely. The product $b(x)\,dx$ is the load acting on a small segment of material near the point $x$. When the beam-column buckles, this point moves downwards an amount given by the expression in the curly braces; this expression is the same one as we had before except that it gives the motion at an arbitrary point $x$ as opposed to at the end of the beam-column. Thus the term in the square brackets represents the potential energy of the distributed load acting on a segment of length $dx$ at $x$ of the beam-column. To account for the entire distributed load we need to integrate this expression from 0 to $L$. With a simple end-load $P$, the contribution to the potential energy due to the load is

$$-P\int_0^L \frac{1}{2}(v')^2\,dx. \tag{12.86}$$

What has been done is a replacement of $L$ with $x$ and $P$ with $b(x)dx$ and the summation of this contribution from 0 to $L$.

---

**Example 12.7**

*Buckling due to self-weight.* Consider a tall narrow tree loaded only by gravitationally forces. For a given material density and effective cross-sectional area, determine how tall the tree can be before it buckles under its own weight. For simplicity, assume the cross-sectional area, $A$, the Young's modulus, $E$, and the second moment of the area, $I$, to all be constants. Use an approximation method.

*Solution*

In this situation the load is a constant distributed load $b(x) = A\rho g = \gamma$, where $\rho$ is the mass density of the column and $g$ is the gravitational constant. Thus $\gamma$ is the weight per unit length of the tree. To solve let us determine the buckling condition for the system in terms of $\gamma$. We will then invert this relation at the end to find a restriction of the height of the tree in terms of $\gamma$.

The potential energy when the distributed load is a constant can be written as:

$$\Pi = \int_0^L \frac{1}{2}EI(v'')^2\,dx - \gamma\int_0^L \left\{\int_0^x \frac{1}{2}(v'(\bar{x}))^2\,d\bar{x}\right\}dx. \tag{12.87}$$

To approximately solve this problem, consider a one-parameter approximation for the solution in the form: $v(x) \approx Cf(x)$, where $C$ is the undetermined coefficient. For this assumed form,

$$\begin{aligned}\Pi(v) \Rightarrow \Pi(C) = &\int_0^L \frac{1}{2}EI(Cf'')^2\,dx \\ &- \gamma\int_0^L \left\{\int_0^x \frac{1}{2}(Cf'(\bar{x}))^2\,d\bar{x}\right\}dx.\end{aligned} \tag{12.88}$$

For stationarity,

$$\frac{d\Pi(C)}{dC} = C\left[\int_0^L EI(f'')^2\,dx - \gamma\int_0^L \left\{\int_0^x (f'(\bar{x}))^2\,d\bar{x}\right\}\,dx\right] = 0. \tag{12.89}$$

For a non-trivial solution, $C \neq 0$, one must have

$$\gamma = \frac{\int_0^L EI(f'')^2\,dx}{\int_0^L \left\{\int_0^x (f'(\bar{x}))^2\,d\bar{x}\right\}\,dx}. \tag{12.90}$$

Selecting $f(x) = x^2$, for our functional form, yields

$$\gamma = \frac{\int_0^L EI(2)^2\,dx}{\int_0^L \left\{\int_0^x (2\bar{x})^2\,d\bar{x}\right\}\,dx} \tag{12.91}$$

$$= \frac{4EIL}{\int_0^L \left\{\frac{4}{3}x^3\right\}\,dx} \tag{12.92}$$

$$= \frac{4EIL}{\frac{1}{3}L^4} \tag{12.93}$$

$$= \frac{12EI}{L^3} \tag{12.94}$$

For $\gamma < 12EI/L^3$ there will be no buckling; thus the height restriction is

$$L^3 < \frac{12EI}{\gamma} = \frac{12EI}{\rho g A} = \frac{12Er^2}{\rho g}, \tag{12.95}$$

where for the last equality we have introduced the radius of gyration $r^2 = I/A$ of the cross-section.

**Remarks:**

(1) From the result, we see that denser trees are necessarily shorter. Likewise, a tree can grow taller if it increases its radius of gyration.

(2) We can assess the accuracy of our computation, since there is a known reference solution. Up to four digits this is[4]

$$\gamma_{\text{exact}} = 7.837\frac{EI}{L^3}. \tag{12.96}$$

Comparing, we find that the relative error for our simple approximation is $|(12 - 7.837)/(7.837)| \approx 53\%$, which is not too good. Adding additional polynomial terms would greatly improve the accuracy.

(3) As an alternative choice one can pick $f(x) = 1 - \cos\left(\frac{\pi}{2}\frac{x}{L}\right)$. This yields

[4] This solution can be computed from the governing ordinary differential equation. However, it requires knowledge of Bessel functions of fractional order.

$$\gamma = 8.2979 \frac{EI}{L^3}, \tag{12.97}$$

which results in a relative error of approximately 6%.

(4) Every tree you see must know this relation!

### 12.5.2 Deflection behavior for beam-columns with combined axial and transverse loads

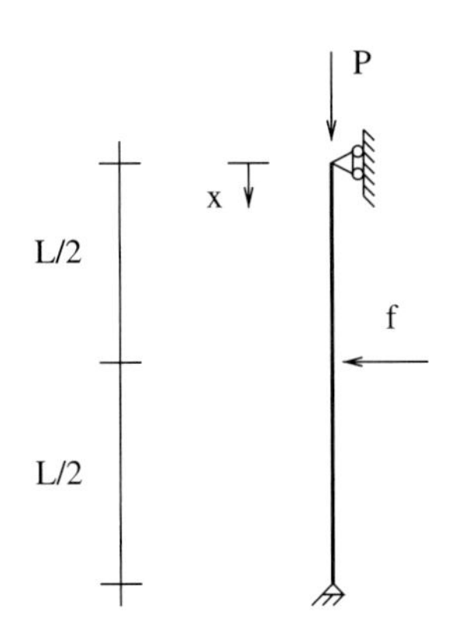


**Fig. 12.22** Beam-column with axial and transverse loads.

Energy methods of the type we have just examined are also useful for other *buckling-like* problems. Consider, for example, a simply supported beam with a load $f$ applied at mid-span that is also subject to an axial compression $P$; see Fig. 12.22. The total potential energy for this mechanical system is

$$\Pi = \int_0^L \frac{1}{2} EI(v'')^2\, dx - P \int_0^L \frac{1}{2}(v')^2\, dx - fv\left(\frac{L}{2}\right). \tag{12.98}$$

If $f = 0$, we are looking at a classical buckling problem; viz., the beam remains straight until a critical load is reached, after which, the beam bends suddenly. The critical load for the configuration shown is $P_{\text{cr}} = \pi^2 EI/L^2$. In the case where $f \neq 0$, the beam no longer has a buckling response where there are trivial *and* non-trivial solutions. Instead, the system responds with only a single solution that has *buckling-like* behavior.

The stationary conditions for the potential energy of the system will still give us the equilibrium equations for $v(x)$. To be concrete, let us focus on approximate solutions of the form $v(x) \approx C \sin(\pi x/L)$. If we introduce this expression into the potential energy eqn (12.98), then we find that

$$\Pi_{\text{total}} = \frac{1}{4} EI \left(\frac{\pi}{L}\right)^4 C^2 L - P \frac{1}{4}\left(\frac{\pi}{L}\right)^2 C^2 L - fC. \tag{12.99}$$

The stationary condition yields

$$\frac{d\Pi}{dC} = \frac{1}{2} EI \left(\frac{\pi}{L}\right)^4 CL - P\frac{1}{2}\left(\frac{\pi}{L}\right)^2 CL - f \tag{12.100}$$

$$= C\left\{\frac{1}{2} EI\left(\frac{\pi}{L}\right)^4 L - P\frac{1}{2}\left(\frac{\pi}{L}\right)^2 L\right\} - f = 0, \tag{12.101}$$

and thus we find that

$$C = \frac{f}{\dfrac{EI\pi^4}{2L^3} - P\dfrac{\pi^2}{2L}} \tag{12.102}$$

$$= \frac{f\ 2L/\pi^2}{\dfrac{EI\pi^2}{L^2} - P} \tag{12.103}$$

$$= \frac{2L}{\pi^2} \frac{f}{P_{\text{cr}} - P} \,. \tag{12.104}$$

Putting everything back together, we have an approximate solution of the form:

$$v(x) \approx \frac{2L}{\pi^2} \frac{f}{P_{\text{cr}} - P} \sin\left(\pi \frac{x}{L}\right). \tag{12.105}$$

**Remarks:**

(1) For fixed values of $P$, the transverse deflection is seen to be a linear function of $f$. However, the magnitude of the deflection for a fixed values of $f$ increases hyperbolically as $P$ approaches $P_{\text{cr}}$ of the Euler column.

(2) The response is quite similar to what we saw when we considered eccentrically loaded beam-columns.

(3) This result illustrates a reasonable rule of thumb: The deflection response of a beam subjected to transverse loads is amplified by a factor of $1/(P_{\text{cr}} - P)$ in the presence of compressive axial loads. The value of $P_{\text{cr}}$ is the one appropriate for the same kinematic boundary conditions without any transverse loads.

## Chapter summary

- Equilibrium of conservative systems: $d\Pi/d\theta = 0$
  - Non-trivial solution conditions lead to buckling loads – eigenvalue problems, eigenvectors give buckling modes
- Stability of conservative systems:
  - $d^2\Pi/d\theta^2 > 0$ – stable
  - $d^2\Pi/d\theta^2 = 0$ – neutral
  - $d^2\Pi/d\theta^2 < 0$ – unstable
- Beam-column: $EIv'''' + Pv'' = q$
  - Pin-pin (Euler load): $P_{\text{cr}} = \pi^2 EI/L^2$
  - Pin-clamped $P_{\text{cr}} = (4.493)^2 EI/L^2$
  - Clamped-free $P_{\text{cr}} = \pi^2 EI/4L^2$
  - Clamped-clamped $P_{\text{cr}} = 4\pi^2 EI/L^2$
- Eccentrically loaded structures display buckling-like response as an upper bound.
- Axial loads amplify transverse deflections by $\approx 1/(P_{\text{cr}} - P)$.
- For approximate solutions, choose $\widetilde{\mathcal{S}} \subset \mathcal{S}$ and solve eigenvalue problem that arises from the stationarity conditions of the potential energy.

# Exercises

(12.1) The structure shown is composed of two hinged rigid bars, a displacement spring, and a torsional spring. Find the buckling load and sketch the associated buckling shape. Assume $k_a = 2$ kN/m, $k_b = 1$ kN · m/rad, and $L = 1$ m.

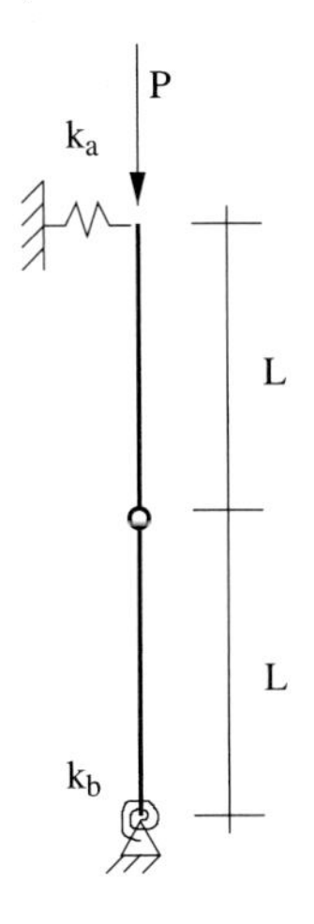


(12.2) For the two-degree-of-freedom system made up of rigid bars and torsional springs, as shown,
(a) find the total potential energy $\Pi$, assuming $\theta_1$ and $\theta_2$ small;
(b) derive the equation governing the critical buckling loads by applying the principle of stationary potential energy.

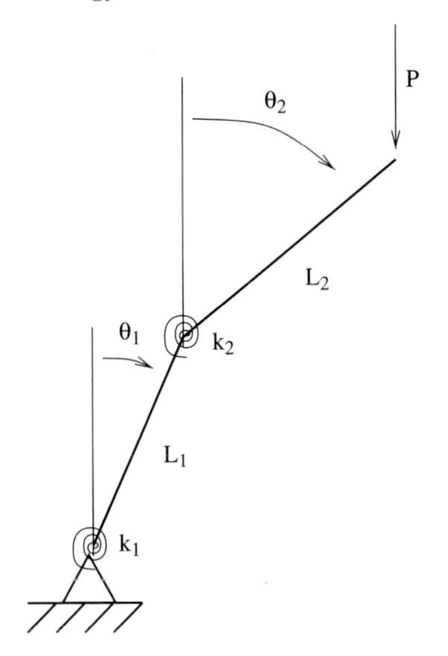


(12.3) The linkage shown is made of three rigid bars, three torsional springs, and one extensional spring. Set up the system of equations that one would have to solve in order to determine the critical load; write your answer in matrix form and indicate in words the remaining steps that would be needed to solve the exercise. *Do not solve the equations.*

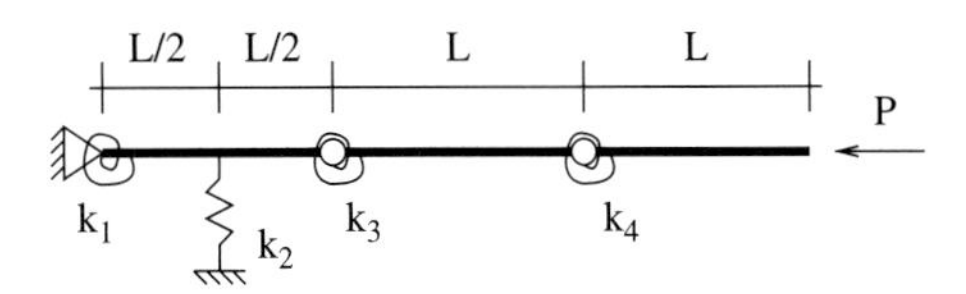


(12.4) Consider the three (rigid) bar system shown where $k = 100$ kN/m and $L = 0.3$ m.

(a) Find the three buckling loads and their associated buckling modes/shapes. *Accurately sketch/plot* the buckling modes.

(b) Which of the three is the critical mode shape?

(c) If the spring constant nearest the left support is quadrupled in value, what is the new critical load and mode shape? *Accurately sketch/plot* the critical mode.

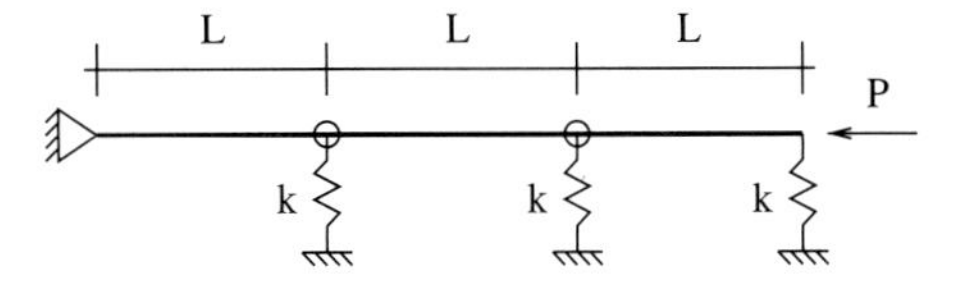


(12.5) As the load $P$ is increased on the structure shown, the rigid inverted-T will displace uniformly downwards. At a certain load $P$ the structure will experience a rotational instability. Determine the load $P$ at which this occurs. Assume that the vertical and horizontal segments have length $L_v$ and $L_h$, respectively. (Hint: The system has 2 degrees of freedom, $\theta$ and $\Delta$.)

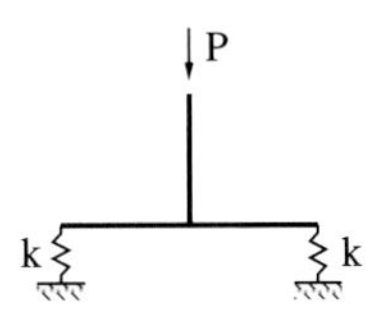


Rigid welded bars

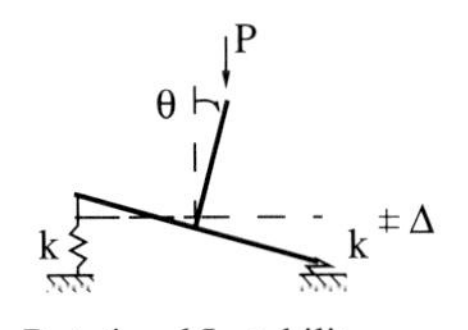


Rotational Instability

(12.6) Similar to Exercise 12.5, find the buckling load for the rigid inverted-T now supported by three springs of equal stiffness.

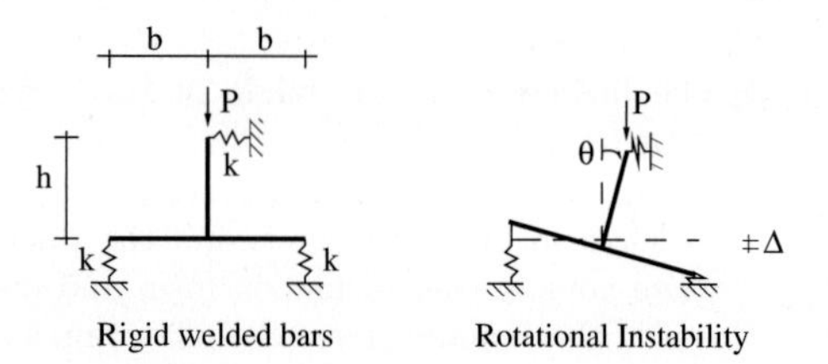


(12.7) Shown is a structure composed of two rigid links of length $L = 2$ m that are joined by a torsional spring with spring constant $c = 2$ kN · m/rad. The top of the structure is supported by a flexible support with spring constant $k = 1$ kN/m. Determine the critical load of the structure and sketch the deflected shape just after collapse.

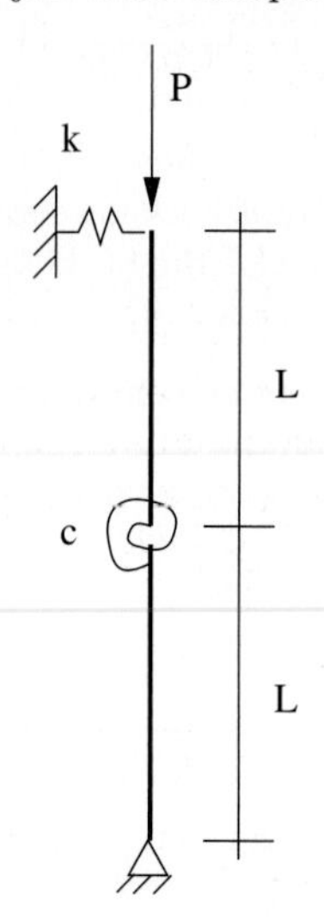


(12.8) The system shown is composed of two rigid bars connected by a (long) slider. This system displays a critical buckling load *in tension.* Find an expression for the critical load in terms of the bar lengths and the torsional spring constant.

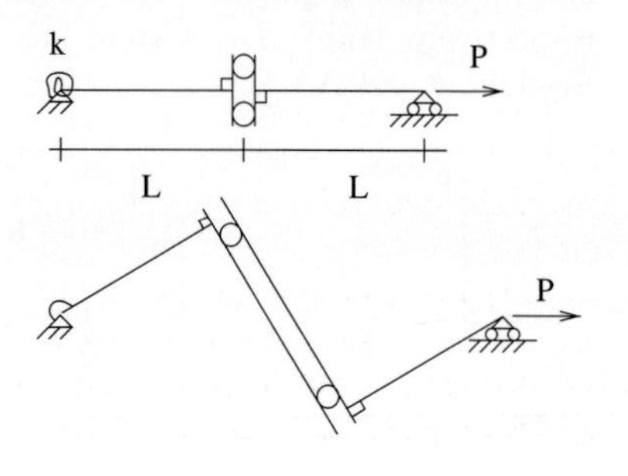


(12.9) Derive the buckling load for a cantilevered column with an axial end-load; i.e. determine the buckling load for a free-clamped column.

(12.10) Derive the buckling load for a column which is free to shorten but has restrained end-deflections and restrained end-rotations; i.e. determine the buckling load for a clamped-clamped column.

(12.11) Consider the linear elastic beam shown. Set up the determinant condition for finding the buckling load. Show that it is satisfied by the Euler solution; i.e. this system has the same buckling load as the pin-pin case.

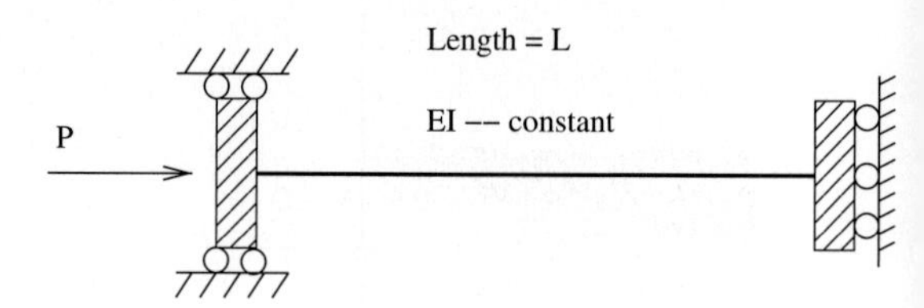


(12.12) A free-standing (cantilever) rectangular post of cross-sectional dimensions $a$ and $b$ ($b > a$) and height $L$ is made of lumber with compressive crushing strength $\sigma_B$ and Young's modulus $E$. Find $L$ such that the post is equally likely to fail by crushing and by buckling.

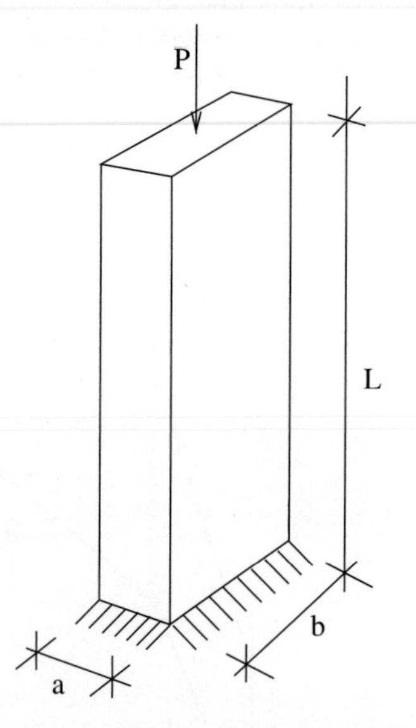


(12.13) Explain in words the essential elements of a buckling analysis. Make sure to discuss how one checks the stability of an equilibrium solution.

(12.14) A toothbrush is an interesting example of a force-limiting system that employs buckling in its basic design. When you brush your teeth, your hand applies a loading that essentially sets up a reaction force on the brush-head from the teeth which is distributed (approximately) in an even manner over the bristles of the brush as an axial load. Idealize the brush-head as a regular array of $N$ bristles spread evenly over an area $w$ by $l$, and determine an expression for the maximum force one can apply to the brush-head. Assume each individual bristle has a length $h$, radius $r$, and modulus $E$.

(12.15) A truss is loaded with a vertical force at the top pin. Determine the buckling load and sketch all the possible ways the system could buckle in the plane under the first buckling load.

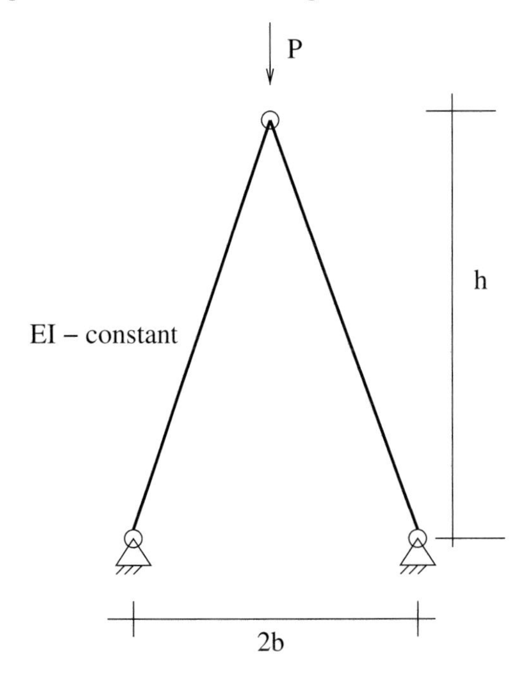


(12.16) Consider the beam-column shown with constant $EI$. Find an expression for the deflection of the beam-column, when $q(x) = q_o$, a constant.

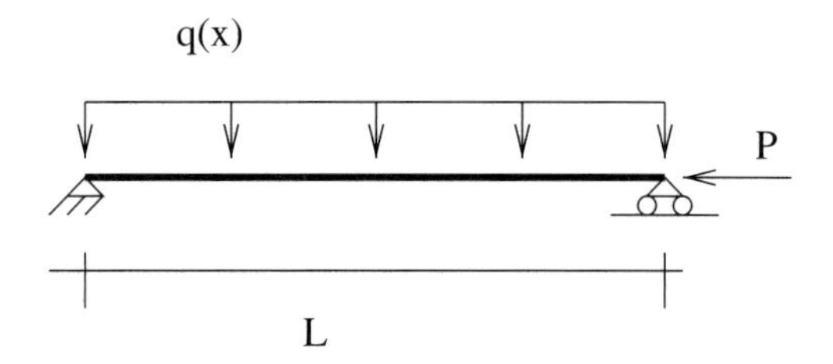


(12.17) Thin-walled pressure vessels made of very flexible materials have a torsional buckling failure mode that is associated with compressive normal stresses. Assuming that the pressure vessel shown below cannot support any compressive normal stresses (in the plane of the vessel walls), find the maximum allowable torque that one can apply to the pressure vessel.

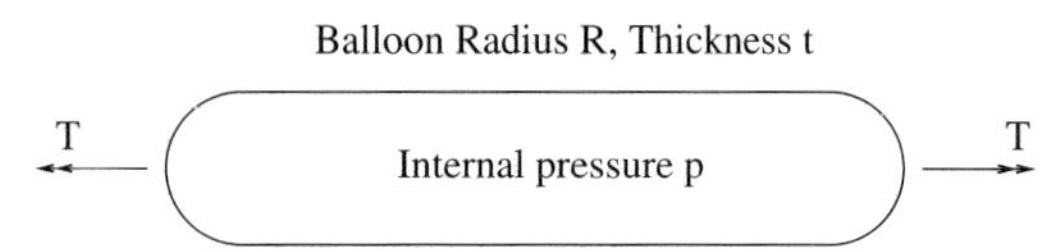


(12.18) The elastic bar shown has a linearly varying coefficient of thermal expansion $\alpha(x) = C + Dx$. What temperature change is needed to cause the bar to buckle. Assume that the constants $C$ and $D$ are both positive; also assume that $h > b$.

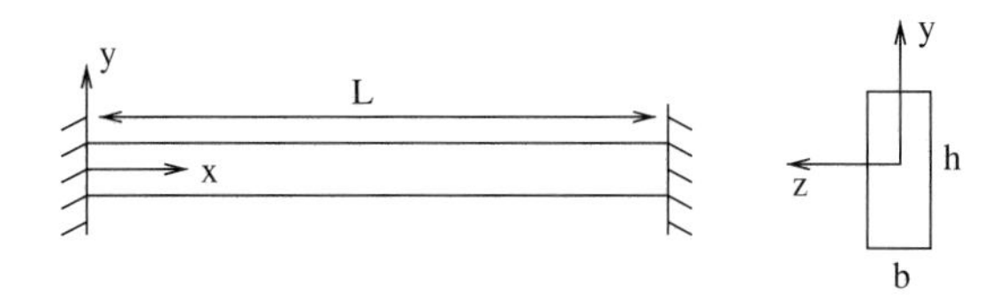


(12.19) Solve Exercise 12.10 using an approximate stationary potential energy method. Use a one-parameter solution space.

(12.20) Solve Exercise 12.11 using an approximate stationary potential energy method. Use a one-parameter solution space.

(12.21) Consider a beam-column with length $L = 1.5$ m and a $1.2 \times 1.2$ cm$^2$ square cross-section. The column is pinned at $x = 0$ and is supported by a pin-roller at $x = L$ – i.e. it is simply supported. Further, it is supported at $x = L/4$ by a linear spring with spring constant $k = 0.5$ N/mm. The column is subjected to an axial compressive force $P$ at the pin-roller support. Estimate the critical load using an approximate potential energy method with a single parameter. Assume $E = 200\ \mathrm{kN/mm^2}$.

(12.22) Consider the system in Exercise 12.21 except that the axial compressive load is now applied at $x = 3L/4$ instead of at $x = L$. Find the critical load using an approximate potential energy method. Provide an answer accurate to at least two digits. (Hint: A small computer program is helpful for this problem.)

(12.23) Consider a beam supported by a Winkler foundation. The beam is 100 ft long with a Young's modulus of $E = 30 \times 10^6$ psi and a cross-sectional area moment of inertia $I = 77.4\ \mathrm{in}^4$. Assume a (continuously distributed) foundation stiffness $k = 100\ \mathrm{lb/in}^2$ and find the axial buckling load (with small deformation assumptions) and buckling mode. To solve this exercise assume an approximation of the form $v(x) = \sum_{i=1}^{n} C_i f_i(x)$ where $f_i = \sin(i\pi x/L)$. Increase $n$ as appropriate and provide accurate plots of the solution.

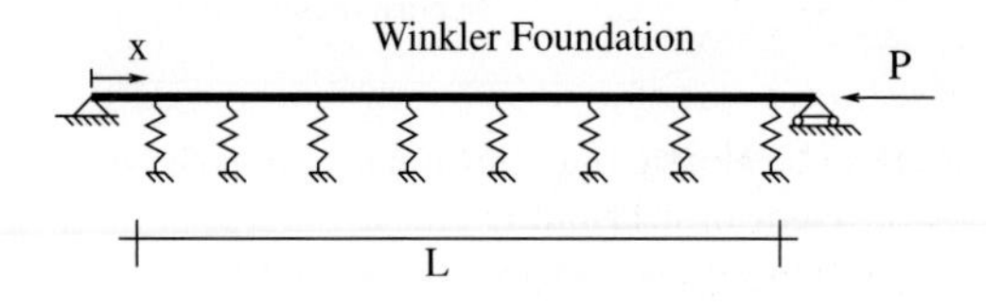


(12.24) Consider a beam supported by a partial distributed spring foundation with spring foundation constant $k$ [F/L$^2$] for $x \in (a, b)$. Using a single term approximation of the form $v(x) = Cx(x - L)$, find an approximation for the system's buckling load.

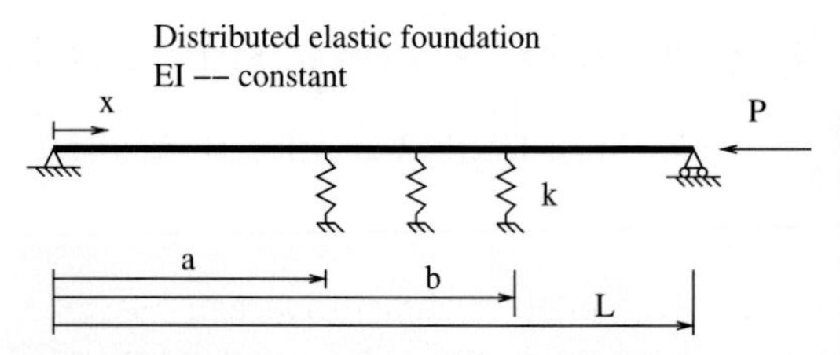


(12.25) A pin-pin beam-column with constant bending stiffness $EI$ has a rigid bar welded to one end, as shown. The rigid bar is subjected to a point load, and the beam-column supports a uniform distributed load plus an applied moment at $x = 0$. Determine an approximate solution for the deflection of the beam. A single-term parameterization is sufficient for this exercise.

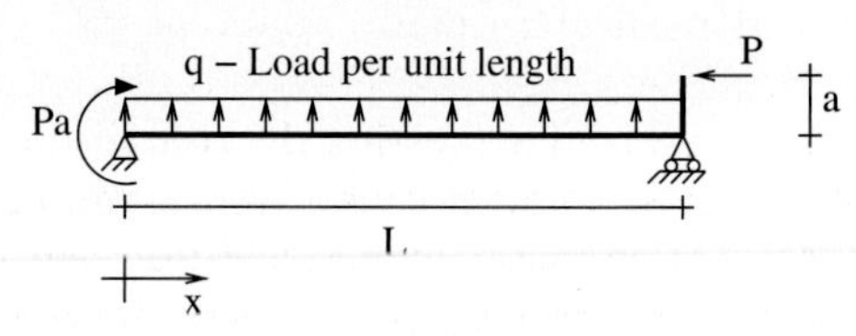

# Virtual Work Methods: Virtual Displacements

# 13



Up to this point we have looked at several different types of energy methods for the formulation and solution of different classes of mechanical problems. We started in Chapter 10 with the concept of virutal work and the method/principle of virtual forces. Then, in Chapter 11 we introduced the principle of stationary potential energy followed by the principle of stationary complementary potential energy. One observation that we made was that the principle of virtual forces was equivalent to the principle of stationary complementary potential energy (for conservative systems). They both were seen to be alternative ways of expressing the fundamental kinematic relations that governed a mechanical problem. Stationary potential energy, on the other hand, was seen to be an alternative way of expressing the equilibrium relations for a conservative mechanical system. In this chapter we will examine one last energy concept: the *principle of virtual displacements*. The principle or method of virtual displacements will be seen to be the virtual work counterpart to the principle of stationary potential energy. It is an alternative way of expressing the equilibirum equations for a mechanical system and, as with the principle of virtual forces, it applies to both conservative and non-conservative systems.

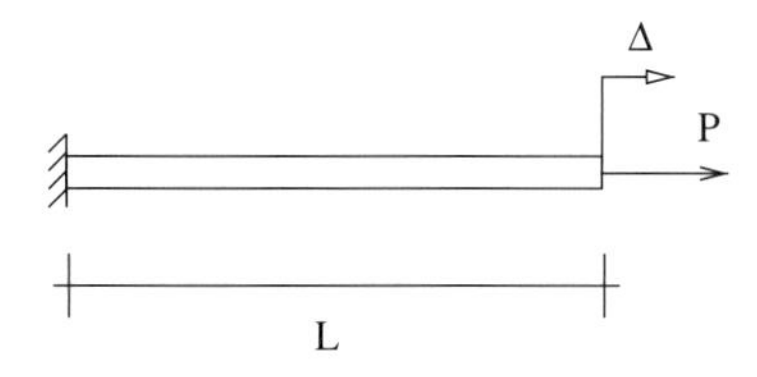


**Fig. 13.1** Bar with an axial end-load.

## 13.1 The virtual work theorem: Virtual displacement version

In Chapter 10 we introduced the virtual work theorem. In its most elementary form it states:

External Virtual Work = Internal Virtual Work

To exploit the theorem we introduced the concept of virtual forces, and this allowed us to generate concrete expressions for internal and external virtual work. Here we wish to look an alternative definition for virtual work that is based upon virtual displacements instead of virtual forces.

To keep the presentation concrete, consider the linear elastic bar shown in Fig. 13.1. The bar is subjected to an end-load $P$, and suppose we would like to determine the deflection $\Delta$ at the end of the bar in terms of the applied load using the method of virtual displacements.

[1] We will denote virtual kinematic quantities with an over-bar. It is also very common to denote virtual quantities with a prepended $\delta$; i.e. one often writes $\delta u$ in place of $\bar{u}$.

To do so we will consider a *virtual displacement* field for our bar $\bar{u}(x)$.[1] This virtual motion need not have anything to do with the actual motion of our bar; its selection is entirely our choice. For our example, let us assume $\bar{u}(x) = \bar{1}x^2$.

The external virtual work (Ext. V.W.) in this context is *defined* as the virtual motion at the point of application of the load times the real load. Thus for our example,

$$\begin{aligned} \text{Ext. V.W.} &= \text{real force} \times \text{virtual displacement} \\ &= P\bar{u}(L) = P\bar{1}L^2. \end{aligned} \tag{13.1}$$

Similarly, the internal virtual work (Int. V.W.) is *defined* as the integral of the real stresses times the virtual strains, where the virtual strains are the strains associated with the virtual displacements ($\bar{\varepsilon} = d\bar{u}/dx$). For our example system we have

$$\text{Int. V.W.} = \int_V \sigma\bar{\varepsilon}\, dV \tag{13.2}$$

$$= \int_L A\sigma\bar{\varepsilon}\, dx \tag{13.3}$$

$$= \int_L R\bar{\varepsilon}\, dx = \int_L R \times \bar{1} \times 2x\, dx. \tag{13.4}$$

If we put this last expression for the internal virtual work together with our expression for the external virtual work in the virtual work theorem, then we find that

$$\int_L R \times \bar{1} \times 2x\, dx = P\bar{1}L^2. \tag{13.5}$$

Since we know that the strains are constant between point loads, we can express the internal force as $R(x) = AE\Delta/L$, which results in

$$AE\frac{\Delta}{L}\bar{1} \times 2\frac{L^2}{2} = P\bar{1}L^2 \tag{13.6}$$

$$\Delta = \frac{PL}{AE}. \tag{13.7}$$

**Remarks:**

(1) Our result is far from astounding as we have seen it many times before, but it does give some confidence in our definitions of internal and external virtual work for the method of virtual displacements.

(2) As before, we will decorate virtual quantities (those not associated with our real system) with an over-bar. The unit value $\bar{1}$ is optional.

(3) Virtual work quantities in the method of virtual displacements will always be virtual kinematic quantities times an energetically conjugate force-like quantity – displacements times forces, strains times stresses, rotations times moments (equivalently torques). For

forces and moments, the virtual work will be assumed positive when the force or moment is in the same direction as the virtual displacement or virtual rotation.

(4) Internal and external virtual work quantities in the method of virtual displacements are additive quantities just like they were in the method of virtual forces. Thus when dealing with systems carrying loads in multiple ways, one can simply add up all the external and internal work quantities and set them equal to each other.

(5) The effective use of the method of virtual displacements can require a bit of creativity in the selection of the virtual displacement field. The essential point is that one needs to select a virtual motion that will help one isolate the part of the system's response that is of interest. In statics for example, where the question is usually associated with determination of support forces or internal forces, it is quite common to select virtual displacements that are composed of rigid motions over most of a system and contain jumps at the points of interest. Note that rigid motions will lead to zero virtual strains and an easy evaluation of the internal virtual work.

(6) In the example, it should be noted that $\bar{u}(0)$ was zero. If it had not been, then there would have been an added contribution from the support force to the external virtual work – viz., $-R(0) \times \bar{u}(0)$.

## 13.2 The virtual work expressions

As with the principle of virtual forces, one needs virtual work expressions for a variety of loading cases, if one is to effectively tackle a wide class of problems. The needed expressions are quite similar to those we used with the principle of virtual forces. We simply need to swap which quantities are virtual and which ones are real.

### 13.2.1 External work expressions

The loading cases of interest are point loads and distributed loads. For point forces we will use expressions of the form

$$\text{Ext. V.W.} = P \times \bar{u}, \tag{13.8}$$

where $P$ is a given real force and $\bar{u}$ is a virtual displacement at the point of application of the force in the direction in which $P$ acts. For point moments, the appropriate expression is

$$\text{Ext. V.W.} = M \times \bar{\theta}, \tag{13.9}$$

where $M$ is a given real moment (or torque) and $\bar{\theta}$ is a virtual rotation at the point of application of the moment about the axis of the moment. In the case of distributed loads we can employ these two definitions on small segments of the system and then add up the result. For an axially loaded bar with distributed load $b(x)$, the total applied load on a small segment of length $dx$ would be $b(x)\,dx$, and the contribution to the external virtual work for this segment would be $b(x)\bar{u}(x)dx$. Summing over the entire bar yields the expression for the total external work as

$$\text{Ext. V.W.} = \int_0^L b(x)\bar{u}(x)\,dx. \tag{13.10}$$

In a similar fashion, for beams one has

$$\text{Ext. V.W.} = \int_0^L q(x)\bar{v}(x)\,dx \tag{13.11}$$

and for torsion rods

$$\text{Ext. V.W.} = \int_0^L t(z)\bar{\phi}(z)\,dz. \tag{13.12}$$

### 13.2.2 Axial rods

If an element in a mechanical system of length $L$ is loaded with axial forces then the internal virtual work in the element for the method of virtual displacements is given by

$$\text{Int. V.W.} = \int_V \sigma_{xx}\bar{\varepsilon}_{xx}\,dV \tag{13.13}$$

$$= \int_V \frac{R}{A}\bar{\varepsilon}\,dV \tag{13.14}$$

$$= \int_0^L R\bar{\varepsilon}\,dx. \tag{13.15}$$

If and only if the system is elastic, one can also write

$$\text{Int. V.W.} = \int_0^L \varepsilon AE\bar{\varepsilon}\,dx. \tag{13.16}$$

---

**Example 13.1**

*Bar with two axial forces.* Let us revisit Example 11.1, in which the bar shown in Fig. 13.2 was to be analyzed to determine the relation between the applied forces ($P_1$ and $P_2$) and the resulting displacements ($\Delta_1$ and $\Delta_2$).

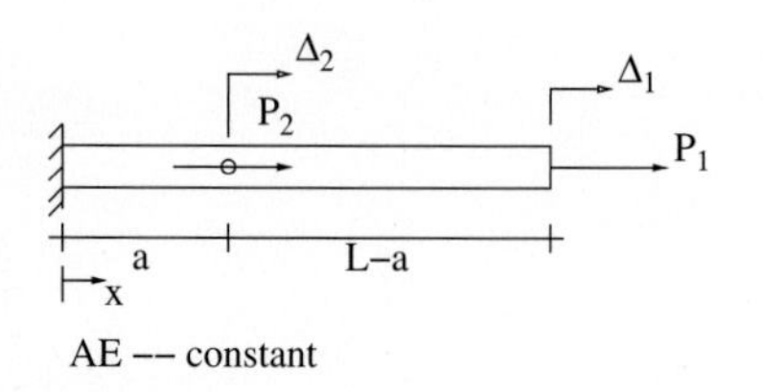


**Fig. 13.2** Bar with two axial forces.

*Solution*

Use the principle of virtual displacements. One first needs expressions for the internal and external virtual work. The external virtual work (accounting also for support forces) is

$$\text{Ext. V.W.} = P_2\bar{u}(a) + P_1\bar{u}(L) - R(0)\bar{u}(0). \tag{13.17}$$

Assuming the bar to be linear elastic, we can employ eqn (13.16) for the internal virtual work. Since we know that the strains are constant in each segment between the loads, we can write

$$\text{Int. V.W.} = \int_0^a \left(\frac{\Delta_2}{a}\right) AE\bar{\varepsilon}\, dx + \int_a^L \left(\frac{\Delta_1 - \Delta_2}{L-a}\right) AE\bar{\varepsilon}\, dx. \quad (13.18)$$

To proceed further, we now need to pick a virtual displacement field in a manner that will allow us to isolate the information in which we are interested. Let us start by selecting the field $\bar{u}(x) = \langle x - a \rangle$. In this case $\bar{\varepsilon} = H(x-a)$, and the principle of virtual displacements tells us that

$$P_2\bar{u}(a) + P_1\bar{u}(L) - R(0)\bar{u}(0) =$$

$$\int_0^a \left(\frac{\Delta_2}{a}\right) AE\bar{\varepsilon}\, dx + \int_a^L \left(\frac{\Delta_1 - \Delta_2}{L-a}\right) AE\bar{\varepsilon}\, dx \quad (13.19)$$

$$P_1 = \left(\frac{\Delta_1 - \Delta_2}{L-a}\right) AE. \quad (13.20)$$

This is one relation between the loads and the displacements. To have a complete description we need a second equation. To generate a second relation we will choose a second virtual displacement field. The selection of a second virtual motion needs to be done in a way that generates an expression involving $P_2$. One possible choice is $\bar{u}(x) = x - \langle x - a \rangle$, which gives $\bar{\varepsilon} = 1 - H(x-a)$. The virtual work theorem then tells us that

$$P_2 + P_1 = \frac{\Delta_2}{a} AE. \quad (13.21)$$

Equations (13.20) and (13.21) form a system of two equations that allow one to compute the loads if the displacements are known or, vice versa, the displacements if the loads are known.

**Remarks:**

(1) Every choice of a virtual displacement field in the principle of virtual displacements generates a single equation that characterizes the response of the system. In this example we needed two relations to have a complete system of equations, and thus we needed to apply the principle of virtual displacements twice.

(2) It should be observed that each equation generated in this example by the principle of virtual displacements is a type of force equilibirum equation. Later, when we give a proof of the principle of virtual displacements for the axially loaded bar, we will see that it is nothing more than a clever reformulation of the governing differential equation of equilibrium in the bar.

(3) Any other two choices of the virtual displacement field would have been perfectly valid for this problem. Alternative choices would likely have led to equations that superficially appeared different

but upon numerical solution would have yielded the same result. Note, however, that choosing the second virtual field as a simple scaling of the first, say $\bar{u}_2 = C\bar{u}_1$, where $C$ is a constant, would not be helpful, as the equation generated with $\bar{u}_2$ would not be independent from the one generated by $\bar{u}_1$.

(4) For the selections made we had the property that $\bar{u}(0) = 0$. By doing this we avoided having to find a third equation to determine the support force $R(0)$.

### 13.2.3 Torsion rods

If an element in a mechanical system of length $L$ is loaded with a torque, then the internal virtual work in the element for the principle of virtual displacements is given by

$$\text{Int. V.W.} = \int_V \sigma_{z\theta}\bar{\gamma}_{z\theta}\, dV \tag{13.22}$$

$$= \int_V \tau r \frac{d\bar{\phi}}{dz}\, dV \tag{13.23}$$

$$= \int_0^L \frac{d\bar{\phi}}{dz}\left[\int_A \tau r\, dA\right] dz \tag{13.24}$$

$$= \int_0^L T\frac{d\bar{\phi}}{dz}\, dz. \tag{13.25}$$

If and only if the system is elastic, one can also write

$$\text{Int. V.W.} = \int_0^L \frac{d\phi}{dz} GJ \frac{d\bar{\phi}}{dz}\, dz. \tag{13.26}$$

#### Example 13.2

*Statically indeterminate rod with a point torque.* Let us revisit Example 11.2 and determine the relation between the applied torque, $T_1$, and the resulting rotation at the point of application, $\theta_1$; see Fig. 13.3.

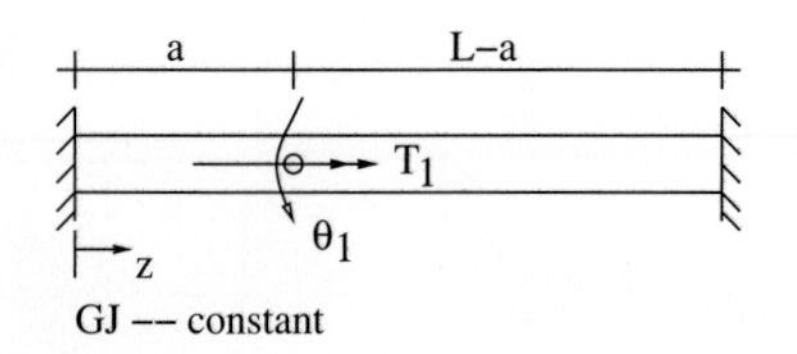


**Fig. 13.3** Rod with a single-point torque.

*Solution*

The external virtual work for this system (including the support torques) is

$$\text{Ext. V.W.} = T_1\bar{\phi}(a) + T(L)\bar{\phi}(L) - T(0)\bar{\phi}(0). \tag{13.27}$$

We note that between the applied load and the supports the rotation field is linear, and thus the twist rates are constant in each segment. This holds, since there are no distributed torques on the rod. This allows us to write the internal virtual work as

$$\text{Int. V.W.} = \int_0^a \left(\frac{\theta_1}{a}\right) GJ\frac{d\bar{\phi}}{dz}\,dz + \int_a^L \left(\frac{-\theta_1}{L-a}\right) GJ\frac{d\bar{\phi}}{dz}\,dz. \quad (13.28)$$

Let us now select as our virtual rotation field

$$\bar{\phi}(z) = z - \langle z-a\rangle\frac{L}{L-a}. \quad (13.29)$$

This function is zero at the two supports and otherwise as simple as possible – i.e. linear. If we plug this into the expressions for internal and external virtual work and set them equal to each other, we find

$$T_1 = GJ\frac{\theta_1}{a} + GJ\frac{\theta_1}{L-a}. \quad (13.30)$$

**Remarks:**

(1) Note that the static indeterminacy did not create any difficulties for the principle of virtual displacements.

(2) Our choice of a virtual rotation field that is zero at the supports conveniently eliminated the need to determine the support torques for the bar.

(3) The choice of a piecewise linear function was a matter of preference which led to simple integrands. Other choices would have given the same final result. The principle of virtual displacements holds for any choice of the virtual motion.

(4) Note that the final expression derived from the principle in this setting is a type of torque equilibrium equation for the bar.

### 13.2.4 Bending of beams

If an element in a mechanical system of length $L$ is loaded in bending, then the internal virtual work (ignoring possible shear effects) in the element is given for the case of virtual displacements as

$$\text{Int. V.W.} = \int_V \sigma_{xx}\bar{\varepsilon}_{xx}\,dV \quad (13.31)$$

$$= \int_V \sigma(-y)\bar{\kappa}\,dV \quad (13.32)$$

$$= \int_0^L \bar{\kappa}\left[\int_A -y\bar{\sigma}\,dA\right]dx \quad (13.33)$$

$$= \int_0^L M\bar{\kappa}\,dx. \quad (13.34)$$

If and only if the system is elastic, one can also write

$$\text{Int. V.W.} = \int_0^L \kappa EI\bar{\kappa}\,dx. \quad (13.35)$$

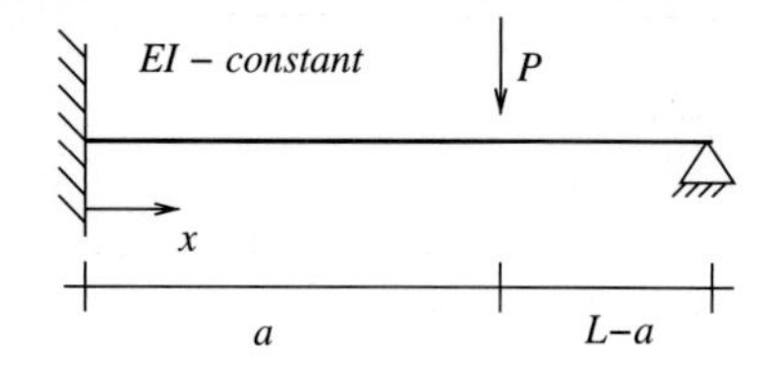


**Fig. 13.4** Indeterminate beam with point load.

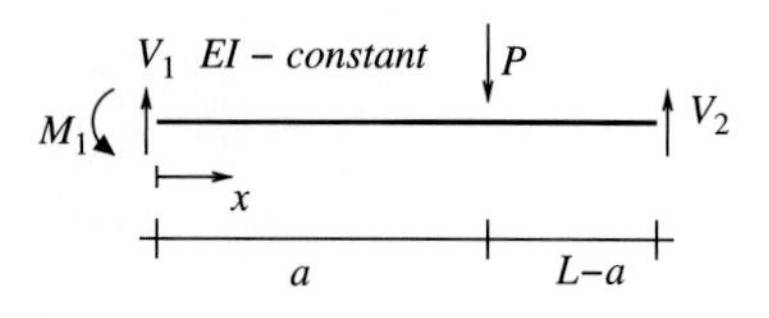


**Fig. 13.5** Indeterminate beam with point load: free-body diagram.

### Example 13.3

*External equilibrium of a statically indeterminate beam.* Consider the beam shown in Fig. 13.4 and determine a set of equations that relate the support forces to the applied load.

*Solution*

Using the free-body diagram shown in Fig. 13.5, we can write the virtual work theorem as:

$$V_1\bar{v}(0) + M_1\bar{\theta}(0) + V_2\bar{v}(L) - P\bar{v}(a) = \int_0^L M\bar{\kappa}\,dx. \tag{13.36}$$

To find a relation between the forces we can choose $\bar{v}_1 = 1$. This implies that $\bar{\theta}_1 = \bar{\kappa}_1 = 0$, from which it follows that

$$V_1 + V_2 - P = 0. \tag{13.37}$$

To find a relation involving the support moment we need a virtual deflection that has non-zero rotation at $x = 0$. A suitable choice is $\bar{v}_2 = x$. This implies $\bar{\theta}_2 = 1$ and $\bar{\kappa}_2 = 0$, from which it follows that

$$M_1 + V_2L - Pa = 0. \tag{13.38}$$

**Remarks:**

(1) Since we were interested in relations having only to do with the external loads, we chose virtual deflections that generated zero virtual curvature and hence no internal virtual work.

(2) The two relations generated are recognized to be the equilibrium of forces in the vertical direction and equilibrium of moments about the left support.

## 13.3 Principle of virtual displacements: Proof

Up to this point we have been applying the principle of virtual displacements to simple problems and observing that the results are consistent with results from earlier chapters. The observation has also been made that the principle leads to equilibrium-like equations. In this section we will give some proof-of-concept proofs to show that the principle is nothing more that a rewriting of the governing differential equation of equilibrium for a given problem. This is similar to the fact that the principle of virtual forces is just a rewriting of the governing kinematic expression for a given problem.

### 13.3.1 Axial bar: Proof

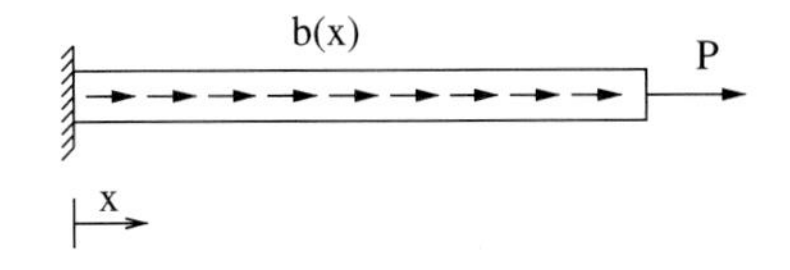


**Fig. 13.6** Bar with axial distributed load and end-load.

Let us consider the system shown in Fig. 13.6 and try to derive the principle of virtual displacements from the govering equations for the bar which consist of the equilibrium equation $dR/dx + b = 0$, the kinematic equation $\varepsilon = du/dx$, and the constitive relation as appropriate for the material. Since we have noted already that the results of applying the principle of virtual displacements leads to equilibrium-like equations, let us start with the equilibrium equation:

$$\frac{dR}{dx} + b = 0. \tag{13.39}$$

First multiply eqn (13.39) by an arbitrary function $\bar{u}(x)$, which we will choose to call a virtual displacement. This results in

$$\frac{dR}{dx}\bar{u} + b\bar{u} = 0. \tag{13.40}$$

We can now integrate eqn (13.40) over the domain to give

$$\int_0^L \frac{dR}{dx}\bar{u}\,dx + \int_0^L b\bar{u}\,dx = 0. \tag{13.41}$$

Using the product rule of differentiation, the first integral can be re-written as

$$\int_0^L \frac{dR}{dx}\bar{u}\,dx = \int_0^L \frac{d}{dx}(R\bar{u}) - R\frac{d\bar{u}}{dx}\,dx \tag{13.42}$$

$$= R\bar{u}\Big|_0^L - \int_0^L R\bar{\varepsilon}\,dx. \tag{13.43}$$

In the last step we introduced the definition $\bar{\varepsilon} = d\bar{u}/dx$. Using this result in eqn (13.41) and the fact that $R(L) = P$ gives

$$P\bar{u}(L) - R(0)\bar{u}(0) + \int_0^L b\bar{u}\,dx = \int_0^L R\bar{\varepsilon}\,dx, \tag{13.44}$$

which in nothing more than external virtual work equals internal virtual work – i.e. the virtual work theorem for the case of virtual displacements.

**Remarks:**

(1) The proof given is, of course, specific to the specific case examined; however, the result is in fact quite general.

(2) From the proof, one should observe that the final result only depended upon the equilibirum equation for the real system. Nowhere did we employ the kinematic equation for the real bar, nor did we employ the constitutive relation for the real bar. Thus the final result is nothing more than an alternative statement of equilibrium for the bar.

(3) The steps used in the derivation were strictly algebraic and calculus-based, and were independent of any particular choice of virtual displacement. For this reason the steps can be reversed,

and one sees that the principle of virtual displacements holds if and only if the equilibrium equation for the bar is satisfied.

### 13.3.2 Beam bending: Proof

As a second proof-of-concept consider a beam with distributed load $q(x)$. Proceeding as with the axial bar, let us start with the two governing differential equations for force and moment equilibrium

$$\frac{dV}{dx} + q = 0 \tag{13.45}$$

$$\frac{dM}{dx} + V = 0 \tag{13.46}$$

and combine them into a single equation

$$\frac{d^2M}{dx^2} = q. \tag{13.47}$$

Now multiply both sides of this equation by an arbitrary function $\bar{v}(x)$ (the virtual deflection) and integrate over the domain. Employing the definitions of $\bar{\theta} = d\bar{v}/dx$ and $\bar{\kappa} = d\bar{\theta}/dx$ results in

$$\int_0^L q\bar{v}\,dx = \int_0^L \frac{d^2M}{dx^2}\bar{v}\,dx \tag{13.48}$$

$$= \int_0^L \frac{d}{dx}\left(\frac{dM}{dx}\bar{v}\right) - \frac{dM}{dx}\bar{\theta}\,dx \tag{13.49}$$

$$= \frac{dM}{dx}\bar{v}\Big|_0^L - \int_0^L \frac{dM}{dx}\bar{\theta}\,dx \tag{13.50}$$

$$= \frac{dM}{dx}\bar{v}\Big|_0^L - \left(\int_0^L \frac{d}{dx}(M\bar{\theta}) - M\bar{\kappa}\,dx\right) \tag{13.51}$$

$$= \frac{dM}{dx}\bar{v}\Big|_0^L - M\bar{\theta}\Big|_0^L + \int_0^L M\bar{\kappa}\,dx. \tag{13.52}$$

Moving the leading terms on the right-hand side to the left gives the final result:

$$\begin{aligned} &M(L)\bar{\theta}(L) - M(0)\bar{\theta}(0) \\ &+ V(L)\bar{v}(L) - V(0)\bar{v}(0) \\ &+ \int_0^L q\bar{v}\,dx = \int_0^L M\bar{\kappa}\,dx; \end{aligned} \tag{13.53}$$

viz., external virtual work equals internal virtual work.

**Remarks:**

(1) In the derviation, the only information that we used about the real system were the equations of equilibrium. Thus the principle

of virtual work is again seen to only be a special way of expressing the equilibirum equations for a mechanical system.

(2) The virtual deflection of the beam was arbitrary. The virtual work theorem holds for all choices of the virtual deflection field, and all the mathematical steps are reversible.

## 13.4 Approximate methods

In the examples which we have seen so far using the method of virtual displacements, the complete behavior of the system of interest could easily be characterized by a finite number of unknowns. In Example 13.1 we had $\Delta_1$ and $\Delta_2$, in Example 13.2 it was $\theta_1$, and in Example 13.4 it was $M_1$, $V_1$, and $V_2$. In a wide class of problems this will be true; however, in a larger class of problems this will not be case. The situation is rather similar to what was seen in Chapter 11 with respect to solving problems via the principle of stationary potential energy, and in a similar fashion we can also utilize the principle of virtual displacements to compute approximate solutions.

We will introduce the basic concepts via the problem illustrated in Fig. 13.7; this is the same problem we started with in Chapter 11 when we introduced approximate solutions via stationary potential energy. The question at hand is to determine the deflection of the beam. The principle of virtual displacements for this problem states

$$-M(0)\bar{v}'(0) - V(0)\bar{v}(0) - P\bar{v}(L) = \int_0^L v'' EI\bar{v}''\, dx. \tag{13.54}$$

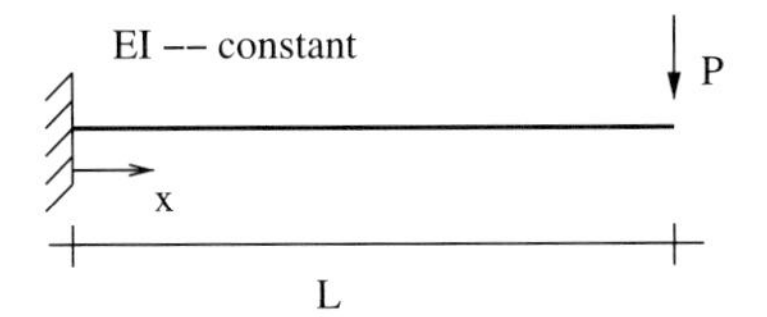


**Fig. 13.7** Cantilever beam with a dead-load.

The unknown in which we are interested is $v(x)$, the beam's deflection, and we would like to use eqn (13.54) to determine it. As before, we note that $v(x)$ is not just any function, since it must satisfy the kinematic boundary conditions. In other words, we should only consider functions in the set

$$\mathcal{S} = \{v(x) \mid v(0) = 0 \text{ and } v'(0) = 0\}. \tag{13.55}$$

As before, the set of functions $\mathcal{S}$ is known as the space of trial solutions – the true equilibrium displacement field lies somewhere within this set of functions. Note that for the actual solution $v(x)$, eqn (13.54) holds true for any virtual displacement $\bar{v}(x)$. The virtual displacements are said to be in the set of functions $\mathcal{V}$ defined as

$$\mathcal{V} = \{\bar{v} \mid \text{(no restrictions)}\}. \tag{13.56}$$

This set is sometimes called the space of virtual displacements or the space of test functions.

Just as with Ritz's method, we will simplify the situation by only considering possible solutions from a subset of $\mathcal{S}$ that can be defined by a finite number of parameters. For example, if we we look for an approximate solution in the set $\widetilde{\mathcal{S}}_N \subset \mathcal{S}$ which has $N$ unknown parameters, then we will need to generate $N$ equations to determine the $N$ parameters. To do so, we will need to consider $N$ different virtual displacement fields. In the Bubnov–Galerkin approximation scheme, these functions are selected from a set $\widetilde{\mathcal{V}}_N \subset \mathcal{V}$, where $\widetilde{\mathcal{V}}_N$ contains functions of the same form as those in $\widetilde{\mathcal{S}}_N$.

To be concrete, consider the subsets

$$\widetilde{\mathcal{S}}_1 = \{v(x) \mid v(x) = Cx^2, \quad C \in \mathbb{R}\} \subset \mathcal{S} \tag{13.57}$$

and

$$\widetilde{\mathcal{V}}_1 = \{\bar{v}(x) \mid \bar{v}(x) = \bar{C}x^2, \quad \bar{C} \in \mathbb{R}\} \subset \mathcal{V}. \tag{13.58}$$

The set $\widetilde{\mathcal{S}}_1$ is composed of a set of parabolas parameterized by the parameter $C$ which is an arbitrary real number and likewise for $\widetilde{\mathcal{V}}_1$.[2] To determine the approximate solution we need to determine $C$. We do this by examining the consequences of requiring the virtual work expression eqn (13.54) to hold true for all $\bar{v} \in \widetilde{\mathcal{V}}_1$. If we plug our approximations into the virtual work expression, we find

[2] $\widetilde{\mathcal{S}}_1$ and $\widetilde{\mathcal{V}}_1$ are one-dimensional sets; one can specify any element in each set by giving a value to a single scalar parameter.

$$-P\bar{C}L^2 = \int_0^L 4C \times EI \times \bar{C}\, dx. \tag{13.59}$$

This can be rearranged as

$$\bar{C}\left[4C \times EIL + PL^2\right] = 0. \tag{13.60}$$

Since we require this to be true for all values of $\bar{C}$, this tells us that the the expression in the square brackets must be zero. In other words, that $C = -PL/4EI$. Thus we find an approximation for the deflection as

$$v(x) \approx -\frac{PLx^2}{4EI}. \tag{13.61}$$

**Remarks:**

(1) The solution we have arrived at is an approximation to the true solution to this problem. The approximation can be improved by adding further terms (and parameters) to our approximation space.

(2) Our final result is in fact identical to our result from Section 11.7. For conservative problems, the Bubnov–Galerkin method will always yield the same result as Ritz's method.

(3) In this example our selection of functions eliminated the (unknown) support reactions from the final equations. This is typically desirable, and will work out automatically as long as the given kinematic boundary conditions are zero.

---

### Example 13.4

*Bar with a constant distributed load.* Consider a bar with one end fixed, the other free, and loaded by a constant distributed load $b(x) = b_o$, as shown in Fig. 13.8. Find an approximate solution for $u(x)$ the bar's displacement.

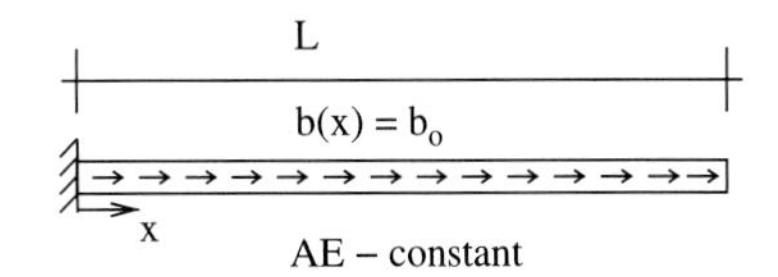


**Fig. 13.8** Linear elastic bar with constant distributed load.

*Solution*
Let us look for an approximate linear solution. The virtual work theorem tells us that

$$-R(0)\bar{u}(0)+\int_0^L b_o\bar{u}\,dx=\int_0^L \varepsilon AE\bar{\varepsilon}\,dx. \tag{13.62}$$

For approximation spaces let us select

$$\widetilde{\mathcal{S}}_1=\{u(x)\mid u(x)=Cx,\quad C\in\mathbb{R}\}\subset\mathcal{S} \tag{13.63}$$

and

$$\widetilde{\mathcal{V}}_1=\{\bar{u}(x)\mid \bar{u}(x)=\bar{C}x,\quad \bar{C}\in\mathbb{R}\}\subset\mathcal{V}. \tag{13.64}$$

With these selections, $\varepsilon=C$ and $\bar{\varepsilon}=\bar{C}$, and we find that

$$\bar{C}\left[b_o\frac{L^2}{2}-CAEL\right]=0. \tag{13.65}$$

Requiring this later result to be true for all functions in $\widetilde{\mathcal{V}}_1$ implies

$$C=\frac{b_oL}{2AE} \tag{13.66}$$

and that our approximate defelction is given by $u(x)\approx\frac{b_oL}{2AE}x$.

**Remarks:**

(1) Figure 13.9 compares the exact quadratic solution to the approximate one. The approximation is seen to be quite reasonable for a single parameter solution.
(2) If we had used $\widetilde{\mathcal{S}}_1$ in the principle of stationary energy for this problem, we would have come to the same result, as the problem is conservative.

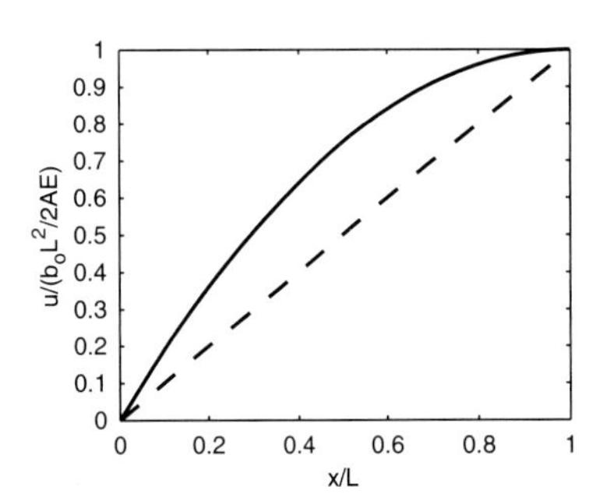


**Fig. 13.9** Exact (solid line) versus approximate (dashed line) solution for Example 13.4.

---

In the case where a single term provides insufficient accuracy for a given problem, one can add additional terms to the approximation spaces. To be concrete, consider the problem of Example 13.4, but this time with an arbitrary distributed load $b(x)$. For a better approximation consider

$$\widetilde{\mathcal{S}}_N=\left\{u(x)\;\middle|\;u(x)=\sum_{i=1}^{N}C_if_i(x),\quad \{C_i\}_{i=1}^N\in\mathbb{R}\right\}\subset\mathcal{S}, \tag{13.67}$$

where the functions $f_i(x)$ are known and have the property that $f_i(0)=0$. For example, they could be the polynomials $f_i(x)=x^i$. In the Bubnov–Galerkin method we then select the space of virtual displacement to have the same form; i.e.

$$\widetilde{\mathcal{V}}_N = \left\{ \bar{u}(x) \;\middle|\; \bar{u}(x) = \sum_{j=1}^{N} \bar{C}_j f_j(x), \quad \{\bar{C}_j\}_{j=1}^{N} \in \mathbb{R} \right\} \subset \mathcal{V}. \tag{13.68}$$

Inserting into the principle of virtual displacements, eqn (13.62), gives

$$\sum_{j=1}^{N} \bar{C}_j \left[ \int_0^L b(x) f_j(x)\, dx - \sum_{i=1}^{N} \left( \int_0^L f_j'(x) AE f_i'(x)\, dx \right) C_i \right] = 0. \tag{13.69}$$

If we require this expression to be true for any function in $\widetilde{\mathcal{V}}_N$, i.e. for arbitrary parameters $\bar{C}_j$, then the term in the square brackets must be zero:

$$\sum_{i=1}^{N} \left( \int_0^L f_j'(x) AE f_i'(x)\, dx \right) C_i = \int_0^L b(x) f_j(x)\, dx. \tag{13.70}$$

If we define the $N \times N$ matrix $\boldsymbol{K}$ to have components

$$K_{ji} = \int_0^L f_j'(x) AE f_i'(x)\, dx \tag{13.71}$$

and the vector $\boldsymbol{F}$ of length $N$ to have components

$$F_j = \int_0^L b(x) f_j(x)\, dx, \tag{13.72}$$

then we see that we have a system of $N$ linear equations in the $N$ unknowns $C_i$; i.e. the problem reduces to finding the solution to the following system of matrix equations:

$$\boldsymbol{KC} = \boldsymbol{F}, \tag{13.73}$$

where $\boldsymbol{C}$ is a vector with components $C_i$.

**Remarks:**

(1) The entire discussion of Section 11.7 about approximation properties and their improvement also applies to the principle of virtual displacements.

(2) The use of an $N$ function approximation in the setting given here is the basis of the well-known finite element method. The only difference is the specific choice of approximation functions. The finite element method is essentially a careful and precise specification of the selection/construction of the approximating functions.

---

**Example 13.5**

*Two-term approximation with non-zero kinematic boundary conditions.* Consider the pin-clamped beam shown in Fig. 13.10 and determine an approximate expression for the beam's deflection using the principle of virtual displacements.

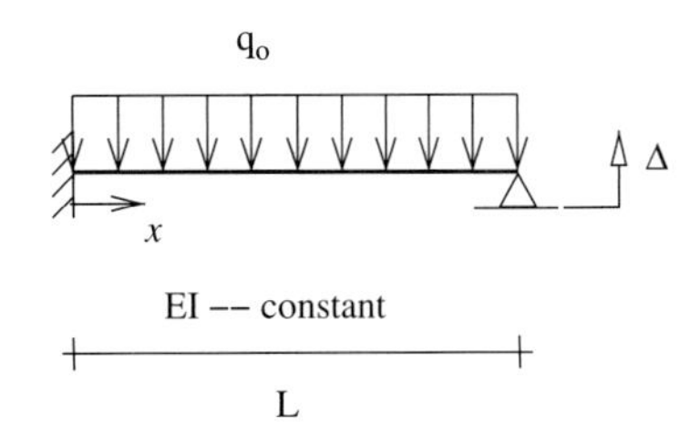


**Fig. 13.10** Pin-clamped beam with constant distributed load and imposed non-zero kinematic boundary condition.

*Solution*

In this problem their is a non-zero kinematic boundary condition. To appropriately handle this we will need to carefully set up our space of solutions and space of virtual deflections. For the space of trial solutions we will select a two-parameter space

$$\widetilde{\mathcal{S}}_2 = \left\{ v(x) \;\middle|\; v(x) = \sum_{i=1}^{2} C_i f_i(x) + g(x) \right\} \subset \mathcal{S}, \tag{13.74}$$

where

$$f_1(x) = \left(\frac{x}{L}\right)^2 \frac{x-L}{L}, \tag{13.75}$$

$$f_2(x) - \left(\frac{x}{L}\right)^3 \frac{x-L}{L}, \tag{13.76}$$

$$g(x) = \Delta\left(\frac{x}{L}\right)^2. \tag{13.77}$$

Note that the $f_i(x)$ satisfy the kinematic boundary conditions at the wall and are zero at the end with the imposed displacement. The function $g(x)$ additionally satisfies the non-zero kinematic boundary condition at the right-end of the beam. The space of virtual deflections is constructed to have the same form as the space of trial solutions minus the terms associated with the non-zero kinematic boundary conditions. Thus we chose

$$\widetilde{\mathcal{V}}_2 = \left\{ \bar{v}(x) \;\middle|\; \bar{v}(x) = \sum_{j=1}^{2} \bar{C}_j f_j(x) \right\} \subset \mathcal{V}. \tag{13.78}$$

The principle of virtual displacements for this problem is given by

$$\int_0^L -q_o \bar{v}(x)\, dx = \int_0^L v''(x) EI \bar{v}''(x)\, dx. \tag{13.79}$$

Inserting our approximate forms yields

$$\sum_{j=1}^{2} \bar{C}_j \left[ \int_0^L g'' EI f_j'' + q_o f_j \, dx + \sum_{i=1}^{2} \left( \int_0^L f_j'' EI f_i'' \, dx \right) C_i \right] = 0. \tag{13.80}$$

For this to be zero for all functions in $\widetilde{\mathcal{V}}_2$, the term in the square brackets must be zero. This is a system of linear equations in the form $\boldsymbol{KC} = \boldsymbol{F}$, where the matrix $\boldsymbol{K}$ has components

$$K_{ji} = \int_0^L f_j'' EI f_i'' \, dx \tag{13.81}$$

and the vector $\boldsymbol{F}$ has components

$$F_j = -\int_0^L g''EIf_j'' + q_o f_j \, dx. \qquad (13.82)$$

Using our assumed forms, the governing system of equations is given by

$$\frac{EI}{L^3}\begin{bmatrix} 4 & 4 \\ 4 & \frac{24}{5} \end{bmatrix}\begin{pmatrix} C_1 \\ C_2 \end{pmatrix} = \begin{pmatrix} \frac{q_o L}{12} - \frac{2\Delta EI}{L^3} \\ \frac{q_o L}{20} + \frac{2\Delta EI}{L^3} \end{pmatrix}. \qquad (13.83)$$

Solving this system of linear equations gives:

$$C_1 = \frac{q_o L^4}{16EI} - \frac{\Delta}{2} \qquad (13.84)$$

$$C_2 = -\frac{q_o L^4}{24EI}. \qquad (13.85)$$

**Remarks:**

(1) This problem is modestly simple and can be solved using the differential equations governing beam deflection. In comparing the results, one finds that our approximate solution is exact. The Bubnov–Galerkin method returns the exact solution should it lie in $\widetilde{\mathcal{S}}_N$.

(2) For more complex problems with a larger number of terms one can automate the generation of the components of the matrix $\boldsymbol{K}$ and the vector $\boldsymbol{F}$, as well as their solution.

### Example 13.6

*Non-conservative load.* The beam shown in Fig. 13.11 is subjected to a force that remains perpendicular to the beam even as it rotates. Such a force is known as a follower force, and is a type of non-conservative load. Use the principle of virtual displacements to find an approximate expression for the beam's deflection.

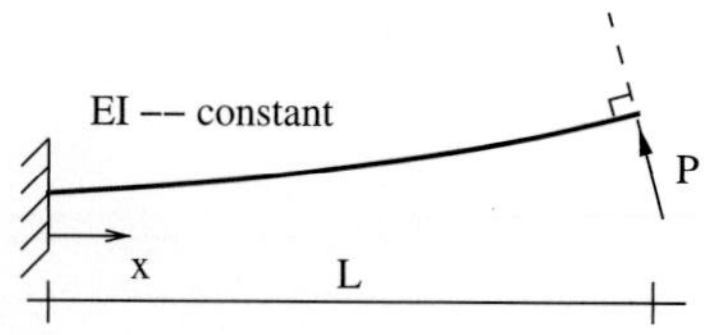

**Fig. 13.11** Cantilever beam subjected to a follower force.

*Solution*

The definitions for the internal virtual work do not change due to the presence of the follower load. The only issue that requires attention is the fact that the vertical component of the force will be $P\cos(v'(L))$. Thus the virtual work theorem will read

$$P\cos(v'(L))\,\bar{v}(L) = \int_0^L v''EI\bar{v}''\,dx. \qquad (13.86)$$

In what follows we will approximate $\cos((v'(L))$ as $1-(v'(L))^2/2$, and for approximate function spaces we will pick the forms $v \approx Cx^2$ and

$\bar{v} = \bar{C}x^2$. Inserting the approximations into the virtual work theorem, and requiring it to hold true for all $\bar{C}$, yields

$$(2PL^4)C^2 + (4EIL)C - PL^2 = 0. \tag{13.87}$$

Solving for $C$ gives

$$C = \frac{-4EIL \pm \sqrt{(4EIL)^2 + 8P^2L^6}}{4PL^4}. \tag{13.88}$$

There are two solutions; however, the solution associated with the negative sign in front of the radical leads to the non-physical result that the beam moves in the opposite direction of the load. Thus we discard that possibility, and find that

$$C = \frac{-4EIL + \sqrt{(4EIL)^2 + 8P^2L^6}}{4PL^4} \tag{13.89}$$

$$= \frac{EI}{PL^3}\left[-1 + \sqrt{1 + \frac{1}{2}\left(\frac{PL^2}{EI}\right)^2}\right]. \tag{13.90}$$

**Remarks:**

(1) This problem illustrates two important features of the principle of virtual work: (a) it applies to non-linear problems, and (b) it applies to non-conservative problems. The load in this case was not conservative, and thus the use of Ritz's method is precluded.

(2) Note that if the load is very small, then the radical can be expanded in a Taylor series to show that $C \approx PL/4EI$, which is the result we had for this level of approxiation when we treated the case of a dead-loaded cantilever. So for small forces, follower loads and dead-loads give the same response as one would intuitively expect.

## Chapter summary

- Virtual work theorem: Ext. V.W. = Int. V.W.
- The principle of virtual displacements is an alternative way to express the equilibrium relations for a mechanical system.
- Ext. V.W. (virtual displacements): $P\bar{u}$ or $M\bar{\theta}$
- Int. V.W. (virtual displacements):
  - Axial forces: $\int_L R\bar{\varepsilon}\,dx$, (elastic) $\int_L \varepsilon AE\bar{\varepsilon}\,dx$

- Torsion: $\int_L T\bar{\phi}'\,dz$, (elastic) $\int_L \phi' GJ\bar{\phi}'\,dz$
- Bending: $\int_L M\bar{\kappa}\,dx$, (elastic) $\int_L \kappa EI\bar{\kappa}\,dx$

- Bubnov–Galerkin scheme: Select $\widetilde{\mathcal{S}} \subset \mathcal{S}$ to have same functional form as $\widetilde{\mathcal{V}} \subset \mathcal{V}$

# Exercises

(13.1) Find an exact solution to Exercise 11.1 from Chapter 11 using the principle of virtual displacements.

(13.2) Find an exact solution to Exercise 11.2 from Chapter 11 using the principle of virtual displacements.

(13.3) Find an exact solution to Exercise 11.3 from Chapter 11 using the principle of virtual displacements.

(13.4) For the truss given in Exercise 10.11 of Chapter 10, find all the support reactions using the principle of virtual displacements.

(13.5) Write the virtual work theorem for the following system. Make sure to define the solution space $\mathcal{S}$; use no restrictions on the test function space $\mathcal{V}$.

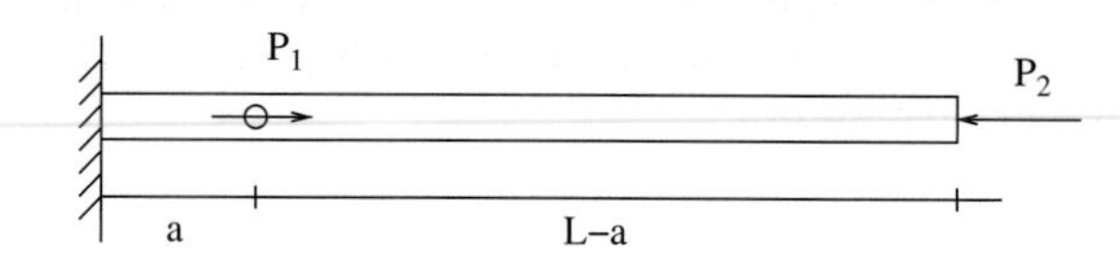


(13.6) Write the virtual work theorem for the following system. Make sure to define the solution space $\mathcal{S}$; use no restrictions on the test function space $\mathcal{V}$.

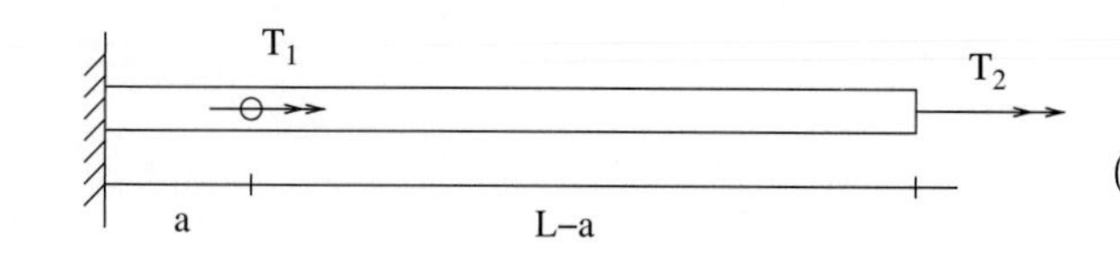


(13.7) Write the virtual work theorem for the following system. Make sure to define the solution space $\mathcal{S}$; use no restrictions on the test function space $\mathcal{V}$.

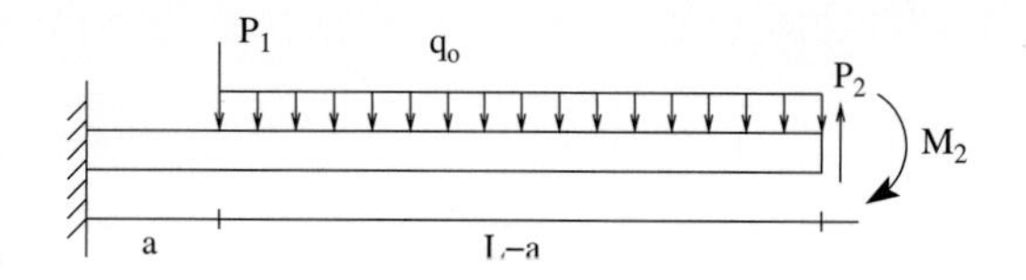


(13.8) Write the virtual work theorem for the following system. Make sure to define the solution space $\mathcal{S}$; use a test function space $\mathcal{V}$ that eliminates support reactions. (Hint: Virtual work expressions are additive like real work.)

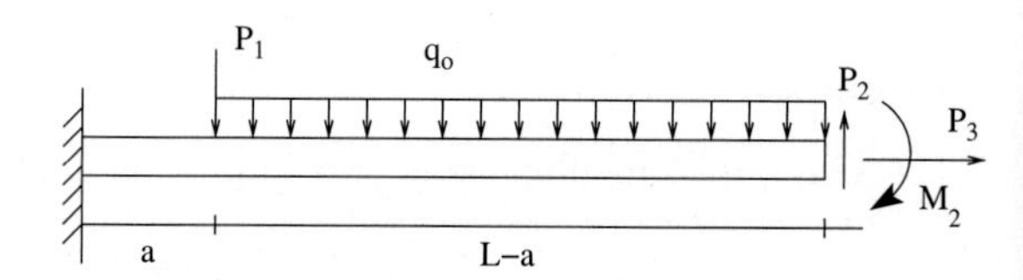


(13.9) For the beam shown write the virtual work theorem. Use a test function space that eliminates support reactions.

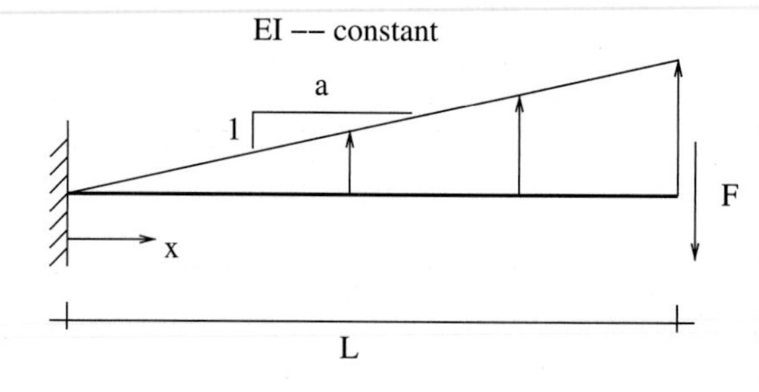


(13.10) Consider an elastic beam of length $L$ and bending stiffness $EI$ subject to a transverse load $q(x)$ and boundary conditions $\theta(0) = \theta_o$, $V(0) = V_o$, $V(L) = V_L$, and $\theta(L) = \theta_L$. Formulate the principle of virtual displacements for this system using a test function space that eliminates support reactions.

(13.11) Starting from the differential equation of equilibrium for a torsion bar, derive the virtual work theorem for the case where the left-end of the bar is subject to the torque $T(0) = -\hat{T}$ and the right-end is fixed.

(13.12) Consider a bar subjected to $N$ applied forces $P_i$ acting at locations $x_i = iL/N$ for $i = 1, 2, \ldots, N$. Formulate the principle of virtual work for this system assuming that $R(0) = 0$.

(13.13) For the beam shown, write an expression for the principle of virtual displacements. Clearly define the space of solutions and the space of test functions. Assume the beam is linear elastic with slender prismatic proportions.

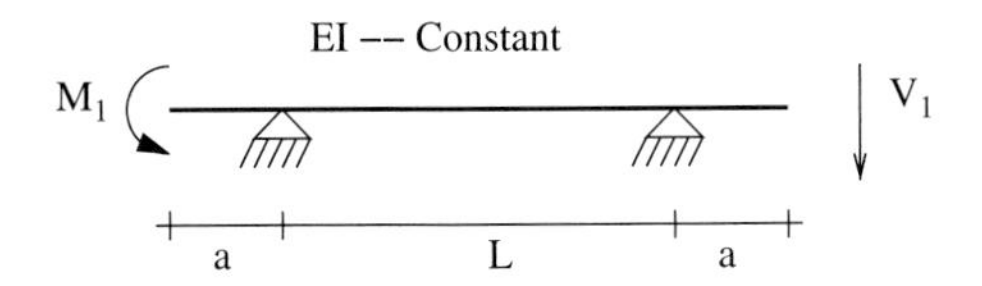


(13.14) For the configuration shown, derive the virtual work equation starting from $d^2M/dx^2 = q$.

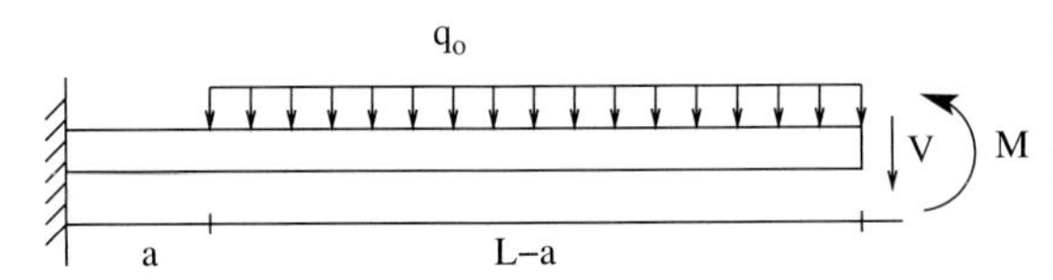


(13.15) Consider the elastic rod shown. Starting from the relevant governing ordinary differential equation, derive the applicable principle of virtual displacements. Be sure to explicitly define your space of trial solutions; use test functions that eliminate support reactions.

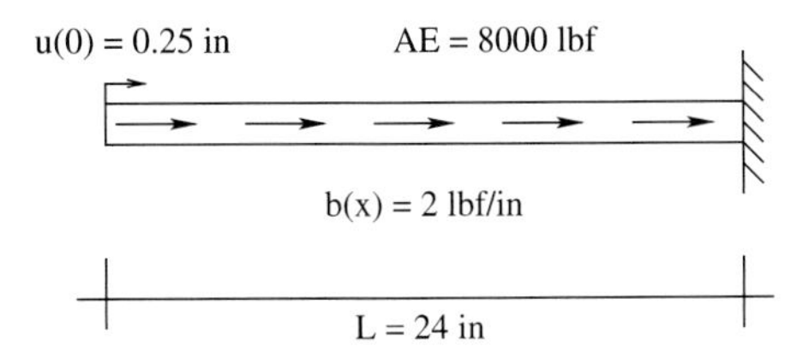


(13.16) A pin-pin beam-column with constant bending stiffness $EI$ has a rigid bar welded to one end as shown. The rigid bar is subjected to a point load and the beam-column supports a uniform distributed load plus an applied moment at $x = 0$. Determine an approximate solution for the deflection of the beam using the principle of virtual displacements. One-dimensional function spaces are sufficient for this exercise.

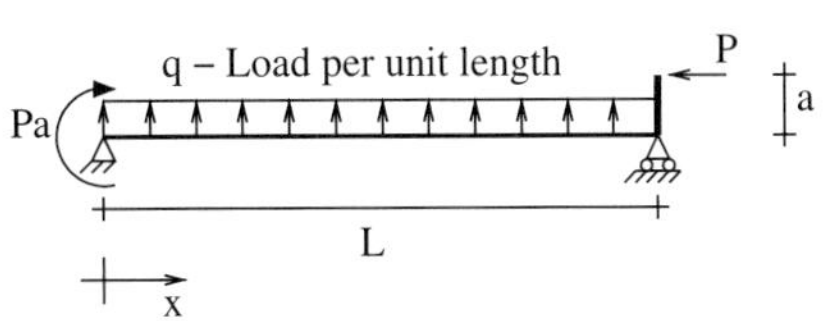


(13.17) Approximately solve Exercise 11.21 from Chapter 11 with the principle of virtual displacements using a single quadratic term.

(13.18) Approximately solve Exercise 11.22 from Chapter 11 using the principle of virtual displacements.

(13.19) Approximately solve Exercise 11.25 from Chapter 11 using the principle of virtual displacements.

(13.20) Approximately solve Exercise 11.32 from Chapter 11 using the principle of virtual displacements.

(13.21) Consider the system shown in Exercise 11.6 from Chapter 11 and approximately solve for the bar's displacement field using a two-parameter approximation with the principle of virtual displacements.

# Appendix A
# Additional Reading

Statics:


(1) Beer, F.P., Johnston, E.R., and Elliot, R.E. (2007). *Vector Mechanics for Engineers. Statics* (8th edn). McGraw-Hill, Boston, MA.

(2) Meriam, J.L. and Kraige, L.G. (2011). *Engineering Mechanics: Statics* (6th edn). Wiley, Hoboken, NJ.


Engineering mathematics:


(1) Kreyszig, E. (2011). *Advanced Engineering Mathematics* (10th edn). Wiley, Hoboken, NJ.

(2) Lay, D.C. (2011). *Linear Algebra and its Applications* (4th edn). Pearson/Addison-Wesley, Boston, MA.

(3) Ottosen, N.S. and Petersson, H. (1992). *Introduction to the Finite Element Method*. Prentice-Hall, New York, NY.

(4) Nagle, R.K., Saff, E.B., and Snider, A.D. (2008). *Fundamentals of Differential Equations and Boundary Value Problems* (5th edn). Pearson/Addison-Wesley, Boston, MA.

(5) Stewart, J. (2007). *Calculus: Early Transcendentals* (6th edn). Thompson, Brooks/Cole, Belmont, CA.

(6) Strang, G. (2006). *Linear Algebra and its Applications* (4th edn). Thompson, Brooks/Cole, Belmont, CA.


Advanced undergraduate solid mechanics:


(1) Ashby, M.F. and Jones, D.R.H. (2012). *Engineering Materials 1: An Introduction to Properties, Applications, and Design* (4th edn). Elsevier Butterworth-Heinemann, Amsterdam.

(2) Boresi, A.P. and Schmidt, R.J. (2003). *Advanced Mechanics of Materials* (6th edn). Wiley, New York, NY.

(3) Sadd, M.H. (2009). *Elasticity: Theory, Applications, and Numerics* (2nd edn). Elsevier/Academic Press, London.

(4) Slaughter, W. (2002). *The Linearized Theory of Elasticity*. Birkhäuser, Boston, MA.

(5) Ugural, A.C. and Fenster, S.K. (2011). *Advanced Mechanics of Materials and Applied Elasticity* (5th edn). Prentice-Hall, Upper Saddle River, NJ.

# Units, Constants, and Symbols

Appendix B

To convert between SI units and USCS units Table B.1 provides a set of helpful conversion relations. Tables B.1–B.5 provide additional useful information. For additional information on units, see the National Institute of Standards *Specifications, Tolerances, and Other Technical Requirements for Weighing and Measuring Devices, National Institute of Standards Handbook 44 – 2008 Edition.* Of particular interest is Appendix C of this handbook, which lists many useful

**Table B.1** Unit conversion table.

| To convert | To | Multiply by |
|---|---|---|
| **Length** | | |
| Inches (in) | Millimeters (mm) | 25.400 |
| Inches (in) | Meters (m) | 0.025400 |
| Feet (ft) | Meters (m) | 0.304800 |
| **Angles** | | |
| Degrees (deg) | Radians (rad) | $\pi/180$ |
| **Area** | | |
| Square inches ($in^2$) | Square meters ($m^2$) | 0.000645 |
| **Force** | | |
| Pounds Force (lbf) | Newtons (N) | 4.448222 |
| **Pressure** | | |
| Pounds per sq. inch (psi) | Pascal ($Pa = N/m^2$) | 6894.757 |
| Thousand pounds per sq. inch (ksi) | Mega-Pascal ($MPa = N/mm^2$) | 6.894757 |
| Million pounds per sq. inch (msi) | Giga-Pascal ($GPa = kN/m^2$) | 6.894757 |

*(cont.)*

**Table B.1** (Cont.)

| | | |
|---|---|---|
| **Mass** | | |
| Pound Mass (lbm) | Kilogram (kg) | 0.45359237 |
| **Torque** | | |
| Inch pound force (lbf-in) | Newton-meter (N-m) | 0.112985 |
| Foot pound force (lbf-ft) | Newton-meter (N-m) | 1.355818 |
| **Power** | | |
| Horsepower (hp = 550 lbf-ft/s) | Watt (W = N-m/s) | 745.700 |

conversion factors between different systems of units. A brief historical discussion of units can be found in Appendix B of this handbook, which is available online at http://ts.nist.gov/WeightsAndMeasures/Publications/H44-08.cfm

**Table B.2** Prefixes used in the SI system.

| Prefix | Symbol | Multiplying factor |
|---|---|---|
| Yotta | Y | $10^{24}$ |
| Zetta | Z | $10^{21}$ |
| Exa | E | $10^{18}$ |
| Peta | P | $10^{15}$ |
| Tera | T | $10^{12}$ |
| Giga | G | $10^{9}$ |
| Mega | M | $10^{6}$ |
| Kilo | k | $10^{3}$ |
| Milli | m | $10^{-3}$ |
| Micro | $\mu$ | $10^{-6}$ |
| Nano | n | $10^{-9}$ |
| Pico | p | $10^{-12}$ |
| Femto | f | $10^{-15}$ |
| Atto | a | $10^{-18}$ |
| Zepto | z | $10^{-21}$ |
| Yocto | y | $10^{-24}$ |

**Table B.3** Useful physical constants.

| | |
|---|---|
| Gravitational constant (g) | 9.80665 ($m/s^2$) |
| Gravitational constant (g) | 32.1740 ($ft/s^2$) |
| Natural Logarithm Base (e) | 2.718281828 |
| Pi ($\pi$) | 3.14159265 (rad) |
| Speed of light (c) | 299792458 (m/s) |
| Standard atmosphere | 101325 (Pa) |

**Table B.4** Greek alphabet.

| Symbol | | Name |
|---|---|---|
| Upper case | Lower case | |
| A | $\alpha$ | alpha |
| B | $\beta$ | beta |
| $\Gamma$ | $\gamma$ | gamma |
| $\Delta$ | $\delta$ | delta |
| E | $\epsilon$, $\varepsilon$ | epsilon |
| Z | $\zeta$ | zeta |
| H | $\eta$ | eta |
| $\Theta$ | $\theta$, $\vartheta$ | theta |
| I | $\iota$ | iota |
| K | $\kappa$ | kappa |
| $\Lambda$ | $\lambda$ | lambda |
| M | $\mu$ | mu |
| N | $\nu$ | nu |
| $\Xi$ | $\xi$ | xi |
| O | o | omicron |
| $\Pi$ | $\pi$, $\varpi$ | pi |
| P | $\rho$, $\varrho$ | rho |
| $\Sigma$ | $\sigma$, $\varsigma$ | sigma |
| T | $\tau$ | tau |
| $\Upsilon$ | $\upsilon$ | upsilon |
| $\Phi$ | $\phi$, $\varphi$ | phi |
| X | $\chi$ | chi |
| $\Psi$ | $\psi$ | psi |
| $\Omega$ | $\omega$ | omega |

**Table B.5** Common Latin abbreviations.

| Abbreviation | Latin | English |
| --- | --- | --- |
| e.g. | *exempli gratia* | for example |
| et al. | *et alii* | and others |
| etc. | *et cetera* | and so on |
| ibid. | *ibidem* | in the same place |
| i.e. | *id est* | that is |
| q.e.d. | *quod erat demonstrandum* | which was to be demonstrated |
| viz. | *videlicet* | namely |

Appendix C

# Representative Material Properties

Table C.1 lists representative material properties for a variety of different materials. The numbers given are approximated for easy recall. Thus interrelations between elastic properties are not preserved nor are strict unit conversions. Allowed stresses are given as initial yield stresses for ductile materials, and fracture stresses for brittle materials. For actual design computations one

**Table C.1** Material property values.

| Material | $E$ | $\nu$ | $G$ | $\sigma_a$ | $\alpha$ |
|---|---|---|---|---|---|
| | (msi) | (-) | (msi) | (ksi) | ($\mu$strain/F) |
| | (GPa) | (-) | (GPa) | (MPa) | ($\mu$strain/C) |
| Alumina | 55 | 0.22 | 22 | 380 | 4.7 |
| $Al_2O_3$ (99.5%) | 375 | 0.22 | 150 | 2600 | 8.4 |
| Aluminum | 11 | 0.33 | 4 | 40 | 14 |
| (6061-T6) | 77 | 0.33 | 30 | 280 | 24 |
| Diamond | 175 | 0.2 | 73 | 170 | 0.66 |
| | 1220 | 0.2 | 510 | 1200 | 1.2 |
| Cast iron | 15 | 0.21 | 6 | 8 | 6.6 |
| | 105 | 0.21 | 43 | 56 | 12 |
| Copper | 17 | 0.34 | 7 | 10 | 9 |
| (99.9%) | 120 | 0.34 | 50 | 70 | 16 |
| Polyamide | 0.47 | 0.4 | 0.17 | 11 | 55 |
| (Nylon 6/6) | 3.25 | 0.4 | 1.2 | 75 | 95 |
| Polycarbonate | 0.33 | 0.38 | 0.12 | 8.5 | 40 |
| (Lexan®) | 2.3 | 0.38 | 0.83 | 60 | 70 |
| Polychloroprene | 250e-6 | 0.46 | 85e-6 | 4 | 130 |
| (Neoprene) | 1.7e-3 | 0.46 | 0.60e-3 | 30 | 230 |
| Polyethelyne | 0.11 | 0.45 | 0.038 | 4 | 100 |
| (HDPE) | 0.8 | 0.45 | .28 | 30 | 180 |
| Steel | 30 | 0.3 | 12 | 35 | 6 |
| (Low Carbon) | 210 | 0.3 | 80 | 250 | 11 |

*(cont.)*

**Table C.1** (Cont.)

| **Material** | $E$ | $\nu$ | $G$ | $\sigma_a$ | $\alpha$ |
|---|---|---|---|---|---|
| | (msi) | (-) | (msi) | (ksi) | ($\mu$strain/F) |
| | (GPa) | (-) | (GPa) | (MPa) | ($\mu$strain/C) |
| Steel | 30 | 0.3 | 12 | 270 | 5.5 |
| (440A tempered) | 210 | 0.3 | 80 | 1900 | 10 |
| Titanium | 15 | 0.33 | 5.6 | 120 | 5 |
| | 105 | 0.33 | 40 | 840 | 10 |
| Tungsten | 50 | 0.28 | 20 | 100 | 2.5 |
| | 350 | 0.28 | 135 | 700 | 4.5 |
| Uranium | 24 | 0.21 | 10 | 30 | 6 |
| (D-38) | 170 | 0.21 | 70 | 210 | 11 |
| Wood | 2 | | | 4 | |
| (Douglas-fir, ‖-grain) | 14 | | | 30 | |
| Zerodur® | 13 | 0.24 | 5.2 | 6 | 0.01 |
| (glass) | 90 | 0.24 | 36 | 42 | 0.02 |

must use reliably measured data associated with the actual material supply, as material properties can vary substantially, depending upon processing parameters.

# Appendix D

# Parallel-Axis Theorem

The area moments of inertia of a given area $A$ are defined as

$$I_z = I_{zz} = \int_A y^2 dA,$$
$$I_y = I_{yy} = \int_A z^2 dA, \quad \text{(D.1)}$$
$$I_{zy} = I_{yz} = \int_A yz dA.$$

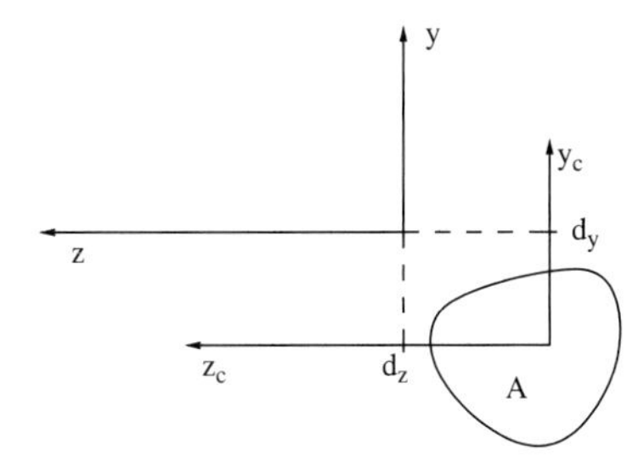


**Fig. D.1** Parallel-axis theorem construction. The axes $y_c$ and $z_c$ represent the centroidal axes.

If the area moments of inertia $(I_{z_c}, I_{y_c}, I_{y_c z_c})$ of an area are known with respect to the centroidal axes of the area, then the parallel-axis theorem tells us that the area moments of inertia with respect to any other set of (parallel) axes are given by:

$$I_z = I_{zz} = \int_A y^2 dA = I_{z_c} + A d_y^2 \quad \text{(D.2)}$$

$$I_y = I_{yy} = \int_A z^2 dA = I_{y_c} + A d_z^2 \quad \text{(D.3)}$$

$$I_{yz} = I_{zy} = \int_A yz dA = I_{y_c z_c} + A d_y d_z, \quad \text{(D.4)}$$

where $d_y$ and $d_z$ are the (centroidal) coordinates of the $y$-$z$ frame (see Fig. D.1). The proof of the theorem follows directly from eqn (D.1) under the substitution $y = y_c - d_y$ and $z = z_c - d_z$, and the fact that $\int_A z_c \, dA = \int_A y_c \, dA = 0$ by definition of the centroid.

# Appendix E

# Integration Facts



In calculus courses one usually learns the mechanics of integration and becomes adept at the computation of given integrals. In engineering, this skill is important, as the computation of a particular integral is often the final step in the development of a solution to a particular problem. However, in engineering one also faces the need to create models, and these models often contain integral terms. In order to construct these models, one needs to have a firm grasp of the meaning of an integral. In this Appendix we first address integration at a conceptual level and then list some important facts – all associated with Riemannian integration.

## E.1 Integration is addition in the limit

At an elementary level, (Riemannian) integration is nothing more than a formal (and very useful) methodology for expressing the result of an infinite sum. Given an (integrable) function $f : [a, b] \to \mathbb{R}$, the integral

$$\int_a^b f(x)\, dx \tag{E.1}$$

*is defined as* the limit value of the Riemannian sum

$$\lim_{N\to\infty} \sum_{k=1}^{N+1} f(x_k)\Delta x_k, \tag{E.2}$$

where $x_0 = a < x_1 < x_2 < \cdots < x_N < b = x_{N+1}$, $\Delta x_k = x_k - x_{k-1}$, and all $\Delta x_k \to 0$ as $N \to \infty$.

The key to understanding integration from a model-building perspective is the understanding of this definition.

In particular, one should observe that the summand in eqn (E.2) represents the value of a function in the neighborhood of a point times a measure of the size of the neighborhood. The summation process then adds all these values over a given region.

**Example E.1**

*Integral of $f(x) = x$.* As a first example let us verify that this definition of the integral gives us what we expect for $f(x) = x$ and $a = 0$ and $b = 3$.

*Solution*

We know from elementary calculus that

$$\int_0^3 x\,dx = \frac{1}{2}x^2\Big|_0^3 = \frac{9}{2}. \tag{E.3}$$

To compute the Riemann sum we need to select values for the $x_k$s. Any selection that satisfies the ordering requirement given above will suffice. Let us assume that $x_k = 3k/(N+1)$; thus $\Delta x_k = 3/(N+1)$. The summand is then

$$f(x_k)\Delta x_k = \frac{3k}{N+1}\frac{3}{N+1}. \tag{E.4}$$

The sum is thus

$$\sum_{k=1}^{N+1} \frac{3k}{N+1}\frac{3}{N+1} = \frac{9}{(N+1)^2}\sum_{k=1}^{N+1} k - \frac{9}{(N+1)^2}\frac{(N+1)(N+2)}{2}. \tag{E.5}$$

In the limit $N \to \infty$, we see that our sum gives us

$$\lim_{N\to\infty}\sum_{k=1}^{N+1} f(x_k)\Delta x_k = \lim_{N\to\infty}\frac{9}{(N+1)^2}\frac{(N+1)(N+2)}{2} = \frac{9}{2}. \tag{E.6}$$

Thus we see that the definition given above does provide the expected result.

**Example E.2**

*Sand-pile.* Let us assume that we have a board upon which is a pile of sand; see Fig. E.1. The height of the sand is given by a function $h(x)$. What is the total weight which the board supports? Assume that the weight density of the sand is $\gamma$ and the width of the board into the page is $w$.

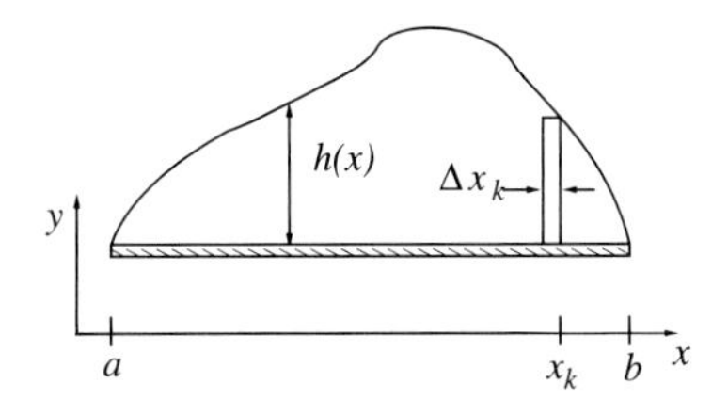


**Fig. E.1** Sand-pile.

*Solution*

If we consider a small segment of the board located at $x_k$ of length $\Delta x_k$, then an approximation to the volume of the sand over this segment is $h(x_k)w\Delta x_k$. Thus, the approximate weight that this segment carries is given by $\gamma h(x_k)w\Delta x_k$. The smaller $\Delta x_k$, the more accurate the estimate. The total weight being supported by the board can be estimated by choosing many such $x_k$s between $a$ and $b$, and adding all the terms. In the limit that one takes an infinite number of such points we will obtain an exact result. But this is simply the Riemann sum

$$\lim_{N\to\infty}\sum_{k=1}^{N+1} w\gamma h(x_k)\Delta x_k. \tag{E.7}$$

Thus the total weight supported by the board is given by the integral

$$\gamma w \int_a^b h(x)\, dx. \tag{E.8}$$

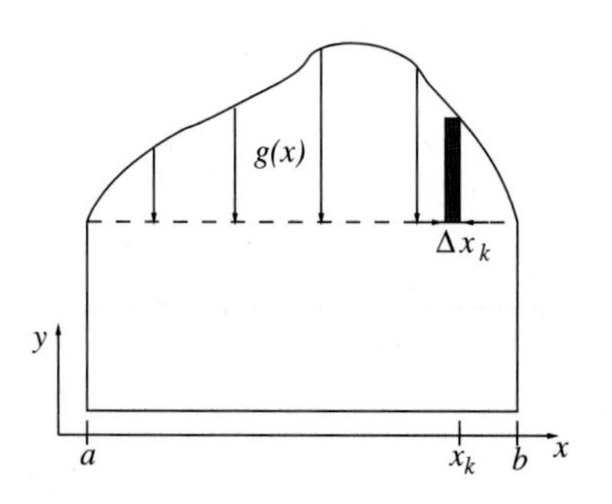


**Fig. E.2** Two-dimensional box.

**Example E.3**

*Mass flow.* Let us assume that we have a two-dimensional box, and that across the top of the box there is a flow of gas; see Fig. E.2. The gas flow is defined by a function $g(x) : [a, b] \to \mathbb{R}$ which gives the mass per unit time per unit length flowing across the top of the box; i.e. the dimensions of $g$ are $[g] = [M/TL]$. We would like to determine an expression for the total mass per unit time that flows across the top of the box.

*Solution*

If we consider the top of the box and look at a small segment at $x_k$ of length $\Delta x_k$, then the flow rate across this segment can be approximated by $g(x_k)\Delta x_k$. The smaller $\Delta x_k$ is, the more accurate is the estimate. The total flow rate across the top of the box can be estimated by choosing many such $x_k$s between $a$ and $b$ and adding up all the terms. In the limit that one takes an infinite number of such points, we will obtain an exact result. But this is simply the Riemann sum

$$\lim_{N\to\infty} \sum_{k=1}^{N+1} g(x_k)\Delta x_k. \tag{E.9}$$

Thus the total mass flow rate across the top of the box is given by the integral

$$\int_a^b g(x)\, dx. \tag{E.10}$$

## E.2 Additivity

It should be observed that if one integrates over a domain $\Omega = (a, b) \cup (c, d)$ where $(a, b) \cap (c, d) = \emptyset$, then the integral

$$\int_\Omega f(x)\, dx = \int_{(a,b)} f(x)\, dx + \int_{(c,d)} f(x)\, dx;$$

i.e., integration is an additive process. This follows directly from our definition of integration in the previous section. This observation is quite helpful when faced with complex integration domains, if one can decompose the domain into a set of simpler domains. This result also holds in higher dimensions.

**Example E.4**

*Exploitation of additivity.* Suppose you wish to integrate $f(x, y) = xy$ over the domain $\Omega$ shown in Fig. E.3. In this case one can breakup the domain, $\Omega$ into two domains, $\Omega_1 = (0, a) \times (0, b)$ and $\Omega_2 = (a, c) \times (e, d)$; i.e. $\Omega = \Omega_1 \cup \Omega_2$ and $\Omega_1 \cap \Omega_2 = \emptyset$. We then have that

$$\int_{\Omega} xy\, dxdy = \int_{\Omega_1} xy\, dxdy + \int_{\Omega_2} xy\, dxdy = \frac{a^2b^2}{4} + \frac{(c^2 - a^2)(d^2 - e^2)}{4}$$

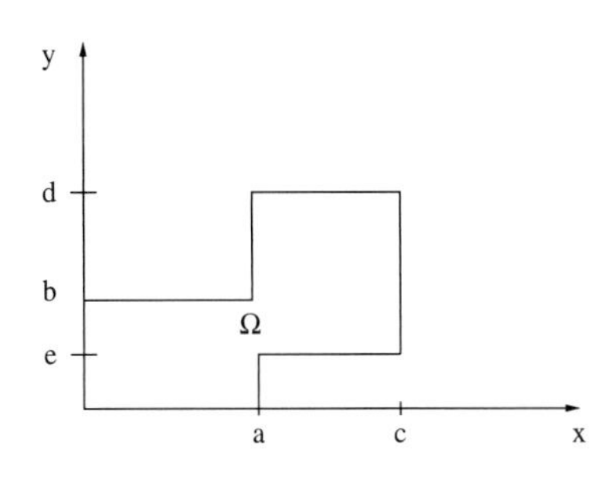


**Fig. E.3** Easily-decomposable domain of integration.

# E.3 Fundamental theorem of calculus

The fundamental theorem of calculus states that for a function $f : [a, b] \to \mathbb{R}$,

$$f(b) - f(a) = \int_a^b \frac{df(x)}{dx}\, dx$$

(when $df/dx$ is integrable). The theorem is used quite often in mechanics when one knows the rate of change of a quantity and one wishes to know the overall change in the quantity over a given interval.

**Example E.5**

*Application of the fundamental theorem of calculus.* Suppose one knows that the rate of change of the internal force in a rod (see Fig. E.4) is given by $dR/dx = -b(x) = -3x$ (assuming consistent units), then one immediately knows that the difference between the forces on the two ends of the rod is

$$R(L) - R(0) = \int_0^L -(3x)\, dx = -\frac{3L^2}{2}.$$

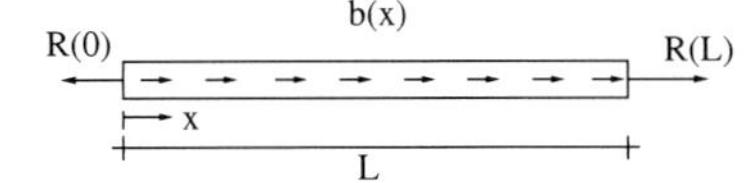


**Fig. E.4** Rod with internal force distribution.

# E.4 Mean value

The mean value (or average value) of a function over a domain $\Omega$ is defined by

$$f_{\text{mean}} = \frac{\int_{\Omega} f\, dA}{\int_{\Omega} 1\, dA},$$

where we have implicitly assumed that the domain is two-dimensional. Translations to one and three dimensions are obvious. The average value of a function has many uses. For example, it is used in defining the centroid of an area or volume. The definition can also be exploited to compute various integrals if one already knows the mean value of an integrand; i.e. one can rewrite the expression as

$$\int_\Omega f \, dA = A_\Omega f_{\text{mean}},$$

where $A_\Omega$ is the area of the domain $\Omega$.

**Example E.6**

*Integration using the mean value.* Consider the computation of the integral of the function

$$f(x,y) = \begin{cases} y & (x,y) \in \Omega_1 \\ k & (x,y) \in \Omega_2 \end{cases}$$

over $\Omega$ in Fig. E.3, where $k$ is a constant. The mean value of $f(x,y)$ over $\Omega_1$ is $b/2$ and the mean value over $\Omega_2$ is $k$. Thus

$$\begin{aligned} \int_\Omega f(x,y)\, dxdy &= \int_{\Omega_1} f(x,y)\, dxdy + \int_{\Omega_2} f(x,y)\, dxdy \\ &= ab\frac{b}{2} + (c-a)(d-e)k. \end{aligned}$$

## E.5 The product rule and integration by parts

The product rule of differentiation states that for two functions $f(x)$ and $g(x)$, the derivative of their product is given by

$$\frac{d}{dx}(fg) = \frac{df}{dx}g + f\frac{dg}{dx}.$$

We can use this result to our advantage in certain integration problems by combining it with the fundamental theorem of calculus. In particular, if we integrate both sides of this expression over a given interval $[a,b]$, then we have:

$$[f(b)g(b) - f(a)g(a)] = \int_a^b \frac{df}{dx} g \, dx + \int_a^b f \frac{dg}{dx} \, dx.$$

This result is normally written by moving one of the integrals to the other side of the equals sign – for example, as:

$$\int_a^b \frac{df}{dx} g \, dx = [f(b)g(b) - f(a)g(a)] - \int_a^b f \frac{dg}{dx} \, dx.$$

This formula, or a small variant of it, is known as the integration by parts rule.

**Example E.7**

*Derivation of a weak form.* Suppose we have the second-order ordinary differential equation

$$\frac{d^2u}{dx^2} + h(x) = 0 \tag{E.11}$$

over the interval $(a, b)$, where $h(x)$ is known, as are the boundary conditions $u(a) = u_a$ and $u(b) = u_b$. The usual problem statement is: find a function $u(x)$ with $u(a) = u_a$ and $u(b) = u_b$ such that it satisfies eqn (E.11).

A very important restatement of this problem is found by multiplying eqn (E.11) by an arbitrary function $\bar{v}(x)$, which has the properties that $\bar{v}(a) = \bar{v}(b) = 0$, and then integrating by parts. This gives:

$$0 = \bar{v}\frac{d^2u}{dx^2} + \bar{v}h \tag{E.12}$$

$$0 = \int_a^b \bar{v}\frac{d^2u}{dx^2} + \bar{v}h\,dx \tag{E.13}$$

$$0 = \int_a^b \frac{d}{dx}\left(\bar{v}\frac{du}{dx}\right) - \frac{d\bar{v}}{dx}\frac{du}{dx} + \bar{v}h\,dx \tag{E.14}$$

$$0 = \underbrace{\left[\bar{v}\frac{du}{dx}\right]_a^b}_{=0} + \int_a^b -\frac{d\bar{v}}{dx}\frac{du}{dx} + \bar{v}h\,dx \tag{E.15}$$

$$0 = \int_a^b \frac{d\bar{v}}{dx}\frac{du}{dx} - \bar{v}h\,dx. \tag{E.16}$$

Equation (E.16) is known as the weak form of the ordinary differential equation[1]. A restatement of the problem is then: find a function $u(x)$ with $u(a) = u_a$ and $u(b) = u_b$ such that eqn (E.16) is satisfied for all functions $\bar{v}(x)$, where $\bar{v}(a) = \bar{v}(b) = 0$. This form of the problem statement is essentially equivalent to the form with the ordinary differential equation modulo a few technical mathematical details associated with the types of function one may use. The weak form of the problem may seem rather complicated, but it has some very important advantages over the standard differential equation form. It is intimately related to the principle of virtual power/work, and is the basis for the finite-element method for solving many engineering problems.

[1] The ordinary differential equation is sometimes known as the strong form.

# E.6 Integral theorems

## E.6.1 Mean value theorem

Given a continuous function $f(\boldsymbol{x})$ over a region $\Omega$, then there exists a point $\hat{\boldsymbol{x}} \in \Omega$ such that

$$\int_{\Omega} f(\boldsymbol{x})\, d\boldsymbol{x} = \text{meas}(\Omega) f(\hat{\boldsymbol{x}}), \tag{E.17}$$

where meas($\Omega$) is the measure of $\Omega$ – its length, area, or volume, dependent upon the dimension of the domain. Note that this theorem states also that the function takes on its mean value at the point $\hat{\boldsymbol{x}}$; i.e. $f(\hat{\boldsymbol{x}}) = f_{\text{mean}}$.

### E.6.2 Localization theorem

Given a continuous function $f(\boldsymbol{x})$ over a region $\Omega$ for which

$$\int_{B} f(\boldsymbol{x})\, d\boldsymbol{x} = 0, \tag{E.18}$$

for all $B \subseteq \Omega$, then $f(\boldsymbol{x}) = 0$ for all $\boldsymbol{x} \in \Omega$.

### E.6.3 Divergence theorem

Given a differentiable vector valued function $\boldsymbol{f}(\boldsymbol{x})$ over a regular[2] region $\Omega$ with boundary $\partial\Omega$ having outward normal field $\boldsymbol{n}(\boldsymbol{x})$, then

[2] Roughly speaking, a regular region is a closed domain with piecewise smooth boundaries.

$$\int_{\partial\Omega} \boldsymbol{f} \cdot \boldsymbol{n}\, dA = \int_{\Omega} \text{div}[\boldsymbol{f}]\, dV. \tag{E.19}$$

For example in two dimensions and Cartesian coordinates,

$$\int_{\partial\Omega} f_x n_x + f_y n_y \, d\Gamma = \int_{\Omega} \frac{\partial f_x}{\partial x} + \frac{\partial f_y}{\partial y}\, dA. \tag{E.20}$$

In the three main coordinate systems the divergence operator acts on a vector as

$$\begin{aligned}
\text{div}[\boldsymbol{f}] &= \frac{\partial}{\partial x}(f_x) + \frac{\partial}{\partial y}(f_y) + \frac{\partial}{\partial z}(f_z) && \text{(E.21)}\\
&= \frac{1}{r}\frac{\partial}{\partial r}(r f_r) + \frac{1}{r}\frac{\partial}{\partial \theta}(f_\theta) + \frac{\partial}{\partial z}(f_z) && \text{(E.22)}\\
&= \frac{1}{r^2}\frac{\partial}{\partial r}\left(r^2 f_r\right) + \frac{1}{r\sin(\phi)}\frac{\partial}{\partial \theta}(f_\theta) && \text{(E.23)}\\
&\quad + \frac{1}{r\sin(\phi)}\frac{\partial}{\partial \phi}(f_\phi \sin(\phi)).
\end{aligned}$$

# Bending without Twisting: Shear Center

In Chapter 8 we treated the problem of bending about one axis when the cross-section of the beam possessed a single plane of symmetry and bending about two axes when the cross-section possessed two planes of symmetry. In all the cases treated the applied transverse loads were assumed to pass through the centroid of the cross-section. Without these assumptions it is possible that the beam will twist in addition to bend. In many situations this is an undesirable effect, and one wishes to know how to place the transverse load such that the beam only bends and does not twist. The net result is that there is a point on each cross-section through which the transverse load must pass in order to avoid twisting. This point is known as the shear center. For circular and rectangular cross-sections this point lies at the centroid of the cross-section, but for more complex cross-sections it does not. An exact analysis of this situation is rather complex, but can be very reasonably estimated and understood as outlined below.

## F.1 Shear center

Consider a beam with an arbitrary cross-section, where we will assume that the $x$-axis is coincident with the centroidal axis of the beam. For simplicity we will further assume that the beam is linear elastic, isotropic, and homogeneous, and loaded only with transverse forces. The stresses on a given cross-section (the tractions) consist of the normal stress $\sigma_{xx}$ and the two shear stresses $\sigma_{xy}$ and $\sigma_{xz}$. The resultants on the cross-section consist of the bending moments, the axial torque, the axial force, and the two shear forces. These are given in terms of the stress components as

$$R = \int_A \sigma_{xx}\, dA\,, \tag{F.1}$$

$$V_y = \int_A \sigma_{xy}\, dA\,, \tag{F.2}$$

$$V_z = \int_A \sigma_{xz}\, dA\,, \tag{F.3}$$

$$T = \int_A -\sigma_{xy} z + \sigma_{xz} y\, dA\,, \tag{F.4}$$

$$M_z = \int_A -\sigma_{xx} y \, dA \,, \tag{F.5}$$

$$M_y = \int_A \sigma_{xx} z \, dA \,. \tag{F.6}$$

To understand the issue of twisting we need to consider the shear stresses on the cross-section in detail. When we bend a beam with transverse forces we will directly generate shear stresses, say, $\sigma_{xy}^{DS}$ and $\sigma_{xz}^{DS}$; i.e. direct shear stresses. If we separately twist the beam we will also have contributions to the shear stresses due to the applied twist, say $\sigma_{xy}^{AT}$ and $\sigma_{xz}^{AT}$. The total shear stresses on the cross-section, in general, will be the sum of these two contributions.

The torque associated with the shear stresses from the applied twist is given by

$$T^{AT} = \int_A -\sigma_{xy}^{AT} z + \sigma_{xz}^{AT} y \, dA \,. \tag{F.7}$$

Since we are assuming that the system is linear elastic, we can also assume that this torque is related to the applied twist rate as:

$$T^{AT} = C \frac{d\phi}{dx} \,, \tag{F.8}$$

where $C$ represents the effective torsional stiffness of the cross-section – i.e. a sort of $(GJ)_{\text{eff}}$. Thus the we can express the total torque on the cross-section as

$$T = C \frac{d\phi}{dx} + \int_A -\sigma_{xy}^{DS} z + \sigma_{xz}^{DS} y \, dA \,. \tag{F.9}$$

If it is desired that $d\phi/dx = 0$ (i.e. the beam does not twist), then there must be a net torque on the cross-section (about the centroidal axis) of

$$T_{\text{no twist}} = \int_A -\sigma_{xy}^{DS} z + \sigma_{xz}^{DS} y \, dA \,. \tag{F.10}$$

**Remarks:**

(1) This final result states that if we wish to bend a beam with shear forces through the centroid of a cross-section without it simultaneously twisting, then in general we will need to also apply a torque about the the centroidal axis of the beam. For circular and rectangular cross-sections this torque turns out to be zero. However, for more general shapes it is often non-zero.

(2) One common way of generating this torque is to shift the applied shear forces away from the centroid, as we will see in the example below.

(3) Note that for the argument presented, one in principle needs to know how to compute the torsional stiffness of a general cross-section. However, in the end, we want to enforce the desire that the twist is zero, and thus we never actually need to be able to compute the constant $C$ above.

**Example F.1**

*Bending a C-section without twisting.* Consider a cantilevered beam with an end-load and cross-section as shown in Fig. F.1(a). We would like to bend the beam with the force $P$ without the beam twisting. What additional torque needs to be applied to the beam so that it does not twist?

*Solution*

Since the cross-section is thin-walled we can determine the direct shear stress distribution using the thin-walled approximation. The area moment of inertia of the cross-section is

$$I = \frac{0.1 \times 10^3}{12} + 2\left[\frac{10 \times 0.1^3}{12} + 0.1 \times 10 \times 5^2\right] = 58.3 \text{ mm}^4 . \tag{F.11}$$

Thus the shear flow at the junction of the web and the flanges is given by

$$q = \frac{PQ}{I} = \frac{P}{I} 5 \times 10 \times 0.1 = 0.0857P \tag{F.12}$$

and the total horizontal shear force in the upper and lower flanges is

$$F_{\text{flange}} = 0.5 \times 10 \times 0.0857P = 0.429P . \tag{F.13}$$

So the required torque about the centroid, to keep the section from twisting while bending, is

$$\begin{aligned} T &= \underbrace{2 \times 5 \times F_{\text{flange}}}_{\text{Flange Contribution}} + \underbrace{P \times z_c}_{\text{Web Contribution}} \\ &= (4.29 + z_c)P , \end{aligned} \tag{F.14}$$

where $z_c$ is the distance of the centroid from the web; see Fig. F.1(b).

**Remarks:**

(1) The torque $T = (4.29 + z_c)P$ needs to be applied about the centroidal axis if we want to prevent the beam from twisting while we bend it with a shear force of magnitude $P$.

(2) One way to effect this is to shift the point of application of the shear force to the left by an amount $z_c + e$. The shift has been written in this way so that $e$ represents the magnitude of the shift from the web – a convenient reference point, since we have not yet computed the location of the centroid. The requirement for a statically equivalent force system will be that

$$\begin{aligned} (z_c + e)P &= (4.29 + z_c)P \\ e &= 4.29 \text{ mm} ; \end{aligned} \tag{F.15}$$

see Fig. F.1(b)(right). Note that this shift of the load lies outside the cross-section of the beam. In practice one would overcome this point by welding a small plate onto the beam so that the load could be properly applied; see Fig. F.1(c).

(3) Any point along the line shifted a distance of 4.29 mm from the web will suffice as shown in Fig. F.1(b). If the beam is also bent with a

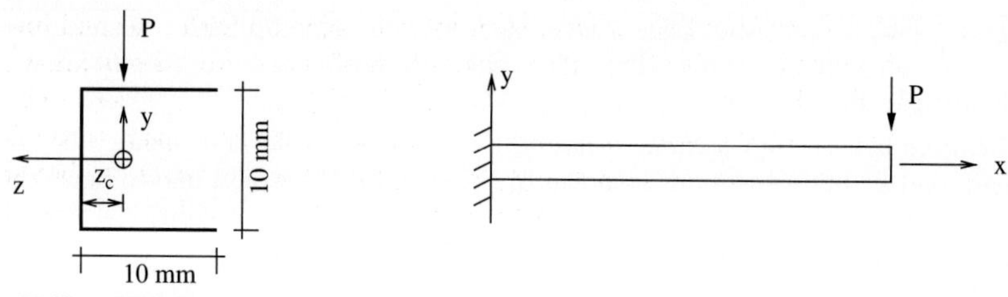


(a) Cantilever beam with C-section.

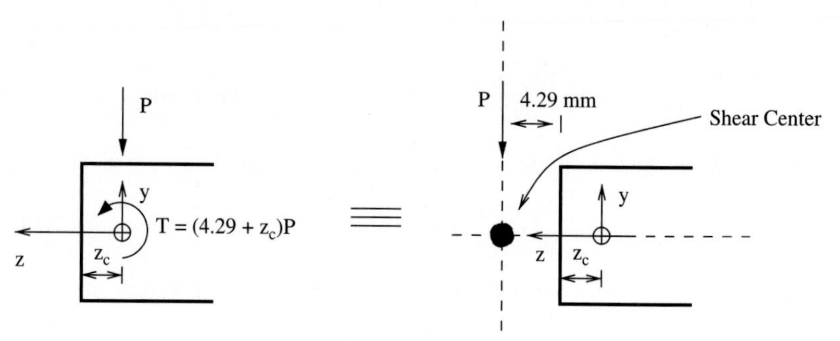


(b) Two loading cases that generate bending without twist.

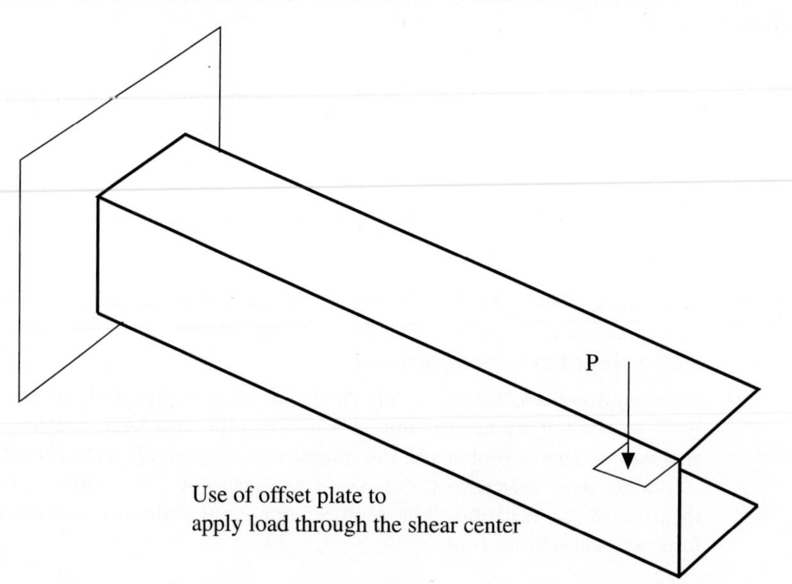


(c) One technique to effect the needed torque using only the given shear force.

**Fig. F.1** C-section loaded with a single transverse shear force avoiding twisting.

shear force about the $y$-axis, then we will come up with a second line of points of application. The intersection of these two lines is known as the shear center (as shown in Fig. F.1(b)(right)). For zero-twist bending of the beam, all transverse shear forces must pass through this point if additional torques are not applied to the beam. It should be noted that the shear center always lies along lines of symmetry of the cross-section.

# Index

Page references in *italics* refer to exercises. Page references in **bold** refer to tables/figures.